Advances in Vascular Medicine

David Abraham, Clive Handler,
Michael Dashwood and Gerry Coghlan (Eds.)

Advances in Vascular Medicine

Editors
David Abraham
University College London
London
UK
dabraham@medsch.ucl.ac.uk

Clive Handler
University College London
London
UK
chandler@medsch.ucl.ac.uk

Michael Dashwood
University College London
London
UK
m.dashwood@medsch.ucl.ac.uk

Gerry Coghlan
University College London
London
UK
Gerry.coghlan@royalfree.nhs.uk

ISBN 978-1-4471-5763-2 ISBN 978-1-84882-637-3 (eBook)
DOI 10.1007/978-1-84882-637-3
Springer London Dordrecht Heidelberg New York

British Library Cataloguing in Publication Data
A catalogue record for this book is available from the British Library

Printed on acid-free paper

Springer is part of Springer Science+Business Media (www.springer.com)

Preface

Understanding the many complex cellular and molecular mechanisms underlying human vascular diseases is essential in improving the treatment of this important and wide-ranging group of diseases that affect a large proportion of the world population. This book is based on lectures presented at an International Vascular Biology Workshop held in London and chaired by Professor Dame Carol Black. The contents are complemented by some invited chapters, all written by world experts in areas of basic science and clinical medicine highly relevant to vascular biology and disease. We are particularly grateful to Professor Arshed Quyyumi, Professor of Medicine and Cardiology at Emory University, who with his research group and clinical colleagues, has provided a substantial contribution to this book. In common with our previous book – *Vascular Complications in Human Disease: Mechanisms and Consequences* published by Springer in 2008, our aim with this book is to highlight some of the established relationships between basic science and clinical medicine, and to outline new and exciting fields of research and practice in vascular biology and pathobiology.

There are two sections: Basic Science of Vascular Biology and Clinical Aspects of Vascular Biology. In the first section, dealing with basic science, we have included three important growth areas: "Genetics and Gene Therapy" cover approaches to gene therapy and delivery systems, "Animal Models to Study Vascular Disease" with chapters on animal models of scleroderma, animal models of atherosclerosis, and finally on the endothelin system. The final section on basic science titled "Molecules and Mediators and Therapeutic Applications" encompasses the role of endothelin in systemic sclerosis, and other aspects of the genetics and biology of endothelium and vascular function and includes a chapter on Cell Therapy for Cardiovascular Diseases and Cell and Molecular Mechanism(s) Underlying Vascular Remodeling. These basic science topics underpin what may further improve the clinical care of patients with vascular diseases.

The first section on clinical aspects of vascular biology is written by our colleagues from Papworth Hospital, currently the only UK center operating on patients with chronic thromboembolic disease associated pulmonary hypertension; this section also includes a chapter on imaging in acute and chronic thromboembolic disease. Vascular disease in connective tissue diseases includes chapters on pulmonary arterial hypertension in connective tissue disease, registry and epidemiological data in systemic sclerosis associated pulmonary arterial hypertension, and a review of vascular disease in systemic sclerosis. The final clinical section on Cardiovascular Disease, includes the important topics of coronary heart disease in women, graft performance in coronary artery surgery, predicting cardiovascular risk and the metabolic syndrome.

Although common basic science strands link the chapters, each chapter stands alone as an authoritative, up-to-date and powerful insight into these important topics of vascular biology. The chapters help the basic scientist understand clinical problems as well as explaining to clinicians the scientific foundations of vascular diseases and allude to possible tracks for future research.

Although we are making progress in understanding some of the basic scientific mechanisms of vascular disease, there is much work to be done. The picture is thus far from complete. We hope that the information and insights contained in this book will be a useful contribution to the literature and help other scientists and clinicians make progress in this exciting field of biomedicine.

London, UK

David Abraham
Clive Handler
Michael Dashwood
Gerry Coghlan

Foreword

Vascular medicine, in the form of vascular surgery, has its origins in attempts to ligate blood vessels to prevent haemorrhage, which is a well-described practice in one of the first extant textbooks of surgery written by the Indian doctor Sushruta over 2,500 years ago. However, reconstructive vascular surgery to repair torn or aneurysmic vessels, or to create anastomoses, only really became routinely feasible in the 19th century when surgery in general benefitted from the introduction of antiseptic and anesthetic procedures. Other areas of vascular medicine are much more recent, and needed for their impetus the realization that blood vessels are far more than structural conduits but have an intrinsic biology and pathology that contribute to a wide range of diseases.

For example, though the atherosclerotic process was accurately described by eminent 19th century pathologists such as Virchow, it is only since after World War II that cardiologists began to take seriously the idea that acute myocardial infarction was a consequence of atherosclerosis in the coronary arteries, leading to the huge upsurge in the last 50 years of novel interventional attempts to remedy the problem – coronary bypass surgery since 1960, angioplasty since the mid 1970s, bare metal stents since the mid 1980s, and drug eluting stents since 2002.

In parallel with developments in intervention, vascular biology has steadily increased our understanding of the cellular and molecular physiology of blood vessels, revealing the characteristic responses of endothelial and smooth muscle cells to insult and injury and defining their active roles in the maintenance of vascular homeostasis. As late as 1960, the endothelium was described as a passive, blood-compatible, semi-permeable membrane leading to Lord Florey's rejection of this view of the endothelium as "little more than a sheet of nucleated cellophane" and the beginnings of endothelial cell biology. This was to lead within 20 years to important discoveries, including the key molecular mechanisms by which endothelium controls leukocyte traffic and other aspects of the inflammatory response, and then the Nobel prize-winning discovery of nitric oxide as a novel endogenous endothelium-derived signalling molecule that regulates vascular tone and platelet function. Another hugely significant body of cell and molecular biology has stemmed from Folkman's careful description of new blood vessel formation (angiogenesis) and his seminal discovery of its importance for tumor growth.

Thus we are currently in an exciting phase, as the molecular discoveries from vascular biology are beginning to be combined with the increasingly sophisticated technical expertise of vascular interventionists and surgeons, leading confidently into an era when a series of novel preventive and regenerative strategies will be applied successfully to vascular medicine. This volume provides a valuable snapshot of several growth points in vascular medicine, both preclinical and clinical, that reflect these strategies and the wide range of diseases where they impact.

King's College, London Jeremy D. Pearson

Contents

Contributors

David J. Abraham
Centre for Rheumatology, Royal Free Hospital, University College London, NW3 2PF, London, UK

Mikael Arbeus
Department of Thoracic and Cardiovascular Surgery, Örebro University Hospital, Örebro, Sweden

François A. Auger
Laboratoire d'Organogénèse Expérimentale (LOEX), Hôpital du Saint-Sacrement du CHA, Department of Surgery, Laval University, Quebec, Canada

Konstantinos Aznaouridis
Division of Cardiology, Department of Medicine, Emory University School of Medicine, 1364 Clifton Rd. NE, Suite D403C, Atlanta, GA, USA

Andrew H. Baker
BHF Glasgow Cardiovascular Research Centre, University of Glasgow, 126 University Place, Glasgow, G12 8TA, UK, ab11f@clinmed.gla.ac.uk

Rosa J. Baron
Department of Pharmacology, University of Glasgow, Glasgow, G12 8QQ, Scotland, UK

Ulrike Benbow
Department of Medicine, Dartmouth Medical School, Hanover, New Hampshire 03755, USA

Andrew R. Bond
Clinical Science at South Bristol, Bristol Heart Institute, Bristol, UK

Charlotte J. Cash
Department of Radiology, Royal Free Hospital, Pond Street, London, NW3 2QG, UK

Joanne R. Cleverley
Department of Radiology, Royal Free Hospital, Pond Street, London, NW3 2QG, UK

Gerry Coghlan
Department of Cardiology, University College London, Royal Free Hospital, London, UK

Robin A. Condliffe
Pulmonary Vascular Disease Unit, Royal Hallamshire Hospital, Sheffield, UK

Michael Dashwood
Clinical Biochemistry, Royal Free and University College Medical School, London, NW3 2QG, UK

Christopher P. Denton
Centre for Rheumatology, University College London, Royal Free Hospital, London, NW3 2PF, UK

Emma Derrett-Smith
Centre for Rheumatology, Royal Free Hospital, London, UK

Saurabh S. Dhawan
Division of Cardiology, Department of Medicine, Emory University School of Medicine, Atlanta, GA, 30322, USA

Oliver Distler
Center of Experimental Rheumatology, Department of Rheumatology, University Hospital Zurich, Zurich, CH-8091 Zurich, Switzerland

Pedro D'Orleans-Juste
Department of Pharmacology, Université de Sherbrooke, 3001, 12th Avenue North, Sherbrooke, QC, J1H 5N4, Canada, pedro.dorleans-juste@usherbrooke.ca

Maria S. Fernández-Alfonso
Unidad de Cartografía Cerebral, Instituto Pluridisciplinar, Universidad Complutense, Juan XXIII, 1, E-28040, Madrid, Spain

Derek Filbey
Department of Transfusion Medicine, Örebro University Hospital, Örebro, Sweden

Deborah Galley
Bristol Heart Institute, University of Bristol, Level 7, Bristol Royal Infirmary, Bristol, BS2 8HW, Avon, UK

Lucie Germain
Laboratoire d'Organogénèse Expérimentale (LOEX), Hôpital du Saint-Sacrement du CHA, Department of Surgery, Laval University, Quebec, Canada

Marta Gil-Ortega
Departamento de Farmacología, Tecnología y Desarrollo Farmacéutico, Facultad de Farmacia, Universidad CEU-San Pablo, Madrid

David G. Harrison
Department of Medicine, Division of Cardiology, Emory University School of Medicine, Atlanta, GA, USA, Atlanta Veterans Affairs Medical Center, Decatur, Georgia, USA

Susi Heiden
Center for Cardiovascular Research/Department of Pharmacology and Toxicology, Charité, Hessische Strasse 3-4, Berlin, 10115, Germany, Department of Nephrology, Charité – Campus Benjamin Franklin, Berlin, Germany

Berthold Hocher
Center for Cardiovascular Research/Department of Pharmacology and Toxicology, Charité, Hessische Strasse 3-4, Berlin, 10115, Germany, berthold.hocher@charite.de

Sarah L. Howat
Cardiovascular Division, King's College London, School of Medicine, Waterloo Campus, SE1 9NH, London, UK

Christopher L. Jackson
Bristol Heart Institute, University of Bristol, Level 7, Bristol Royal Infirmary, Bristol, BS2 8HW, Avon, UK, chris.jackson@bristol.ac.uk

David P. Jenkins
Cardiac Surgery, Papworth Hospital, Cambridge, Cambridgeshire, UK

Dan Lacroix
Laboratoire d'Organogénèse Expérimentale (LOEX), Hôpital du Saint-Sacrement du CHA, Department of Surgery, Laval University, Quebec, Canada

Benjamin D. Mackie
Department of Medicine, Division of Cardiology, Emory University School of Medicine, Atlanta, GA, USA, Atlanta Veterans Affairs Medical Center, Decatur, GA, 30033, USA

Mirko Manetti
Department of Internal Medicine and Rheumatology, Justus-Liebig-University of Giessen, Division of Rheumatology and Clinical Immunology, Kerckhoff-Klinik, Bad Nauheim, Germany, Department of Anatomy, Histology and Forensic Medicine, University of Florence, Florence, Italy

Rachel Masson
BHF Glasgow Cardiovascular Research Centre, University of Glasgow, 126 University Place, Glasgow, G12 8TA, UK, r.masson.1@research.gla.ac.uk

Nicholas W. Morrell
Department of Medicine, Addenbrooke's and Papworth Hospitals, Cambridge, CB2 2QQ, UK

Jonathan R. Murrow
Division of Cardiology, Department of Medicine, Emory University School of Medicine, 1364 Clifton Road, Suite D403C, Atlanta, GA, 30322, USA

Ulf Müller-Ladner
Department of Internal Medicine and Rheumatology, Justus-Liebig-University of Giessen, Division of Rheumatology and Clinical Immunology, Kerckhoff-Klinik, Bad Nauheim, Germany

Elena Neumann
Department of Internal Medicine and Rheumatology, Justus-Liebig-University of Giessen, Division of Rheumatology and Clinical Immunology, Kerckhoff-Klinik, Bad Nauheim, Germany

Stuart A. Nicklin
BHF Glasgow Cardiovascular Research Centre, University of Glasgow, 126 University Place, Glasgow, G12 8TA, UK, san3x@clinmed.gla.ac.uk

Anna Nuttall
Centre for Rheumatology, Royal Free Hospital, London, UK

James S. Owen
Department of Medicine, Royal Free and University College Medical School, Royal Free Campus, Rowland Hill Street, London, NW3 2PF, UK, j.owen@medsch.ucl.ac.uk

Muhiddin A. Ozkor
Division of Cardiology, Department of Medicine, Emory University School of Medicine, Emory University, 1364 Clifton Road, Suite D403C, Atlanta, GA 30322, USA

Ioannis Papaioannou
Department of Medicine, Royal Free and University College Medical School, Royal Free Campus, Rowland Hill Street, London, NW3 2PF, UK, i.papaioannou@medsch.ucl.ac.uk

Jeremy D. Pearson
Cardiovascular Division, King's College London, School of Medicine, Waterloo Campus, SE1 9NH, London, UK, jeremy.pearson@kcl.ac.uk

Joanna Pepke-Zaba
Director of Pulmonary Vascular Disease Unit, Papworth Hospital, Papworth Everard, Cambridge, UK

Thiemo Pfab
Center for Cardiovascular Research/Department of Pharmacology and Toxicology, Charité, Hessische Strasse 3-4, Berlin, 10115, Germany, Department of Nephrology, Charité – Campus Benjamin Franklin, Berlin, Germany

Markella Ponticos
Division of Medicine, Research Department of Inflammation, Centre for Rheumatology and Connective Tissue Diseases, Royal Free and University College Medical School, University College London, Royal Free Campus, Rowland Hill Street, London, NW3 2PF, UK, m.ponticos@medsch.ucl.ac.uk

Arshed A. Quyyumi
Division of Cardiology, Emory University, 1364 Clifton Road, Suite D403C, Atlanta, GA, 30322, USA, aquyyum@emory.edu

Claire L. Schwartz
Clinical Sciences at South Bristol, Bristol Heart Institute, Bristol Royal Infirmary, Bristol, Avon, UK

James R. Seibold
University of Michigan Scleroderma Program, University of Michigan Health System, Ann Arbor, 48128, Michigan, USA

Leslee J. Shaw
Professor of Medicine (Cardiology), Emory University School of Medicine, Atlanta, GA, USA

Xu Shi-wen
Centre for Rheumatology, University College London, Royal Free Hospital, London, NW3 2PF, UK

Beatriz Somoza
Departamento de Farmacología, Tecnología y Desarrollo Farmacéutico, Facultad de Farmacia, Universidad CEU-San Pablo, Madrid

Domingos S.R. Souza
Department of Thoracic and Cardiovascular Surgery, Örebro University Hospital, Örebro, Sweden

Richard J. Stratton
Centre for Rheumatology, Royal Free Hospital, London, UK

Jay Suntharalingam
Consultant Respiratory Physician, Respiratory Department, Royal United Hospital Bath, Combe Park, Bath BA1 3NG, UK

Viola Vaccarino
Emory University School of Medicine, Rollins School of Public Health, Emory University, Atlanta, GA, USA

Karoline von Websky
Center for Cardiovascular Research/Department of Pharmacology and Toxicology, Charité, Hessische Strasse 3-4, Berlin 10115, Germany, Department of Nephrology, Charité – Campus Benjamin Franklin, Berlin, Germany

Nanette K. Wenger
Emory University School of Medicine, Emory Heart and Vascular Center, Grady Memorial Hospital, Atlanta, GA, USA

A. Maziar Zafari
Cardiovascular Training Program, Emory University School of Medicine, 1639 Pierce Drive, 322 WMB, Atlanta, GA, 30322, USA, azafari@emory.edu

Section 1

Part 1
Genetics and Gene Therapy

Chapter 1
Oligonucleotide Therapeutics to Treat Dyslipoproteinemia and Atherosclerosis

Ioannis Papaioannou and James S. Owen

1.1 Plasma Lipoproteins and Dyslipoproteinemia

1.1.1 Delivering Lipid Nutrients

Plasma lipoproteins are large macromolecular particles integral to the transport of lipids between the liver, intestine, and peripheral tissues (Fig. 1.1).[1] Dietary triglycerides and other fat-soluble substances are absorbed from the intestine and packaged into chylomicrons, which are secreted into the bloodstream via the lymphatics. Following lipolysis by endothelium-bound lipoprotein lipase, their energy-rich triglyceride is delivered to peripheral tissues as free fatty-acids, leaving behind cholesterol-enriched chylomicron remnants, which are rapidly cleared by the liver.[2] In contrast, following carbohydrate ingestion or fasting, endogenous triglycerides and also cholesterol are packaged into very-low-density lipoprotein (VLDL) for transport from liver to the periphery.[2,3] As with chylomicron lipolysis, the delivery of triglycerides decreases VLDL volume and changes its composition, forming initially intermediate-density lipoprotein (IDL) and then low-density lipoprotein (LDL) particles. The half-life of LDL, which carries three-quarters of plasma cholesterol, is much longer than chylomicron remnants, and the particles are catabolized by LDL-receptors in both peripheral tissues and the liver.[1]

1.1.2 Reverse Cholesterol Transport

The remaining major class of lipoproteins, high-density lipoprotein (HDL), is involved in reverse cholesterol transport, the pathway by which excess cellular cholesterol is transferred from the periphery to the liver for excretion (Fig. 1.1).[4-6]

I. Papaioannou and J.S. Owen (✉)
Department of Medicine, Royal Free and University College Medical School,
Royal Free Campus, Rowland Hill Street, London, NW3 2PF, UK
e-mail: j.owen@medsch.ucl.ac.uk

D. Abraham et al. (eds.), *Advances in Vascular Medicine*,
DOI 10.1007/978-1-84882-637-3_1,

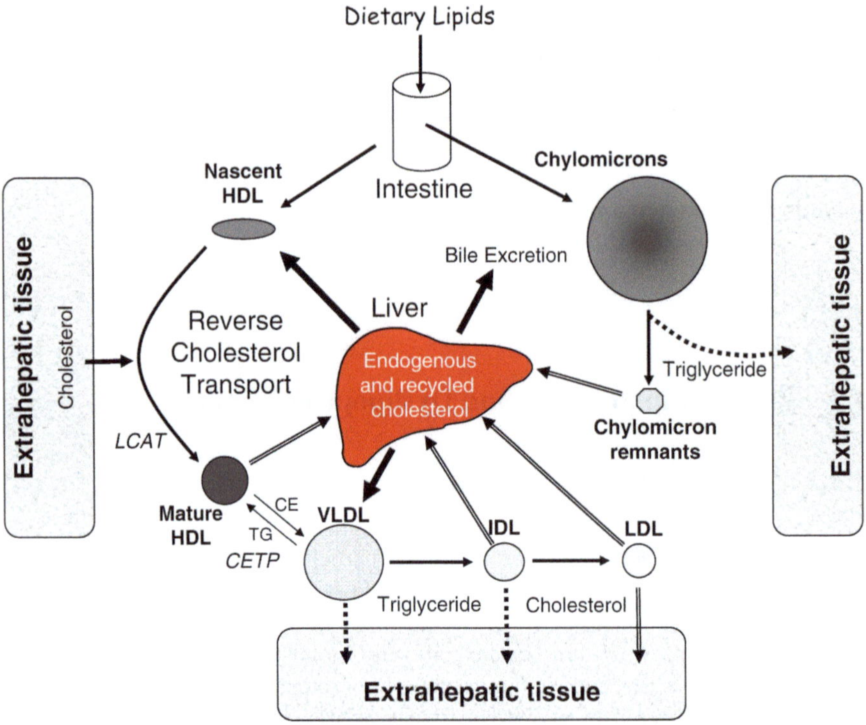

Fig. 1.1 The lipoprotein transport system for plasma lipids. Dietary lipids are absorbed in the intestine and packaged into chylomicrons, which then deliver energy-rich triglycerides to peripheral tissues, particularly muscle and adipose. The resultant remnant particles are cleared by hepatic receptors via ApoE. Endogenous lipids in liver are secreted as very-low-density lipoprotein (VLDL), which undergoes a similar lipolysis to generate cholesterol-rich low-density lipoprotein (LDL) particles that are taken up equally by liver and extrahepatic tissues. Reverse cholesterol transport from periphery to liver occurs when nascent discoidal high-density lipoprotein (HDL) (lipid-poor ApoAI) sequesters cellular cholesterol, which, following esterification to facilitate internal packing and formation of mature spherical HDL, is then transported to liver, for recycling or excretion

Newly synthesized or nascent HDL particles are discoidal and secreted by the enterohepatic system. They sequester free cholesterol from peripheral cells, which is esterified by the plasma enzyme lecithin-cholesterol acyltransferase (LCAT) to form cholesteryl esters that move into a hydrophobic core. Loading of nascent HDL with cholesterol produces mature, spherical HDL particles, which can deliver their cholesteryl ester to liver via scavenger receptors.[4] Alternatively, some cholesteryl esters are transferred to VLDL and LDL in exchange for triglyceride by the cholesteryl ester transfer protein (CETP), and so are eventually taken up by hepatic LDL-receptors.[5,6]

1.1.3 Clinical Relevance

The metabolism and relative concentrations of plasma lipoproteins are of profound clinical importance, as dyslipoproteinemias are common and are major risk factors for cardiovascular disease (CVD), the leading cause of death worldwide. High plasma cholesterol levels in the form of LDL or remnant particles result in deposition of lipids in the arterial wall, ultimately causing atherosclerosis.[7] Moreover, low HDL is an important independent risk factor because of its role in removing excess cholesterol from arterial cells.[1,4,5] Controlling risk through lifestyle modification and/or drugs, such as statins to lower LDL, are major public health goals. Nonetheless, about two-thirds of adverse cardiovascular events continue despite these interventions; increasingly, new therapeutic approaches are required.

1.2 Key Protein Targets within Plasma Lipid Transfer Pathways

1.2.1 Apolipoprotein B (ApoB)

1.2.1.1 Structural and Receptor-Binding Functions

Apolipoprotein B (ApoB) is the principal protein component of VLDL, IDL, LDL, and chylomicrons.[8] It is primarily expressed in liver and intestine, but the two translational products differ. In the intestine, ApoB mRNA is subject to post-transcriptional editing, which results in a premature stop codon.[1,8] Thus, both the full-length protein, ApoB100 (515-kDa), and a carboxy-terminus truncated form, ApoB48 (244-kDa; hence 48% the size of ApoB100), are produced from the same gene. Editing of ApoB mRNA is accomplished by the editosome, a multiprotein complex whose active subunit is APOBEC-1 (ApoB mRNA-editing enzyme catalytic polypeptide 1).[4] Circulating liver-derived VLDL has ApoB100 as its major protein component, which in LDL is the sole protein constituent, while ApoB48 is the structural protein of chylomicrons.[1-3,8]

This difference determines their catabolism: as ApoB48 lacks the C-terminal moiety, it cannot bind to the LDL-receptor and hence chylomicron remnants are cleared via ApoE interaction with hepatic LDL-receptor-related protein (LRP), a fast catabolic step. In contrast, binding of ApoB100 is a slow process and the half-life of LDL is 100-fold greater than chylomicrons. This exacerbates the susceptibility of LDL to oxidation and ingestion by monocyte-macrophages to form "foam cells," the lipid streak of early atherosclerotic lesions.[7] As LDL has a central role in atherosclerotic plaque formation, inhibition of ApoB synthesis is considered a therapeutic target to reduce CVD risk. However, genetic evidence suggests that any suppression of ApoB expression must be carefully controlled to prevent adverse consequences.

1.2.1.2 ApoB Deficiency

Familial hypobetalipoproteinemia (FHBL) is an inherited genetic disease, which naturally reduces levels of functional ApoB.[1,9,10] In most cases of FHBL, the mutation results in production of truncated inactive ApoB, although nontruncating missense mutations with impaired secretion, such as R463W, are also documented.[11] The more severe form of the disease abetalipoproteinemia (ABL), in which there is complete absence of ApoB-containing lipoproteins, is often caused by inactivating mutations in the microsomal trigyceride transfer protein (MTP), a molecular chaperone essential for the correct folding and lipidation of ApoB in the endoplasmic reticulum.[12] ABL results in intestinal fat malabsorption, and patients present with severe neurological and other disorders due to lipid-soluble vitamin deficiency.[12] In FHBL, heterozygous carriers have decreased LDL levels and a high incidence of nonalcoholic fatty liver disease and mild intestinal dysfunction, but are otherwise asymptomatic.[1,10,13] Homozygous carriers, on the other hand, have a more severe phenotype similar to that of ABL.[10,12]

1.2.2 Apolipoprotein AI (ApoAI)

1.2.2.1 ApoAI and HDL are Atheroprotective

Apolipoprotein AI is a 243 amino acid (28-kDa) amphipathic protein produced by liver and intestine, and is the main protein component of HDL. It is secreted into the circulation as a longer propeptide, which is proteolytically processed by cleavage of a hexapeptide to mature ApoAI.[14] ApoAI is essential for formation of HDL and several of its biological functions; thus, impaired ApoAI activity results in diminished reverse cholesterol transport and low HDL levels.[4,5] Genetic evidence reveals that mutations associated with defective ApoAI correlate strongly with increased risk for premature CVD.[14] Complete absence of normal ApoAI, a condition known as analphalipoproteinemia, results in undetectable levels of HDL.[1] Various missense mutations leading to premature termination of the protein, such as codon-2 (Q[-2]X) of the ApoAI propeptide[14] or codon 136 (E136X),[15] are associated with markedly decreased plasma HDL and greatly increased risk of premature CVD. Patients homozygous for Q[-2]X show additional pathology such as neuropathy and, due to subretinal lipid deposition, premature cataracts and retinopathy.[14]

1.2.2.2 $ApoAI_{Milano}$

In 1980, a novel "gain-of-function" ApoAI variant termed $ApoAI_{Milano}$ and arising from a point mutation (C→T; R173C) was identified, which appeared to be atheroprotective.[16] Heterozygous patients with the $ApoAI_{Milano}$ mutation have decreased incidence of CVD, despite paradoxical low plasma HDL. Although anti-atherogenic mechanism(s) of $ApoAI_{Milano}$ are poorly understood, the Cysteine-173 substitution allows intramolecular disulfide bond formation, which confers unique structural and functional properties to the HDL particles.[17,18] Some studies in vitro suggest that $ApoAI_{Milano}$ increases cholesterol efflux, while in experimental animals infusion of $ApoAI_{Milano}$ stabilizes or even regresses established atherosclerotic plaque.[19] Indeed, weekly injections of $ApoAI_{Milano}$/phospholipid complexes for 5 weeks caused regression of coronary atheroma in patients with acute coronary syndrome, validating the potential of HDL-based therapeutics to treat CVD.[20]

1.2.3 Apolipoprotein (ApoE)

Plasma ApoE is a polymorphic glycoprotein of 299 amino acids (34-kDa), primarily secreted by liver (90%) and monocyte-macrophages (10%); it is also synthesized by brain, kidneys, and spleen.[21] ApoE plays a critical role in clearing remnant chylomicrons by targeting them to liver for receptor-mediated endocytosis.[21] Two common isoforms arise from wild-type human ApoE3 by single nucleotide polymorphisms (SNPs) and increase risk of coronary heart disease.[1,21,22] The rarest variant ApoE2 (Arg158Cys; 8% frequency) has defective binding to LDL-receptors and LRP, and in homozygous carriers predisposes to Type III hyperlipoproteinemia.[22] ApoE4 (Cys112Arg; 15% frequency) produces dominant hyperlipidemia and, additionally, is strongly associated with increased susceptibility to Alzheimer's disease.[23]

1.2.4 Proprotein Convertase Subtilisin Kexin Type 9 (PCSK9)

PCSK9 is a serine protease, mainly expressed by liver and intestine, with a key role in the catabolism of the LDL-receptor and other lipoprotein receptors of the same family, such as the VLDL-receptor and ApoE receptor 2 (ApoER2), though not LRP.[1,24,25] Initially, it was thought that PCSK9 degraded LDL receptors intracellularly. However, later studies showed that PCSK9 was also secreted into the circulation where it acts as a ligand for the LDL-receptor and, following their endocytosis, directs the receptor for degradation rather than recycling to the plasma membrane. Its physiological significance is emphasized by the close positive correlation between levels of plasma PCSK9 and total or LDL-cholesterol. Missense "gain-of-function" mutations in the *PCSK9* gene are associated with autosomal dominant hypercholesterolemia (ADH), a rare form of familial hypercholesterolemia in which both the LDL-receptor and the ligand binding domain of ApoB100 are normal.[24] Such mutations, including F216L, lead to increased LDL-receptor catabolism and hence higher plasma levels of total cholesterol and LDL.[24] In contrast, "loss-of-function" mutations increase the amount of LDL-receptor protein, thereby reducing plasma LDL levels[25,26]; indeed, patients homozygous for the C679X mutation, which produces a 14 amino acid truncated PCSK9, have severe hypobetaliproteinemia.[26]

1.3 Oligonucleotide Therapeutics

1.3.1 Overview

Oligonucleotides, either chemically synthesized or generated by selective enrichment strategies, are well-established research tools. Moreover, advances in their production and in vivo delivery have seen them emerge as a novel class of nonsmall molecule therapeutics, and several products are in clinical trial. RNA aptamers and short interfering RNA (siRNA) are at the forefront of clinical development, but other technologies also offer promise.[27-29] Here, we focus on three key types of oligonucleotide therapeutics, which have potential to treat dyslipoproteinemias and atherosclerosis: RNA interference, exon skipping, and oligonucleotide-directed gene editing. The first two, RNA interference[30] and exon skipping,[31] are now established examples of antisense oligonucleotide technology for manipulating gene expression, while gene editing[32] is a radical technique, uniquely suited for introducing small, permanent changes into the genome of the target cells.

1.3.2 RNA Interference

1.3.2.1 Mechanism

Gene silencing or RNA interference is a form of post-translational gene down regulation that is activated in response to double-stranded RNA.[30] Most likely, it evolved as a defense mechanism against parasitic RNA sequences, for example, viruses or transposons, but is now in widespread use as a research tool and as a developing clinical therapeutic. The mechanism of RNA interference has been elucidated (Fig. 1.2).[30] Double-stranded RNA is recognized and processed by Dicer, an enzyme that makes staggered cuts to cleave it into shorter 21 nucleotide double-stranded fragments (short interfering RNA or siRNA). In turn, the siRNA fragments are targets for RNA-induced silencing complex (RISC), a multiprotein cluster, which becomes activated by degrading one of the siRNA strands and incorporating the second. The activated RISC then binds mRNA sequences complementary to the assimilated RNA strand, resulting in mRNA degradation and hence gene downregulation.

This process is commonly exploited experimentally to silence genes in two ways: (1) transfection of synthetic preformed siRNA duplexes; or (2) transfection of a plasmid or viral vector that produces short hairpin RNA (shRNA).[30] Synthetic siRNAs are designed, so their sequence matches only the gene of interest. When transfected into cells, the siRNA is recognized by RISC and used to target and degrade mRNA; this produces a transient and sequence-specific downregulation of gene expression. The siRNA-induced silencing is temporary because the siRNAs and the activated RISCs have finite lifetimes. Vectors producing shRNA are constructed

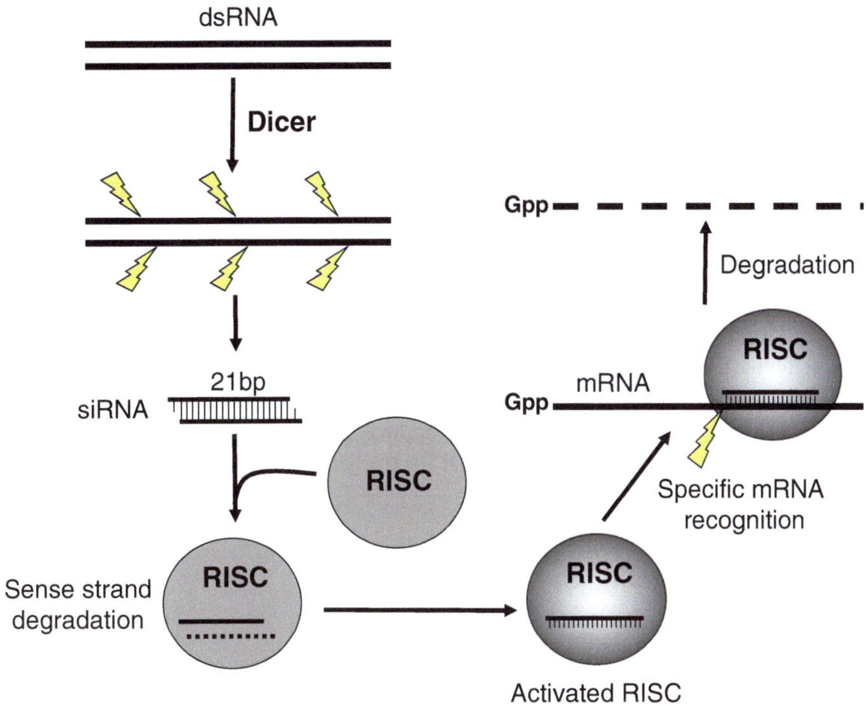

Fig. 1.2 RNA interference. Double-stranded RNA (dsRNA) is recognized by Dicer, an enzyme which chops it into 21–23 base pair fragments with end-overhangs of 1 nucleotide, termed short interfering RNA (siRNA). These are bound by the multienzyme RNA-induced silencing complex (RISC), which is primed by displaying one siRNA strand after degrading its partner. The siRNA strand retained by activated RISC recognizes and binds complementary mRNA targets, which are then destroyed by the enzyme complex. Thus, targeting mRNA by transfecting cells with short dsRNA that has sequence match marks the mRNA for degradation and hence silences gene expression

to express a short RNA sequence that forms a hairpin structure. The double-stranded part of the hairpin is designed to match the target gene sequence. This structure is recognized by Dicer and processed into siRNAs that then downregulate gene expression. As the plasmid or viral vector persists, especially if selection is applied, it can be used for long-term gene silencing.

Use of unmodified (naked) RNA in vivo is limited by its susceptibility to ubiquitous RNases and its inability to cross cellular plasma membranes.[33] Indeed, naked siRNA molecules administered to animals are eliminated rapidly and fail to reach target tissues.[33] This problem was initially circumvented by using RNA with a modified, nuclease-resistant backbone and by conjugation to lipophilic molecules, such as cholesterol,[33] which facilitate entry into cells. Although this approach is relatively effective, newer liposomal formulations of siRNAs are superior and show much promise as potential future therapeutics.[34]

1.3.2.2 Lipoprotein Targets

There are two reports of reducing ApoB expression in vivo by using siRNA. The first was intravenous injection (liver-directed) of cholesterol-conjugated ApoB-specific siRNA into transgenic mice expressing human ApoB100. Significantly lower levels of plasma ApoB100 were seen, which were accompanied by 40–50% reductions in chylomicrons, LDL, and total cholesterol.[33] A later study in cynomolgus monkeys showed that a single injection of a liposomal formulation of ApoB siRNA reduced levels of ApoB mRNA by >80% for a period of 11 days; concomitant reductions of 62 and 82% in plasma total cholesterol and LDL were also noted.[34] An additional target for cholesterol lowering, well-suited to siRNA treatment and a major focus for drug companies, is PCSK9, as its downregulation would increase LDL-receptor protein and hence reduce circulating LDL. Indeed, a study in vivo in nonhuman primates showed siRNA to reduce PCSK9 expression by 70% and plasma LDL by 60%, while leaving HDL or triglyceride levels unchanged.[35]

1.3.3 RNA Splicing and Exon Skipping

1.3.3.1 Mechanism

An interesting property of eukaryotic genes is their discontinuous nature. The protein coding sequence exists in segments called exons that are separated by stretches of noncoding sequence, known as introns.[36] Gene transcription produces a long pre-mRNA molecule containing both exons and introns, which is then converted to mature mRNA (Fig. 1.3). This process, known as RNA splicing, removes introns and joins the exons into one continuous protein-coding sequence.[36]

The mechanism of RNA splicing has been characterized[36,37] in some detail (Fig. 1.4). Splicing is carried out by the spliceosome, a large nucleoprotein complex that comprises all necessary enzymes and factors, and assembles directly on the pre-mRNA molecule. There are five types of sequences that are important in defining introns and exons (see Fig. 1.4): the 5′ and 3′ splice sites, the branch point sequence (BPS), the polypyrimidine tract, and the splicing enhancers/silencers. The 5′ and 3′ splice sites are the intron/exon boundaries and are defined through conserved sequence elements and via binding of the splicing machinery components and accessory proteins during spliceosome assembly. As the spliceosome catalyzes intron removal and joining of the two adjacent exons, it follows that if one of the splice sites is incorrectly chosen, or skipped, then this will drastically change the final mRNA sequence, as intron sequence might be included, or exon sequence deleted (Fig. 1.5).

Where a splice site is skipped, leaving the next suitable splice site along the pre-mRNA to be selected instead, one or more exons can be completely omitted. This can occur naturally, in a process termed alternative splicing (Fig. 1.5), which allows more than one protein product to be produced from the same pre-mRNA. In most cases, alternative splice site selection is regulated by the presence of factors that

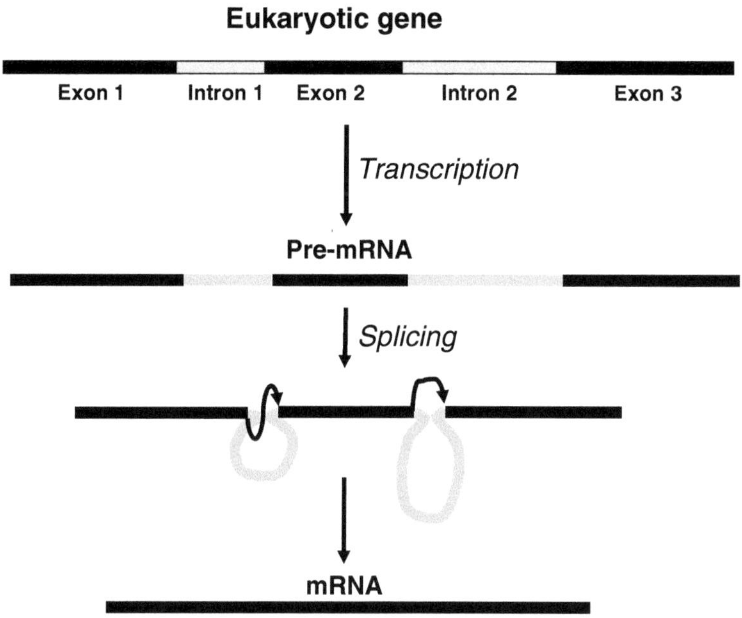

Fig. 1.3 Structure of eukaryotic genes. Protein coding segments (exons) of eukaryotic genes are interspersed with noncoding regions, termed introns. Gene transcription produces a pre-mRNA that contains both introns and exons. Formation of mature mRNA requires excision of each intron and the merging of adjacent exons

bind to splicing enhancers/silencers.[36] However, erroneous disruption of the splicing signals by naturally occurring mutations can give rise to aberrant, often nonfunctional, products with pathological consequences. Recently, short antisense RNA oligonucleotides (ASO) were shown to interfere with the splicing machinery by inhibiting the binding of splicing factors and thereby artificially dictating the choice of splice sites.[31] It is feasible, therefore, to use specific ASOs for blocking selection of a splice site and hence force the splicing machinery to use the next available one, causing an exon to be skipped (Fig. 1.5). This outcome, termed exon skipping, can be exploited to remove exons, which through mutations have deleterious effects on the final gene product.

The most successful application of exon skipping is in Duchenne muscular dystrophy (DMD), a muscle wasting disorder caused by dysfunctional forms of the cytoplasmic structural protein, dystrophin. Usually, this occurs through frameshift mutations, or formation of internal stop codons, which completely disrupt protein function.[31] The central part of dystrophin contains multiple repeats of the same sequence and partially functional dystrophin can be produced by small internal deletions, which skip certain repeats but keep the reading frame intact. This type of defect results in a much milder phenotype, termed Becker muscular dystrophy (BMD).[31] Importantly, the tolerance of dystrophin to internal deletions can be utilized

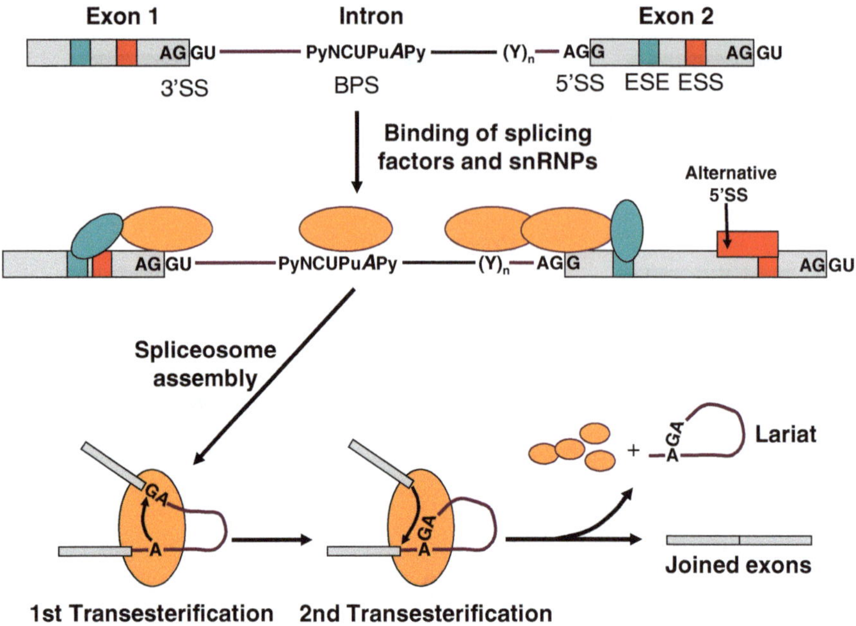

Fig. 1.4 mRNA splicing. This is mediated by the spliceosome, a ribonucleoprotein particle containing several small nuclear RNAs (snRNP, or small nuclear ribonucleoprotein), predominantly U1, U2, U4, U5, and U6, and various accessory proteins, which act as splicing enhancers or silencers. The spliceosome binds five sequence elements within pre-mRNA, which define introns and exon boundaries: the 5′ and 3′ splice sites (5′SS or 3′SS), the branch point sequence (BPS), the polypyrimidine tract, and exon splicing enchancers/silencers (ESE/ESS). This complexity is important, as the relative affinities of the sequence elements for the splicing factors expressed in each cell decides which pairs of splice sites are actually used. Once assembled, the two splice sites and spliceosome catalytic components are brought together, allowing splicing to proceed via a double transesterification reaction. First, the 2′ OH on the ribose ring of the invariant BPS adenine (shown in italic) attacks the phosphodiester bond at the intron–exon boundary in the 5′ splice site (i.e. the bond between the last nucleotide of the upstream exon and the first nucleotide of the intron). This releases the 3′ end of the upstream exon and leaves the 5′ end of the intron bound to the invariant adenine on the BPS, forming a "lariat" structure. The free 3′ end of the upstream exon is now available to attack the phosphodiester bond at the intron–exon boundary in the 3′ splice site. This leaves the upstream and downstream exons joined, while the intron (still as a lariat structure) is released

therapeutically: out-of-frame DMD mutations can be "rescued" to partially functional forms, by skipping appropriate exons. This either removes the mutation completely if it preserves the reading frame or, if skipping changes the reading frame, it restores the reading frame in the remaining sequence. This approach was used in a mouse DMD model to remove mutated exon 23 and produced a partially functional dystrophin, rescuing the animals muscle from degeneration.[38] Indeed, phase I/II clinical trials are underway to assess safety and efficacy of locally injected ASOs directed against exon 51.[39]

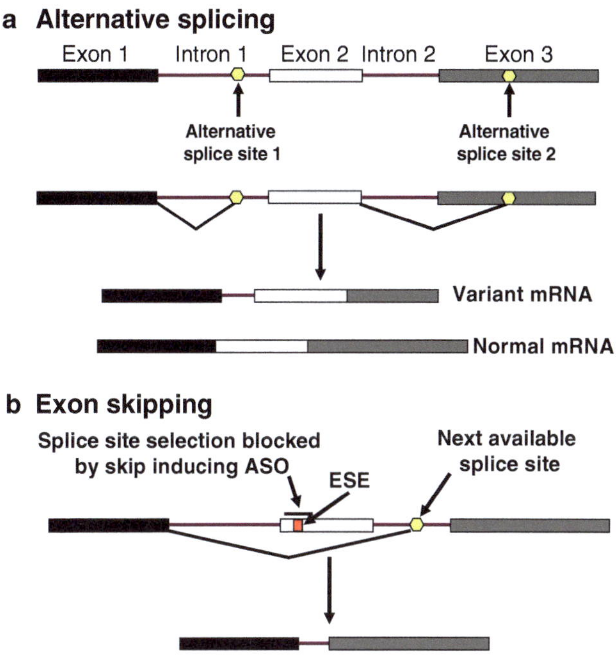

Fig. 1.5 Alternative splicing and exon skipping. (**a**) Molecules of mRNA often have more than one pair of splice sites for each exon. Use of such alternate splice sites may result in exclusion of some exon sequence, inclusion of some intron sequence, or both; this produces, from the same gene, mRNA with significantly different sequences. Notably, by regulation of appropriate splicing factors, this mechanism termed alternative splicing allows different cell types to produce distinct product(s) from a particular gene. (**b**) Splice site selection can be blocked artificially using antisense oligonucleotides (ASO), for example (as shown) by targeting an exonic splicing enhancer (ESE) or silencer (ESS) element. This blocks binding of the ESE or ESS, which in turn reduces the affinity of additional splicing factors for the elements pertaining to this splice site, potentially preventing its selection

1.3.3.2 Lipoprotein Targets

Exon skipping is also being considered as an alternative to ApoB siRNA for treating hypercholesterolemia.[40] Although RNA interference successfully reduced plasma LDL, it is uncertain whether impaired production of chylomicrons might cause liver toxicity and intestinal dysfunction as is sometimes observed in FHBL patients. The strategy is based on the observation that skipping of ApoB exon 27 generates a product, $ApoB87_{SKIP27}$, which is almost identical to $ApoB87_{Padova}$, a naturally occurring dominant ApoB variant from patients with FHBL.[40] Although $ApoB87_{Padova}$ produces functional LDL particles, they have much faster catabolic rates and hence FHBL kindreds have markedly decreased plasma LDL and total cholesterol. At the molecular level $ApoB87_{Padova}$, and a similar variant ApoB89, have frameshift mutations (single nucleotide deletions) in exons 28 and 29, respectively, which produce truncated ApoB with novel C-termini that confer increased LDL-receptor affinity.[40] Importantly, both natural variants retain exon 26, within which lies the nucleotide targeted by the editosome to create ApoB48, thereby allowing normal assembly and secretion of chylomicrons. It has been established that suitably selected ASOs can indeed induce skipping of exon 27 to generate $ApoB87_{SKIP27}$, albeit with a low frequency.[40] However, since the increased catabolism conferred by these variants is dominant, it is predicted that expressing a relatively small amount of $ApoB87_{SKIP27}$ will still increase LDL catabolism without affecting chylomicron production. Although not yet tested in vivo, this use of "gain-of-function" exon skipping is an exciting application of oligonucleotide therapeutics.

1.3.4 Gene Editing

1.3.4.1 Early Difficulties Using RNA–DNA Oligonucleotides

Oligonucleotide-mediated targeted gene editing is a novel and potentially very powerful technology for introducing permanent genetic changes into a cell's genome.[32,41] In a clinical setting, it represents the ultimate gene therapy protocol for inherited diseases: repair to a defective gene would be permanent and existing enhancers and promoters, and cell-specific control and context to regulate gene expression would be retained. Over a decade ago, chimeric RNA–DNA oligonucleotides (RDOs) were reported to induce 30% editing in an episomal target[42] and 50% in a cell model of sickle cell anemia.[43] Encouraged by such reports, we pioneered the technique in CVD, initially converting the dysfunctional ε2 allele in recombinant CHO cells to wild-type ε3 using 68-mer RDOs. The correction was confirmed at both genomic and protein levels.[44] However, on extending this emerging technology to other targets, we began to question its practicality. Thus, while we demonstrated successful conversions of *APOE4* to *APOE3*[45] and *APOAI* to $APOAI_{Milano}$[46] by PCR-based assays, we also noted poor reproducibility and apparently unstable conversions[45]; such concerns were voiced by others.[47,48] Adverse factors included low-quality RDOs, as these hairpin-capped duplex reagents were difficult to synthesize, which meant higher reagent doses and delivery vehicles were needed to effect gene editing; these amplified cytotoxic and pro-apoptotic actions, or induced cell cycle arrest.[45,49,50]

1.3.4.2 Single-Stranded All-DNA Oligonucleotides (ssODN) Give Reproducibility

To resolve such issues, we undertook a back-to-basics approach and targeted cells that stably-express nonfluorescent EGFP due to a point mutation (TAC→TCC; Tyr^{66} Ser).[50] Successful correction produced green cells, which we accurately quantified by flow cytometry. We also switched from problematic RDOs to short (27-mer) single-stranded all-DNA oligonucleotides (ssODNs), as these 2nd generation reagents have high purity and increased reproducibility. Our experiments help clarify certain inconsistent findings and the apparent instability of edited genes, but most importantly stringently validated gene editing as a real and reproducible phenomenon.[51,52]

The basic ssODN technology involves a design that is complementary to the DNA genomic target sequence, except for the desired change, e.g. a 1–3 nucleotide mismatch, deletion, or insertion, and its direct transfection into the nucleus of cells. The change borne within the ssODN is apparently incorporated into the genome through the action of the cell's own DNA repair machinery (Fig. 1.6). Clearly, gene editing has great therapeutic potential to correct harmful missense or frameshift mutations, and additionally might introduce beneficial "gain-of-function" mutations into normal genes. As yet, conversion efficiencies are low, although true correction rates are often uncertain because many studies are proof-of-principle in which multiple copies of reporter gene integrants have been targeted.

1.3.4.3 Mechanism of ssODN-Mediated Gene Editing

The mechanism of gene editing is still largely unknown, as indicated in Fig. 1.6, although a study with biotinylated ssODN suggests that at least part of the ssODN is physically incorporated into the cell's genome.[53] It is also clear that homologous recombination (HR) is involved, as gene editing is stimulated by expression of RAD51, a key protein for HR, and by inhibition of nonhomologous end joining (NHEJ), a pathway that competes with HR.[54] Another important factor is that the 5′ and 3′ ends of the targeting ssODN must be capable of ligation/extension.[55] Altogether, this evidence suggests that the first step in gene editing is homologous base pairing to the target sequence, followed by a partial strand exchange between the ssODN and cellular DNA.

Perhaps surprisingly, the role of the DNA repair system is also poorly defined. Thus, although the mismatch repair system (MMR) corrects small postreplication errors, including mismatches and 1–3 nucleotide insertion or deletion loops,[41] it appears unnecessary for gene editing; indeed, downregulation of key MMR proteins such as MSH2 (mutS homolog 2) can increase editing efficiency.[43] An interesting observation is that cells that have undergone gene editing appear to arrest, at least temporarily, in the G2 phase of the cell cycle.[50,52] It has been suggested that since MSH2 can trigger G2 cell cycle arrest in response to some forms of DNA damage, it may in part be responsible for the arrest, though this is not yet confirmed.[49] On the other hand, gene editing requires the nucleotide excision repair (NER) pathway, a system which corrects larger, bulkier, and more complex lesions by removing a large piece of the DNA strand harboring the lesion, and then filling in the gap using the undamaged strand as a template. The two endonucleases responsible for the initial strand cleavage reactions, XPF and XPG (xeroderma pigmentosum complementation groups F and G), are implicated in gene editing, as their absence significantly suppresses editing efficiency.[56] Conceivably, XPF and XPG help mediate the strand exchange reaction postulated to occur between genomic DNA and the targeting ssODN. The Cockayne syndrome B (CSB) gene product, which is involved in global DNA repair, is also suggested to enhance gene editing and may provide a link between gene editing and transcription.[56]

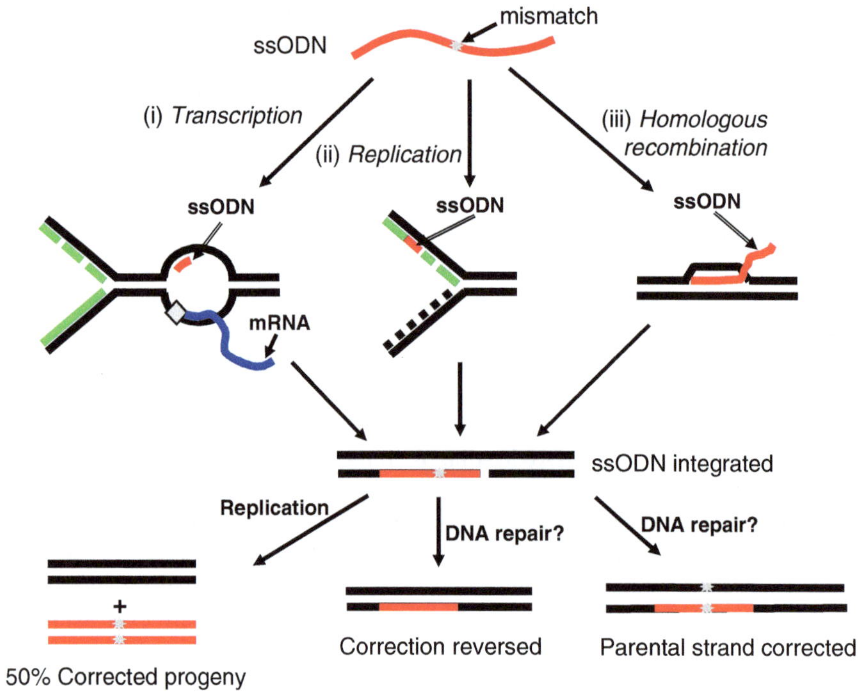

Fig. 1.6 A model of gene editing. Synthetic oligonucleotide (ssODN) is incorporated into the targeted gene via three possible mechanisms: (1) during transcription by binding to the nontranscribed strand, (2) during DNA replication in the replication fork ahead of the DNA polymerase, or (3) directly via the homologous recombination (HR) machinery. Incorporation results in a mismatch as the ssODN contains the desired change to the gene sequence. The mismatch is resolved either by a second round of replication giving rise to 50% corrected progeny, or via DNA repair mechanisms, which can either correct the ssODN (reversing the gene editing event), or correct the parental strand, resulting in a fully edited gene

1.3.4.4 Future Therapeutic Possibilities

Can the efficiency of gene editing be increased? One established approach is to use modified, nuclease-resistant nucleotides; for example, adding phosphorothioate residues to both ends of the ssODN prolongs its half-life and substantially enhances editing efficiency.[52,57] Unfortunately, as demonstrated by recent work from our laboratory, such nuclease-protected ssODNs adversely affect the ability of edited cells to proliferate[52] (see Fig. 1.7); an important consideration as it imparts a strong selective disadvantage to edited cells in the presence of non-edited cells, which can proliferate freely. In addition it may lead to reversion of successful editing events. This is possible because the current model of gene editing proposes that the targeting ssODN is physically integrated into genomic DNA, replacing the original sequence and leaving a mismatched duplex. Such a heteroduplex is thought to persist until the next cell division, when the new round of replication produces one daughter cell homozygous

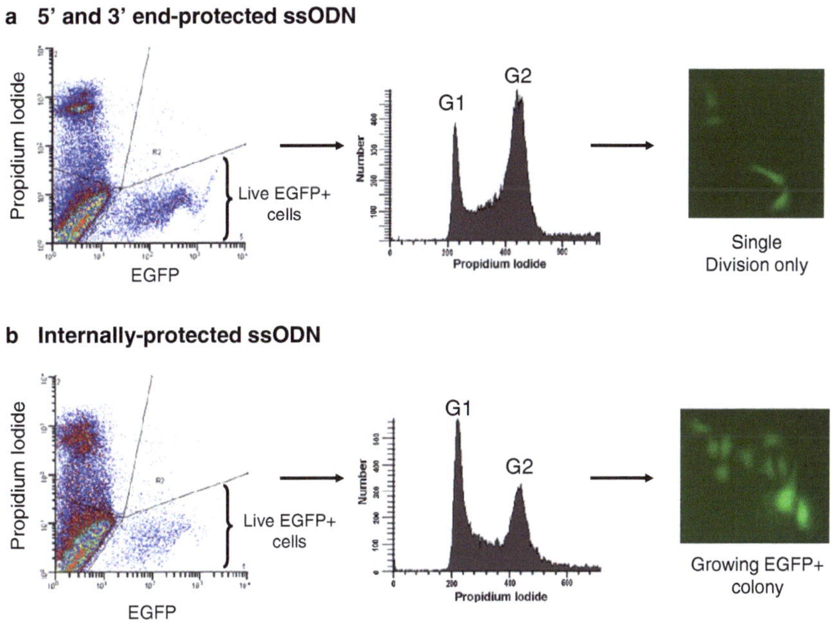

Fig. 1.7 Viability of gene-edited cells is increased by using internally protected oligonucleotides. Use of conventional end-protected ssODNs (three phosphorothioate [PTO] linkages at both 5′ and 3′ ends) to target mutated, nonfluorescent EGFP results in high gene editing efficiency (3–5%). However, it also strongly perturbs the cell cycle, reducing the G1:G2 ratio and only the occasional divided pair of green cells, but no proliferating colonies are seen 64 h post-transfection (**a**). In contrast, if the protecting PTO groups are positioned internally to span the mismatch, the cell cycle is much less disturbed and a high G1:G2 ratio is seen (**b**). Thus, even though the correction efficiency is lower with internally protected ssODN (typically 30–40% of end-protection), the green cells recover faster and growing colonies can be readily identified post-transfection

for the edited sequence and another for the original sequence. However, a delay in cell division could allow recognition of the heteroduplex by DNA repair systems and consequently loss of the editing event.

Despite the controversy and criticism it received along the way, gene editing has now reached a stage of greater maturity. Work from our laboratory and from other groups has consolidated the validity of the technique using ssODNs and established its feasibility for introducing small genetic changes in cultured cells in vitro and ex vivo, and also in creating transgenic mice.[58] Moreover, in a significant advance, we showed that ssODNs, which are internally protected (i.e. having phosphorothioate-modified nucleotides in the center of the ssODN spanning the mismatch), are less harmful and allow active replication of corrected cells[52] (Fig. 1.7).

Ultimately, gene editing can have wide applications to treat dyslipoproteinemias, as most genetic factors that influence plasma lipid levels involve point mutations. As described earlier, these targets include repair of dysfunctional ApoE2 or ApoE4, or generation of atheroprotective $\text{ApopAI}_{\text{Milano}}$. Conversion of wild-type

PCSK9 to the C679X loss-of-function mutant is also an attractive target for reducing plasma LDL levels, while ApoB100 itself is amenable to gene editing, as the simple frameshift mutations causing ApoB89 and $ApoB87_{Padova}$ could be generated by ssODN. Even if the conversions are at low frequency, they are permanent so that a single treatment, or limited number, might produce measurably lower levels of plasma LDL-cholesterol and hence reduce risk of CVD.

The goal for future studies is to release the enormous potential of targeted gene editing for therapeutic gain. One approach in advanced development exploits the observations that double-stranded breaks (DSBs) in genomic DNA are repaired extremely efficiently by the cell's HR machinery and that the template used can be extra-chromosomal DNA. In practice, the DSB is specifically introduced into the target gene using a custom engineered zinc-finger nuclease in the presence of an exogenously delivered homologous DNA template harboring the desired sequence change(s). Repair of the DSB by HR then inadvertently incorporates the genetic changes into the chromosomal target with very high conversion frequencies (perhaps up to 20%).[59,60] However, though offering much promise, including the possibility of introducing targeted inserts of several Kb,[61] there remain important concerns about off-target cleavages and associated genotoxicity, as well as immunogenicity.[60] In contrast, the simple technology of transfecting synthetic oligonucleotides is predicted to have high safety and high fidelity, as increasingly demonstrated by the success of siRNA.

Acknowledgments Dr Papaioannou was supported by a British Heart Foundation project grant (PG/06/015/20305).

References

1. Burnett JR, Hooper AJ. Common and rare gene variants affecting plasma LDL cholesterol. *Clin Biochem Rev*. 2008;29:11–26.
2. Chester A, Scott J, Anant S, Navaratnam N. RNA editing: cytidine to uridine conversion in apolipoprotein B mRNA. *Biochim Biophys Acta*. 2000;1494:1–13.
3. Olofsson SO, Asp L, Boren J. The assembly and secretion of apolipoprotein B-containing lipoproteins. *Curr Opin Lipidol*. 1999;10:341–346.
4. Owen JS, Mulcahy JV. ATP-binding cassette A1 protein and HDL homeostasis. *Atheroscler Suppl*. 2002;3:13–22.
5. Rader DJ. Molecular regulation of HDL metabolism and function: implications for novel therapies. *J Clin Invest*. 2006;116:3090–3100.
6. Rye KA, Barter PJ. Formation and metabolism of prebeta-migrating, lipid-poor apolipoprotein A-I. *Arterioscler Thromb Vasc Biol*. 2004;24:421–428.
7. Lusis AJ. Atherosclerosis. *Nature*. 2000;407:233–241.
8. Fisher EA, Ginsberg HN. Complexity in the secretory pathway: the assembly and secretion of apolipoprotein B-containing lipoproteins. *J Biol Chem*. 2002;277:17377–17380.
9. Willer CJ, Sanna S, Jackson AU, et al. Newly identified loci that influence lipid concentrations and risk of coronary artery disease. *Nat Genet*. 2008;40:161–169.
10. Schonfeld G. Familial hypobetalipoproteinemia: a review. *J Lipid Res*. 2003;44:878–883.
11. Burnett JR, Shan J, Miskie BA, et al. A novel nontruncating APOB gene mutation, R463W, causes familial hypobetalipoproteinemia. *J Biol Chem*. 2003;278:13442–12352.

12. Tarugi P, Lonardo A, Ballarini G, et al. A study of fatty liver disease and plasma lipoproteins in a kindred with familial hypobetalipoproteinemia due to a novel truncated form of apolipoprotein B (APO B-54.5). *J Hepatol.* 2000;33:361–370.
13. Zamel R, Khan R, Pollex RL, Hegele RA. Abetalipoproteinemia: two case reports and literature review. *Orphanet J Rare Dis.* 2008;3:1.
14. Ng DS, Leiter LA, Vezina C, Connelly PW, Hegele RA. Apolipoprotein A-I Q[-2]X causing isolated apolipoprotein A-I deficiency in a family with analphalipoproteinemia. *J Clin Invest.* 1994;93:223–229.
15. Dastani Z, Dangoisse C, Boucher B, et al. A novel nonsense apolipoprotein A-I mutation (apoA-I(E136X)) causes low HDL cholesterol in French Canadians. *Atherosclerosis.* 2006;185:127–136.
16. Franceschini G, Sirtori CR, Capurso A, Weisgraber KH, Mahley RW. A-IMilano apoprotein. Decreased high density lipoprotein cholesterol levels with significant lipoprotein modifications and without clinical atherosclerosis in an Italian family. *J Clin Invest.* 1980;66:892–900.
17. Calabresi L, Sirtori CR, Paoletti R, Franceschini G. Recombinant apolipoprotein A-IMilano for the treatment of cardiovascular diseases. *Curr Atheroscler Rep.* 2006;8:163–167.
18. Favari E, Gomaraschi M, Zanotti I, et al. A unique protease-sensitive high density lipoprotein particle containing the apolipoprotein A-I(Milano) dimer effectively promotes ATP-binding cassette A1-mediated cell cholesterol efflux. *J Biol Chem.* 2007;282:5125–5132.
19. Disterer P, Osman E, Owen JS. Gene therapy for apolipoprotein A-I and HDL–the ultimate treatment for atherosclerosis? In: Abraham D, Handler C, Dashwood M, Coghlan G, eds. *Vascular Complications in Human Disease. Mechanisms and Consequences.* London: Springer; 2007.
20. Nissen SE, Tsunoda T, Tuzcu EM, et al. Effect of recombinant ApoA-I Milano on coronary atherosclerosis in patients with acute coronary syndromes: a randomized controlled trial. *JAMA.* 2003;290:2292–2300.
21. Mahley RW, Rall SC Jr. Apolipoprotein E: far more than a lipid transport protein. *Annu Rev Genomics Hum Genet.* 2000;1:507–537.
22. Mahley RW, Huang Y, Rall SC Jr. Pathogenesis of type III hyperlipoproteinemia (dysbetalipoproteinemia): questions, quandaries, and paradoxes. *J Lipid Res.* 1999;40:1933–1949.
23. Fazekas F, Enzinger C, Ropele S, Schmidt H, Schmidt R, Strasser-Fuchs S. The impact of our genes: consequences of the apolipoprotein E polymorphism in Alzheimer disease and multiple sclerosis. *J Neurol Sci.* 2006;245:35–39.
24. Folsom AR, Peacock JM, Boerwinkle E. Variation in PCSK9, low LDL cholesterol, and risk of peripheral arterial disease. *Atherosclerosis.* 2009;202:211–215.
25. Lambert G, Charlton F, Rye KA, Piper DE. Molecular basis of PCSK9 function. *Atherosclerosis.* 2009;2003:1–7.
26. Cohen J, Pertsemlidis A, Kotowski IK, Graham R, Garcia CK, Hobbs HH. Low LDL cholesterol in individuals of African descent resulting from frequent nonsense mutations in PCSK9. *Nat Genet.* 2005;37:161–165.
27. Wilson C, Keefe AD. Building oligonucleotide therapeutics using non-natural chemistries. *Curr Opin Chem Biol.* 2006;10:607–614.
28. Keefe AD, Schaub RG. Aptamers as candidate therapeutics for cardiovascular indications. *Curr Opin Pharmacol.* 2008;8:147–152.
29. Bhindi R, Fahmy RG, Lowe HC, et al. Brothers in arms: DNA enzymes, short interfering RNA, and the emerging wave of small-molecule nucleic acid-based gene-silencing strategies. *Am J Pathol.* 2007;171:1079–1088.
30. Novina CD, Sharp PA. The RNAi revolution. *Nature.* 2004;430:161–164.
31. Aartsma-Rus A, van Ommen GJ. Antisense-mediated exon skipping: a versatile tool with therapeutic and research applications. *RNA.* 2007;13:1609–1624.
32. Parekh-Olmedo H, Kmiec EB. Progress and prospects: targeted gene alteration (TGA). *Gene Ther.* 2007;14:1675–1680.

33. Soutschek J, Akinc A, Bramlage B, et al. Therapeutic silencing of an endogenous gene by systemic administration of modified siRNAs. *Nature*. 2004;432:173–178.
34. Zimmermann TS, Lee AC, Akinc A, et al. RNAi-mediated gene silencing in non-human primates. *Nature*. 2006;441:111–114.
35. Frank-Kamenetsky M, Grefhorst A, Anderson NN, et al. Therapeutic RNAi targeting PCSK9 acutely lowers plasma cholesterol in rodents and LDL cholesterol in nonhuman primates. *Proc Natl Acad Sci U S A*. 2008;105:11915–11920.
36. Cartegni L, Chew SL, Krainer AR. Listening to silence and understanding nonsense: exonic mutations that affect splicing. *Nat Rev Genet*. 2002;3:285–298.
37. Krämer A. The structure and function of proteins involved in mammalian pre-mRNA splicing. *Annu Rev Biochem*. 1996;65:367–409.
38. Goyenvalle A, Vulin A, Fougerousse F, et al. Rescue of dystrophic muscle through U7 snRNA-mediated exon skipping. *Science*. 2004;306:1796–1799.
39. van Deutekom JC, Janson AA, Ginjaar IB, et al. Local dystrophin restoration with antisense oligonucleotide PRO051. *N Engl J Med*. 2007;357:2677–2686.
40. Khoo B, Roca X, Chew SL, Krainer AR. Antisense oligonucleotide-induced alternative splicing of the APOB mRNA generates a novel isoform of APOB. *BMC Mol Biol*. 2007;8:3.
41. Dekker M, Brouwers C, te Riele H. Targeted gene modification in mismatch-repair-deficient embryonic stem cells by single-stranded DNA oligonucleotides. *Nucleic Acids Res*. 2003;31:e27.
42. Yoon K, Cole-Strauss A, Kmiec EB. Targeted gene correction of episomal DNA in mammalian cells mediated by a chimeric RNA-DNA oligonucleotide. *Proc Natl Acad Sci U S A*. 1996;93:2071–2076.
43. Cole-Strauss A, Yoon K, Xiang Y, et al. Correction of the mutation responsible for sickle cell anemia by an RNA-DNA oligonucleotide. *Science*. 1996;273:1386–1389.
44. Tagalakis AD, Graham IR, Riddell DR, Dickson JG, Owen JS. Gene correction of the apolipoprotein (Apo) E2 phenotype to wild-type ApoE3 by in situ chimeraplasty. *J Biol Chem*. 2001;276:13226–13230.
45. Tagalakis AD, Dickson G, Owen JS, Simons PJ. Gene correction of human apolipoprotein (apo) E4 to apoE3 in vitro using synthetic RNA/DNA oligonucleotides (chimeraplasts). *J Mol Neurosci*. 2005;25:95–104.
46. Manzano A, Mohri Z, Sperber G, et al. Failure to generate atheroprotective apolipoprotein AI phenotypes using synthetic RNA/DNA oligonucleotides (chimeraplasts). *J Gene Med*. 2003;5:795–802.
47. van der Steege G, Schuilenga-Hut PH, Buys CH, Scheffer H, Pas HH, Jonkman MF. Persistent failures in gene repair. *Nat Biotechnol*. 2001;19:305–306.
48. Taubes G. The strange case of chimeraplasty. *Science*. 2002;298:2116–2120.
49. Ferrara L, Kmiec EB. Targeted gene repair activates Chk1 and Chk2 and stalls replication in corrected cells. *DNA Repair (Amst)*. 2006;5:422–431.
50. Olsen PA, Randol M, Krauss S. Implications of cell cycle progression on functional sequence correction by short single-stranded DNA oligonucleotides. *Gene Ther*. 2005;12:546–551.
51. Disterer P, Simons JP, Owen JS. Validation of oligonucleotide-mediated gene editing. *Gene Ther*. 2009;16:824–826.
52. Papaioannou I, Disterer P, Owen JS. Use of internally nuclease-protected single-strand DNA oligonucleotides and silencing the mismatch repair protein, MSH2, enhance replication of corrected cells following gene editing. *J Gene Med*. 2009;11:267–274.
53. Radecke S, Radecke F, Peter I, Schwarz K. Physical incorporation of a single-stranded oligodeoxynucleotide during targeted repair of a human chromosomal locus. *J Gene Med*. 2006;8:217–228.
54. Morozov V, Wawrousek EF. Single-strand DNA-mediated targeted mutagenesis of genomic DNA in early mouse embryos is stimulated by Rad51/54 and by Ku70/86 inhibition. *Gene Ther*. 2007;15:468–472.
55. Olsen PA, Randøl M, Luna L, Brown T, Krauss S. Genomic sequence correction by single-stranded DNA oligonucleotides: role of DNA synthesis and chemical modifications of the oligonucleotide ends. *J Gene Med*. 2005;7:1534–1544.

56. Igoucheva O, Alexeev V, Scharer O, Yoon K. Involvement of ERCC1/XPF and XPG in oligodeoxynucleotide-directed gene modification. *Oligonucleotides*. 2006;16:94–104.
57. Pierce EA, Liu Q, Igoucheva O, et al. Oligonucleotide-directed single-base DNA alterations in mouse embryonic stem cells. *Gene Ther*. 2003;10:24–33.
58. Aarts M, Dekker M, de Vries S, van der Wal A, te Riele H. Generation of a mouse mutant by oligonucleotide-mediated gene modification in ES cells. *Nucleic Acids Res*. 2006;34:e147.
59. Carroll D. Progress and prospects: Zinc-finger nucleases as gene therapy agents. *Gene Ther.*. 2008;15:1463–1468.
60. Cathomen T, Joung JK. Zinc-finger nucleases: the next generation. *Mol Ther*. 2008;16:1200–1207.
61. Moehle EA, Rock JM, Lee YL, et al. Targeted gene addition into a specified location in the human genome using designed zinc finger nucleases. *Proc Natl Acad Sci U S A*. 2007;104:3055–3060.

Chapter 2
Gene Delivery to Cardiovascular Tissue

Rachel Masson, Stuart A. Nicklin, and Andrew H. Baker

Abbreviations

AAV	Adeno-associated virus
Ad	Adenovirus
CABG	Coronary artery bypass graft
CAR	Coxsackie virus and adenovirus receptor
CVD	Cardiovascular disease
EC	Endothelial cells
eNOS	Endothelial nitric oxide synthase
FGF	Fibroblast growth factor
FGFR1	Fibroblast growth factor receptor-1
FH	Familial hypercholesterolemia
HIV	Human immunodeficiency virus
HO-1	Heme oxygenase-1
HSPG	Heparan sulfate proteoglycans
HSV	Herpes simplex virus
iNOS	Inducible nitric oxide synthase
ITR	Inverted terminal repeats
LDL	Low-density lipoprotein
LDLR	Low-density lipoprotein receptor
MMPs	Matrix metalloproteinases
nNOS	Neuronal nitric oxide synthase
NO	Nitric oxide

R. Masson (✉)
BHF Glasgow Cardiovascular Research Centre, University of Glasgow,
126 University Place, Glasgow, G12 8TA, UK
e-mail: r.masson.1@research.gla.ac.uk

D. Abraham et al. (eds.), *Advances in Vascular Medicine*,
DOI 10.1007/978-1-84882-637-3_2, © Springer-Verlag London Limited 2010

NOS	Nitric oxide synthase
NPC	Nuclear pore complex
PDGFR	Platelet-derived growth factor receptor
SERCA2a	Sarco-endoplasmic reticulum calcium ATPase pump
SHR	Spontaneously hypertensive rat
SMC	Smooth muscle cell
TIMPs	Tissue inhibitors of metalloproteinases
VEGF	Vascular endothelial growth factor

2.1 Gene Therapy

Advancement in the understanding of molecular therapeutics has allowed the development of novel treatments to prevent and treat many diseases. Originally conceived for the treatment of inherited monogenic disorders, such as Duchenne's muscular dystrophy and hemophilia, where gene replacement should restore a normal phenotype, gene therapy approaches can now be applied to the treatment of more complex acquired diseases, including cardiovascular diseases (CVDs) and cancers. Before the full potential of gene therapy can be reached, many limitations common to all methods of gene delivery must be overcome. The efficiency of gene transfer will determine how successful the gene therapy application will be. To date, difficulties in achieving sustained gene expression in the target tissue or cell have resulted in limited clinical benefits from gene therapy. The success of gene therapy is restricted by the relative lack of suitable vectors and will depend on the ability of researchers to address a number of still unsolved problems. This can be approached by either the isolation of new viral serotypes that can be developed into vectors or the creation of new vectors by the modification of the existing ones.

2.2 Justification for Gene Therapy for Cardiovascular Disease

CVD remain the leading cause of mortality and morbidity in the western population. An estimated 2.6 million people have CVD in the UK, accounting for over 216,000 deaths in 2004. More than one in three people (37%) die from CVD (www.bhf.org.uk). Despite advances and improvements in treatments, the incidence of CVD continues to increase worldwide. Gene therapy for the treatment of CVD is currently being developed preclinically and tested clinically. Developments in the field of gene therapy have been rapid. Since 1990, over 1,300 clinical trials have been approved worldwide (www.advisorybodies.doh.gov.uk). The majority of clinical trials are for the treatment of cancers (66.5%), with the second biggest field in gene therapy being for the treatment of CVDs (9.1%).

2.3 Therapeutic Genes for Cardiovascular Diseases

With the identification of the genes involved in CVD and the assignment of function to these genes, the potential to translate this information and identify candidate therapeutic genes is enormous. The genes of interest include targets for the treatment of heart failure, such as sarco-endoplasmic reticulum calcium ATPase pump (SERCA2a), targets for treatment of hypertension, including components of the renin–angiotensin system, and targets for the induction of therapeutic angiogenesis, including angiogenic factors, such as the vascular endothelial growth factor (VEGF) and the fibroblast growth factor (FGF).

VEGF production is induced in response to a number of stimuli, such as hypoxia. Its activity can result in a revascularization process, known as therapeutic angiogenesis,[1,2] through the induction of EC growth or proliferation. Being an angiogenic factor, and therefore having the ability to induce the formation of new blood vessels from the existing vascular bed, VEGF is an ideal gene to overexpress in the context of ischemic vascular disease. Direct muscular injection of human VEGF cDNA into patients with ischemic limbs led to an increased blood flow to the limbs and the subsequent healing of ulcers.[3,4] However, some experiments have demonstrated that nonregulated overexpression of pro-inflammatory VEGF can also lead to detrimental effects, including hypotension and arthritis, and so an element of transcriptional control needs to be included. Adeno-associated virus (AAV) vectors expressing the VEGF transgene under the control of hypoxia response elements induced gene expression in ischemic mouse hearts in vivo.[5]

The absence of heme oxygenase (HO)-1 is implicated in the exacerbation of atherosclerosis, demonstrated by the accelerated and more advanced atherosclerotic lesion formation in HO-1-deficient mice.[6] Retroviral-mediated overexpression of HO-1 in the spontaneously hypertensive rat (SHR) resulted in the attenuation of hypertension,[7] while adenoviral-mediated HO-1 gene transfer prevented the development of atherosclerosis in apolipoprotein E deficient mice.[8] Adenoviral-mediated HO-1 overexpression has also resulted in the attenuation of remodeling responses to experimental vascular injury.[9] The many advantageous effects make this gene an important novel target in the treatment of vascular disease.

The potential of gene therapy in the treatment of hypertension has been explored, although this strategy is unlikely to be tested clinically. Nitric oxide (NO) plays an important role in vascular smooth muscle relaxation, and many vascular diseases are influenced by a reduction in NO bioavailability. Gene therapy approaches aim to increase NO bioavailability to improve vascular function. The direct injection of a plasmid expressing human endothelial NO synthase (eNOS) fused to the CMV promoter significantly reduced systemic blood pressure in the SHR, that was prolonged for 5–6 weeks.[10] In vitro, adenoviral-mediated expression of eNOS and inducible nitric oxide synthase (iNOS) had antiproliferative and antiangiogenic effects on endothelial and smooth muscle cells (SMCs).[11-13] Nitric oxide synthase (NOS) is one of the many genes that have beneficial effects on endothelial function and blood pressure. Other vasodilatory promoting genes include atrial natriuretic peptide, human kallikrein, and bradykinin, and are being investigated for their role in the treatment of CVDs.

2.4 Requirements of a Gene Delivery Vector

A multitude of vector systems, viral and nonviral, have been assessed as tools for gene delivery. A generic vector that is suitable for use in all circumstances is unlikely; gene expression is required in different target tissues for varying lengths of time for different conditions. To avoid eliciting host immune responses, a lack of immunogenicity is desirable and would allow for vector readministration. The induction of immune response is a limiting factor, particularly for adenovirus serotype 5 (Ad5) vectors, which target dendritic cells and some monocytes. The removal of virulence genes in viral vectors helps to limit host defenses.[14,15] Vectors capable of sustained transgene expression would avert the problems of vector readministration; however, some gene therapy applications only require transient transgene expression. Vectors must be producible on a large scale resulting in high vector concentrations. To date, no vector possesses all these qualities, although many steps are being made to overcome these hurdles. Each vector system has its own advantages and disadvantages, depending on its intended use.

For cardiovascular gene delivery, vectors with the ability to transduce cells of the vasculature or of the myocardium are being developed. To increase specificity of cardiovascular gene delivery vectors, methods of tropism alteration and incorporation of cell-specific promoters can be applied.[16-18] Vector tropisms need to be modified to allow efficient and selective transgene expression in vascular cells in vivo.

2.5 Ex Vivo and In Vivo Gene Delivery for CVD

Gene delivery approaches are based on two major concepts: ex vivo and in vivo delivery. In ex vivo cell-based gene therapy, autologous cells or tissue are harvested from a patient, incubated with the vector carrying the desired therapeutic gene, and then reintroduced into the patient. Genetically modified cells will express the transgene, usually at high levels. Owing to the lack of effective pharmacological interventions, this method is being developed for gene therapy of vein graft failure during coronary artery bypass graft (CABG) surgery. CABG surgery is performed on patients with significant atherosclerotic narrowing and blockages of the arteries. CABG allows for the incubation of the graft vessel with a gene therapy vector prior to coronary grafting. Late vein graft failure is a common clinical problem[19,20] and occurs due to thrombosis or neointima formation and accelerated atherosclerosis, a process in which a role for matrix-degrading matrix metalloproteinases (MMPs) and neuronal nitric oxide synthase (nNOS), amongst others, has been implicated. Tissue inhibitor of metalloproteinase-3 (TIMP-3) has been shown to inhibit MMP activity and promote apoptosis, thus inhibiting the progression of neointima formation associated with late vein graft failure in human and pig model systems.[21] Adenovirus-mediated overexpression of nNOS-induced beneficial effects on vein graft remodeling and improved endothelial function,[22] demonstrating the potential of this technique.

Transgene expression in nontarget tissue is limited by this ex vivo method by the removal of excess virus prior to engraftment.

Ex vivo gene delivery has also been utilized in the treatment of familial hypercholesterolemia (FH), in which patients have a deficiency of low-density lipoprotein receptors (LDLRs). For this approach, autologous hepatocytes are harvested, transduced with recombinant retroviruses expressing LDLR, and then transplanted back into the patient. This technique has been validated in rabbit models of FH[23] and in patients,[24] both showing persistent and significantly reduced levels of low-density lipoprotein (LDL) cholesterol. However, ex vivo approaches are limited to largely invasive surgical procedures and to tissues and cells that can easily be removed from the body and then reimplanted. Thus, its clinical applications are severely limited. In vivo gene delivery may be able to help overcome this limitation, although faces many challenges of its own.

For in vivo gene delivery, the vector is either administered directly into diseased tissue within a patient, or is systemically delivered and targeted to the site of action by the vector. Local delivery will ensure relatively efficient transduction of target cells unattainable by systemic administration and avoids the need for the delivery vector to cross endothelial barriers, thus resulting in high vector levels in the target tissue.[25] The route of administration has a major influence on the ability of the vector to transduce various cells and tissues. Delivery methods encompass direct injection into the tissue of interest, catheter-mediated gene transfer techniques,[26] or perfusion.[27] Intramyocardial injection of rAAV2 vectors was used to achieve beneficial therapeutic effects in rat ischemia/reperfusion models and demonstrated highly selective transduction of myocardial tissue.[28] Infusion-perfusion catheters have been used in the context of restenosis prevention. In this case, either adenovirus expressing human vascular endothelial growth factor 165 ($hVEGF_{165}$) or plasmid–liposome complexes containing the $hVEGF_{165}$ gene were delivered directly into the artery. However, in both groups there was no significant change in the lumen diameter or clinical restenosis rate when compared with the control group.[29] A surgical technique to improve gene delivery efficiency involves treating the heart with permeability agents in vivo. Simultaneous clamping of all vessels to/from the heart is followed by continuous retrograde perfusion of the heart through a catheter positioned in the aortic root.[30] This technique eliminates excess virus, which ultimately reduces peripheral tissue infection. Local delivery can however result in leakage of transgene expression into nontarget tissues.[31,32]

Systemic delivery is the ultimate goal of gene therapy as it is, in concept, a simple and noninvasive route of delivery. However, the challenge with this approach is that the body has evolved many highly specific systems to remove foreign particles and pathogens from the bloodstream. Many vectors for systemic gene transfer remain ineffective at delivering genes to the vasculature and myocardium, as a result of liver sequestering after vector administration. There is a trend for viral vectors to display tropism for nonvascular tissues. Liver sequestration is a major limitation of Ad vectors, which are mainly based on serotype 5.[33,34] This hepatic tropism limits the use of systemic delivery to gene therapy for liver disorders. Advances in vector technology and development are helping to overcome this major barrier.

Some AAV serotypes have been recently shown to efficiently cross the blood vessel barrier and as such can be intravenously injected.[35,36] The major limitation of these vectors is that other noncardiac organs may also be targeted. Transductional and transcriptional targeting strategies can be used to improve transgene expression and cell specificity.

2.6 Nonviral Vectors

Nonviral vectors account for approximately 25% of the clinical trials currently in operation (www.wiley.co.uk/genmed/clinical). The simplest form of the vector is naked plasmid DNA encoding for the gene of interest and can be directly injected into the target tissue. Nonviral vector gene delivery is highly inefficient with levels of transduction being significantly less than those achieved by viral vector gene delivery. Nonviral vectors have no specific mechanism with which to cross cell membranes or traffic the injected DNA into the host cell nucleus.[37,38] Strategies to improve vector delivery can be categorized into two general groups: (1) the association of the DNA with other molecules, and (2) the application of physical energy to aid cell entry through the cell membrane (Table 2.1). The major problems of nonviral vector delivery include the interactions of the vector–DNA complex with blood plasma proteins and nontarget cells, and entrapment within endosomes from which

Table 2.1 Characteristics of nonviral gene delivery techniques

Method of gene transfer		Advantages	Disadvantages
Physical	Hydrodynamic injection	Potent gene transfer to liver	Restricted to the liver
	Bioballistic (gene gun)	High transfection efficiency	Shallow penetration of DNA into the tissue Short duration of gene transfer Dependent on cell line used
	Ultrasound	Low invasiveness Nontoxic	Relatively short duration of gene expression
Chemical	Liposomes	Large capacity for DNA (>20 kb) Lack of immunogenicity Broad tropism	Low transfer efficiency in comparison to viral vectors Poor efficiency in transduction of nondividing cells
	Polycation DNA complexes	Safe in vivo High transduction efficiency in vitro	Instability Cleared rapidly from blood stream Nonspecific interactions with other proteins
	Peptide DNA complexes	Low toxicity Low immunogenicity	Conjugation reactions may reduce biological activities of the proteins and peptides

the vector must escape. Once inside the target cell, the challenge of resisting nonspecific cytoplasmic degradation and passage through the physical barrier of the nuclear envelope must be faced.[38] Additionally, plasmid DNA that reaches the nucleus remains extrachromosomal and is usually lost during breakdown of the nuclear envelope at mitosis.[39] Recent studies have thus focused on the development of specially designed vectors with reduced affinity for intracellular proteins and cellular surfaces[40,41] and on mimicking viral properties that will allow the nonviral vector to be maintained and replicate in the target cells. As plasmids contain no proteins to interact with cellular receptors, physical methods of gene delivery can be applied to bring the vector into closer proximity with the cell membrane or to temporarily disrupt the cell membrane, making it permeable to the DNA. Potentially, the use of nonviral vectors offers several advantages over the use of viral vectors including ease and thrift of mass-production, lessened immunogenicity, and a lower risk of unwanted transgene expression in nontarget tissues. However, clinical applications of nonviral vectors remain impeded by the low efficiency of transfection and transient transgene expression. Producing sustained gene expression and potentiating the efficiency of delivery remains a goal of nonviral gene therapy applications.

2.7 Viral Vectors

Viruses have evolved highly specialized mechanisms to enable them to insert their genomes into target cells, making them an ideal candidate to deliver therapeutic genes. In a direct comparison of gene transfer vectors for myocardial gene transfer, recombinant (E1-/E3-) adenovirus, recombinant AAV, and recombinant (ICP27-) HSV all exhibited robust transgene expression, while uncomplexed and complexed naked DNA displayed very limited expression.[42] The efficiency of viral vectors can be attributed to the viral proteins that interact selectively with cell surface receptors and potentially in the trafficking of the virus to the nucleus.[43,44] However, low-level expression of viral genes often evokes an adaptive immune response, and as such the host would destroy the vector-transduced cell.[45] Ad vectors in particular evoke strong immune responses and on administration, can activate an innate immune response mediated by the viral particle itself.[45] This type of immune response is not specific and is aimed at clearing the body of foreign particles, being the first line of defense. Rapid clearance of the vector by cellular elements of the innate immune response involves Kupffer cells,[46] activation of the classical arm of the complement pathway,[47] and an inflammatory response. Adaptive cellular responses are subsequently induced, which activates cytotoxic T-lymphocytes (CTLs).[48] B-cells are activated during the humoral response, which can result in the production of neutralizing antibodies, thereby eliminating the option of vector readministration. By removing genes necessary for viral replication to provide space in which to insert foreign genes, viruses can be manipulated to express foreign genes in any cells that the virus transduces. This also minimizes host immune responses through removal of the adaptive arm of the immune response. Recombinant vectors are thus replication

Table 2.2 Characteristics of viral vectors for use in gene therapy

Vector	Ability to integrate	Transgene capacity	Tropism	Immune response activation	Reference
Retrovirus	Yes	9 kb	Dividing cells only	Minimal	51
Lentivirus	Yes	7–9 kb	Dividing and nondividing cells. Ideal for endothelial cells	Minimal	57,76
Herpes simplex virus-1	No	152 kb	Dividing and nondividing cells. Natural tropism for neuronal cells	Minimal	83
Adenovirus	No	36 kb	Dividing and nondividing cells	Strong	45
Adeno-associated virus	Yes	4.6 kb	Dividing and nondividing cells	Minimal	145,178

deficient, and to produce them, the replication genes must be provided in *trans*, either integrated into the genome of the packaging cell line or on a plasmid.

In principle, any virus can be used as a vector. There are five main classes of clinically applicable viral vectors being studied for cardiovascular applications; retroviruses, lentiviruses, HSV, adenoviruses (Ad), and AAV, a summary of which can be seen in Table 2.2. These five vector classes can be further subcategorized according to whether the vector genome integrates into the host chromosome or exists extra-chromosomally.[49] Integrating vectors are associated with an increased risk of insertional mutagenesis,[50] although careful engineering may be applied to minimize these risks. For example, the engineering of vectors that integrate into predetermined sites could allow long-term transgene expression while preventing the detrimental effects through inappropriate integration.[51] Since each vector system has its own unique set of properties, one vector may be preferential above another in a particular setting and will determine its range of uses in gene therapy.

2.7.1 Retrovirus

Retroviruses were the first viral vectors to be used in human gene therapy[52] and approximately 25% of the world's gene therapy clinical trials use retroviruses as their platform vector (www.wiley.co.uk/genmed/clinical). Retroviruses can be further subdivided into oncoretroviruses, lentiviruses, and spumaviruses, all of which are being developed for gene therapy applications to varying extents. Retroviruses are small enveloped RNA viruses, which replicate via an integrated DNA intermediate by the actions of the enzyme reverse transcriptase. The viral genome is approximately 10 kb, comprising at least three genes: *gag* (group-specific antigens), *pol* (reverse transcriptase), and *env* (the viral envelope protein). These viral genes are flanked by long terminal repeats (LTRs), which are required for integration into the host genome and control viral gene expression. The genome also contains a packaging sequence that allows it to be distinguished from the host-cell RNA.[53]

Retroviral vectors have all their viral genes removed and replaced with the transgene of interest, thus rendering them replication-incompetent.[54] Despite their wide use as gene delivery vectors, the small genome of retroviruses allows for only 9 kb of foreign sequence to be inserted. Production of high-titer preparations required for gene therapy applications is problematic. Retroviruses are associated with low-efficiency gene transfer owing to their inability to deliver genes to nondividing cells.[55] Thus, their utility as gene delivery vectors for vascular applications is severely limited as they are not able to infect nondividing vascular cells. These inefficiencies have led to the development of lentiviral vectors, which are capable of infecting both dividing and nondividing quiescent cells.[56-58]

The genome of retroviruses integrates into the host's genome leading to the potential for long-term transgene expression. However, integration is not site-specific and subsequently this vector has many safety concerns associated with it. Random insertion of an LTR sequence adjacent to a cellular proto-oncogene can lead to inappropriate expression of a protein involved in cellular regulation. Random insertional mutagenesis could also disrupt tumor suppressor genes, potentially leading to dysregulation and a malignancy. In 2000, a clinical trial carried out in France to treat children with severe combined immunodeficiency-X1, illustrated the oncogenic potential of retroviral vectors.[59] This study was based on ex vivo transfer of the γc gene into CD34+ cells using a defective gamma Moloney retrovirus-derived vector. After 10 months, the therapy was found to provide sustained full correction of disease phenotype demonstrating the unique potential of gene therapy. However, by 2003, two patients had developed a serious adverse complication consisting of uncontrolled leukemia-like clonal lymphocyte proliferation,[50] with a third case of leukemia-like illness being reported in 2005.[60] Two of the three patients were found to have retrovirus integration within or in close proximity to the LM02 proto-oncogene promoter, which is associated with childhood leukemia. This integration resulted in the inappropriate upregulation of the proto-oncogene and proved fatal in one of the patients.[61] However, the beneficial outcomes in the remaining patients are not to be overlooked. To date, 17 out of 20 patients in both the Paris and London clinical trials have had their immune system restored and has remained functional for over 7 years.[62] One adverse effect has recently been reported in the UK-based clinical trial (www.news.bbc.co.uk/1/hi/health).

2.7.2 Lentivirus

Lentiviruses are a subclass of retroviruses that are increasing being used in gene therapy. In particular, they are being developed for the treatment of neurodegenerative disorders, because of their ability to efficiently transduce cells of the nervous system.[63,64] The lentiviruses used are usually derived from human immunodeficiency virus-1 (HIV-1) and so raise many potential clinical safety concerns. The vector integrated into the genome randomly. To improve the biosafety of these vectors, significant modification to the HIV-1 genome can be made.[65] Deletion of accessory genes *tat, vif, vpr, vpu,* and *nef* produces minimal vectors that contain only genes necessary for replication and packaging, thus minimizing deleterious effects.[65] Development of nonhuman lentiviral-based systems, including simian,[66] feline,[67,68] and bovine immunodeficiency viruses,[69,70] has also been given attention to increase the safety profile of these vectors.

Lentiviruses have a relatively large packaging capacity of up to 8 kb and an ability to infect a wide range of cells. They are also minimally immunogenic having been shown to sustain gene expression for several months[71] without detectable pathology.[72-74] Gene transfer through lentiviruses is relatively stable, as the transgene integrates into the host genome and is copied along with the host genome every time the cell divides. One of the most appealing features of these vectors is that unlike other retroviruses, lentiviruses can infect nondividing cells, being able to enter the nucleus without mitosis.[75,76] This ability makes these vectors ideal for targeting the endothelium, which is largely composed of nondividing cells. Lentivirus transduction of both primary human saphenous vein endothelial cells (EC) and SMC was shown to be efficient and without toxicity,[57] but there are relatively few studies to date. Lentivirus-based vectors have been also shown to be successful at transducing adult cardiomyocytes of a transplanted heart,[77] and the hearts of SHR in a study of cardiac physiology.[78]

Recently, a new generation of lentiviral vectors has been produced with enormous potential. These are in the form of nonintegrating lentiviral vectors. By introducing mutations into highly conserved acidic residues in the viral integrase gene, catalytic site or chromosome binding site, vectors can be rendered integration defective without interrupting viral DNA synthesis or accumulation in the nucleus.[79-81] Efficient sustained transgene expression in vivo is attainable with nonintegrating lentiviral vectors as has been demonstrated in muscle[81] and in rat ocular and brain tissue at levels high enough to improve retinal degeneration in an appropriate disease model.[82]

2.7.3 Herpes Simplex Virus (HSV)

HSV type 1 is an enveloped double-stranded DNA virus containing an icosahedral-shaped capsid surrounded by a layer of proteins referred to as tegument. It has a relatively large genome of 150 kb, which facilitates large foreign DNA inserts of up to 30–40 kb.[83] HSV is able to infect a broad range of cell types including nondividing cells. Natural viral infection can take the form of a cycle of lytic replication or can enter a latent state in which the viral genome persists without the expression of any viral proteins, possibly for the life of the host. Latently infected neurons function normally and do not illicit an immune response.[84] HSV-1 has many key features making it a highly desirable vector for gene delivery. First, it has a large transgene capacity, which is provided by the deletion of genes superfluous for viral replication.

However, because its genome does not integrate, HSV vectors are unlikely to be suitable for the treatment of conditions requiring long-term gene expression. Because of its natural tropism for neuronal cells, it has become a promising vector for the treatment of neurological disorders such as Parkinson's disease.[85] HSV vectors have also emerged as promising vectors in cancer therapies in the form of replication-selective oncolytic vectors.[86-88] These vectors fail to replicate efficiently in healthy cells and will replicate in cancer cells only, destroying them through oncolysis.

2.7.4 Adenovirus

Adenoviruses are nonenveloped dsDNA viruses with an icosahedral capsid consisting of three main structural proteins, hexon, fiber, and penton base, and several minor capsid proteins. Their genomes range in size from 26 to 45 kb. Adenoviruses were first isolated from tonsils and adenoid tissue[89] and are infectious human viruses, which often cause mild infection of the gastrointestinal, upper respiratory tract and eye. Most adenoviral infections are self-limiting being efficiently counteracted by the host's immune system. Deletion of the virulent genes during vector production may help in reducing the pathogenesis of these viruses.

Adenoviral vectors, most commonly Ad5 and adenovirus serotype 2 (Ad2), are a popular choice in gene therapy and such a status has led to much information about them becoming widely available. As such, adenovirus is well characterized and can be easily genetically altered and grown to high titers. They have a high capacity for the insertion of foreign DNA allowing up to 36 kb (helper-dependent Ads) to be accommodated. They were initially deemed promising vectors for cardiovascular gene therapy applications as they were shown to transduce human vascular cells in vitro[90] and in vivo.[91,92] Adenoviral vectors exhibit a tropism for many human cells and can infect quiescent as well as dividing cells,[93] an important characteristic for the transduction of vascular EC and SMC, which have low mitotic rates, even in diseased states.[94] Adenovirus replicates episomally, thus reducing the risk of random integration into the host genome. However, because Ad vectors are nonintegrating, it means that their genomes are lost in proliferating cells, and so transgene expression will be transient, although this may be advantageous in certain clinical applications. Transient gene expression coupled with hepatic tropism is a major limiting factor for adenoviral vectors and has led to their use in niche areas such as vein grafting, where gene transfer can be carried out ex vivo.[21,95]

The major inadequacy of adenoviral vectors is their high immunogenicity. Many individuals produce neutralizing antibodies and memory T cells directed at Ad proteins after exposure to the vectors. This is a result of the expression of viral genes, which trigger a cascade of humoral and innate immune responses.[48] This is a significant problem, as gene expression is consequently short-lived[96] and vector readministration is less effective.[97] In view of this, current studies focus on strategies to eliminate host immune responses,[14,98] and also on engineering vectors with increased transduction of cardiovascular cells. This can be achieved in several ways, one of which involves the abolition of the natural tropism of the virus and subsequently endowing it with a new tropism for the target cell type.[99-101]

2.7.4.1 Ad Vector Development

To reduce the immunogenicity of Ad vectors and create additional space for the insertion of new genetic material, Ad has been altered in several ways to remove unnecessary parts of the genome (Fig. 2.1). Expression of adenovirus proteins occurs in phases – early and late. The adenovirus genome contains five early transcription units (E1A, E1B, E2, E3, E4), two early delayed (intermediate) transcription units, and five late units (L1–L5), and encodes over 70 gene products.[102] The genome is flanked by inverted terminal repeats (ITRs) of 100–140 bp in size that serve as replication origins. Early genes (E1A and E1B) are involved in gene expression regulation and their activation leads to the expression of viral late genes (involved in the expression of structural proteins) and the production of infectious viral particles. The foreign gene can be inserted into the region occupied by either E1 or E3 genes with one or both being deleted in the vector construct. In the first generation Ad vector, the E1 (E1A and E1B) gene is replaced by the gene of interest and the resultant defective virus is propagated in cell lines, such as 293 cells,[103] which provide the early gene products in *trans*. The progeny virus cannot replicate in normal cells and on introduction into the host, it will infect cells and express the foreign gene, but no progeny virus will be produced. As the E3 region of the genome is dispensable in viral replication, many first-generation vectors will also have all or part of the E3 region deleted. Despite these deletions, first-generation vectors still express wild-type late viral genes at low levels and trigger a CTL immune response,[104] resulting in a short duration of transgene expression.

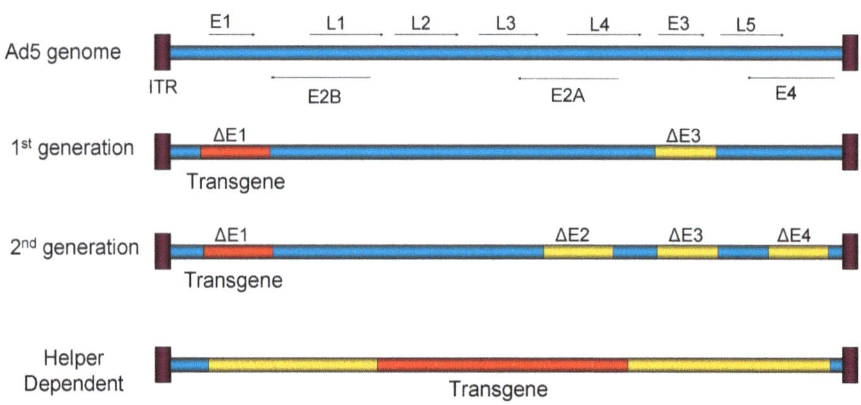

Fig. 2.1 Adenovirus serotype 5 (Ad5) vector development. Adenovirus 5 genome and maps of first-, second-, and third-generation adenoviral vectors showing regions of the genome deleted to facilitate transgene insertion. Adapted from ref.[186]

Second-generation Ad vectors have the E2 and/or E4 regions deleted from their genomes in addition to the E1±E3 deletion. However, second-generation vectors were not found to reduce inflammation of humoral immune response to adenovirus in rabbit models in comparison with first-generation vectors, and most disappointingly did not increase longevity of transgene expression.[96]

Helper virus-dependent or gutless vectors have essential regions of the viral genome (L1, L2, VA, and TP) deleted and rely on the provision of essential viral functions from a helper virus. The gutless adenovirus only keeps the two ITRs and the packaging signal from the wild-type adenovirus required for DNA replication and packaging.[105-108] By deleting most of the viral genomes, it is possible to accommodate up to 37 kb of insert DNA into defective vectors. In vivo studies have shown substantially longer transgene expression with helper-dependent vectors[14,98] sustained up to a year in baboons. However, an innate immune response is still activated against the vectors.[48] Recently, however, Barcia et al demonstrated that helper-dependent Ad vectors mediated sustained transgene expression for up to 1 year in the brains of mice preimmunized against adenovirus.[15] This highlights the potential of these vectors in the treatment of chronic diseases, as the immune system was unable to inhibit transgene expression.

2.7.4.2 Vector Capsid Engineering

There are more than 50 different serotypes of adenovirus, classified into six groups (A–F) based on biochemical and immunological properties. These viruses infect different cell types through the utilization of different primary cellular receptors and thus have a wide tissue tropism range. Most adenoviruses, except subgroup B and the short fiber of subgroup F, use the coxsackie virus and adenovirus receptor (CAR).[109,110] The two-step mechanism of Ad5 infection is well characterized, making it possible to reengineer it to alter its tropism. Ad5 virus interacts in vitro with CAR by means of the knob domain of the capsid fiber, bringing the capsid into close proximity with integrins. After attachment, the RGD motif in the penton base at the N-terminus of the fiber interacts with coreceptors $\alpha_v\beta_3/\alpha_v\beta_5$ integrins.[111] Adenovirus is then internalized by receptor-mediated endocytosis and released by endosomal acidification in fiber-free form to the cytosol, before trafficking to the nucleus. CAR is relatively ubiquitous, resulting in the infection of undesirable tissues as well as target tissues. Ad5 can transduce EC,[90] coronary arteries,[92] the heart[112] and at lower efficiency vascular SMC.[113] This is reflective of the distribution of CAR expression, with high CAR expression leading to high transduction efficiency. Indeed, after systemic injection in the rat and mouse models, Ad5 virions preferentially accumulated in the liver and spleen.[33,114] This highlights the need to substantially alter Ad5 tropism to retarget it to alternative sites, for example the brain, kidney, and heart vasculature.

Genetic strategies to alter adenovirus tropism can either focus on pseudotyping the Ad5 fiber with that of another serotype, or on modification of the existing receptor binding mechanisms. Native hepatic tropism can be altered by mutating the virus in areas integral to cellular receptor binding. The identification and mutation of the residues in the knob involved in CAR binding[115] has allowed the production of modified Ad5.[116]

2.7.4.3 Adenoviral Retargeting by Pseudotyping

The adenovirus fiber protein mediates primary binding of adenovirus to its receptor. Vector retargeting may be achieved through the use of chimeric vectors that incorporate the entire fiber or part of the fiber from a different serotype in place of its own. This could potentially ablate the virus's natural tropism by removal of both the CAR- and heparan sulfate proteoglycan (HSPG)-binding sites and bestow a new tropism upon the vector. Several adenovirus serotypes have shown increased transduction of specific tissues. Proof of the concept of chimeric vectors was first shown in 1996 with the production of functional adenoviral vectors in which the fiber was composed of the tail and shaft domains of Ad5 and the knob domain of serotype 3.[117] Alterations in adenoviral tropism were achieved through primary binding via the Ad3 receptor with subsequent internalization steps achieved via domains of the penton base of Ad5.

Ad5 vectors pseudotyped with serotype 37 and 19p fibers have demonstrated a lack of native tropism for mouse, rat, and human hepatocytes in vitro and demonstrated greatly reduced transduction of liver after systemic injection into rats.[118] Further genetic modifications can allow the development of targeted and thus more efficient vectors. Isolated targeting peptides can be genetically incorporated into the HI loop of the fiber of Ad19p between amino acids 331 and 332. Kidney targeting peptides HTTHREP and HITSLLS, which were identified through in vivo phage display, were incorporated into Ad19p-pseudotyped vectors. These peptide-modified vectors were shown in vitro and in vivo, after systemic administration, to display a significant increase in selective renal targeting with higher levels of transduction than the unmodified Ad19p vectors.[119]

2.7.4.4 Nongenetic Targeting

A simple way of altering vector tropism without genetic modification is the coating of the viral particle with a bispecific antibody. One domain of the bispecific molecule binds to the virus capsid, while the other domain binds to a novel receptor thus acting as a molecular bridge. This concept has been used in vitro to enhance Ad-mediated transduction of human umbilical vascular EC,[120] and in vivo to redirect Ad vectors to a new cellular receptor after systemic delivery.[121,122] Although the addition of a protein adapter enhances the affinity of Ad vectors for their targets, it also increases the difficulty of crossing the barrier from laboratory to clinic as there are more components to be considered and reproduced without batch variation.

2.7.4.5 Retargeting Detargeted Vectors by Ligand Insertion

The insertion of targeting peptides into the fiber gene of Ad5 can provide new tropism to detargeted vectors. The exposed HI loop has been identified as a preferred insertion site for peptides[123,124] without detriment to virion assembly or fiber trimerization. As the fiber is present at a frequency of 36 copies per virion, the vector can display the targeting peptide a maximum of 36 times. Foreign peptides have also been successfully incorporated in the hypervariable region 5 surface loop of the hexon of Ad vectors.[125] In this region, peptides can be displayed at a copy number of 720. However, in a direct comparison of peptide-modified fiber and hexon vectors, hexon-mediated targeting failed to change the tropism of the vectors.[126]

Recent work has shown the application of the phage display technology to identify sequences with desired biological properties, and subsequently introduced these sequences in the retargeting site of the vector.[127-129] One disadvantage of these small targeting peptides is their often weak binding affinity for their targets. The concept of phage display of exogenous peptides was first conceived in 1985, and is simply the display of peptides or proteins on the surface of bacteriophage.[130] The technology of phage display has since been developed and is now used in a wide range of applications, including the rapid isolation of novel peptides with the ability to bind to defined target molecules in vitro or in vivo.[131] For use in cardiovascular applications, phage display could potentially identify ligands, which are specific for the vasculature. Highly efficient and selective peptides can be isolated through the process known as biopanning, which can be carried out in vitro and in vivo. Successive rounds of biopanning enrich the pool of phage with clones that specifically bind the target.

The distinct disadvantage of using in vitro biopanning is that the question remains as to whether the ligands isolated in vitro will display the same specificity in vivo. Phage libraries can be directly introduced into live animals, to select for targeting peptide sequences. However, targeting peptides identified in animal models may not always be applicable and achieve the same level of targeting in humans. In 2002, the first in vivo screening of a peptide library in a patient was carried out.[132] Isolated motifs from tissue biopsies showed high similarity to ligands for cell-surface proteins of the human vasculature. This method has since been used in stage IV cancer patients to identify tumor-targeting ligands.[133] This study displays how this method can be directly applicable in a clinical setting.

2.7.5 *Adeno-Associated Virus (AAV)*

AAV vectors have developed rapidly over the past decade and have become promising vectors for several genres of gene therapy. RAAV2 vectors have been extensively researched and are the most characterized and predominantly used of the AAV vectors. The potential of these vectors in cardiovascular gene delivery was first shown through rAAV-mediated expression of the cytoprotective gene HO-1 in rat myocardium.[28] The safety and efficiency of these vectors was further proven through rAAV-mediated myocardial gene transfer in mice. Transgene expression was observed 1 year postinfusion with no significant inflammatory response or adverse affects on LV systolic function.[134] In a study by Xiao et al,[135] the introduction of rAAV vectors expressing the *lacZ* gene into the muscles of immunocompetent mice resulted in persistent gene expression for more than 1.5 years.[135] AAV vectors are thus minimally pathogenic and possess the ability to mediate long-term transgene expression, and so could prove useful in clinical situations where prolonged transgene expression is desirable. Stable transgene expression is a prerequisite for vectors to treat inherited disorders and would be desirable in the treatment of many acquired CVDs, which progressively worsen over time. However, the progress of AAV vectors has been hampered by their poor transduction of many target tissues.

RAAV vectors evoke little innate immune response, with only transient infiltration of neutrophils and chemokines.[136] Immune response against the virus appears to be restricted to the generation of antibodies specific for the viral capsid protein.[137] AAV vectors are inefficient transducers of antigen presenting cells such as macrophages and dendritic cells, which are believed to be necessary in the production of cellular immune responses.[138] However, recently the duration of transgene expression in the liver mediated by rAAV2 vectors was found to be limited to 8 weeks.[139] Upon further investigation, it was suggested that transduced hepatocytes were destroyed by the activation of T-cells against the capsid of rAAV2.[139,140] Direct comparison of T-cell responses activated against the capsids of rAAV serotypes 2, 7, and 8 revealed little evidence of T-cell activation against rAAV7 and 8 and postulated a potential role for heparin binding in directing immune response against the capsid proteins.[140] Thus, utilization of alternative serotypes that do not use HSPG as their receptor for cell entry may help to avoid this limitation.

In AAV vectors, the viral DNA, except the ITRs, has been eliminated to allow for foreign DNA insertion. This adds a safety feature that will reduce host immune responses directed at viral gene expression and eliminate the possibility of the generation of replication competent pseudo-wild-type AAV. One important safety concern with AAV vectors is the potential for germ-line transmission. Intramyocardial injection of AAV vectors expressing *lacZ* into Sprague–Dawley rats resulted in the detection of *lacZ* expression and β-galactosidase activity in the testes at 6 months postinfusion.[141] In a similar study, Arruda et al found that while vector DNA could be detected in the gonad of rat, mouse, rabbit, and dog, no AAV vector sequences could be detected in the semen.[142] Another major safety concern lies among reports of high incidences of hepatic carcinomas after rAAV vector infusion into mice.[143] Carcinomas that developed in these mice were subsequently found to contain AAV vector proviruses at a specific chromosomal locus,[144] implicating insertional mutagenesis by AAV vectors as a causative factor. These findings raise questions of rAAV vector safety.

2.7.5.1 AAV Biology

AAVs are small 4.7-kb linear single-stranded DNA nonenveloped viruses. Their genomes are organized in similar ways, being extremely simple in composition and containing only two large open reading frames (ORFs) flanked by ITRs of approximately 145 bp, which are required for viral genome replication and packaging (Fig. 2.2). The two ORFs encode two genes, rep (replication) and cap (capsid), which are, respectively, involved in gene expression regulation and structure. Four multifunctional rep isoforms with molecular masses of 78, 68, 52, and 40 kDa are encoded by the 5′ ORF and are transcribed from two different promoters. The rep proteins are involved in specific DNA-binding, helicase, and site-specific endonuclease and modulation of transcription of viral genome promoters. The 3′ ORF encodes three capsid proteins (VP1, VP2, and VP3) through alternate splicing of the cap gene. All three proteins use the same stop codon, and so VP2 and VP3 are successive amino-terminal truncated forms of VP1. The three proteins interact together to form a capsid with icosahedral symmetry. When used as gene delivery vectors, the rep & cap genes, which make up 96% of the genome, are replaced by the transgene. Recombinant vectors are produced by supplying these deleted genes in *trans*. The resultant vectors are less likely to evoke a host immune response. The small size of the AAV virion is responsible for the limited DNA packaging capacity and is a major disadvantage of AAV vectors. Transgenes can be packaged as long as they are not significantly larger (119% maximum capacity) or smaller than the wild-type genome.[145] Without these limits, the resultant vectors are severely defective with regard to producing infectious virions. One method to overcome this limitation is the trans-splicing of larger genes between two independent AAV vectors that will be coadministered.[146] This technique utilizes the ability of AAV genomes to combine, although results in lower transgene expression as a result of the complexity of the system. However, further development may increase the utility of AAV vectors allowing them to appeal to a wide range of applications.

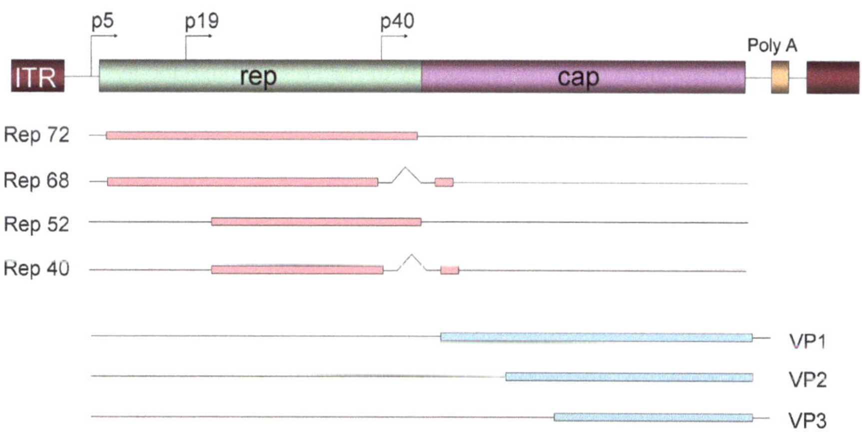

Fig. 2.2 Genome organization of adeno-associated virus (AAV). The AAV genome is a 4.7 kb single-stranded linear DNA genome and is made up of 2 genes, rep and cap, with two flanking inverted terminal repeats (ITRs). Three different promoters drive transcription: P5, P19, and P40. Four transcripts encode nonstructural genes (Rep72, 68, 52, and 40) and three transcripts encode structural proteins (VP1, 2, and 3)

2.7.5.2 AAV Replication

AAVs are helper-dependent viruses with a bi-phasic life cycle. They cannot replicate autonomously, instead requiring coinfection with an unrelated virus, such as Ad or HSV, to complete its life cycle. In the absence of coinfection, AAV can undergo latent infection as an episome or may integrate its viral DNA into the host genome[147] in human chromosome 19 by site-specific recombination directed by the viral *rep* function.[148] AAV genomes can be excised from the host genome in the presence of helper factors and can lead to a productive infection cycle.[149] It is important to note that rAAV vectors lack the integration function as their viral genes have been removed. Advances in AAV vector production have eliminated the need for helper adenovirus infection.[150] Instead, to be packaged into functional vectors, genomes must be provided with all rep, cap, and helper functions *in trans* on exogenous plasmids.[150,151]

There are several common stages for replication of all AAV vectors that must be carried out for successful transgene expression. The first step in infection is the attachment of the vector to the cell surface receptor, and in the case of AAV vectors will require the use of coreceptors to assist in internalization. The virus must then be internalized into the cell by the process of receptor-mediated endocytosis. The vector is subsequently trafficked from early endosomes to late endosomal compartments.[152] It must then escape the endosome to be released into the cytosol and undergo nuclear translocation. Endosomal processing is thought to be an essential step for AAVs, exemplified by the fact that AAV2 directly injected into the cytosol fails to reach the nucleus.[44] After endosomal release, which may occur through weak acidification of the vesicle, AAV rapidly traffics to the nucleus and accumulates in the perinuclear region.[153] Nuclear translocation was initially thought to occur through the virus slowly penetrating the nuclear pore complex (NPC) into the nucleus, with the majority of the virus remaining in perinuclear compartments.[153] However, entry into the nucleus has since been shown to occur independently of the NPC through the use of agents that block NPC function.[154] It is unknown whether viral uncoating to release the genome occurs within or outside the nucleus. However, capsid proteins[155] and the necessary machinery for virion uncoating[154] have been identified within the nucleus, suggesting that nuclear virion uncoating may be a reality, although direct evidence is lacking. The single-stranded DNA genome is converted to double-stranded DNA within the nucleus and is then the transcription template. After entry into the host cell nucleus, the virus can either establish a lytic or lysogenic life cycle, which is determined by the presence or absence of helper virus. The efficiency of all these replication steps will determine the overall efficiency of the vector.

2.7.5.3 AAV Serotypes and Receptors

To date, over 100 AAV genetic variants have been isolated.[156] Twelve known serotypes of AAV have been identified, all displaying a variety of tissue tropisms and receptor-binding characteristics (Table 2.3). The sequence identities among the

Table 2.3 Adeno-associated virus (AAV) serotypes and their varying tropisms and receptors

Serotype	Tropism	Receptor
AAV1	Skeletal muscle[179] cardiac tissue[112]	α2–3 linked or α2–6 linked sialic acid
AAV2	Broad tropism – muscle, brain, retina, liver, lung.	HSPG, αVβ5 integrin, fibroblast or hepatocyte growth factor receptors, 37/67-kDa laminin receptor
AAV3	Cochlear inner hair cells[180]	heparin, heparan sulfate, and FGFR-1, 37/67-kDa laminin receptor
AAV4	Ependymal cells[181]	α2–3 O-linked sialic acid
AAV5	Neurons[182], dendritic cells[183]	PDGFR, α2–3 N-linked sialic acid
AAV6	Skeletal muscle, cardiac tissue[35]	α2–3 linked or α2–6 linked sialic acid
AAV7	Skeletal muscle[184]	Unknown
AAV8	Liver[184]	37/67-kDa laminin receptor
AAV9	Liver, skeletal muscle, cardiac tissue[36]	37/67-kDa laminin receptor
AAV10	Liver, heart, skeletal muscle, lung, kidney, uterus[185]	Unknown
AAV11	Muscle, kidney, spleen, lung, heart, stomach[185]	Unknown

different serotypes are high with a general homology in nucleotide sequence of approximately 80%. The greatest divergence in sequence can be observed in the capsid proteins, especially in regions thought to lie on the utmost exterior of the virion.[157] This may account for the differing natural tropisms of these viruses. The pattern of transgene expression has been demonstrated to be affected by the serotype of AAV,[158] which may be due, in part, to viral receptor distribution, as receptor binding is the primary step in viral infection. The discrepancies in tissue tropisms between serotypes are likely as a result of different mechanisms of uptake into a target cell. To comprehend the differences in transduction efficiencies of the different serotypes, it is important to understand the full mechanism of the initial AAV binding and internalization steps.

AAV2 has a wide host range and utilizes HSPG as an attachment receptor,[159] and at least three different coreceptors including αVβ5 integrin,[160] and the fibroblast or hepatocyte growth factor receptors.[161] AAV3 has been shown to bind to heparin, heparan sulfate, and fibroblast growth factor receptor-1 (FGFR-1), making its array of receptors similar to those of AAV2.[162] Competition assays identified that closely related serotypes AAV1 and AAV6 use either α2–3 linked or α2–6 linked sialic acid as primary receptors when transducing numerous cell types.[163] Platelet-derived growth factor receptor (PDGFR) has been identified as a coreceptor for AAV5, with the in vivo tropism of AAV5 correlating with the distribution of PDGFR.[164] AAV5 also requires α2–3 sialic acid for binding and transduction.[165] AAV4 shares the requirement of AAV5 for sialic acid; however, the difference between these two vectors lies in linkage specificity; AAV4 requires O-linked sialic acid, whereas AAV5 requires N-linked sialic acid, offering an explanation for tropism differences.[166] A 2-yeast hybrid screen with subsequent functional studies revealed the 37/67-kDa laminin receptor (LamR) as important in binding and transduction of AAV8.[167]

It was also shown to be important in the binding of AAV2, -3, and -9. AAV10 and -11 have not yet been fully characterized.

AAV2 vectors have been quite disappointing in the area of cardiovascular gene therapy due to inefficiencies in transduction of both myocardial cells and EC. Direct comparison of Ad5 and AAV2 for transduction of vascular cells has revealed the poor tropism of AAV2 for EC.[57] Transduction of vascular EC has been shown to be inefficient with AAV2 vectors resulting in virion degradation by the proteasome during the trafficking process.[168] Although no AAV serotype appears substantially more efficient than AAV2 in transduction of the vascular endothelium, other EC have been transduced by alternate serotypes. AAV6-based vectors demonstrate a higher transduction efficiency of airway epithelia than AAV2,[169] illustrating the potential of exploiting naturally occurring serotypes. Thus, alternate serotypes with naturally occurring tropism differences can be exploited as potential gene therapy vectors to see if they offer an enhanced tropism for cardiovascular tissues. AAV serotypes 1 and 6 have shown preferential transduction of the musculature.

2.7.5.4 AAV Transcapsidation

RAAV vectors are based on the AAV2 genome and onto which the capsid proteins from a different serotype have been pseudotyped. Capsid proteins from most serotypes have been successfully cross-packaged with ITRs from AAV2. Several studies have been carried out to compare the transduction efficiencies of the ever increasing array of alternate serotype AAV vectors. In a study by Du et al,[170] the capacity of AAV serotypes 1–5 for in vitro myocardial transduction was tested.[170] This study demonstrated the differing capacities of the alternative serotypes, and identified AAV1 as having the highest enhanced ability to transduce adult human cardiomyocytes. In another study that compared the efficiency of recombinant vectors of eight different serotypes in transducing rat myocardium in vivo, AAV1, 6, and 8 demonstrated the highest efficiency in transducing rat hearts in vivo.[112] It is difficult to compare between AAV serotype studies as no standard for titering AAV has been set up, and different routes of administration and different aged animals have been used. However, general trends can be observed, demonstrating that AAV serotypes 1, 6, 8, and 9 show high levels of cardiac transduction.

2.7.5.5 Retargeting AAV Vectors

Although several serotypes of AAV have been identified, several cell types remain nonpermissive to AAV infection. Retargeting vectors may encompass these nonpermissive cells into AAVs vast repertoire, and may improve the efficiency of transduction of cells already permissive to infection. Retargeting of AAV vectors has mainly been applied to AAV2 vectors, and has been achieved in vitro through two main strategies. These are (1) the use of bi-functional antibodies[171] and (2) the genetic modification of the capsid through the insertion of targeting peptides.[172] Vector binding is enhanced by the use of bi-specific antibodies. During this process, one arm of the antibody binds to the surface of the cell of interest, and the other arm to the AAV capsid structure. Bartlett et al[171] achieved AAV2-mediated transduction of nonpermissive human megakaryocytic cells through the interaction of a bispecific F $(ab)_2$ antibody with both the cell surface receptor $\alpha_{IIb}\beta3$, and the viral capsid. This facilitated the binding and internalization of the vector via an alternative receptor and represents the potential to improve the binding and transduction profile of AAV2. This technique has been used to redirect AAV binding by insertion of an immunoglobulin binding domain to couple it to various antibodies to mediate altered receptor binding.[173] However, this relies on a very stable interaction between the antibody and the vector.

The AAV capsid protein is important in the initial stages of viral infection and primarily interacts with the cell surface receptor. The capsid protein determines the tissue tropisms of the virus through its selective interactions. Short peptide sequences can be cloned into the capsid gene to change or expand the vector tropism and can even be used to disrupt the native tropism. Targeting peptides may be derived from phage-display techniques previously described. To be successful, the peptide insertion should have minimal effects on subsequent vector assembly, packaging, and infectivity. Several suitable sites for insertion of targeting peptides into the AAV2 capsid have been identified and evaluated for tolerance to insertions and mutations; peptides may be inserted at the optimal position of 587 in the AAV2 capsid to be displayed on the surface of the virion.[172,174] Genetic incorporation of peptides into the AAV capsid has been used to enhance transduction of human EC[129] and to alter tropism toward cells expressing the CD13 receptor[128] and human luteinizing receptor (LH-R).[175]

A variant of this technique is the use of AAV libraries, which are similar in concept to phage libraries. A random peptide is inserted into the AAV2 capsid sequence in a position that allows it to be displayed on the surface of the virion, while at the same time ablating HSPG binding. Chimeric capsid AAV libraries are screened to identify vectors that exclusively transduce a particular target cell or tissue type. This technique was first developed by Müller et al,[176] who used the AAV library to identify vectors that could transduce human coronary artery EC more readily than nonendothelial control cells. Others have used this approach to identify AAV vectors that efficiently transduce acute myeloid leukemia cell lines,[177] a cell type that no other vectors have been found to efficiently transduce. AAV libraries allow the selection of vectors with targeting peptides that have been identified while already in the AAV2 capsid. This eliminates the possibility of the targeting peptide losing its specificity when incorporated into the vector.

References

1. Josko J, Gwozdz B, Jedrzejowska-Szypulka H, Hendryk S. Vascular endothelial growth factor (VEGF) and its effect on angiogenesis. *Med Sci Monit.* 2000;6(5):1047–1052.
2. Lee M, Rentz J, Bikram M, Han S, Bull DA, Kim SW. Hypoxia-inducible VEGF gene delivery to ischemic myocardium using water-soluble lipopolymer. *Gene Ther.* 2003;10(18):1535–1542.

3. Baumgartner I, Pieczek A, Manor O, et al. Constitutive expression of phVEGF165 after intramuscular gene transfer promotes collateral vessel development in patients with critical limb ischemia. *Circulation*. 1998;97(12):1114–1123.
4. Shyu KG, Chang H, Wang BW, Kuan P. Intramuscular vascular endothelial growth factor gene therapy in patients with chronic critical leg ischemia. *Am J Med*. 2003;114(2):85–92.
5. Su H, Arakawa-Hoyt J, Kan YW. Adeno-associated viral vector-mediated hypoxia response element-regulated gene expression in mouse ischemic heart model. *Proc Natl Acad Sci U S A*. 2002;99(14):9480–9485.
6. Yet SF, Layne MD, Liu X, et al. Absence of heme oxygenase-1 exacerbates atherosclerotic lesion formation and vascular remodeling. *Faseb J*. 2003;17(12):1759–1761.
7. Sabaawy HE, Zhang F, Nguyen X, et al. Human heme oxygenase-1 gene transfer lowers blood pressure and promotes growth in spontaneously hypertensive rats. *Hypertension*. 2001;38(2):210–215.
8. Juan SH, Lee TS, Tseng KW, et al. Adenovirus-mediated heme oxygenase-1 gene transfer inhibits the development of atherosclerosis in apolipoprotein E-deficient mice. *Circulation*. 2001;104(13):1519–1525.
9. Tulis DA, Durante W, Liu X, Evans AJ, Peyton KJ, Schafer AI. Adenovirus-mediated heme oxygenase-1 gene delivery inhibits injury-induced vascular neointima formation. *Circulation*. 2001;104(22):2710–2715.
10. Lin KF, Chao L, Chao J. Prolonged reduction of high blood pressure with human nitric oxide synthase gene delivery. *Hypertension*. 1997;30(3 pt 1):307–313.
11. Sato J, Nair K, Hiddinga J, et al. eNOS gene transfer to vascular smooth muscle cells inhibits cell proliferation via upregulation of p27 and p21 and not apoptosis. *Cardiovasc Res*. 2000;47(4):697–706.
12. Kibbe MR, Li J, Nie S, et al. Inducible nitric oxide synthase (iNOS) expression upregulates p21 and inhibits vascular smooth muscle cell proliferation through p42/44 mitogen-activated protein kinase activation and independent of p53 and cyclic guanosine monophosphate. *J Vasc Surg*. 2000;31(6):1214–1228.
13. Cooney R, Hynes SO, Duffy AM, Sharif F, O'Brien T. Adenoviral-mediated gene transfer of nitric oxide synthase isoforms and vascular cell proliferation. *J Vasc Res*. 2006;43(5):462–472.
14. Morral N, O'Neal W, Rice K, et al. Administration of helper-dependent adenoviral vectors and sequential delivery of different vector serotype for long-term liver-directed gene transfer in baboons. *Proc Natl Acad Sci U S A*. 1999;96(22):12816–12821.
15. Barcia C, Jimenez-Dalmaroni M, Kroeger KM, et al. One-year expression from high-capacity adenoviral vectors in the brains of animals with pre-existing anti-adenoviral immunity: clinical implications. *Mol Ther*. 2007;12:2154–2163.
16. Wickham TJ. Targeting adenovirus. *Gene Ther*. 2000;7(2):110–114.
17. Reynolds PN, Nicklin SA, Kaliberova L, et al. Combined transductional and transcriptional targeting improves the specificity of transgene expression in vivo. *Nat Biotechnol*. 2001;19(9):838–842.
18. Barnett BG, Tillman BW, Curiel DT, Douglas JT. Dual targeting of adenoviral vectors at the levels of transduction and transcription enhances the specificity of gene expression in cancer cells. *Mol Ther*. 2002;6(3):377–385.
19. Campeau L, Enjalbert M, Lesperance J, Vaislic C, Grondin CM, Bourassa MG. Atherosclerosis and late closure of aortocoronary saphenous vein grafts: sequential angiographic studies at 2 weeks, 1 year, 5 to 7 years, and 10 to 12 years after surgery. *Circulation*. 1983;68(3 pt 2):II1–II7.
20. Davies MG, Hagen PO. Pathobiology of intimal hyperplasia. *Br J Surg*. 1994;81(9):1254–1269.
21. George SJ, Lloyd CT, Angelini GD, Newby AC, Baker AH. Inhibition of late vein graft neointima formation in human and porcine models by adenovirus-mediated overexpression of tissue inhibitor of metalloproteinase-3. *Circulation*. 2000;101(3):296–304.
22. West NE, Qian H, Guzik TJ, et al. Nitric oxide synthase (nNOS) gene transfer modifies venous bypass graft remodeling: effects on vascular smooth muscle cell differentiation and superoxide production. *Circulation*. 2001;104(13):1526–1532.

23. Chowdhury JR, Grossman M, Gupta S, Chowdhury NR, Baker JR Jr, Wilson JM. Long-term improvement of hypercholesterolemia after ex vivo gene therapy in LDLR-deficient rabbits. *Science*. 1991;254(5039):1802–1805.
24. Grossman M, Rader DJ, Muller DW, et al. A pilot study of ex vivo gene therapy for homozygous familial hypercholesterolaemia. *Nat Med*. 1995;1(11):1148–1154.
25. Schwarz ER, Speakman MT, Patterson M, et al. Evaluation of the effects of intramyocardial injection of DNA expressing vascular endothelial growth factor (VEGF) in a myocardial infarction model in the rat – angiogenesis and angioma formation. *J Am Coll Cardiol*. 2000;35(5):1323–1330.
26. Sasano T, Kikuchi K, McDonald AD, Lai S, Donahue JK. Targeted high-efficiency, homogeneous myocardial gene transfer. *J Mol Cell Cardiol*. 2007;42(5):954–961.
27. Bridges CR, Burkman JM, Malekan R, et al. Global cardiac-specific transgene expression using cardiopulmonary bypass with cardiac isolation. *Ann Thorac Surg*. 2002;73(6):1939–1946.
28. Melo LG, Agrawal R, Zhang L, et al. Gene therapy strategy for long-term myocardial protection using adeno-associated virus-mediated delivery of heme oxygenase gene. *Circulation*. 2002;105(5):602–607.
29. Hedman M, Hartikainen J, Syvanne M, et al. Safety and feasibility of catheter-based local intracoronary vascular endothelial growth factor gene transfer in the prevention of postangioplasty and in-stent restenosis and in the treatment of chronic myocardial ischemia: phase II results of the Kuopio Angiogenesis Trial (KAT). *Circulation*. 2003;107(21):2677–2683.
30. O'Donnell JM, Lewandowski ED. Efficient, cardiac-specific adenoviral gene transfer in rat heart by isolated retrograde perfusion in vivo. *Gene Ther*. 2005;12(12):958–964.
31. Champion HC, Georgakopoulos D, Haldar S, Wang L, Wang Y, Kass DA. Robust adenoviral and adeno-associated viral gene transfer to the in vivo murine heart: application to study of phospholamban physiology. *Circulation*. 2003;108(22):2790–2797.
32. Ikeda Y, Gu Y, Iwanaga Y, et al. Restoration of deficient membrane proteins in the cardiomyopathic hamster by in vivo cardiac gene transfer. *Circulation*. 2002;105(4):502–508.
33. Huard J, Lochmuller H, Acsadi G, Jani A, Massie B, Karpati G. The route of administration is a major determinant of the transduction efficiency of rat tissues by adenoviral recombinants. *Gene Ther*. 1995;2(2):107–115.
34. Mizuguchi H, Koizumi N, Hosono T, et al. CAR- or alphav integrin-binding ablated adenovirus vectors, but not fiber-modified vectors containing RGD peptide, do not change the systemic gene transfer properties in mice. *Gene Ther*. 2002;9(12):769–776.
35. Blankinship MJ, Gregorevic P, Allen JM, et al. Efficient transduction of skeletal muscle using vectors based on adeno-associated virus serotype 6. *Mol Ther*. 2004;10(4):671–678.
36. Pacak CA, Mah CS, Thattaliyath BD, et al. Recombinant adeno-associated virus serotype 9 leads to preferential cardiac transduction in vivo. *Circ Res*. 2006;99(4):e3–e9.
37. Lechardeur D, Sohn KJ, Haardt M, et al. Metabolic instability of plasmid DNA in the cytosol: a potential barrier to gene transfer. *Gene Ther*. 1999;6(4):482–497.
38. Johnson-Saliba M, Jans DA. Gene therapy: optimising DNA delivery to the nucleus. *Curr Drug Targets*. 2001;2(4):371–399.
39. Niidome T, Huang L. Gene therapy progress and prospects: nonviral vectors. *Gene Ther*. 2002;9(24):1647–1652.
40. Ogris M, Brunner S, Schuller S, Kircheis R, Wagner E. PEGylated DNA/transferrin-PEI complexes: reduced interaction with blood components, extended circulation in blood and potential for systemic gene delivery. *Gene Ther*. 1999;6(4):595 605.
41. Kursa M, Walker GF, Roessler V, et al. Novel shielded transferrin-polyethylene glycol-polyethylenimine/DNA complexes for systemic tumor-targeted gene transfer. *Bioconjug Chem*. 2003;14(1):222–231.
42. Wright MJ, Wightman LM, Lilley C, et al. In vivo myocardial gene transfer: optimization, evaluation and direct comparison of gene transfer vectors. *Basic Res Cardiol*. 2001;96(3):227–236.
43. Roelvink PW, Mi Lee G, Einfeld DA, Kovesdi I, Wickham TJ. Identification of a conserved receptor-binding site on the fiber proteins of CAR-recognizing adenoviridae. *Science*. 1999;286(5444):1568–1571.

44. Ding W, Zhang L, Yan Z, Engelhardt JF. Intracellular trafficking of adeno-associated viral vectors. *Gene Ther.* 2005;12(11):873–880.
45. McConnell MJ, Imperiale MJ. Biology of adenovirus and its use as a vector for gene therapy. *Hum Gene Ther.* 2004;15(11):1022–1033.
46. Worgall S, Wolff G, Falck-Pedersen E, Crystal RG. Innate immune mechanisms dominate elimination of adenoviral vectors following in vivo administration. *Hum Gene Ther.* 1997;8(1):37–44.
47. Cichon G, Boeckh-Herwig S, Schmidt HH, et al. Complement activation by recombinant adenoviruses. *Gene Ther.* 2001;8(23):1794–1800.
48. Muruve DA. The innate immune response to adenovirus vectors. *Hum Gene Ther.* 2004;15(12):1157–1166.
49. Thomas CE, Ehrhardt A, Kay MA. Progress and problems with the use of viral vectors for gene therapy. *Nat Rev Genet.* 2003;4(5):346–358.
50. Hacein-Bey-Abina S, von Kalle C, Schmidt M, et al. A serious adverse event after successful gene therapy for X-linked severe combined immunodeficiency. *N Engl J Med.* 2003;348(3):255–256.
51. Kay MA, Glorioso JC, Naldini L. Viral vectors for gene therapy: the art of turning infectious agents into vehicles of therapeutics. *Nat Med.* 2001;7(1):33–40.
52. Nabel EG, Plautz G, Nabel GJ. Site-specific gene expression in vivo by direct gene transfer into the arterial wall. *Science.* 1990;249(4974):1285–1288.
53. Verma IM, Somia N. Gene therapy – promises, problems and prospects. *Nature.* 1997;389(6648):239–242.
54. Young LS, Searle PF, Onion D, Mautner V. Viral gene therapy strategies: from basic science to clinical application. *J Pathol.* 2006;208(2):299–318.
55. Miller DG, Adam MA, Miller AD. Gene transfer by retrovirus vectors occurs only in cells that are actively replicating at the time of infection. *Mol Cell Biol.* 1990;10(8):4239–4242.
56. Lewis P, Hensel M, Emerman M. Human immunodeficiency virus infection of cells arrested in the cell cycle. *EMBO J.* 1992;11(8):3053–3058.
57. Dishart KL, Denby L, George SJ, et al. Third-generation lentivirus vectors efficiently transduce and phenotypically modify vascular cells: implications for gene therapy. *J Mol Cell Cardiol.* 2003;35(7):739–748.
58. Tsui LV, Kelly M, Zayek N, et al. Production of human clotting Factor IX without toxicity in mice after vascular delivery of a lentiviral vector. *Nat Biotechnol.* 2002;20(1):53–57.
59. Cavazzana-Calvo M, Hacein-Bey S, de Saint Basile G, et al. Gene therapy of human severe combined immunodeficiency (SCID)-X1 disease. *Science.* 2000;288(5466):669–672.
60. Couzin J, Kaiser J. Gene therapy. As Gelsinger case ends, gene therapy suffers another blow. *Science.* 2005;307(5712):1028.
61. Hacein-Bey-Abina S, Von Kalle C, Schmidt M, et al. LMO2-associated clonal T cell proliferation in two patients after gene therapy for SCID-X1. *Science.* 2003;302(5644):415–419.
62. Cavazzana-Calvo M, Fischer A. Gene therapy for severe combined immunodeficiency: are we there yet? *J Clin Invest.* 2007;117(6):1456–1465.
63. Mitrophanous K, Yoon S, Rohll J, et al. Stable gene transfer to the nervous system using a non-primate lentiviral vector. *Gene Ther.* 1999;6(11):1808–1818.
64. Wong LF, Azzouz M, Walmsley LE, et al. Transduction patterns of pseudotyped lentiviral vectors in the nervous system. *Mol Ther.* 2004;9(1):101–111.
65. Kim VN, Mitrophanous K, Kingsman SM, Kingsman AJ. Minimal requirement for a lentivirus vector based on human immunodeficiency virus type 1. *J Virol.* 1998;72(1):811–816.
66. Fischer-Lougheed JY, Tarantal AF, Shulkin I, et al. Gene therapy to inhibit xenoantibody production using lentiviral vectors in non-human primates. *Gene Ther.* 2007;14(1):49–57.
67. Browning MT, Schmidt RD, Lew KA, Rizvi TA. Primate and feline lentivirus vector RNA packaging and propagation by heterologous lentivirus virions. *J Virol.* 2001;75(11):5129–5140.
68. Lin YL, Noel D, Mettling C, et al. Feline immunodeficiency virus vectors for efficient transduction of primary human synoviocytes: application to an original model of rheumatoid arthritis. *Hum Gene Ther.* 2004;15(6):588–596.

69. Takahashi K, Luo T, Saishin Y, et al. Sustained transduction of ocular cells with a bovine immunodeficiency viral vector. *Hum Gene Ther*. 2002;13(11):1305–1316.
70. Molina RP, Ye HQ, Brady J, et al. A synthetic Rev-independent bovine immunodeficiency virus-based packaging construct. *Hum Gene Ther*. 2004;15(9):865–877.
71. Zhang XY, La Russa VF, Bao L, Kolls J, Schwarzenberger P, Reiser J. Lentiviral vectors for sustained transgene expression in human bone marrow-derived stromal cells. *Mol Therapy*. 2002;5(5 pt 1):555–565.
72. Naldini L, Blomer U, Gage FH, Trono D, Verma IM. Efficient transfer, integration, and sustained long-term expression of the transgene in adult rat brains injected with a lentiviral vector. *Proc Natl Acad Sci U S A*. 1996;93(21):11382–11388.
73. Abordo-Adesida E, Follenzi A, Barcia C, et al. Stability of lentiviral vector-mediated transgene expression in the brain in the presence of systemic antivector immune responses. *Human Gene Therapy*. 2005;16(6):741–751.
74. Azzouz M, Ralph GS, Storkebaum E, et al. VEGF delivery with retrogradely transported lentivector prolongs survival in a mouse ALS model. *Nature*. 2004;429(6990):413–417.
75. Uchida N, Sutton RE, Friera AM, et al. HIV, but not murine leukemia virus, vectors mediate high efficiency gene transfer into freshly isolated G0/G1 human hematopoietic stem cells. *Proc Natl Acad Sci U S A*. 1998;95(20):11939–11944.
76. Zennou V, Petit C, Guetard D, Nerhbass U, Montagnier L, Charneau P. HIV-1 genome nuclear import is mediated by a central DNA flap. *Cell*. 2000;101(2):173–185.
77. Zhao J, Pettigrew GJ, Thomas J, et al. Lentiviral vectors for delivery of genes into neonatal and adult ventricular cardiac myocytes in vitro and in vivo. *Basic Res Cardiol*. 2002;97(5): 348–358.
78. Diez-Freire C, Vazquez J, de Adjounian MF Correa, et al. ACE2 gene transfer attenuates hypertension-linked pathophysiological changes in the SHR. *Physiol Genomics*. 2006;27(1): 12–19.
79. Leavitt AD, Robles G, Alesandro N, Varmus HE. Human immunodeficiency virus type 1 integrase mutants retain in vitro integrase activity yet fail to integrate viral DNA efficiently during infection. *J Virol*. 1996;70(2):721–728.
80. Engelman A. In vivo analysis of retroviral integrase structure and function. *Adv Virus Res*. 1999;52:411–426.
81. Apolonia L, Waddington SN, Fernandes C, et al. Stable gene transfer to muscle using non-integrating lentiviral vectors. *Mol Ther*. 2007;15(11):1947–1954.
82. Yanez-Munoz RJ, Balaggan KS, MacNeil A, et al. Effective gene therapy with nonintegrating lentiviral vectors. *Nat Med*. 2006;12(3):348–353.
83. Latchman DS. Gene delivery and gene therapy with herpes simplex virus-based vectors. *Gene*. 2001;264(1):1–9.
84. Jacobs A, Breakefield XO, Fraefel C. HSV-1-based vectors for gene therapy of neurological diseases and brain tumors: part I. HSV-1 structure, replication and pathogenesis. *Neoplasia*. 1999;1(5):387–401.
85. Burton EA, Glorioso JC, Fink DJ. Gene therapy progress and prospects: Parkinson's disease. *Gene Ther*. 2003;10(20):1721–1727.
86. Kirn D, Martuza RL, Zwiebel J. Replication-selective virotherapy for cancer: biological principles, risk management and future directions. *Nat Med*. 2001;7(7):781–787.
87. Han ZQ, Assenberg M, Liu BL, et al. Development of a second-generation oncolytic Herpes simplex virus expressing TNFalpha for cancer therapy. *J Gene Med*. 2007;9(2):99–106.
88. Liu BL, Robinson M, Han ZQ, et al. ICP34.5 deleted herpes simplex virus with enhanced oncolytic, immune stimulating, and anti-tumour properties. *Gene Ther*. 2003;10(4):292–303.
89. Rowe WP, Huebner RJ, Gilmore LK, Parrott RH, Ward TG. Isolation of a cytopathogenic agent from human adenoids undergoing spontaneous degeneration in tissue culture. *Proc Soc Exp Biol Med*. 1953;84(3):570–573.
90. Lemarchand P, Jaffe HA, Danel C, et al. Adenovirus-mediated transfer of a recombinant human alpha 1-antitrypsin cDNA to human endothelial cells. *Proc Natl Acad Sci U S A*. 1992;89(14):6482–6486.

91. Lemarchand P, Jones M, Yamada I, Crystal RG. In vivo gene transfer and expression in normal uninjured blood vessels using replication-deficient recombinant adenovirus vectors. *Circ Res*. 1993;72(5):1132–1138.
92. French BA, Mazur W, Geske RS, Bolli R. Direct in vivo gene transfer into porcine myocardium using replication-deficient adenoviral vectors. *Circulation*. 1994;90(5):2414–2424.
93. Berkner KL. Development of adenovirus vectors for the expression of heterologous genes. *Biotechniques*. 1988;6(7):616–629.
94. Gordon D, Reidy MA, Benditt EP, Schwartz SM. Cell proliferation in human coronary arteries. *Proc Natl Acad Sci U S A*. 1990;87(12):4600–4604.
95. Turunen P, Puhakka HL, Heikura T, et al. Extracellular superoxide dismutase with vaccinia virus anti-inflammatory protein 35K or tissue inhibitor of metalloproteinase-1: combination gene therapy in the treatment of vein graft stenosis in rabbits. *Hum Gene Ther*. 2006;17(4): 405–414.
96. Wen S, Schneider DB, Driscoll RM, Vassalli G, Sassani AB, Dichek DA. Second-generation adenoviral vectors do not prevent rapid loss of transgene expression and vector DNA from the arterial wall. *Arterioscler Thromb Vasc Biol*. 2000;20(6):1452–1458.
97. Yang Y, Li Q, Ertl HC, Wilson JM. Cellular and humoral immune responses to viral antigens create barriers to lung-directed gene therapy with recombinant adenoviruses. *J Virol*. 1995;69(4):2004–2015.
98. Schiedner G, Morral N, Parks RJ, et al. Genomic DNA transfer with a high-capacity adenovirus vector results in improved in vivo gene expression and decreased toxicity. *Nat Genet*. 1998;18(2):180–183.
99. Biermann V, Volpers C, Hussmann S, et al. Targeting of high-capacity adenoviral vectors. *Hum Gene Ther*. 2001;12(14):1757–1769.
100. Dmitriev I, Krasnykh V, Miller CR, et al. An adenovirus vector with genetically modified fibers demonstrates expanded tropism via utilization of a coxsackievirus and adenovirus receptor-independent cell entry mechanism. *J Virol*. 1998;72(12):9706–9713.
101. Haviv YS, Blackwell JL, Kanerva A, et al. Adenoviral gene therapy for renal cancer requires retargeting to alternative cellular receptors. *Cancer Res*. 2002;62(15):4273–4281.
102. Mizuguchi H, Kay MA, Hayakawa T. Approaches for generating recombinant adenovirus vectors. *Adv Drug Deliv Rev*. 2001;52(3):165–176.
103. Graham FL, Smiley J, Russell WC, Nairn R. Characteristics of a human cell line transformed by DNA from human adenovirus type 5. *J Gen Virol*. 1977;36(1):59–74.
104. Yang Y, Nunes FA, Berencsi K, Furth EE, Gonczol E, Wilson JM. Cellular immunity to viral antigens limits E1-deleted adenoviruses for gene therapy. *Proc Natl Acad Sci U S A*. 1994;91(10):4407–4411.
105. Mitani K, Graham FL, Caskey CT, Kochanek S. Rescue, propagation, and partial purification of a helper virus-dependent adenovirus vector. *Proc Natl Acad Sci U S A*. 1995;92(9):3854–3858.
106. Kochanek S, Clemens PR, Mitani K, Chen HH, Chan S, Caskey CT. A new adenoviral vector: Replacement of all viral coding sequences with 28 kb of DNA independently expressing both full-length dystrophin and beta-galactosidase. *Proc Natl Acad Sci U S A*. 1996;93(12):5731–5736.
107. Parks RJ, Chen L, Anton M, Sankar U, Rudnicki MA, Graham FL. A helper-dependent adenovirus vector system: removal of helper virus by Cre-mediated excision of the viral packaging signal. *Proc Natl Acad Sci U S A*. 1996;93(24):13565–13570.
108. Ng P, Beauchamp C, Evelegh C, Parks R, Graham FL. Development of a FLP/frt system for generating helper-dependent adenoviral vectors. *Mol Ther*. 2001;3(5 pt 1):809–815.
109. Bergelson JM, Cunningham JA, Droguett G, et al. Isolation of a common receptor for Coxsackie B viruses and adenoviruses 2 and 5. *Science*. 1997;275(5304):1320–1323.
110. Tomko RP, Xu R, Philipson L. HCAR and MCAR: the human and mouse cellular receptors for subgroup C adenoviruses and group B coxsackieviruses. *Proc Natl Acad Sci U S A*. 1997;94(7):3352–3356.
111. Wickham TJ, Mathias P, Cheresh DA, Nemerow GR. Integrins alpha v beta 3 and alpha v beta 5 promote adenovirus internalization but not virus attachment. *Cell*. 1993;73(2):309–319.
112. Palomeque J, Chemaly ER, Colosi P, et al. Efficiency of eight different AAV serotypes in transducing rat myocardium in vivo. *Gene Ther*. 2007;14(13):989–997.

113. Ohno T, Gordon D, San H, et al. Gene therapy for vascular smooth muscle cell proliferation after arterial injury. *Science*. 1994;265(5173):781–784.
114. Koeberl DD, Alexander IE, Halbert CL, Russell DW, Miller AD. Persistent expression of human clotting factor IX from mouse liver after intravenous injection of adeno-associated virus vectors. *Proc Natl Acad Sci U S A*. 1997;94(4):1426–1431.
115. Kirby I, Davison E, Beavil AJ, et al. Mutations in the DG loop of adenovirus type 5 fiber knob protein abolish high-affinity binding to its cellular receptor CAR. *J Virol*. 1999;73(11):9508–9514.
116. Jakubczak JL, Rollence ML, Stewart DA, et al. Adenovirus type 5 viral particles pseudotyped with mutagenized fiber proteins show diminished infectivity of coxsackie B-adenovirus receptor-bearing cells. *J Virol*. 2001;75(6):2972–2981.
117. Krasnykh VN, Mikheeva GV, Douglas JT, Curiel DT. Generation of recombinant adenovirus vectors with modified fibers for altering viral tropism. *J Virol*. 1996;70(10):6839–6846.
118. Denby L, Work LM, Graham D, et al. Adenoviral serotype 5 vectors pseudotyped with fibers from subgroup D show modified tropism in vitro and in vivo. *Hum Gene Ther*. 2004;15(11):1054–1064.
119. Denby L, Work LM, Seggern DJ, et al. Development of renal-targeted vectors through combined in vivo phage display and capsid engineering of adenoviral fibers from serotype 19p. *Mol Ther*. 2007;15(9):1647–1654.
120. Nettelbeck DM, Miller DW, Jerome V, et al. Targeting of adenovirus to endothelial cells by a bispecific single–chain diabody directed against the adenovirus fiber knob domain and human endoglin (CD105). *Mol Ther*. 2001;3(6):882–891.
121. Printz MA, Gonzalez AM, Cunningham M, et al. Fibroblast growth factor 2-retargeted adenoviral vectors exhibit a modified biolocalization pattern and display reduced toxicity relative to native adenoviral vectors. *Hum Gene Ther*. 2000;11(1):191–204.
122. Reynolds PN, Zinn KR, Gavrilyuk VD, et al. A targetable, injectable adenoviral vector for selective gene delivery to pulmonary endothelium in vivo. *Mol Ther*. 2000;2(6):562–578.
123. Krasnykh V, Dmitriev I, Mikheeva G, Miller R, Belousova N, Curiel DT. Characterisation of an adenovirus vector containing a heterologous peptide epitope in the HI loop of the fiber knob. *J Virol*. 1998;72(3):1844–1852.
124. Dmitriev I, Krasnykh V, Miller CR, et al. An adenovirus vector with genetically modified fibers demonstrates expanded tropism via utilisation of a coxsackievirus and adenovirus receptor-independent cell entry mechanism. *J Virol*. 1998;72:9706–9713.
125. Vigne E, Mahfouz I, Dedieu JF, Brie A, Perricaudet M, Yeh P. RGD inclusion in the hexon monomer provides adenovirus type 5-based vectors with a fiber knob-independent pathway for infection. *J Virol*. 1999;73(6):5156–5161.
126. Campos SK, Barry MA. Comparison of adenovirus fiber, protein IX, and hexon capsomeres as scaffolds for vector purification and cell targeting. *Virology*. 2006;349(2):453–462.
127. Engelstadter M, Bobkova M, Baier M, et al. Targeting human T cells by retroviral vectors displaying antibody domains selected from a phage display library. *Hum Gene Ther*. 2000;11(2):293–303.
128. Grifman M, Trepel M, Speece P, et al. Incorporation of tumor-targeting peptides into recombinant adeno-associated virus capsids. *Mol Ther*. 2001;3(6):964–975.
129. Nicklin SA, Buening H, Dishart KL, et al. Efficient and selective AAV2-mediated gene transfer directed to human vascular endothelial cells. *Mol Ther*. 2001;4(3):174–181.
130. Smith GP. Filamentous fusion phage: novel expression vectors that display cloned antigens on the virion surface. *Science*. 1985;228(4705):1315–1317.
131. Johnsson K, Ge L. Phage display of combinatorial peptide and protein libraries and their applications in biology and chemistry. *Curr Top Microbiol Immunol*. 1999;243:87–105.
132. Arap W, Kolonin MG, Trepel M, et al. Steps toward mapping the human vasculature by phage display. *Nat Med*. 2002;8(2):121–127.
133. Krag DN, Shukla GS, Shen GP, et al. Selection of tumor-binding ligands in cancer patients with phage display libraries. *Cancer Res*. 2006;66(15):7724–7733.
134. Woo YJ, Zhang JC, Taylor MD, Cohen JE, Hsu VM, Sweeney HL. One year transgene expression with adeno-associated virus cardiac gene transfer. *Int J Cardiol*. 2005;100(3):421–426.

135. Xiao X, Li J, Samulski RJ. Efficient long-term gene transfer into muscle tissue of immunocompetent mice by adeno-associated virus vector. *J Virol.* 1996;70(11):8098–8108.
136. Zaiss AK, Liu Q, Bowen GP, Wong NC, Bartlett JS, Muruve DA. Differential activation of innate immune responses by adenovirus and adeno-associated virus vectors. *J Virol.* 2002;76(9): 4580–4590.
137. Bessis N, GarciaCozar FJ, Boissier MC. Immune responses to gene therapy vectors: influence on vector function and effector mechanisms. *Gene Ther.* 2004;11(suppl 1):S10–S17.
138. Jooss K, Yang Y, Fisher KJ, Wilson JM. Transduction of dendritic cells by DNA viral vectors directs the immune response to transgene products in muscle fibers. *J Virol.* 1998;72(5):4212–4223.
139. Manno CS, Pierce GF, Arruda VR, et al. Successful transduction of liver in hemophilia by AAV-Factor IX and limitations imposed by the host immune response. *Nat Med.* 2006;12(3):342–347.
140. Vandenberghe LH, Wang L, Somanathan S, et al. Heparin binding directs activation of T cells against adeno-associated virus serotype 2 capsid. *Nat Med.* 2006;12(8):967–971.
141. Pachori AS, Melo LG, Zhang L, Loda M, Pratt RE, Dzau VJ. Potential for germ line transmission after intramyocardial gene delivery by adeno-associated virus. *Biochem Biophys Res Commun.* 2004;313(3):528–533.
142. Arruda VR, Fields PA, Milner R, et al. Lack of germline transmission of vector sequences following systemic administration of recombinant AAV-2 vector in males. *Mol Ther.* 2001;4(6):586–592.
143. Donsante A, Vogler C, Muzyczka N, et al. Observed incidence of tumorigenesis in long-term rodent studies of rAAV vectors. *Gene Ther.* 2001;8(17):1343–1346.
144. Donsante A, Miller DG, Li Y, et al. AAV vector integration sites in mouse hepatocellular carcinoma. *Science.* 2007;317(5837):477.
145. Hermonat PL, Quirk JG, Bishop BM, Han L. The packaging capacity of adeno-associated virus (AAV) and the potential for wild-type-plus AAV gene therapy vectors. *FEBS Lett.* 1997;407(1):78–84.
146. Yan Z, Zhang Y, Duan D, Engelhardt JF. Trans-splicing vectors expand the utility of adeno-associated virus for gene therapy. *Proc Natl Acad Sci U S A.* 2000;97(12):6716–6721.
147. Cheung AK, Hoggan MD, Hauswirth WW, Berns KI. Integration of the adeno-associated virus genome into cellular DNA in latently infected human Detroit 6 cells. *J Virol.* 1980;33(2):739–748.
148. Kotin RM, Menninger JC, Ward DC, Berns KI. Mapping and direct visualization of a region-specific viral DNA integration site on chromosome 19q13-qter. *Genomics.* 1991;10(3):831–834.
149. Berns KI, Pinkerton TC, Thomas GF, Hoggan MD. Detection of adeno-associated virus (AAV)-specific nucleotide sequences in DNA isolated from latently infected Detroit 6 cells. *Virology.* 1975;68(2):556–560.
150. Xiao X, Li J, Samulski RJ. Production of high-titer recombinant adeno-associated virus vectors in the absence of helper adenovirus. *J Virol.* 1998;72(3):2224–2232.
151. Grimm D, Kern A, Rittner K, Kleinschmidt JA. Novel tools for production and purification of recombinant adenoassociated virus vectors. *Hum Gene Ther.* 1998;9(18):2745–2760.
152. Douar AM, Poulard K, Stockholm D, Danos O. Intracellular trafficking of adeno-associated virus vectors: routing to the late endosomal compartment and proteasome degradation. *J Virol.* 2001;75(4):1824–1833.
153. Bartlett JS, Wilcher R, Samulski RJ. Infectious entry pathway of adeno-associated virus and adeno-associated virus vectors. *J Virol.* 2000;74(6):2777–2785.
154. Hansen J, Qing K, Srivastava A. Infection of purified nuclei by adeno-associated virus 2. *Mol Ther.* 2001;4(4):289–296.
155. Sanlioglu S, Benson PK, Yang J, Atkinson EM, Reynolds T, Engelhardt JF. Endocytosis and nuclear trafficking of adeno-associated virus type 2 are controlled by rac1 and phosphatidylinositol-3 kinase activation. *J Virol.* 2000;74(19):9184–9196.
156. Gao G, Vandenberghe LH, Alvira MR, et al. Clades of adeno-associated viruses are widely disseminated in human tissues. *J Virol.* 2004;78(12):6381–6388.

157. Gao G, Alvira MR, Somanathan S, et al. Adeno-associated viruses undergo substantial evolution in primates during natural infections. *Proc Natl Acad Sci U S A*. 2003;100(10):6081–6086.
158. Rabinowitz JE, Rolling F, Li C, et al. Cross-packaging of a single adeno-associated virus (AAV) type 2 vector genome into multiple AAV serotypes enables transduction with broad specificity. *J Virol*. 2002;76(2):791–801.
159. Summerford C, Samulski RJ. Membrane-associated heparan sulfate proteoglycan is a receptor for adeno-associated virus type 2 virions. *J Virol*. 1998;72(2):1438–1445.
160. Summerford C, Bartlett JS, Samulski RJ. AlphaVbeta5 integrin: a co-receptor for adeno-associated virus type 2 infection. *Nat Med*. 1999;5(1):78–82.
161. Qing K, Mah C, Hansen J, Zhou S, Dwarki V, Srivastava A. Human fibroblast growth factor receptor 1 is a co-receptor for infection by adeno-associated virus 2. *Nat Med*. 1999;5(1):71–77.
162. Blackburn SD, Steadman RA, Johnson FB. Attachment of adeno-associated virus type 3H to fibroblast growth factor receptor 1. *Arch Virol*. 2006;151(3):617–623.
163. Wu Z, Miller E, Agbandje-McKenna M, Samulski RJ. Alpha2, 3 and alpha2, 6 N-linked sialic acids facilitate efficient binding and transduction by adeno-associated virus types 1 and 6. *J Virol*. 2006;80(18):9093–9103.
164. Di Pasquale G, Davidson BL, Stein CS, et al. Identification of PDGFR as a receptor for AAV-5 transduction. *Nat Med*. 2003;9(10):1306–1312.
165. Walters RW, Yi SM, Keshavjee S, et al. Binding of adeno-associated virus type 5 to 2, 3-linked sialic acid is required for gene transfer. *J Biol Chem*. 2001;276(23):20610–20616.
166. Kaludov N, Brown KE, Walters RW, Zabner J, Chiorini JA. Adeno-associated virus serotype 4 (AAV4) and AAV5 both require sialic acid binding for hemagglutination and efficient transduction but differ in sialic acid linkage specificity. *J Virol*. 2001;75(15):6884–6893.
167. Akache B, Grimm D, Pandey K, Yant SR, Xu H, Kay MA. The 37/67-kilodalton laminin receptor is a receptor for adeno-associated virus serotypes 8, 2, 3, and 9. *J Virol*. 2006;80(19):9831–9836.
168. Nicklin SA, Von Seggern DJ, Work LM, et al. Ablating adenovirus type 5 fiber-CAR binding and HI loop insertion of the SIGYPLP peptide generate an endothelial cell-selective adenovirus. *Mol Ther*. 2001;4(6):534–542.
169. Halbert CL, Allen JM, Miller AD. Adeno-associated virus type 6 (AAV6) vectors mediate efficient transduction of airway epithelial cells in mouse lungs compared to that of AAV2 vectors. *J Virol*. 2001;75(14):6615–6624.
170. Du L, Kido M, Lee DV, et al. Differential myocardial gene delivery by recombinant serotype-specific adeno-associated viral vectors. *Mol Ther*. 2004;10(3):604–608.
171. Bartlett JS, Kleinschmidt J, Boucher RC, Samulski RJ. Targeted adeno-associated virus vector transduction of nonpermissive cells mediated by a bispecific F(ab′gamma)2 antibody. *Nat Biotechnol*. 1999;17(2):181–186.
172. Wu P, Xiao W, Conlon T, et al. Mutational analysis of the adeno-associated virus type 2 (AAV2) capsid gene and construction of AAV2 vectors with altered tropism. *J Virol*. 2000;74(18):8635–8647.
173. Ried MU, Girod A, Leike K, Buning H, Hallek M. Adeno-associated virus capsids displaying immunoglobulin-binding domains permit antibody-mediated vector retargeting to specific cell surface receptors. *J Virol*. 2002;76(9):4559–4566.
174. Girod A, Ried M, Wobus C, et al. Genetic capsid modifications allow efficient re-targeting of adeno-associated virus type 2. *Nat Med*. 1999;5(9):1052–1056.
175. Shi W, Arnold GS, Bartlett JS. Insertional mutagenesis of the adeno-associated virus type 2 (AAV2) capsid gene and generation of AAV2 vectors targeted to alternative cell-surface receptors. *Hum Gene Ther*. 2001;12(14):1697–1711.
176. Muller OJ, Kaul F, Weitzman MD, et al. Random peptide libraries displayed on adeno-associated virus to select for targeted gene therapy vectors. *Nat Biotechnol*. 2003;21(9):1040–1046.
177. Perabo L, Buning H, Kofler DM, et al. In vitro selection of viral vectors with modified tropism: the adeno-associated virus display. *Mol Ther*. 2003;8(1):151–157.
178. Xiao W, Chirmule N, Berta SC, McCullough B, Gao G, Wilson JM. Gene therapy vectors based on adeno-associated virus type 1. *J Virol*. 1999;73(5):3994–4003.

179. Hauck B, Xiao W. Characterization of tissue tropism determinants of adeno-associated virus type 1. *J Virol*. 2003;77(4):2768–2774.
180. Liu Y, Okada T, Sheykholeslami K, et al. Specific and efficient transduction of Cochlear inner hair cells with recombinant adeno-associated virus type 3 vector. *Mol Ther*. 2005;12(4): 725–733.
181. Davidson BL, Stein CS, Heth JA, et al. Recombinant adeno-associated virus type 2, 4, and 5 vectors: transduction of variant cell types and regions in the mammalian central nervous system. *Proc Natl Acad Sci U S A*. 2000;97(7):3428–3432.
182. Alisky JM, Hughes SM, Sauter SL, et al. Transduction of murine cerebellar neurons with recombinant FIV and AAV5 vectors. *Neuroreport*. 2000;11(12):2669–2673.
183. Xin KQ, Mizukami H, Urabe M, et al. Induction of robust immune responses against human immunodeficiency virus is supported by the inherent tropism of adeno-associated virus type 5 for dendritic cells. *J Virol*. 2006;80(24):11899–11910.
184. Gao GP, Alvira MR, Wang L, Calcedo R, Johnston J, Wilson JM. Novel adeno-associated viruses from rhesus monkeys as vectors for human gene therapy. *Proc Natl Acad Sci U S A*. 2002;99(18):11854–11859.
185. Mori S, Wang L, Takeuchi T, Kanda T. Two novel adeno-associated viruses from cynomolgus monkey: pseudotyping characterization of capsid protein. *Virology*. 2004;330(2):375–383.
186. Alba R, Bosch A, Chillon M. Gutless adenovirus: last-generation adenovirus for gene therapy. *Gene Ther*. 2005;12(suppl 1):S18–S27.

Part 2
Animal Models to Study Vascular Disease

Chapter 3
Animal Models of Scleroderma: From Cellular and Molecular Mechanisms to Novel Antifibrotic Strategies

Mirko Manetti, Elena Neumann, Oliver Distler, and Ulf Müller-Ladner

3.1 Introduction

Systemic sclerosis (SSc; scleroderma) is a complex connective tissue disorder that leads to fibrosis of the skin and various internal organs including the heart, lung, kidney, and gastrointestinal tract.[1] Although fibrosis is the main pathological hallmark of SSc, autoimmunity, inflammation, and widespread small-vessel vasculopathy characteristically precede the excessive synthesis and deposition of extracellular matrix (ECM), which ultimately disrupts the physiologic structure of the affected tissues and leads to dysfunction of the affected organs.[2,3] Fibrosis results from a complex interplay among endothelial cells, inflammatory cells, immune cells, and fibroblasts activated by and inducing a number of mediators.[2,3]

Several cytokines, chemokines, and growth factors have been shown to be strongly associated with the pathogenesis of SSc, including the transforming growth factor β (TGFβ), the connective tissue growth factor (CTGF), the platelet-derived growth factor (PDGF), interleukin-4 (IL-4), monocyte chemoattractant protein-1 (MCP-1/CCL2), and endothelin-1 (ET-1).[2-4] Furthermore, increasing evidence suggests that an altered balance between Th1 and Th2 cytokines toward a Th2-polarized immune response plays a pivotal role in the pathogenesis of fibrotic diseases.[5] The resulting cellular microenvironment ultimately leads to a constitutive activation of fibroblasts, which transdifferentiate into contractile myofibroblasts, produce high amount of types I, III, VI, and VII collagen, fibronectins, and proteoglycans, and secrete growth factors and fibrogenic cytokines that perpetuate the fibrotic process in an autocrine or paracrine manner.[2,3] Several alterations in the downstream signaling pathways of TGFβ, and in particular, a dysregulation in the expression and activation of Smad transcription factors have been reported in SSc

M. Manetti (✉)
Department of Anatomy, Histology and Forensic Medicine, University of Florence, viale G. B. Morgagni 85, 50134 Florence, Italy;
Department of Internal Medicine and Rheumatology, Division of Rheumatology and Clinical Immunology, Justus-Liebig-University of Giessen, Kerckhoff-Klinik, Bad Nauheim, Hessen, Germany
e-mail: mirkomanetti@yahoo.it

D. Abraham et al. (eds.), *Advances in Vascular Medicine*,
DOI 10.1007/978-1-84882-637-3_3, © Springer-Verlag London Limited 2010

fibroblasts.[3,6,7] Of note is the observation that these cellular alterations showed to be stable over multiple generations in vitro. In addition, recent observations lead to the suggestion that Smad-independent TGFβ signaling pathways, as well as TGFβ-independent transduction, and epigenetic regulation in SSc fibroblasts also might contribute to their persistent dysfunction, thus playing an important role in inducing or maintaining the SSc phenotype.[6,8-11]

Despite substantial progress during the past years in uncovering the molecular and cellular pathogenetic mechanisms of SSc, its etiology still remains unclear. In addition, there is no therapeutic approach available that is able to reverse or slow down the progression of skin fibrosis and, thus, to substantially modify the natural progression of the disease.

Animal models are important tools for a better understanding of the mechanisms that trigger and sustain fibrosis and to exploit potential therapeutic interventions for SSc. So far, various induced or naturally occurring in vivo models have been generated and investigated in detail and, although not exclusively, murine models, such as bleomycin-induced scleroderma, sclerodermatous graft-versus-host disease (Scl GvHD), and types 1 and 2 tight skin (Tsk1 and Tsk2) mouse models, have been studied most extensively.[12] The large body of genetic information available for mice represents the principal advantage of murine models in comparison with other species. Therefore, the generation of novel models by genetic manipulations such as gene knockout or differential gene expression in selective cell types using transgenesis, was possible.[12-15] Nevertheless, the University of California at Davis (UCD)-200/206 chicken model is another important animal model for SSc, which recapitulates several aspects of the human disease, such as a prominent vascular involvement with inflammation and widespread fibrosis.[12]

Although all animal models display fibrotic skin alterations resembling those in SSc patients, a model that shows all the pathogenetic components and histologic and biochemical features of human scleroderma is currently not available.[12,16] However, animal models recapitulate selected aspects of SSc phenotype and have contributed significantly to advance the understanding of the pathogenetic mechanisms of human SSc (Table 3.1). Hence, a growing body of studies provides evidence that such model systems may be used as platforms to test promising novel therapeutic interventions for SSc.

In this review, the different animal models and their recent contribution to the knowledge of the pathogenesis and potential treatment of human scleroderma are summarized.

3.2 Bleomycin-Induced Scleroderma Model

Since its development by Yamamoto et al about 10 years ago,[17] the bleomycin-induced scleroderma model has been extensively used by several groups, providing substantial and valuable information about the pathogenesis of human SSc.[18] Because of its reproducibility and easy induction, this model is used with increasing frequency to investigate the role of cellular mediators and signaling pathways in fibrosis, and to test novel antifibrotic strategies in vivo.

Table 3.1 Aspects of systemic sclerosis (SSc) pathogenesis reproduced in animal models[a]

Model	Skin fibrosis	Inflam mation	Auto immunity	Vasculo pathy	Further features
Murine bleomycin	•	•	–	–	Lung fibrosis
Scl GvHD	•	•	–	–	Lung fibrosis
Modified Scl GvHD	•	•	•	•	Fibrosis in kidneys and small intestine, but not in lungs. Vascular alterations in skin and kidneys
Tsk1	•	–	•	–	Skin tethering, subcutaneous hyperplasia
Tsk2	•	•	•	–	
TβRIIΔk transgenic	•	•	–	–	Lung fibrosis
TβRI^{CA}/Cre-ER transgenic	•	–	–	•	Vasculopathy in lungs and kidneys
MRL/lpr-IFNγR$^{-/-}$	•	•	•	•	Fibrosis in lungs and other internal organs
RLX$^{-/-}$	•	–	–	–	Pulmonary, cardiac, and renal fibrosis
Chicken UCD-200/206	•	•	•	•	Widespread visceral vasculopathy, inflammation and fibrosis, rheumatoid factors and distal polyarthritis

[a]*Scl GvHD* sclerodermatous graft-versus-host disease; *Tsk1 and Tsk2* types 1 and 2 tight skin mice; *TβRIIΔk* transgenic mice expressing a kinase-deficient TβRII selectively on fibroblasts; *TβRI^{CA}/Cre-ER* transgenic mice expressing a postnatally constitutively active TβRI in fibroblasts; *MRL/lpr-IFNγR$^{-/-}$* MRL/lpr mouse strains lacking IFNγ receptor; *RLX$^{-/-}$* relaxin gene knockout mice; *UCD200/206* University of California at Davis chicken lines 200/206; • present; – absent

In this model, repeated subcutaneous injections of the anticancer drug bleomycin induce skin as well as pulmonary fibrosis in several mouse strains[18] (Fig. 3.1). Moreover, in contrast to the human disease, in bleomycin-treated mice there is no evidence of microvasculopathy and autoimmunity. Cutaneous changes are localized to the injected skin area and remain at least for 6 weeks after the last bleomycin treatment.[18] The histological features observed in the involved skin of the mice include a gradual increase in dermal thickness caused by a severe dermal fibrosis characterized by deposition of dense collagen bundles with mononuclear cell infiltrates, mimicking the cutaneous changes of human scleroderma[17,18] (Fig. 3.1). In the lung, fibrosis with thickened alveolar walls and the presence of cellular infiltrates are commonly observed at early time points.[17,18]

In this model, the development of fibrosis appears to be triggered by excessive oxidative stress mediated by reactive oxygen species (ROS) and the subsequent acute inflammatory reaction.[18] Of note, oxidative stress has also been implicated in the pathogenesis of human SSc.[3,19] The early cutaneous inflammatory infiltrates consist mainly of CD4+ T cells, macrophages, eosinophils, and mast cells, the latter displaying a marked degranulation. The inflammatory process precedes the onset

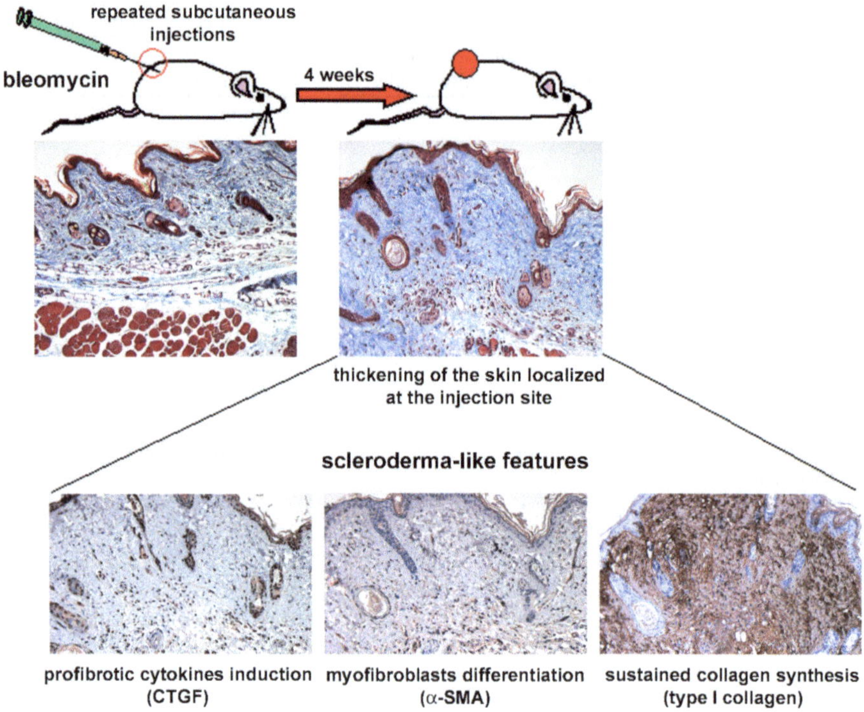

Fig. 3.1 Bleomycin-induced scleroderma model. Repeated subcutaneous injections of bleomycin induce skin fibrosis in several mouse strains. Cutaneous changes are evident after 4 weeks of treatment, are localized at the injected skin area, and remain at least for 6 weeks after the last bleomycin injection. *Upper panels*: Histological sections of control (*left*) and bleomycin-injected (*right*) mouse skin showing a marked increase in dermal thickness caused by a severe dermal fibrosis characterized by deposition of dense collagen bundles with mononuclear cell infiltrates after bleomycin treatment (Masson's trichrome staining: *red* color, cytoplasm and muscle fibers; *blue* color, collagen). *Lower panels*: Serial histological sections of bleomycin-injected mouse skin immunostained with specific antibodies for connective tissue growth factor (CTGF), as an example of profibrotic cytokine, α-smooth muscle actin (α-SMA), a marker of myofibroblasts, and type I collagen. The immunoreactivity was detected using biotinylated secondary antibodies and the avidin–biotin peroxidase complex followed by color development with 3-amino-9-ethylcarbazole substrate (*red* staining). Sections were counterstained with hematoxylin (*blue* staining) (original magnification, 100×)

of dermal fibrosis, closely resembling the sequence of histopathological changes observed in human SSc.[3,18] Moreover, there is evidence of ROS-mediated endothelial cell damage and adhesion molecule expression, and of excessive Fas/Fas ligand-mediated apoptosis of mononuclear cells, which occur in parallel with the induction of dermal fibrosis in this animal model.[18,20-22]

It is well known that excessive ECM deposition in SSc is the result of a complex imbalance between matrix synthesis and degradation.[2] Similarly, increased production of type I collagen, upregulation of TGFβ in infiltrating cells and fibroblasts, differentiation of α-smooth muscle actin (α-SMA)-expressing myofibroblasts, as well as increased expression of tissue inhibitor of matrix metalloproteinases 1 (TIMP-1) and plasminogen activator inhibitor 1 (PAI-1) were observed in the skin of bleomycin-treated mice.[18,23-25]

The bleomycin model also exhibits SSc-like early and sustained activation of the profibrotic Smad-dependent TGFβ signaling with elevated levels and nuclear localization of phospho-Smad2/3, and the downregulation of endogenous TGFβ signaling inhibitor Smad7 in lesional fibroblasts.[26] Interestingly, blockade of TGFβ activity using specific antibodies reduced the development of dermal fibrosis and the number of myofibroblasts in the lesional skin of bleomycin-treated mice.[18,24] Moreover, bleomycin administration reduced collagen gene expression and induced a mild fibrogenic response in genetically deleted Smad3 null mice (Smad3$^{-/-}$) in comparison with wild-type animals, suggesting a pivotal role of Smad-dependent TGFβ signaling in the pathogenesis of the murine scleroderma-like phenotype.[27]

In addition to TGFβ, the overexpression of other profibrotic cytokines and chemokines, which are known to be strongly involved in the pathogenesis of human SSc, such as CTGF, PDGF, MCP-1/CCL2, and IL-4, is also found in this model.[18] In particular, MCP-1 and its receptor (CCR2) are strongly expressed in inflammatory cellular infiltrates at early stages following bleomycin treatment and in the fibroblasts of sclerotic skin at later stages.[28] Of note, the application of neutralizing antibodies against MCP-1 led to reduced cutaneous fibrosis.[28] Moreover, mice with MCP-1 deficiency (MCP-1 null mice, MCP-1$^{-/-}$) showed a substantial diminished induction of skin fibrosis following bleomycin administration.[29] These data underline the prominent role for MCP-1 in the pathogenesis of bleomycin-induced skin fibrosis, mimicking an important molecular aspect of human SSc pathogenesis. Furthermore, MCP-1 is known to mediate its profibrotic effects in SSc fibroblasts through the release of IL-4 from T cells.[4] In addition, the Th2-polarized immune response leads to an increase of type 2 cytokines, such as TGFβ, IL-4, and IL-13.[5] The potent profibrotic cytokines IL-4 and IL-13 are dysregulated in bleomycin-induced cutaneous fibrosis as well as in a mouse model of pulmonary fibrosis, and IL-13 transgenic mice show increased levels of TGFβ1 and sustained development of tissue fibrosis.[30-32] Moreover, Aliprantis et al reported that mice deficient for the transcription factor T-bet, which is selectively expressed in T cells and regulates type 1 immunity, display increased sensitivity to bleomycin-induced dermal fibrosis. Here, T-bet appears to exert its antifibrotic effect by repressing IL-13.[33] Other recent observations showed that IL-13 triggers its profibrotic response in a TGFβ-dependent as well as -independent fashion.[34] Thus, IL-13 is regarded as an attractive potential target for developing novel therapeutic strategies for SSc.

The relative role of the adaptive and innate immune system in the pathogenesis of bleomycin-induced fibrosis is unclear and the published data are inconsistent. It has been reported that both nude mice, and mice immunodepleted of CD4+ and CD8+ T cells after treatment with specific antibodies, show attenuated bleomycin-mediated pulmonary disease.[18] Conversely, severe combined immunodeficient (SCID) mice have been described to develop both bleomycin-induced pulmonary fibrosis and dermal sclerosis comparable with those of wild-type mice.[35,36] Dermal sclerosis was also inducible in nude mice.[37] In a very recent study, the bleomycin-induced scleroderma model has been reproduced in RAG2$^{-/-}$ knockout (RAG2 KO) mice, which are deficient for the recombinase gene necessary for antigen receptor rearrangement, and therefore totally lack mature T and B cells.[33] RAG2 KO mice developed skin sclerosis comparable with that of wild-type mice, indicating that the bleomycin-induced scleroderma phenotype is not mediated by the adaptive immune system, but must involve innate immune cells.[33]

The bleomycin-induced scleroderma model has been used with increasing frequency to investigate the effects of various molecules both as preventive agents and as a treatment for established fibrosis.

Yamamoto et al reported that systemic administration of anti-TGFβ antibodies along with local subcutaneous injections of bleomycin resulted in a significant reduction of dermal fibrosis and inflammatory cell infiltration.[18] However, because of the short half-life of the antibody, repeated injections were required to achieve the described antifibrotic effects.[18] In a recent study, the topical application of a peptide inhibitor for TGFβ1 (P144) has been shown to ameliorate established skin fibrosis in bleomycin-treated mice.[38] These effects were mediated through the suppression of collagen synthesis, CTGF expression, Smad2/3 phosphorylation in fibroblasts, and differentiation of α-SMA-expressing myofibroblasts at sites of fibrosis, whereas mast cell and mononuclear cell infiltration was not affected.[38]

Transfection of human hepatocyte growth factor (HGF) cDNA into the skeletal muscle prevented and ameliorated the symptoms of dermal sclerosis and lung fibrosis induced by subcutaneous bleomycin injections.[39] Of note, the treatment reduced TGFβ1 expression in mononuclear inflammatory cells and fibroblasts.[39] Thus, the authors proposed that gene therapy using HGF cDNA transfection could be regarded as a potential therapeutic approach for the treatment of both skin and pulmonary fibrosis in SSc.

Systemic administration of interferon-γ (IFN-γ), which is an inducer of Th1 immune responses and is known to act as a potent inhibitor of collagen synthesis and Smad-dependent TGFβ signaling, reduced but did not suppress dermal fibrosis in the bleomycin-induced scleroderma model, most likely by counteracting the profibrotic Th1–Th2 imbalance.[18,40] Furthermore, the potential therapeutic utility of IFN-γ for the treatment of skin lesions in localized scleroderma has already been evaluated in clinical trials, achieving only modest results after long-term treatment.[41]

Kimura et al recently showed that a novel synthetic small molecule, SKL-2841, which exerts a dual antichemotactic activity for MCP-1 and macrophage inflammatory protein-1β (MIP-1β) significantly suppresses the inflammatory response in the acute phase and skin fibrosis in the chronic phase of the bleomycin-induced scleroderma model.[42] Considering the pivotal role of MCP-1 in the pathogenesis of SSc, this compound needs further investigation to clarify its mode of action in vivo and might be useful for the treatment of skin fibrosis in human scleroderma.

Using the bleomycin-induced murine model of scleroderma and in vitro studies on human dermal fibroblasts, the role of adenosine A_{2A} receptors in the pathogenesis of dermal fibrosis could be demonstrated.[43] Adenosine A_{2A} receptor-knockout mice were protected from developing bleomycin-induced dermal fibrosis.

Pharmacologic blockade with a specific antagonist of adenosine A_{2A} receptors substantially attenuated dermal fibrosis in wild-type mice.[43] These results show that adenosine A_{2A} receptors could be novel therapeutic targets in the treatment of dermal fibrosis in SSc.

Very recently, the in vitro and in vivo potential of imatinib mesylate as antifibrotic agent for the treatment of dermal fibrosis in SSc has been investigated.[44] Imatinib mesylate is an ATP analog tyrosine kinase inhibitor. It blocks the binding site of the ATP pocket of the kinase c-Abl and inhibits another membrane tyrosine kinase, c-kit, and the fusion protein Bcr-Abl. Its inhibitory action selectively interferes with both TGFβ and PDGF signaling, thus making the drug a promising candidate as inhibitor of fibrosis. Imatinib mesylate significantly inhibits the production of ECM by SSc fibroblasts under basal condition and under PDGF or TGFβ stimulation.[44] Simultaneous systemic administration of imatinib mesylate in biologically relevant concentrations prevented dermal fibrosis induced by bleomycin without any toxic side effects.[44] These results indicate that imatinib mesylate may be a potential antifibrotic agent in SSc. Further studies using animal models are needed to assess its efficacy as treatment for established fibrosis in vivo in addition to its potential as a preventive agent.

In another recent study, the antifibrogenic effects of histone deacetylase inhibitor trichostatin A (TSA) in human SSc skin fibroblasts in vitro and in the bleomycin-induced scleroderma model in vivo were analyzed.[11] Intraperitoneal injection of TSA after repeated subcutaneous bleomycin treatments showed substantial antifibrotic efficacy in vivo. The histological features included the prevention of dermal thickening and the presence of less densely packed dermal collagen bundles in comparison with the TSA-untreated mice. Furthermore, no toxic effects were observed in TSA-treated mice.[11] As there is growing evidence that epigenetic histone modifications regulate the expression of fibrogenic molecules and that epigenetic mechanisms are involved in SSc pathogenesis, the study supported the possibility that TSA and related inhibitory compounds could be useful as pharmacologic strategy to prevent the development and progression of fibrosis in early scleroderma.[3,10,11]

3.3 Sclerodermatous Graft-Versus-Host Disease Model

Chronic GvHD resembling scleroderma showing fibrotic involvement of the skin and internal organs is a serious complication, which can occur after bone marrow hematopoietic cell transplantation from donors mismatched for minor histocompatibility loci. In addition, the hypothesis that fetal microchimerism may play a role in the pathogenesis of SSc further supports the notion of the utility of a GvHD murine model to study the pathogenetic mechanisms of SSc.[2]

The murine Scl GvHD was first proposed as a model to study human scleroderma by Jaffee and Claman in 1983.[45] Several years later, McCormick et al generated Scl GvHD using B10.D2 bone marrow and spleen cell transplantation across minor histocompatibility loci into sublethally irradiated BALB/c mice[46] (Fig. 3.2).

The cellular and molecular changes and the sequence of events that characterize the development of the murine scleroderma-like phenotype have been described in detail.[46,47] A prominent inflammation and the subsequent fibrotic response appear to be the principal characteristics of murine Scl GvHD without evidence of microvascular injury or production of autoantibodies. In Scl GvHD mice, infiltrating cutaneous mononuclear cells and increased TGFβ1 mRNA levels precede the increase of type I collagen mRNA expression and protein synthesis, with subsequent remarkable skin thickening and lung fibrosis at early time points.[46,47] These features closely resemble those of early and rapidly progressive human scleroderma. Increased expression of C-C chemokines such as MCP-1, macrophage inflammatory protein-1α (MIP-1α/CCL3), and RANTES (CCL5) also precedes the onset of skin and pulmonary fibrosis.[47] In the lesional skin, the inflammatory infiltrates consist mainly of CD11b+ monocytes/macrophages, and T cells. In particular, monocytes displaying upregulated activation markers appear to be the predominant cell population infiltrating not only the skin, but also the lungs.[46] Furthermore, the mononuclear cells infiltrating the skin in Scl GvHD are of donor origin, as demonstrated by PCR analysis of Y chromosome sequences when female BALB/c mice are transplanted with B10.D2 male cells, thus mimicking the microchimerism, which has been detected in the lesional skin of SSc female patients.[46]

Interestingly, neutralizing antibodies to TGFβ have been shown to prevent both skin and lung fibrosis by blocking the recruitment of monocytes/macrophages and T cells in the skin and by abrogating the upregulation of TGFβ1.[46] Furthermore, the same group reported that latency-associated peptide, which is released from latent form of TGFβ is effective in preventing skin fibrosis and thickening in Scl GvHD. In contrast, it is not able to suppress mononuclear cell activation and influx into the skin.[48] Very recently, cutaneous gene expression analysis using DNA microarrays showed that several inflammatory cytokines, chemokines, growth factors, and cell adhesion molecules are upregulated in the lesional skin of Scl GvHD mice, closely resembling the expression pattern of human scleroderma involved skin.[49]

These findings lead to the suggestion that murine Scl GvHD is TGFβ and chemokine dependent and might be regarded as a valuable platform to analyze novel antifibrotic interventions such as chemokine targeting in early scleroderma.

In 2004, Ruzek et al developed a modified model of graft-versus-host-induced SSc. Here, spleen cells from B10.D2 donor mice were injected intravenously into BALB/c RAG2$^{-/-}$ knockout (RAG2 KO) recipient mice, which are deficient in mature T and B cells[50] (Fig. 3.2). This model exhibits several aspects of the human disease, including dermal thickening, particularly in the extremities, progressive fibrosis involving internal organs such as kidney and small intestine but not lungs, and a prominent mononuclear cell infiltration.[50] Mainly monocytes/macrophages, CD4+ and CD8+ T cells infiltrate the affected tissues. Early immune activation and SSc-specific circulating autoantibodies such as antinuclear antibodies (ANAs) and anti-topoisomerase I antibodies (anti-Scl-70) could be detected. In addition, there is evidence of vascular alterations with a prominent vasoconstriction and altered expression of vascular markers such as ET-1 and α-SMA in skin and kidney.[50] In this model, TGFβ levels are elevated in skin and kidney. The administration of a monoclonal antibody (1D11), which

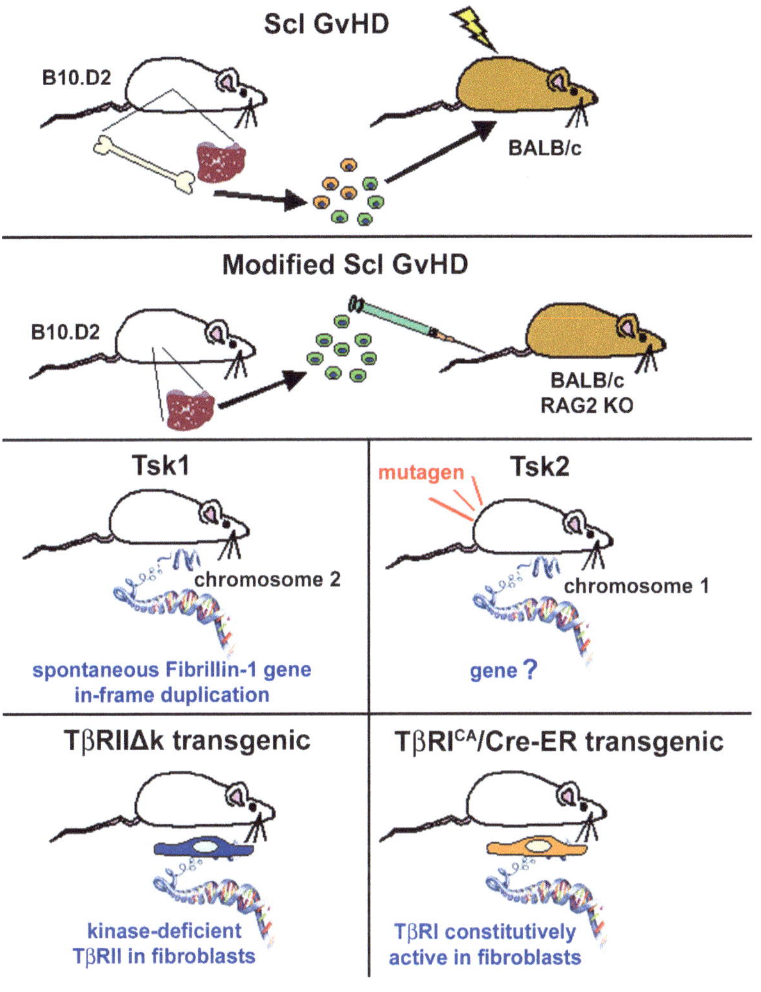

Fig. 3.2 Induced, spontaneous, and transgenic murine models for scleroderma. *Scl GvHD* sclerodermatous graft-versus-host disease model: transplantation of bone marrow and spleen cells from wild-type B10.D2 mice into sublethally irradiated BALB/c mice. Modified Scl GvHD: intravenous injection of spleen cells from wild-type B10.D2 mice into RAG2 knockout (RAG2 KO) BALB/c mice. *Tsk1* type 1 tight skin mouse: spontaneous mutation in the gene encoding fibrillin-1 on the mouse chromosome 2. *Tsk2* type 2 tight skin mouse: unknown mutagen-induced mutation on the mouse chromosome 1. TβRIIΔk: transgenic mice expressing a kinase-deficient TβRII selectively on fibroblasts. $T\beta RI^{CA}$/Cre-ER: transgenic mice expressing a postnatally constitutively active TβRI in fibroblasts

neutralizes all isoforms of TGFβ was able to reduce dermal thickening and ECM accumulation as well as changes in vessel morphology in the affected organs.[50] Because of the numerous similarities to the human disease, the modified GvHD model should be considered for further investigations aimed to achieve novel insights into the pathogenetic mechanisms and potential treatment of SSc, in particular of the diffuse SSc form, which is characterized by most prominent multiorgan involvement.

3.4 Tight Skin (Tsk) Mice

3.4.1 Tsk1

The type 1 tight skin (Tsk1) mouse represents a naturally occurring disease model in which a partial in-frame duplication within the coding region of the fibrillin-1 gene on chromosome 2 causes the SSc-like phenotype[51] (Fig. 3.2). This mutation is inherited as an autosomal dominant trait. Homozygous embryos die in utero whereas neonatal heterozygous animals develop a tight skin phenotype, which is apparent at 2–3 weeks of age.[51,52] Tsk1 mice spontaneously develop a fibrotic response, which probably originates deeply from the fascia and is characterized by diffuse thickening and tethering of the skin that is firmly bound to the subcutaneous tissue.[52,53] The initial descriptions of the tight skin phenotype included an excessive growth of connective tissue and skeleton with thickened dermis and altered ultrastructure of dermal collagen.[52,53] Subsequently, several groups reported that in contrast to human SSc, which is characterized by thickening and sclerosis of the dermis, Tsk1 mice rather show increased hypodermal thickness and severe subcutaneous hyperplasia while the dermis appears unaffected.[53,54] Vascular involvement has not been reported in Tsk1 mice and the histological changes detected in the lungs resemble those of human emphysema rather than interstitial fibrosis.[53,54] Nevertheless, the genetic background and other important features, such as evidence of autoimmunity,[55] make the Tsk1 a useful model for SSc which has been extensively studied in the past years.

Fibrillin-1 is a structural protein of connective tissue microfibrils, which has been implicated in the pathogenesis of scleroderma. Studies on Choctaw Native Americans, a population with a higher prevalence of scleroderma than the general population, show a genetic linkage, implicating the chromosomal region containing the fibrillin-1 encoding gene as a region of SSc susceptibility.[56,57] Furthermore, increased numbers of extracellular microfibrils have been detected in SSc skin lesions.[58] Microfibrils produced in vitro by SSc fibroblasts showed an altered structure, and circulating autoantibodies against fibrillin-1 have been found in patients with scleroderma.[56,59] Similarly, the mutation that is responsible for the tight skin phenotype in mice leads to a mutant fibrillin protein that has been shown to alter the organization of the ECM and to increase the deposition of other ECM proteins.[53,60,61] Fibrillin-1 is supposed to have a role in controlling the bioavailability of TGFβ by interactions with latent TGFβ binding proteins. Therefore, it has been suggested that the mutant fibrillin protein might lead to increased release of this profibrotic cytokine in Tsk1 mice.[62]

Increased ECM production, including collagen types I, III, and VI, elevated numbers of procollagen α1(I) and α1(III)-expressing fibroblasts, and upregulated TGFβ mRNA expression have been reported during the rapid postnatal development of the skin in Tsk1 mice.[53] The levels of collagen transcripts decrease with age. TGFβ is not elevated in the skin of adult Tsk1 mice.[53] Denton et al reported an increased in vivo activity of a reporter transgene harboring a fibroblast-specific enhancer of procollagen α2(I) and demonstrated an increased expression of procollagens α1(I) and α2(I) in Tsk1 mice.[63] It has been suggested that the development of the tight skin phenotype is due to the activation of TGFβ signaling, and that TGFβ and IL-4 play a pivotal role in the pathogenesis of fibrosis in this model. Explanted fibroblasts from the lesional skin of Tsk1 mice showed a sustained increase in procollagen α1(I) transcript and collagen protein under IL-4 and TGFβ stimulation.[64] Moreover, other studies suggested that the tight skin phenotype is abrogated by the disruption of the IL-4 axis, most likely through the consequent deregulation of TGFβ signaling.[65,66] Interestingly, the deletion of the IL-4 gene has been shown to rescue the Tsk1/Tsk1 homozygous mice that usually die during embryonic development.[66] CD4+ T cells are the main source of IL-4, and Tsk1 mice lacking CD4+ T cells did not develop skin thickening.[67] In contrast, the tight skin phenotype could be reproduced in SCID mice.[68,69] Administration of anti-IL-4 antibodies to neonatal Tsk1 mice prevented skin thickening.[70] Intravenous immunoglobulin therapy decreased the secretion of IL-4 and TGFβ from splenocytes and abrogated fibrosis in this model.[71] Dermal or systemic administration of halofuginone, which inhibits TGF-β-induced α1(I) procollagen gene expression and Smad3 phosphorylation, prevented thickening of the skin in Tsk1 mice.[72,73] Furthermore, a recent study showed that transfection with human recombinant HGF significantly reduced hypodermal thickness and IL-4 and TGFβ expression in the skin of Tsk1 mice, although did not ameliorate the pulmonary disease.[74] Using high-density gene arrays, Ong et al demonstrated a substantial similarity in the gene expression profile between neonatal Tsk1 fibroblasts and neonatal wild-type fibroblasts activated by recombinant TGFβ1.[75] This analysis also allowed the identification of several mediators such as monocyte chemoattractant protein 3 (MCP-3/CCL7), which are overexpressed in Tsk1 fibroblasts but had not previously been implicated in the pathogenesis of the tight skin phenotype and fibrosis.[75] Conversely, the transcript profiling analysis performed on Tsk1 skin by Baxter et al did not show any evidence of increased activity of TGFβ-induced signaling pathways. Many genes known to be involved in human fibrosis were also not upregulated in this model.[54]

Nevertheless, the large body of evidence of autoimmunity in Tsk1 mice indicates that such model may be useful to clarify the controversial role, which has been assigned to B cells in the development and progression of the human disease. Several groups reported that Tsk1 mice produce various autoantibodies similar to those found in SSc patients, such as anti-topoisomerase I, anti-RNA polymerase I, and anti-fibrillin-1 antibodies.[55] Interestingly, it has also been demonstrated that there is a correlation between skin thickness, soluble collagen content, and the serum level of anti-topoisomerase I antibodies.[55] Peripheral blood B cells showed elevated CD19 expression and activated CD19-mediated signal transduction in Tsk1 mice, leading to chronic B cell activation, which might have important implications for autoantibody production.[76-78] The loss of CD19 significantly decreased skin fibrosis, and the levels of circulating anti-topoisomerase I antibodies correlated with CD19 expression in

Tsk1 mice.[76] In a recent study, Hasegawa et al investigated the effects of B cell depletion on tight skin phenotype using an anti-CD20 monoclonal antibody for immunotherapy.[79] Interestingly, B cell depletion in neonatal Tsk1 mice has been shown to significantly suppress the development of skin thickening, autoantibody production, and the imbalance between Th1 and Th2 cytokine expression. In contrast, it was not effective in improving the pathological signs in adult mice with established disease.[79] These data suggest that B cells play a major role in the initiation of autoimmunity and fibrotic disease in Tsk1 mice, but are not required for the maintenance of the tight skin phenotype. Since no immunotherapy to date has been proven effective in the treatment of skin sclerosis in SSc patients, an early B cell-targeted intervention prior to the onset of skin fibrosis might be a potential therapeutic strategy.

Moreover, other recent studies showed that blockade of CD40–CD40 ligand (CD40L) interaction and antagonism of the B-cell-activating factor belonging to the tumor necrosis factor family (BAFF) remarkably attenuate the development of skin fibrosis as well as autoantibody production in the Tsk1 mouse model.[80,81] Thus, the critical immunoregulatory CD40–CD40L and BAFF–BAFF receptor axes could be regarded as attractive therapeutic targets for the treatment of skin fibrosis and autoimmunity in SSc patients.

3.4.2 *Tsk2*

The type 2 tight skin (Tsk2) mouse is an animal model resulting from a mutagen-induced mutation on the mouse chromosome 1[82] (Fig. 3.2). This mutation is inherited as an autosomal dominant trait. Only heterozygous animals survive and develop the Tsk2 phenotype, which is characterized by several histological and biochemical cutaneous abnormalities similar to those in the skin of SSc patients.[82] Histological examination of the skin showed a marked dermal thickening and excessive deposition of dense collagen bundles in the dermis and in the subdermal adipose tissue.[82] Biochemical analyzes on Tsk2 explanted fibroblasts showed increased procollagen $\alpha 1$(I) and $\alpha 3$(I) gene expression and collagen synthesis when compared with wild-type fibroblasts. In addition, elevated collagen gene expression was detected in the skin of the animals.[82-84] Tsk2 mice display a prominent mononuclear cell infiltration in the lower dermis and subcutaneous adipose tissue, a feature that is not displayed in Tsk1 mouse.[82] A recent study reported autoimmunity in Tsk2 mice, including the presence of several autoantibodies against nuclear antigens similar to those detected in SSc patients, such as anti-topoisomerase I, anti-centromere, and anti-ribonucleoprotein antibodies.[84] Interestingly, the circulating levels of anti-centromere antibodies significantly correlated with the increase of dermal thickness. These mice also showed autoantibodies, which are not associated with SSc, such as anti-dsDNA antibodies that are specific for systemic lupus erythematosus.[84] Therefore, Tsk2 mouse may be useful to better understand the complex interrelationship between autoimmunity and fibrosis in the pathogenesis of scleroderma and can be considered as in vivo model to study autoimmunity in different human disorders.

3.5 Genetically Modified and Transgenic Mouse Models

Genetically modified mice harboring manipulations or disruptions of pivotal signaling pathways allow the determination of the cellular and molecular mechanisms, which trigger key pathogenetic events of diseases. In the recent years, mutant or transgenic murine models have been generated in the attempt to create an animal model, which closely recapitulates the complex pathogenesis and the pathological features of human SSc.

Because of the proposed central role of TGFβ in SSc, the recently developed animal models include mice characterized by perturbation in TGFβ signaling following genetic modifications or deletion of key downstream mediators of TGFβ transduction.

Using transgenesis, Denton et al generated mice expressing a kinase-deficient type II TGFβ receptor (TβRIIΔk) selectively on fibroblasts[14] (Fig. 3.2). This genetic modification results in the development of dermal and pulmonary fibrosis in vivo. Further in vitro studies on explanted dermal murine fibroblasts provided evidence of TGFβ signaling alteration and SSc-like molecular features in this model.[14] The expression of the mutant receptor has been shown to lead to sustained activation of the TGFβ ligand-receptor axis through the kinase activity of the TGFβ receptor type 1 (TβRI, ALK5). TβRIIΔk transgenic fibroblasts showed increased synthesis of type I collagen, altered production of matrix metalloproteinases, elevated expression of CTGF, and myofibroblast transdifferentiation with enhanced ability to contract collagen gel matrices.[14]

Very recently, Sonnylal et al established a mouse model in which the TGFβ pathway is constitutively activated postnatally in fibroblasts.[15] These transgenic mice harbor an inducible constitutively active mutant TβRI and a Cre-ER transgene that is driven by a fibroblast-specific Col1a2 enhancer (TβRI^{CA}/Cre-ER mice) (Fig. 3.2). After birth, the administration of 4-hydroxytamoxifen induces the expression of fibroblast-restricted constitutively active TβRI, leading to pronounced and generalized fibrosis of the dermis, and widespread vasculopathy in internal organs with fibrotic thickening of the small vessel wall in lung and kidney.[15] Moreover, primary explanted skin fibroblasts showed enhanced myofibroblast differentiation as well as increased gene expression of several ECM proteins and downstream targets of TGFβ signaling, such as procollagen α1(I), fibronectin, CTGF, and increased Smad2/3 phosphorylation.[15] Since this murine phenotype recapitulates several histologic and biochemical features of human scleroderma and is characterized by sustained TGFβ signaling activation, it should be a valuable model to test novel antifibrotic strategies, such as therapies targeting TGFβ activation or downstream mediators of TGFβ transduction, including CTGF and Smad transcription factors. In addition, further characterization of the structural vasculopathy and future modifications of this murine model are likely to result in an even more comprehensive clinical scleroderma phenotype.

The same group has recently generated transgenic mice that overexpress CTGF in fibroblastic cells, and these mice also appear to show generalized skin fibrosis.[15] The characterization of this new animal model should aid in understanding the still unknown molecular mechanisms of CTGF function in the pathogenesis of SSc.

Le Hir et al reported that MRL/lpr mouse strains lacking IFNγ receptor (MRL/lpr-IFNγ$R^{-/-}$) spontaneously develop an inflammatory and fibrotic syndrome, which displays several features of human SSc, including abnormal deposition of dermal collagen bundles, a prominent mononuclear cell infiltration and fibrosis in skin, lungs, and other internal organs, production of autoantibodies, and microvascular alterations.[85]

Samuel et al recently proposed the relaxin gene knockout ($RLX^{-/-}$) mouse as a model of systemic sclerosis.[13] $RLX^{-/-}$ mice showed an age-related progression of skin thickening and dermal fibrosis, with increased type I and III collagen deposition and marked collagen overexpression in explanted lesional fibroblasts. These mice also consistently developed an age-related progression of pulmonary, cardiac, and renal fibrosis. Interestingly, the administration of relaxin effectively and completely reverted dermal fibrosis at early onset of the disease but did not improve the established sclerosis and excessive dermal scarring in later stages.[13]

3.6 UCD-200/206 Chicken

The UCD chicken lines 200/206 are an animal model particularly useful to study the vascular changes, which characterize human SSc.[86] The UCD-200/2006 chickens spontaneously develop an inherited scleroderma-like disease exhibiting the entire spectrum of SSc manifestations, including vascular occlusion, severe perivascular lymphocytic infiltration of the skin and viscera, fibrosis of skin and internal organs, the presence of circulating autoantibodies against nuclear antigens, anti-cardiolipin antibodies, anti-endothelial cell antibodies (AECAs), rheumatoid factors, and distal polyarthritis.[87] Studies on skin lesions from UCD-200/206 chickens have highlighted the role of endothelial dysfunction as an early acute disease feature. These studies demonstrated that circulating AECAs lead to endothelial cell injury and that AECA-induced endothelial cell apoptosis is a primary event in the pathogenesis of the SSc-like disease in this animal model.[88,89] Similar findings of endothelial cell apoptosis in early disease stages have been observed in skin biopsies from SSc patients.[88,89] These findings were further supported by the observation that AECA-positive serum from UCD-200 chickens injected into normal chicken embryos results in AECA binding to microvascular endothelium in vivo and a significant increase in endothelial cell apoptosis.[90] A further study focused on endothelial cell apoptosis, mononuclear cell infiltration, and collagen deposition in the visceral organs of UCD chickens.[91] In this study, apoptotic endothelial cells were found mostly in esophagus, lungs, and kidneys, but not in heart or liver, in UCD-200 chickens at the initial stage of the disease. This study also showed that the esophagus was the most affected organ, with a prominent mononuclear cell infiltration and increased collagen deposition. These observations support the hypothesis that endothelial cell apoptosis initiates the disease process, followed by mononuclear cell infiltration and fibrosis.

3.7 Concluding Remarks and Future Perspectives

The complexity and the still limited knowledge of the molecular mechanisms that trigger the pathogenesis of SSc avert the development of a prototypic animal model that would include all the hallmarks of the human disease. To date, several internal organ manifestations that are the most frequent causes of morbidity and mortality in SSc patients, such as gastrointestinal tract dysfunction and mainly gastro-esophageal disease, pulmonary arterial hypertension (PAH), and SSc-associated interstitial lung disease (SSc-ILD), could not be reproduced in an experimental approach. Furthermore, the only available animal model that appears to recapitulate the entire spectrum of vascular changes of human SSc is an avian model, which is difficult to further characterize and to use for molecular biology studies because of the almost completely unknown genetic background.

Nevertheless, some of the key features of the disease have been recapitulated in a number of animal models, which have been extensively studied by many groups to decipher the cellular and molecular pathogenetic events that drive the onset and the progression of fibrosis in human SSc. Based on the large body of knowledge achieved up to date, further in-depth investigations using established and new mouse models and the development of novel genetically determined experimental approaches, that more closely resemble human SSc, will provide new insight into disease pathways and potential therapeutic targets. The prudent extrapolation of the achieved results and the evaluation of their potential use in therapeutic approaches finally will allow the translation of basic research into more effective treatment of scleroderma patients.

References

1. LeRoy EC, Black C, Fleischmajer R, et al. Scleroderma (systemic sclerosis): classification, subsets and pathogenesis. *J Rheumatol.* 1988;15:202–205.
2. Jimenez SA, Derk CT. Following the molecular pathways toward an understanding of the pathogenesis of systemic sclerosis. *Ann Intern Med.* 2004;140:37–50.
3. Varga J, Abraham D. Systemic sclerosis: a prototypic multisystem fibrotic disorder. *J Clin Invest.* 2007;117:557–567.
4. Distler JH, Jüngel A, Caretto D, et al. Monocyte chemoattractant protein 1 released from glycosaminoglycans mediates its profibrotic effects in systemic sclerosis via the release of interleukin-4 from T cells. *Arthritis Rheum.* 2006;54:214–225.
5. Wynn TA. Fibrotic disease and the T(H)1/T(H)2 paradigm. *Nat Rev Immunol.* 2004;4:583–594.
6. Ishida W, Mori Y, Lakos G, et al. Intracellular TGF-β receptor blockade abrogates Smad-dependent fibroblast activation in vitro and in vivo. *J Invest Dermatol.* 2006;126:1733–1744.
7. Mori Y, Chen SJ, Varga J. Expression and regulation of intracellular SMAD signaling in scleroderma skin fibroblasts. *Arthritis Rheum.* 2003;48:1964–1978.
8. Pannu J, Gardner H, Shearstone JR, Smith E, Trojanowska M. Increased levels of transforming growth factor beta receptor type I and up-regulation of matrix gene program: a model of scleroderma. *Arthritis Rheum.* 2006;54:3011–3021.
9. Shi-Wen X, Rodriguez-Pascual F, Lamas S, et al. Constitutive ALK5-independent c-Jun N-terminal kinase activation contributes to endothelin-1 overexpression in pulmonary fibrosis: evidence of an autocrine endothelin loop operating through the endothelin A and B receptors. *Mol Cell Biol.* 2006;26:5518–5527.

54. Baxter RM, Crowell TP, McCrann ME, Frew EM, Gardner H. Analysis of the tight skin (Tsk1/+) mouse as a model for testing antifibrotic agents. *Lab Invest.* 2005;85:1199–1209.
55. Saito S, Kasturi K, Bona C. Genetic and immunologic features associated with scleroderma-like syndrome of TSK mice. *Curr Rheumatol Rep.* 1999;1:34–37.
56. Wallis DD, Tan FK, Kessler R, et al. Fibrillin 1 abnormalities in dermal fibroblast cultures from first degree relatives of patients with systemic sclerosis (scleroderma). *Arthritis Rheum.* 2004;50:329–331.
57. Tan FK, Stivers DN, Foster MW, et al. Association of microsatellite markers near the fibrillin 1 gene on human chromosome 15q with scleroderma in a Native American population. *Arthritis Rheum.* 1998;41:1729–1737.
58. Fleischmajer R, Jacobs L, Schwartz E, Sakai LY. Extracellular microfibrils are increased in localized and systemic scleroderma skin. *Lab Invest.* 1991;64:791–798.
59. Tan FK, Arnett FC, Antohi S, et al. Autoantibodies to the extracellular matrix microfibrillar protein, fibrillin-1, in patients with scleroderma and other connective tissue diseases. *J Immunol.* 1999;163:1066–1072.
60. Saito S, Nishimura H, Brumeanu TD, et al. Characterization of mutated protein encoded by partially duplicated fibrillin-1 gene in tight skin (TSK) mice. *Mol Immunol.* 1999;36:169–176.
61. Lemaire R, Farina G, Kissin E, et al. Mutant fibrillin 1 from tight skin mice increases extracellular matrix incorporation of microfibril-associated glycoprotein 2 and type I collagen. *Arthritis Rheum.* 2004;50:915–926.
62. Isogai Z, Ono RN, Ushiro S, et al. Latent transforming growth factor β-binding protein 1 interacts with fibrillin and is a microfibril-associated protein. *J Biol Chem.* 2003;278:2750–2757.
63. Denton CP, Zheng B, Shiwen X, et al. Activation of a fibroblast specific enhancer of the Proα2(I) collagen gene in tight-skin mice. *Arthritis Rheum.* 2001;44:712–722.
64. McGaha TL, Le M, Kodera T, et al. Molecular mechanisms of interleukin-4-induced up-regulation of type I collagen gene expression in murine fibroblasts. *Arthritis Rheum.* 2003;48:2275–2284.
65. McGaha TL, Saito S, Phelps RG, et al. Lack of skin fibrosis in tight skin (TSK) mice with targeted mutation in the interleukin-4R alpha and transforming growth factor-beta genes. *J Invest Dermatol.* 2001;116:136–143.
66. Kodera T, McGaha TL, Phelps R, Paul WE, Bona CA. Disrupting the IL-4 gene rescues mice homozygous for the tight-skin mutation from embryonic death and diminishes TGF-β production by fibroblasts. *Proc Natl Acad Sci USA.* 2002;99:3800–3805.
67. Wallace VA, Kondo S, Kono T, et al. A role for CD4+ T cells in the pathogenesis of skin fibrosis in tight skin mice. *Eur J Immunol.* 1994;24:1463–1466.
68. Siracusa LD, McGrath R, Fisher JK, Jimenez SA. The mouse tight skin (Tsk) phenotype is not dependent on the presence of mature T and B lymphocytes. *Mamm Genome.* 1998;9:907–909.
69. Dodig TD, Mack KT, Cassarino DF, Clark SH. Development of the tight-skin phenotype in immune-deficient mice. *Arthritis Rheum.* 2001;44:723–727.
70. Ong D, Wong C, Roberts CR, Teh HS, Jirik FR. Anti-IL-4 treatment prevents dermal collagen deposition in the tight-skin mouse model of scleroderma. *Eur J Immunol.* 1998;128:2619–2629.
71. Blank M, Levy Y, Amital H, Shoenfeld Y, Pines M, Genina O. The role of intravenous immunoglobulin therapy in mediating skin fibrosis in tight skin mice. *Arthritis Rheum.* 2002;46: 1689–1690.
72. Levi-Schaffer F, Nagler A, Slavin S, Knopov V, Pines M. Inhibition of collagen synthesis and changes in skin morphology in murine graft-versus-host disease and tight skin mice. Effect of halofuginone. *J Invest Dermatol.* 1996;106:84–88.
73. McGaha TL, Phelps RG, Spiera H, Bona C. Halofuginone, an inhibitor of type-I collagen synthesis and skin sclerosis, blocks transforming-growth-factor-β-mediated Smad3 activation in fibroblasts. *J Invest Dermatol.* 2002;118:461–470.
74. Iwasaki T, Imado T, Kitano S, Sano H. Hepatocyte growth factor ameliorates dermal sclerosis in the tight-skin mouse model of scleroderma. *Arthritis Res Ther.* 2006;8:R161.
75. Ong VH, Evans LA, Shiwen X, et al. Monocyte chemoattractant protein 3 as a mediator of fibrosis. Overexpression in systemic sclerosis and the type 1 tight-skin mouse. *Arthritis Rheum.* 2003;48:1979–1991.

76. Saito E, Fujimoto M, Hasegawa M, et al. CD19-dependent B lymphocyte signaling thresholds influence skin fibrosis and autoimmunity in the tight-skin mouse. *J Clin Invest.* 2002;109: 1453–1462.
77. Tedder TF, Poe JC, Fujimoto M, Haas KM, Sato S. The CD19-CD21 signal transduction complex of B lymphocytes regulates the balance between health and autoimmune disease: systemic sclerosis as a model system. *Curr Dir Autoimm.* 2005;8:55–90.
78. Asano N, Fujimoto M, Yazawa N, et al. B lymphocyte signaling established by the CD19/CD22 loop regulates autoimmunity in the tight-skin mouse. *Am J Pathol.* 2004;165:641–650.
79. Hasegawa M, Hamaguchi Y, Yanaba K, et al. B-lymphocyte depletion reduces skin fibrosis and autoimmunity in the tight-skin mouse model for systemic sclerosis. *Am J Pathol.* 2006;169:954–966.
80. Komura K, Manabu F, Yanaba K, et al. Blockade of CD40-CD40 ligand interactions attenuates skin fibrosis and autoimmunity in the tight-skin mouse. *Ann Rheum Dis.* 2008; doi: 10.1136/ard.2007.073387.
81. Matsushita T, Fujimoto M, Hasegawa M, et al. BAFF antagonist attenuates the development of skin fibrosis in tight-skin mice. *J Invest Dermatol.* 2007;127:2772–2780.
82. Christner PJ, Peters J, Hawkins D, Siracusa LD, Jimenez SA. The tight skin 2 mouse. An animal model of scleroderma displaying cutaneous fibrosis and mononuclear cell infiltration. *Arthritis Rheum.* 1995;38:1791–1798.
83. Christner PJ, Yufit T, Peters J, McGrath R, Conway RF, Jimenez SA. Transcriptional activation of $\alpha 1$(III) procollagen gene in Tsk2/+ dermal fibroblasts. *Biochem Biophys Res Commun.* 2003;303:406–412.
84. Gentiletti J, McCloskey L, Artlett CM, Peters J, Jimenez SA, Christner PJ. Demonstration of autoimmunity in the tight skin-2 mouse: a model for scleroderma. *J Immunol.* 2005;175:2418–2426.
85. Le Hir M, Martin M, Haas C. A syndrome resembling human systemic sclerosis (scleroderma) in MRL/lpr mice lacking interferon-gamma (IFNγ) receptor (MRL/lprγR–/–). *Clin Exp Immunol.* 1999;115:281–287.
86. Gershwin ME, Abplanalp H, Castles JJ, et al. Characterization of a spontaneous disease of white leghorn chickens resembling progressive systemic sclerosis (scleroderma). *J Exp Med.* 1981;153:1640–1659.
87. Sgonc R. The vascular perspective of systemic sclerosis: of chickens, mice and men. *Int Arch Allergy Immunol.* 1999;120:169–176.
88. Sgonc R, Gruschwitz MS, Dietrich H, Recheis H, Gershwin ME, Wick G. Endothelial cell apoptosis is a primary pathogenetic event underlying skin lesions in avian and human scleroderma. *J Clin Invest.* 1996;98:785–792.
89. Sgonc R, Gruschwitz MS, Boeck G, Sepp N, Gruber J, Wick G. Endothelial cell apoptosis in systemic sclerosis is induced by antibody-dependent cell-mediated cytotoxicity via CD95. *Arthritis Rheum.* 2000;43:2550–2562.
90. Worda M, Sgonc R, Dietrich H, et al. In vivo analysis of the apoptosis-inducing effect of anti-endothelial cell antibodies in systemic sclerosis by the chorionallantoic membrane assay. *Arthritis Rheum.* 2003;48:2605–2614.
91. Nguyen VA, Sgonc R, Dietrich H, Wick G. Endothelial injury in internal organs of University of California at Davis line 200 (UCD 200) chickens, an animal model for systemic sclerosis (scleroderma). *J Autoimmun.* 2000;14:143–149.

Chapter 4
Animal Models of Atherosclerosis

Christopher Jackson, Ulrike Benbow, Andrew Bond, Deborah Galley, and Claire Schwartz

4.1 Introduction

Atherosclerosis has been studied in animals for exactly a century, starting with the pioneering studies of Ignatowski (reviewed in ref.).[1] The motivation for those early studies is unclear, as atherosclerosis was not then recognized as a major disease. However, the increasing prevalence of cardiovascular disorders within the global burden of disease[2] has given the desire to understand and treat atherosclerosis ever greater impetus. Animal models have an important part to play in this, in two main regards: increasing our understanding of the pathophysiology of atherosclerosis and developing new treatments for the disease. This chapter takes the view that most of the atherosclerosis research done in animal models until now has served the first purpose well, but the second rather poorly. To support this contention, it will be necessary to consider how atherosclerosis starts and develops.

4.2 Natural History of Atherosclerosis

Atherosclerosis is a disease of the intimal lining of the larger arteries. It is most prevalent in those vessels closest to the heart in terms of the time taken for blood to travel to them: the coronary arteries, the aortic arch and its branches, the abdominal aorta, and the iliac arteries.

An important feature of atherosclerosis is that it is a focal disease, characterized by lesions that arise at discrete locations. As they grow, they may coalesce and give the appearance of a single large damaged surface, but this is misleading. How atherosclerosis is initiated is a matter of dispute, as this is impossible to study in humans. It is possible to derive hypotheses based on certain common features of

---

C. Jackson (✉)
Bristol Heart Institute, University of Bristol, Level 7, Bristol Royal Infirmary, Marlborough Street, Bristol, BS2 8HW, Avon, UK
e-mail: chris.jackson@bristol.ac.uk

D. Abraham et al. (eds.), *Advances in Vascular Medicine*,
DOI 10.1007/978-1-84882-637-3_4,

lesion location, which include disturbed blood flow (hence a predilection for branch points), and the presence of pre-existing accumulations of smooth muscle cells in the intima. The first easily recognized pathological change is the appearance of small collections of lipid-filled cells called foam cells (Fig. 4.1). These are derived either from macrophages or from smooth muscle cells and suggest that lipoprotein trafficking into and out of the vessel wall is compromised. Since this would require at least some impairment of endothelial function, it is reasonable to suggest that endothelial dysfunction is a very early feature of atherogenesis.

The accumulation of foamy cells eventually becomes big enough to be seen with the naked eye, and is then termed a fatty streak. This gradually develops a cap formed from smooth muscle cells, structural proteins, and proteoglycans. Despite existing in an environment conducive to foam cell formation, these smooth muscle cells do not become foamy. Further elaboration of proteins such as collagen and elastin by smooth muscle cells results in the fibrous lesion (Fig. 4.2), the next step in plaque evolution after the fatty streak.

Death of cells within the plaque, particularly foam cells, results in the formation of pools of extracellular lipid, and the plaque is now said to be complex (Fig. 4.3). The plaque may remain in the complex state, or may progress further. Continued deposition of extracellular lipid, loss of cells through apoptosis, degradation of structural proteins, and mechanical stress and strain may together sufficiently weaken the plaque to the point where it splits open, an event usually called a rupture (Fig. 4.4). Plaque rupture results in the inner components of the atherosclerotic

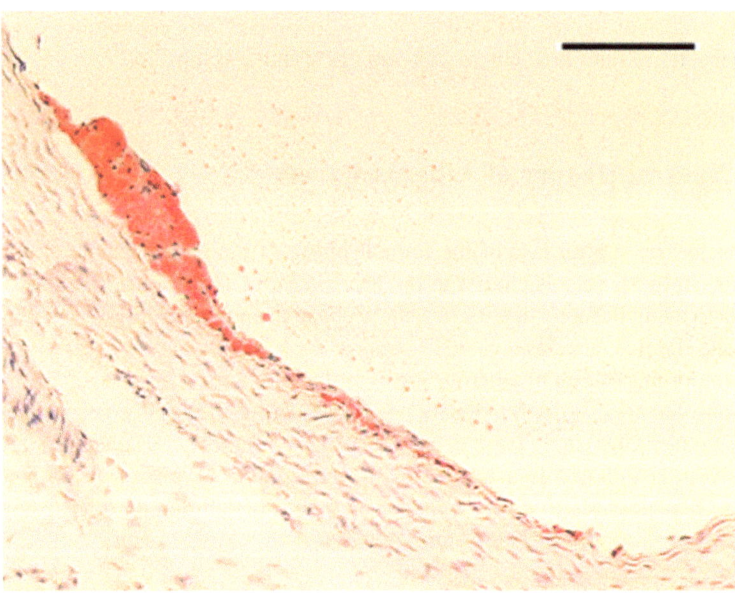

Fig. 4.1 A fatty streak lesion in the thoracic aorta of a male apolipoprotein E (apoE) knockout mouse fed high-fat diet for 4 weeks. Stained for lipid with Oil Red O. Nuclei are counter-stained with hematoxylin. Scale bar = 100 μm. Reproduced with permission from ref.[37]

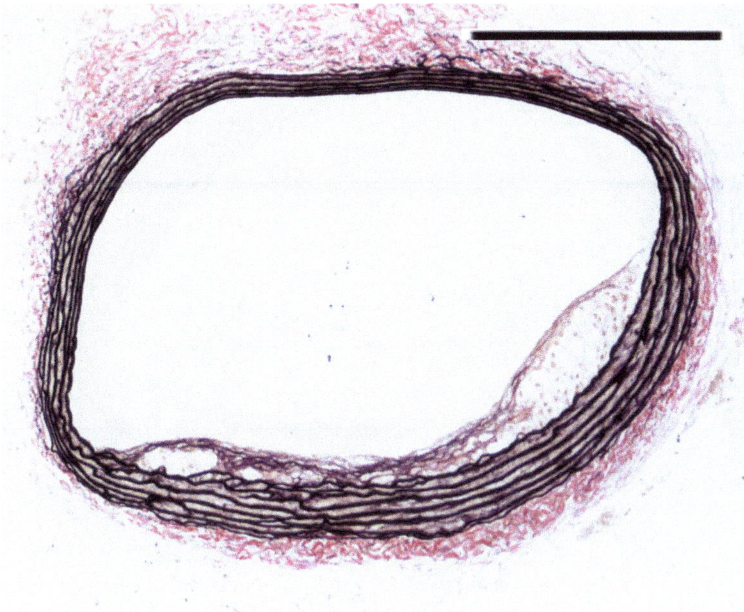

Fig. 4.2 A fibrous plaque in the proximal brachiocephalic artery of a male apoE knockout mouse fed high-fat diet for 6 weeks. Stained for elastin. Scale bar = 200 μm. Reproduced with permission from ref. [37]

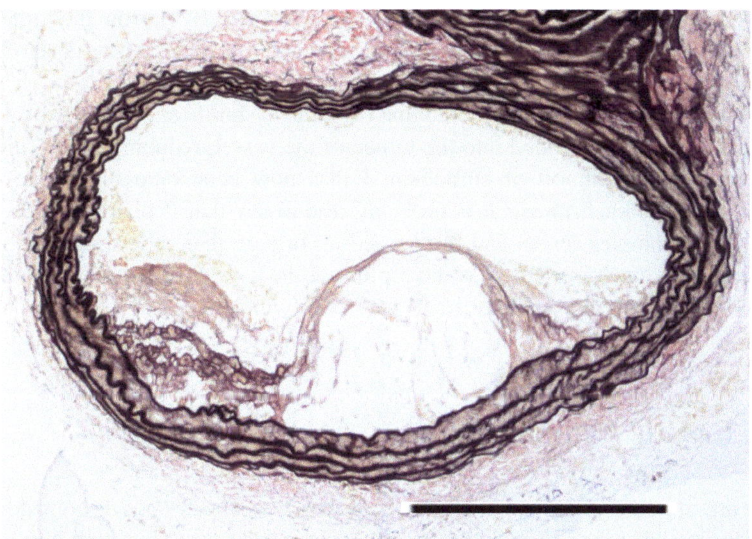

Fig. 4.3 A complex lesion in the proximal brachiocephalic artery of a male apoE knockout mouse fed high-fat diet for 8 weeks. Stained for elastin. Scale bar = 200 μm. Reproduced with permission from ref. [37]

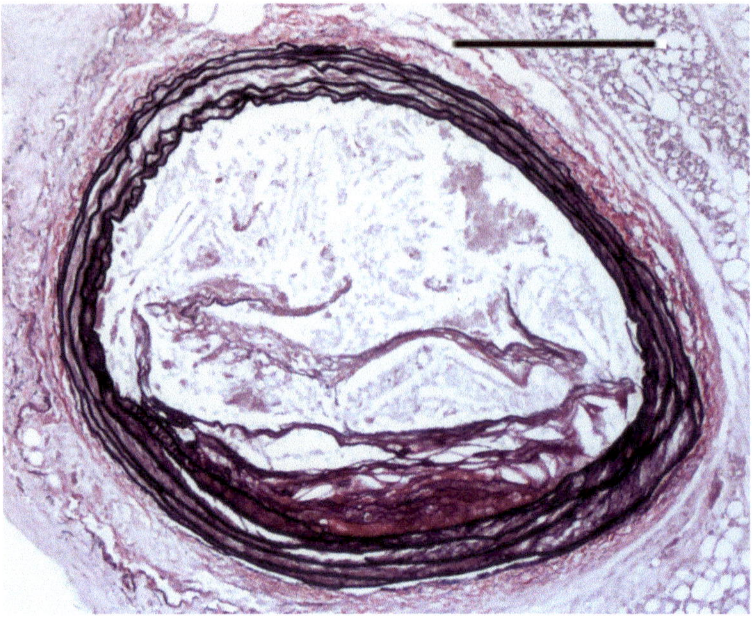

Fig. 4.4 An acute atherosclerotic plaque rupture in the proximal brachiocephalic artery of a male apoE knockout mouse fed high-fat diet for 8 weeks. Stained for elastin. Scale bar = 200 μm. Reproduced with permission from ref. [37]

lesion coming into contact with the blood, and thus thrombosis. This is usually resolved without symptom but, in some cases, will critically narrow the lumen and precipitate end-organ ischemia. More often than not, this manifests as a heart attack or a stroke.

For many years, it was considered that the clinical significance of atherosclerosis was that its lesions obtruded into the lumen of the vessel, reducing blood flow and increasing the likelihood of embolism. It has now been recognized that many plaques pose no such threat; it is probably true to say that if plaques reached the stage of the complex lesion and progressed no further, then atherosclerosis would not be of very much interest. It is the rupture of the complex plaque that is the key pathological event, but one that is still very poorly understood.

4.3 General Considerations

There are of course many animal models in which atherosclerotic lesions develop, involving a wide range of species and manipulations. It is not our intention here to recapitulate all of these, as they have been very expertly reviewed elsewhere (see, for example, refs.) [1] and [3]. Instead, some general points about the use and interpretation of animal models of atherosclerosis will be made.

The first thing to say is that all biological studies, from purified molecules to clinical trials, involve the use of models. Animal models, however, are sometimes treated with a special degree of skepticism and caution; it is often demanded that the animal form of the disease being studied recapitulates the human disease with great morphological fidelity. This is a criterion that is far more stringent than those used for other model systems, and it is quite wrong. An ideal model will faithfully reproduce the biochemistry of the disease, but this need not necessarily result in identical morphology. Provided that the processes that lead to lesion formation are closely congruent between animal and human, the model will yield accurate and useful insights. This is a particularly important point to make in the case of atherosclerosis, a disease that takes decades to develop in humans. An animal model of atherosclerosis must necessarily involve acceleration of the disease, which may mean that processes that are sequential in humans occur in parallel in the animal. This is likely to distort the morphology, but the biochemistry may still be accurately modeled.

4.4 Selecting an Animal Model

The suggestion that it is only rupture of the plaque that is pathologically significant has important implications for how to select animal models. If we assume that pharmacological therapy of atherosclerosis is only going to be instituted in individuals with demonstrable risk, and accept that atherosclerotic lesions begin to appear early in life, then it will be necessary to target just those processes that occur later in plaque development. We will need to study them and test new treatments in animal models of advanced and, crucially, unstable atherosclerosis. On the other hand, the basic scientific understanding of the disease certainly can be advanced by studies of the early stages of plaque initiation in animals.

In most model systems, animals are challenged with some hyperlipemic stimulus, such as dietary or genetic manipulation, and develop fatty streak lesions. This is the case, for example, when normal rabbits are fed a cholesterol-supplemented diet. Lesions are particularly prevalent in the thoracic aorta, but eventually form in most of the large arteries. They take the form of collections of foam cells – these collections can grow quite large – but are uncomplicated lesions that do not develop morphological complexity and do not rupture. What can we learn from such a model?

In terms of the basic processes underlying atherosclerosis, we can learn rather a lot. The changes in endothelial function that permit lipid insudation into the vessel wall, the homing of monocytes to sites of lipid accumulation, the processing of the lipids by monocytes, and the differentiation of those cells into macrophages and macrophage-derived foam cells are all grist to the investigator's mill. However, none of this yields useful information about events in advanced and unstable plaques. Drugs that inhibit, say, lipid accumulation by macrophages will effectively limit plaque growth in such a model, but are very unlikely to show clinical benefit in terms of preventing an individual who has had one heart attack from having another.

This demonstration that animal models must be selected with the purpose of the investigation in mind may seem like a long-winded way to make rather an obvious point, but it is a point that has often been ignored during pre-clinical studies of anti-atherosclerosis drugs. The usual approach has been to take a naïve animal and then to test an intervention administered during the same period as the atherosclerotic provocation is applied. Drugs that interfere with plaque initiation and early development are effective in such models, but do little to influence clinically important advanced unstable atherosclerosis and thus are destined to almost certain failure during clinical development.

4.5 Models of Plaque Initiation and Early Plaque Development

Modeling atherogenesis is of basic scientific interest but does not have much bearing on drug development, for the reasons outlined above. Therefore, the main consideration has to be the precise process that one is trying to investigate. For example, investigation of endothelial dysfunction is very likely to produce misleading results in a model wherein endothelial perturbation is used to provoke lesion formation, such as the administration of high doses of homocysteine[4] or bacterial endotoxin.[5] With such obvious exceptions, the selection of an animal model for this kind of basic study is mainly an issue of cost and convenience, with consideration of things like availability of reagents and the quantity of tissue that can be harvested also needing to be taken into consideration. We do not intend to go into these models in any more detail.

4.6 Models of Unstable Atherosclerosis

There are two kinds of animal models of unstable atherosclerosis: those that do develop unstable atherosclerosis, and those that do not. Surprisingly, most published studies make use of the latter. It has to be admitted that there was in the past some justification for this, because there were only one or two animal models in which plaques actually ruptured. Since 2001 though, a simple mouse model has been available.[6] The continued use of animal models in which plaques do not rupture raises interesting questions regarding the criteria that can be used to evaluate plaque stability.

4.6.1 Models in Which Atherosclerotic Lesions Do Not Rupture

Direct evaluation of instability is clearly impossible in a model in which plaques never rupture, so indirect methods are used instead. Essentially, this involves measuring parameters that are thought to affect lesion stability, such as plaque cellularity, the cell types present, structural protein content, lipid content, fibrous cap thickness, and so on. An intervention that results in an increase in the fractional content of collagen, for example, is interpreted as causing an increase in plaque stability even though the plaques in that particular animal model do not rupture. How do we know that this approach yields useful data?

The logic that is used is that these are the features that differ between human nonruptured and ruptured plaques. Data obtained at postmortem show that reduced collagen content, reduced smooth muscle cell content, increased lipid content, a thin fibrous cap, and so forth are common features of ruptured plaques.[7] In other words, if the change caused by the intervention is also manifested in human plaques, then they will become more stable and less liable to rupture. The problem with this approach stems from its *post hoc* nature. The changes in an animal plaque that are said to predict an increase in stability have been selected because they are opposite to those that show a probabilistic association with plaque rupture in humans. Because 90% of plaque ruptures in humans heal without acute clinical consequence,[8] the factors associated with ruptured human plaques have been deduced from only the other 10% and it is open to question how representative they are. Despite this, it is probably reasonable to say that the use of indirect markers of plaque stability in animal models, which do not suffer from ruptured plaques, is acceptable at present, subject to the caveats expressed below. However, this approach really needs to be validated by the use of a drug treatment that has been shown to improve plaque stability in humans.

4.6.1.1 Study Design Considerations in Models Where Rupture Does Not Occur

One approach that is widely used and published is to study the effects of an intervention on indirect markers of plaque stability, such as plaque collagen content, using a study design where the interventional treatment commences at the same time as the atherosclerotic stimulus. This is a deeply flawed study design that must be avoided: it is confounded by the combination of effects on atherogenesis with effects on plaque composition. An intervention that inhibits, say, monocyte recruitment into the vessel wall will limit atherosclerosis and the plaques will therefore necessarily contain less collagen, fewer smooth muscle cells, more lipid, and so on. This does not mean that the intervention has some specific effect on plaque stability, though such studies are frequently or even usually interpreted this way. This is misleading and wrong. If a model system that does not undergo plaque rupture is to be used, then really it is necessary to start the interventional treatment after plaques have already developed and become established. Changes to indirect markers of plaque stability under these circumstances may indeed reflect a genuine effect.

4.6.2 Models in Which Atherosclerotic Lesions Do Rupture

4.6.2.1 Detection and Quantification of Plaque Rupture

Plaque rupture in humans has been defined as "fibroatheroma with cap disruption; luminal thrombus communicates with the underlying necrotic core."[9] Does this mean that if a disrupted plaque is observed without an accompanying thrombus in an animal model, then the model is not actually generating plaque ruptures? Such events would be mechanistically indistinguishable from plaque ruptures that do result in luminal thrombus, and the break in the fibrous cap would be identical. Even in human studies, plaque rupture has been identified without thrombus formation: for example, where the presence of old ruptures that have repaired is inferred from buried fibrous caps and layering.[8] Separating the consequences of plaque rupture from its genesis in this way is of particular concern when we are considering animal models.

If luminal thrombus is retained as a key diagnostic feature of plaque rupture, animal models of plaque rupture would have to mirror not just the pathophysiological mechanisms of rupture but also to display human-like thrombosis. This kind of restrictive approach is not justified.[10] In particular, and most pertinently in the mouse, the requirement for luminal thrombus is incorrect. Murine brachiocephalic arteries are much smaller than human coronaries, so that the cross-sectional area of a fully occlusive thrombus is about 50-fold less in mice than humans. The volume of even a large thrombus in the mouse is likely to be at least 200-fold less than in humans and its surface area will be about 30-fold less. The fibrinolytic system in mice differs significantly from that in humans, as the plasma level of plasminogen activator inhibitor-1 (PAI-1) is 5–12.5-fold lower in mice than in humans, whereas fibrinogen and tissue-type plasminogen activator (tPA) concentrations are similar.[11] PAI-1 is the major determinant of the rate of lysis of platelet-rich arterial thrombi by pharmacological concentrations of tPA.[12] Furthermore, plasma levels of thrombin-activatable fibrinolysis inhibitor (TAFI) in the mouse are 2–7-fold lower than in humans.[13,14] Activated TAFI can impair fibrinolysis by removing carboxy-terminal lysine residues from fibrin, which act as binding sites for plasminogen and tPA.[13] Thus, the fibrinolytic balance in mice appears to be shifted toward enhanced lysis. Some human coronary thrombi may persist for months,[15] but even if we assume equal rates of fibrinolysis, then mouse plaques will be gone within a few days. If we further assume that the interval between episodes of plaque rupture in mice is of the order of weeks, then the chance of terminating an animal during the period when the thrombus is still present may be as little as 5%, even if luminal thrombus formation is an invariable consequence of murine plaque rupture. It is therefore clear that the presence of luminal thrombosis should not be regarded as a defining characteristic for plaque rupture.

The tear in the fibrous cap will also be much smaller in a murine plaque rupture. Observations of serial sections of ruptured plaques in the mouse brachiocephalic artery suggest that these defects are rarely more than 60 μm in length,[10] whereas in human coronary arteries the average length of such a defect is 1.9 mm.[16] Given similar rates of healing, the breach in the human fibrous cap will be detectable for 30 times as long.

The foregoing makes it clear that it is not reasonable to apply the existing clinical definitions of plaque rupture to mice and other small animal models. So how can we find, recognize, and quantify plaque rupture in animals? If it is true that there is rapid lysis of thrombi and healing of tears in the fibrous caps of mice, acute plaque ruptures of the type usually shown to illustrate the phenomenon in humans would only occasionally be seen in small animals. It is more likely that an animal plaque will be interrogated after a rupture has occurred and the healing process has already started, so the issue is how to recognize a previous, but now healing or fully healed plaque rupture.

Laminar structures are seen in the intima of advanced murine lesions that are rich in elastin and are populated by strongly α-smooth muscle actin-positive cells, taken to be smooth muscle cells[10] (Fig. 4.5). These appearances are highly suggestive of remnants of previous fibrous caps that have ruptured and have been incorporated into the growing lesion as it develops. When such buried fibrous caps are seen in humans, they are interpreted as indicative of previous, nonfatal, healed plaque rupture.[8,17] A number of lines of evidence support a similar interpretation in mice[10]: buried fibrous caps form only at sites where plaque ruptures occur; they are associated with fibrin deposition; and they can be modulated independently of changes in plaque size. This suggests that buried fibrous caps either represent sites of previous plaque rupture in mice, or occur in parallel with plaque rupture, and are

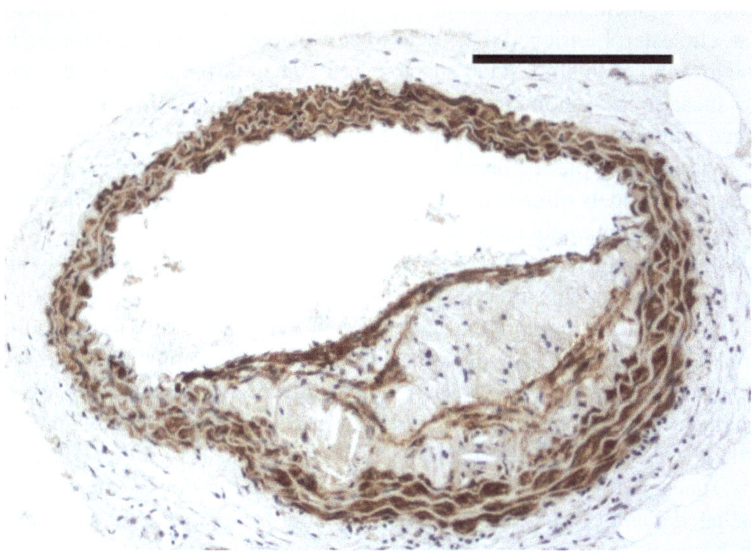

Fig. 4.5 A healed plaque rupture, in the form of a buried fibrous cap, in the proximal brachiocephalic artery of a male apoE knockout mouse fed high-fat diet for 8 weeks. Stained for α-smooth muscle actin. Scale bar = 200 μm. Reproduced with permission from ref. [37]

therefore useful indicators of plaque rupture that can be used to quantify the phenomenon.

4.6.2.2 Mouse Models of Plaque Rupture

The apolipoprotein E (apoE) knockout mouse, first reported in 1992, spontaneously develops atherosclerotic lesions in the thoracic aorta.[18,19] However, these aortic lesions do not generally undergo plaque rupture. To induce plaque rupture in apoE knockouts, compression injury has been applied to the abdominal aorta at sites of atheromatous lesions using blunt forceps.[20] Although this results in a number of areas of thrombosis not associated with plaque, 10 of 32 animals showed evidence of thrombus-associated plaque using this technique. There was histological evidence of intraplaque hemorrhage within disrupted plaques, and plaque-associated luminal thrombus.

Physical and chemical triggers have also been employed, such as using a photochemical reaction to induce thrombus formation.[21]Another approach has been to use apoE knockout mice with carotid atheromatous lesions induced by an externally placed silastic collar,[22] and then to transfect these with an adenovirus expressing the tumor-suppressor protein p53.[23] One day following transfection, increased apoptosis is evident in the cell of the fibrous cap and cap thinning is seen at later time points. These attenuated fibrous caps undergo rupture in 40% of animals after pressor challenge with phenylephrine.[23] Recently, Cheng et al[24] have developed a device that induces low shear stress proximally and oscillatory shear stress distally in a straight segment of the common carotid artery. In apoE knockout mice fed a high-fat, cholesterol-enriched diet for 9 weeks, plaques developed reproducibly in both the low and oscillating shear stress regions. The morphology of the plaques in the low shear stress region closely mimicked human thin cap fibro-atheroma, and prolonged angiotensin II infusion induced intraplaque hemorrhage exclusively in plaques with phenotypic features associated with instability. The particular advantage that these models offer is that vulnerable lesions develop at sites that are very accessible to local manipulation and instrumentation.

Recently, investigations have begun to focus on the brachiocephalic artery of apoE knockout mice. This artery is a short communicating vessel originating at the aortic arch and bifurcating into the right common carotid and right subclavian arteries. Unlike aortic lesions in apoE knockout mice, brachiocephalic plaques closely resemble those of human atherosclerosis. In studies of mice aged 42–54 weeks fed a standard chow diet (some of which received supplementary estrogens), intraplaque hemorrhage was found in 75% of animals, and there was also fibrotic conversion of necrotic zones and loss of fibrous cap.[25] Many of these lesions also demonstrated endothelial denudation or absence of endothelium and macrophage infiltration. Although fibrous cap thinning was observed, neither plaque rupture nor thrombosis was identified. The threefold novel findings of an acellular necrotic core, erosion of this mass through to the lumen, and the presence of intraplaque hemorrhage do however suggest a resemblance to plaque erosion with thrombosis.

This model has been developed further using apoE knockout mice with a strain background that is a mixture of C57BL/6 and 129. These mice, fed a diet containing 21% lard and 0.15% cholesterol starting at the age of 6–8 weeks, show progressive, pronounced atheromatous lesions within the brachiocephalic artery after as little as 5 weeks.[26] These lesions are similar to human coronary vulnerable plaque lesions in that they have a high lipid content and a relatively thin fibrous cap. Most intriguingly however, and in contrast to previous studies, these animals suffer acute plaque rupture with thinning and discontinuity of the fibrous cap and intrusion of blood into the lesion.[6,27] Those animals in which acute rupture is seen also have a higher number of buried fibrous caps within the plaque, strongly suggestive of prior episodes of silent plaque rupture. These observations make this model particularly pertinent to human vulnerable plaque.[8,17] The lesions exhibiting acute plaque rupture are also relatively larger and are associated with a greater rate of luminal occlusion.[27] This model of atherosclerosis development, which induces lesions with many of the characteristics of vulnerable plaque, and most importantly demonstrates acute plaque rupture, has been viewed as a major advance.[28]

Double knockout mice, usually combining apoE deficiency with other pathophysiologically relevant gene knockouts, are also proving increasingly useful. Hypercholesterolemic apoE/LDL receptor double knockout mice fed a diet of 21% fat and 0.15% cholesterol, exposed to mental stress (air being forced into the holding cage, with stress recorded as increased heart rate) or hypoxia (from 21% oxygen stepwise down to 10%), undergo ischemia and myocardial infarction.[29] The animals exhibit complex coronary lesions containing fibrin and extracellular crystalline cholesterol, but neither acute plaque rupture nor plaque-associated thrombosis has been reported.

A similar phenotype to the apoE/LDL receptor double knockout mouse is provided by the double knockout for apoE and the high-density lipoprotein scavenger receptor class B type I (SR-BI).[30] These animals develop aortic sinus atherosclerotic lesions from as early as 4 weeks when fed standard chow and then go on to develop extensive coronary artery disease. There is marked fibrin deposition and cholesterol cleft formation, and the animals suffer increased mortality. Actual evidence of acute plaque rupture is not yet available, though the extensive fibrin staining suggests that intraplaque hemorrhage occurs. Moreover, these animals develop multiple ischemic myocardial infarctions and progressive left ventricular impairmcnt.

4.6.2.3 Rat Models of Plaque Rupture

Dahl salt-sensitive hypertensive rats, transgenic for human cholesteryl ester transfer protein, have been proposed as a useful model of plaque vulnerability.[31] These rats develop age-dependent hypertriglyceridemia, hypercholesterolemia, and decreased HDL levels. They develop atherosclerotic lesions in the aorta and coronary vessels, and suffer myocardial infarctions and premature death. The coronary lesions have some analogies with human vulnerable plaques, in that they contain large globular lipid deposits and a fibrocellular cap. Although plaque rupture has not been observed to date, this anatomic similarity, and the increased incidence of myocardial infarction in these animals, suggest that this system may have potential as a model of coronary artery plaque rupture.

4.6.2.4 Rabbit Models of Plaque Rupture

A rabbit model of induced plaque rupture was described in the 1960s, in which atherosclerosis was induced by cholesterol feeding and then the animals were challenged by injection of Russell's viper venom followed by the vasopressor histamine.[32] This experimental system has more recently been developed further, with the aim of shortening the preparatory period and increasing the yield of thrombosis.[33] Four groups of animals were used, each with a different preparatory regimen. Group I consisted of normal rabbits fed on a normal diet for 8 months. Group II were fed a 1% cholesterol diet in a 2 months on, 2 months off pattern for 8 months. Group III underwent balloon-induced injury to the aorta, and were fed a normal diet for 8 months. Group IV also underwent balloon injury, but were fed according to the intermittent 1% cholesterol diet protocol for 8 months. At the end of the preparatory regimens, Russell's viper venom was injected intraperitoneally followed, after 30 min, by intravenous injection of histamine. This was repeated 24 h later and then, after a further 24 h, the rabbits were terminated.

In the Group I controls, the aorta and iliofemoral arteries were histologically normal. In Group II (cholesterol-fed), there was significant foam cell infiltration of the intima but only 3 of 13 rabbits developed thrombus following triggering. Rabbits in Groups III (balloon injury) and IV (cholesterol-fed plus balloon injury) had extensive plaque formation. All 5 rabbits in Group III, and 10 out of 14 rabbits in Group IV, developed platelet-rich thrombi.

With regard to plaque rupture, some animals in the cholesterol-fed groups showed evidence of fibrous cap thinning in areas of plaque that had overlying thrombus. However, this was a rare finding (less than 0.5% of lesions examined) and in the majority thrombus formation was not associated with obvious plaque rupture. There was also evidence of focal endothelial ulceration in some samples from Groups II and IV, without grossly visible thrombus but with platelets, fibrin, and red blood cells in the bases of the lesions. This model shows that vulnerable plaques can be produced in rabbits by cholesterol feeding and balloon injury. It also demonstrates that arterial thrombosis can be triggered pharmacologically, and that the extent of thrombosis in this setting is increased by the presence of atheroma. Furthermore, thin/fissured fibrous caps overlying lipid pools were present. Unfortunately, these lesions were rare and the authors estimated that cholesterol feeding for 2 years would probably be required to produce a significant number of lesions of this type, and in any case it is not clear if the thromboses were the consequence of plaque erosion or of true plaque rupture.

In a development of this principle, Rekhter and colleagues described a system in which a balloon catheter is allowed to become embedded within the plaque: inflation of the balloon disrupts the plaque and triggers thrombosis.[34] The balloon inflation pressure required to induce plaque rupture can be used as a measure of plaque tensile strength, and the resultant plaque fissuring allows plaque-related thrombosis to be examined. The model also allows for delivery of drug to the lesion site, enabling testing of potential plaque-stabilizing strategies. Its drawback is its unknown but probably limited relevance to spontaneous plaque rupture.

4.6.2.5 Pig Models of Plaque Rupture

Some novel approaches have been examined in the pig. Among these is the creation of lipid-rich lesions within a coronary artery by direct injection of material into the media.[35] This approach has a number of benefits, including the fact that the site and size of lesion can be controlled. Perhaps most importantly, the lesion composition can also potentially be manipulated, such that the contribution to vulnerability of the various plaque components can be assessed and plaque-stabilizing therapies examined.

The Rapacz pig, which suffers inherited hypercholesterolemia, develops spontaneous plaque rupture and hemorrhage at the site of coronary lesions after about a year.[36] Although this is in many ways an ideal large animal model, there is very restricted availability of these pigs, limiting its utility.

4.7 Conclusions

1. All biological studies involve the use of models, and animal models are not special in this regard. They have their advantages and shortcomings, and the search for a perfect recapitulation of human atherosclerosis in an animal model is both pointless and fruitless.
2. The real problem with atherosclerosis is that the plaques can rupture, and applied studies should be focused on this fact.
3. Animal models in which atherosclerotic plaques do not rupture can be useful, because indirect measures of plaque stability are available. However, these indirect measures have not been fully validated and must still be interpreted with caution.
4. Study designs that involve starting an intervention at the same time as the atherosclerotic stimulus do not provide trustworthy data about plaque stability because they are confounded by possible effects on atherogenesis.
5. Animal models in which plaques do actually rupture are available and should see increasing use. Intervention studies in such models, where treatment is started once the plaques have already become established and unstable, are particularly useful.

Acknowledgments The author is grateful to Dr. Jason Johnson for permission to reproduce the figures, for useful scientific discussions, and for critical reading of the manuscript.

References

1. Gross DR. *Animal Models in Cardiovascular Research. Developments in Cardiovascular Medicine*. vol. 153. Dordrecht: Kluwer; 1994.
2. Mathers C, Loncar D. Projections of global mortality and burden of disease from 2002 to 2030. *PLoS Med*. 2006;3:e442.
3. Roberts JC, Straus R, Cooper MS, eds. *Comparative Atherosclerosis*. New York: Hoeber Medical Division; 1965.
4. Harker LA, Ross R, Slichter SJ, Scott CR. Homocystine-induced arteriosclerosis. The role of endothelial cell injury and platelet response in its genesis. *J Clin Invest*. 1976;58:731–741.
5. Kunz J. Proliferation, migration and cell renewal within the arterial wall. *Acta Histochem Suppl*. 1983;27:233–243.
6. Johnson JL, Jackson CL. Atherosclerotic plaque rupture in the apolipoprotein E knockout mouse. *Atherosclerosis*. 2001;154:399–406.
7. Falk E. Why do plaques rupture? *Circulation*. 1992;86:III30–III42.
8. Burke AP, Kolodgie FD, Farb A, et al. Healed plaque ruptures and sudden coronary death: evidence that subclinical rupture has a role in plaque progression. *Circulation*. 2001;103:934–940.
9. Virmani R, Kolodgie FD, Burke AP, Farb A, Schwartz SM. Lessons from sudden coronary death: a comprehensive morphological classification scheme for atherosclerotic lesions. *Arterioscler Thromb Vasc Biol*. 2000;20:1262–1275.
10. Jackson CL, Bennett MR, Biessen EA, Johnson JL, Krams R. Assessment of unstable atherosclerosis in mice. *Arterioscler Thromb Vasc Biol*. 2007;27:714–720.
11. Tsakiris DA, Scudder L, Hodivala-Dilke K, Hynes RO, Coller BS. Hemostasis in the mouse (Mus musculus): a review. *Thromb Haemost*. 1999;81:177–188.
12. Zhu Y, Carmeliet P, Fay WP. Plasminogen activator inhibitor-1 is a major determinant of arterial thrombolysis resistance. *Circulation*. 1999;99:3050–3055.
13. Bouma BN, Meijers JC. Thrombin-activatable fibrinolysis inhibitor (TAFI, plasma procarboxypeptidase B, procarboxypeptidase R, procarboxypeptidase U). *J Thromb Haemost*. 2003;1:1566–1574.
14. te Velde EA, Wagenaar GT, Reijerkerk A, et al. Impaired healing of cutaneous wounds and colonic anastomoses in mice lacking thrombin-activatable fibrinolysis inhibitor. *J Thromb Haemost*. 2003;1:2087–2096.
15. Takano M, Inami S, Ishibashi F, et al. Angioscopic follow-up study of coronary ruptured plaques in nonculprit lesions. *J Am Coll Cardiol*. 2005;45:652–658.
16. Friedman M. The pathogenesis of coronary plaques, thromboses, and hemorrhages: an evaluative review. *Circulation*. 1975;52:III34–III40.
17. Mann J, Davies MJ. Mechanisms of progression in native coronary artery disease: role of healed plaque disruption. *Heart*. 1999;82:265–268.
18. Plump AS, Smith JD, Hayek T, et al. Severe hypercholesterolemia and atherosclerosis in apolipoprotein E-deficient mice created by homologous recombination in ES cells. *Cell*. 1992;71:343–353.
19. Zhang SH, Reddick RL, Piedrahita JA, Maeda N. Spontaneous hypercholesterolemia and arterial lesions in mice lacking apolipoprotein E. *Science*. 1992;258:468–471.
20. Reddick RL, Zhang SH, Maeda N. Aortic atherosclerotic plaque injury in apolipoprotein E deficient mice. *Atherosclerosis*. 1998;140:297–305.
21. Eitzman DT, Westrick RJ, Xu Z, Tyson J, Ginsburg D. Hyperlipidemia promotes thrombosis after injury to atherosclerotic vessels in apolipoprotein E-deficient mice. *Arterioscler Thromb Vasc Biol*. 2000;20:1831–1834.
22. von der Thusen JH, van Berkel TJ, Biessen EA. Induction of rapid atherogenesis by perivascular carotid collar placement in apolipoprotein E-deficient and low-density lipoprotein receptor-deficient mice. *Circulation*. 2001;103:1164–1170.
23. von der Thusen JH, van Vlijmen BJ, Hoeben RC, et al. Induction of atherosclerotic plaque rupture in apolipoprotein E–/– mice after adenovirus-mediated transfer of p53. *Circulation*. 2002;105:2064–2070.

24. Cheng C, Tempel D, van Haperen R, et al. Atherosclerotic lesion size and vulnerability are determined by patterns of fluid shear stress. *Circulation*. 2006;113:2744–2753.
25. Rosenfeld ME, Polinsky P, Virmani R, Kauser K, Rubanyi G, Schwartz SM. Advanced atherosclerotic lesions in the innominate artery of the ApoE knockout mouse. *Arterioscler Thromb Vasc Biol*. 2000;20:2587–2592.
26. Johnson J, Carson K, Williams H, et al. Plaque rupture after short periods of fat feeding in the apolipoprotein E-knockout mouse: model characterization and effects of pravastatin treatment. *Circulation*. 2005;111:1422–1430.
27. Williams H, Johnson JL, Carson KG, Jackson CL. Characteristics of intact and ruptured atherosclerotic plaques in brachiocephalic arteries of apolipoprotein E knockout mice. *Arterioscler Thromb Vasc Biol*. 2002;22:788–792.
28. Bennett MR. Breaking the plaque: evidence for plaque rupture in animal models of atherosclerosis. *Arterioscler Thromb Vasc Biol*. 2002;22:713–714.
29. Caligiuri G, Levy B, Pernow J, Thoren P, Hansson GK. Myocardial infarction mediated by endothelin receptor signaling in hypercholesterolemic mice. *Proc Natl Acad Sci U S A*. 1999;96:6920–6924.
30. Braun A, Trigatti BL, Post MJ, et al. Loss of SR-BI expression leads to the early onset of occlusive atherosclerotic coronary artery disease, spontaneous myocardial infarctions, severe cardiac dysfunction, and premature death in apolipoprotein E-deficient mice. *Circ Res*. 2002;90:270–276.
31. Herrera VM, Didishvili T, Lopez LV, et al. Hypertension exacerbates coronary artery disease in transgenic hyperlipidemic Dahl salt-sensitive hypertensive rats. *Mol Med*. 2001;7:831–844.
32. Constantinides P, Chakravarti RN. Rabbit arterial thrombosis production by systemic procedures. *Arch Pathol*. 1961;72:197–208.
33. Abela GS, Picon PD, Friedl SE, et al. Triggering of plaque disruption and arterial thrombosis in an atherosclerotic rabbit model. *Circulation*. 1995;91:776–784.
34. Rekhter MD, Hicks GW, Brammer DW, et al. Animal model that mimics atherosclerotic plaque rupture. *Circ Res*. 1998;83:705–713.
35. Granada JF, Moreno PR, Burke AP, Schulz DG, Raizner AE, Kaluza GL. Endovascular needle injection of cholesteryl linoleate into the arterial wall produces complex vascular lesions identifiable by intravascular ultrasound: early development in a porcine model of vulnerable plaque. *Coron Artery Dis*. 2005;16:217–224.
36. Prescott MF, McBride CH, Hasler-Rapacz J, Von Linden J, Rapacz J. Development of complex atherosclerotic lesions in pigs with inherited hyper-LDL cholesterolemia bearing mutant alleles for apolipoprotein B. *Am J Pathol*. 1991;139:139–147.
37. Johnson JL. *The Role of Matrix Metalloproteinases in an Animal Model of Atherosclerotic Plaque Rupture*. Bristol Heart Institute, University of Bristol; 2005.

5.1 The Transgenic Animal Models

A model in which the human ET-1 gene under the control of its natural promoter was transferred into the germline of mice was established independently by two research groups.[2,3] Both groups observed similar results: the overexpression of the ET-1 gene creates a phenotype with characteristic, age-dependent pathologic alterations of the kidney while blood pressure remains normal. The most evident effects of the elevated plasma and tissue concentrations of ET-1 are the development of renal interstitial fibrosis, renal cysts, and glomerulosclerosis, leading to a progressive decrease in glomerular filtration rate in aged mice. In these transgenic models, the profibrotic effect of ET-1 is most ostentatious, whereas hypertension can only be induced by salt-loading. This was unexpected since previous studies showed ET-1 to be a potent vasoconstrictor when administered intravenously.[4-7]

A bolus injection of the nitric oxide synthesis inhibitor L-NAME induces an exaggerated hypertensive response in ET-1 transgenic mice as well as in ET-2 transgenic rats.[8,9] This suggests that the nitric oxide system as a vasodilator system is involved.

In an animal model of crossbreds of ET-1 transgenic mice ($ET^{+/+}$) and mice in which the gene for the endothelial nitric oxide synthase (eNOS) was knocked out,[10] this assumption was supported. Compared to the normotensive wild-type and $ET^{+/+}$, the crossbred $ET^{+/+}$; eNOS KO have a significantly elevated blood pressure. This is also seen in $ET^{+/+}$; inducible NOS KO crossbreds,[11] which corroborates the hypothesis that the NO system counter-regulates the ET overexpression in the ET-1 transgenic model leading to a normotensive phenotype.

The ET-1 transgenic model also brought out some interesting findings about the role of ET in the etiology of pulmonary diseases. The development of pulmonary fibrosis and chronic lymphocytic inflammation in the absence of pulmonary hypertension was observed in the transgenic animals,[12] suggesting ET-1 to be involved in the pathogenesis of lung fibrosis and bronchiolitis.

Endothelium-restricted overexpression of human ET-1 does not result in elevated blood pressure; however, altered vascular structure and function in resistance vessels can be observed. This is a sign of blood pressure-independent effects of endothelium-generated ET-1 on vasculature.[13]

Conditional cardiac overexpression of ET-1 induces inflammation and dilated cardiomyopathy in mice, indicating that ET-1 acts as a proinflammatory molecule in the heart.[14]

In rats overexpressing the human ET-2 gene under control of its natural promoter, a blood pressure-independent fibrotic remodeling of the kidney is seen, but is mainly restricted to the glomeruli. This is most probably due to the preferential expression of the transgene within the glomeruli in ET-2 transgenic animals,[15] whereas the transgene is ubiquitously expressed within the entire kidney of ET-1 transgenic mice.[2]

ET-3 plays an important role in the development of epidermal melanocytes.[16] A tetracycline-responsive melanocyte-specific lineage transgenic system was created and used to conditionally regulate the overexpression of ET-3 in different stages of melanocyte development. Investigations showed that ET-3 interacts with precursors and differentiated melanocytes.[17]The activation of the transgene in mice leads to a hyperpigmented phenotype both in embryonic development and mature mice.

5.2 The Knockout Animal Models

Homozygous ET-1 KO mice die within 15–30 min after birth owing to craniofacial abnormalities resulting in an inability to breathe normally.[18] Furthermore, they display malformations of cardiovascular system,[19] thymus, and thyroid.[20] These affected tissues and organs are predominantly formed by cranial neural crest-derived ectomesenchymal cells. No abnormalities in other organs such as the lung, kidney, and central nervous system can be observed.

Similar craniofacial and cardiovascular malformations are seen in mice with disruption of the ETAR gene. The important role of ET-1 in skeletal development during intrauterine life was recently confirmed by using prepro-ET-1-lacZ-transgenic mice.[21]

The examination of expression patterns of ETARs and ET-1 suggests that ET-1/ETAR interaction is essential in cranial bone and connective tissue formation as well as in the development of the heart and its outflow tract.[22] Heterozygous ET-1 KO mice, which have lower serum levels of ET-1 than wild-type mice, paradoxically develop elevated blood pressure.[18] Enhanced basal and hypercapnia-induced renal sympathetic nerve activity in the heterozygous ET-1 KO mice seem to be involved in elevation of blood pressure.[23,24] Some other causes that might contribute to blood pressure elevation in heterozygous ET-1 KO mice have been excluded, such as salt sensitivity or respiratory abnormalities.[25,26]

Collecting duct (CD)-specific ET-1 KO mice do not have gross morphologic abnormalities. Plasma ET-1 levels are not affected; however, urinary excretion of ET-1 is reduced. CD ET-1 KO mice are hypertensive but show no differences in body weight, urine volume, creatinine clearance, sodium and potassium excretion, urine and plasma osmolality, plasma aldosterone concentration, and renin activity. CD ET-1 KO mice are salt-sensitive and exhibit reduced sodium excretion during the first 3 days of high-salt diet. Amiloride and furosemide are able to prevent sodium retention and exacerbation of hypertension. However, they do not reduce blood pressure in CD ET-1 KO mice on a normal sodium diet.[27]

Plasma vasopressin (AVP) concentrations are substantially reduced in CD ET-1 KO mice, despite all other aspects of water metabolism being similar. The response to continuous infusion of 1-deamin-8-D-arginine vasopressin is stronger in CD ET-1 KO mice. There seems to be an increased renal sensitivity to the effects of AVP suggesting that ET-1 acts as a physiological autocrine regulator of AVP action in the CD.[28]

Thus, CD-derived ET-1 seems to be an important physiological regulator of renal salt and water handling and systemic blood pressure.

CD-specific KO of the ETAR has no effect on blood pressure or urinary sodium excretion in mice, independently of salt intake. Those mice have increased plasma AVP levels but do not show differences of renal water excretion under baseline conditions. However, they have a more rapid decrease in urine osmolality following an acute water loading. During exogenous AVP infusion, CD ETAR KO mice increase urine osmolality similar to WT mice but have a more rapid subsequent fall in urine osmolality during sustained AVP administration. The lower AVP responsiveness in CD ETAR KO mice is contradictory to the results seen in CD ET-1 KO mice. These animals have elevated AVP sensitivity, and impaired ability to excrete an acute water load. This implies that ET-1 must exert its influence on vasopressin through CD ETBR and/or through paracrine effects.[29]

At 6 months of age, mice with disruption of the ET-1 gene in cardiomyocytes (CM ET-1 KO) have a significantly lower fractional shortening and develop a dilated left ventricle (LV). During the next month, the LV systolic function of CM ET-1 KO mice declines and ultimately they die at much younger ages (median life expectancy of CM ET-1 KO: 11 months, WT: 2 years). Histological characterization of the CM ET-1 KO hearts reveal dilated heart chambers with heterogeneity of myocyte size, increased fibrosis, and raised apoptosis.

At the age of 2 months, aortic banding leads to dilated cardiomyopathy in CM ET-1 KO mice but to LV hypertrophy in WT mice.[30]

Transcriptional and Western blot analyzes suggest that increased apoptosis in CM ET-1 KO is mediated by enhanced activity of tumor necrosis factor (TNF). CM ET-1 KO hearts also have diminished NF-κB activity, amounting to diminution of downstream inhibitors of TNF signaling.[30]

CM ET-1 KO mice are resistant to the hypertrophic stress of treatment with thyroid hormone. While mice with intact ET-1 gene develop a pronounced LV hypertrophy, CM ET-1 KO mice show only a marginal increase in LV mass.

These findings indicate that the interaction of locally produced ET-1 with thyroid hormone is essential in the development of thyroid hormone induced myocardial hypertrophy.[31]

There are no ET-2 KO models described in the published literature. ET-3 KO mice exhibit aganglionic megacolon and pigmentary disorders of the skin and choroidal layer of the retina. These findings indicate that ET-3 is essential in the development of two neural crest-derived cell lineages, epidermal and choroidal melanocytes, and enteric ganglion neurons.[16,30,32]

Blood pressure and heart rate of infant ET-3 KO mice are not different from those in age-matched WT mice. These results suggest that ET-3 is not involved in cardiovascular regulation at least in early life before they die of complications of the megacolon.[33]

Phenotypes of animals with natural mutations or with a targeted disruption of the ETBR gene are similar to that of ET-3 KO animals.[16,34] This underlines the importance of ET-3/ETBR-mediated signaling in the development of melanocytes and enteric neurons.

A naturally occurring rodent model of the described phenotype is the spotting lethal (sl/sl) rat, which carries a deletion of the ETBR gene.

To establish a viable animal model, it was necessary to support a normal enteric nervous system development. ETBR-deficient sl/sl rats were rescued by breeding them with rats harboring a dopamine-beta-hydroxylase (DBH)-ETBR transgene. This transgene leads to the development of a normal enteric nervous system. The resulting transgenic rats (DBH-ETBR; ETBRsl/sl) are healthy but present with total absence of ETBR in all nonadrenergic tissues including the kidneys.[35] Those ETBR-deficient rats are normotensive on a sodium-deficient diet, mildly hypertensive on a standard diet, and exhibit severe hypertension on a high-sodium diet.[36]

Their endothelial function is impaired, but it was shown that it is independent of the salt-enriched diet and therefore not responsible for the hypertension.[37]

In young sl/sl rats, fractional sodium excretion is markedly reduced. When treated with the specific epithelial sodium channel blocker amiloride, the animals show increased excretion of sodium.[38] This seems to be due to a lack of the inhibitory property of the ETBR on the epithelial sodium channel activity. These data indicate that a reduced renal ETBR activity might contribute to a salt-sensitive hypertension.

The effect of the ETBR on the progression of diabetic nephropathy was analyzed by rendering the transgenic animals diabetic with streptozotocin. These animals develop severe hypertension and show an enhanced functional renal impairment when compared with diabetic WT animals. Further analyzes suggested the elevated plasma levels of ET-1 to be responsible for the hypertension and not the renin–angiotensin system or the NO system.[39]

In a model of endothelial cell-specific ETBR KO, the animals showed endothelial dysfunction but no hypertension in response to a high-salt diet.[40] This points out that the salt-induced hypertension seen in the model of rescued ETBR-deficient rats is not mediated by endothelial cells.

A CD-specific ETBR KO causes salt-sensitive hypertension and sodium retention.[41] Taken together with the results from the CD-specific ET-1 KO, these findings point out an ET-1/ETBR axis, which has influence on systemic blood pressure at least partially mediated through autocrine inhibition of CD sodium reabsorption. Yet, it is not only the renal ETBR, which is crucial for the hypertension, since ETBR-deficient mice with transplanted normal kidneys are hypertensive as well.[42]

The highly specific ET-converting-enzymes (ECEs) are membrane-bound metalloproteases, which catalyze the cleavage of a Trp21 bond in the precursor big-ETs resulting in active ETs. Two isoenzymes ECE-1 and ECE-2 have been identified,[43,44] and their role in the ET system was investigated by creating KO models.

Most of the ECE-1 KO animals die in utero because of an embryonic lethal phenotype of cardiac abnormalities. Those severe defects of patterning of the great vessels and formation of the outflow tract are not found to this extent in other KO models of the ET system.[45]

The surviving mice have craniofacial abnormalities strikingly similar to those of the ET-1 KO and ETAR KO animals. The animals also present with an absence of epidermal melanocytes and enteric neurons of the distal gut, the phenotype known from the ET-3 KO and ETBR KO mice.[18,22] This combined phenotype corroborates the assumption that ECE-1 by activating both ET-1 and ET-3 in vivo has an influence on the development of distinct subsets of neural crest cell lineages.

The tissue levels of ET-1, ET-2, and ET-3 were measured as whole embryo extracts and compared with those of WT animals. ET-3 is markedly reduced, whereas the levels of ET-1 and ET-2 are still about 50% of the WT mice. However, biologically active ET seems not to be sufficiently produced at the sites crucial for normal embryonic development.

Unlike the ECE-1 KO mice, animals with a homozygous KO of ECE-2 show no embryonic developmental disorders. They are healthy, fertile, and have a normal life span.[46]

A cross-breeding of the two KO lines brought out animals with a lack of both ECE genes.[46] These ECE-1 KO; ECE-2 KO animals show a similar but more profound embryonic ECE-1 KO phenotype with the same amount of ET-1 and ET-2 as in the ECE-1 KO, suggesting yet another protease to be also responsible for the activation of ETs.

5.3 Conclusion

Taken together, results from the different transgenic and KO models disclose that the ET system plays a major role in embryonic development. Two ET system-dependent neural crest-driven developmental pathways become obvious: one of them being an ET-1/ETAR axis, responsible for normal cardiac and cranial development[47]; the other seems to be a ET-3/ETBR-mediated signaling pathway. Mutations within this axis are associated with disruptions in epidermal melanocytes and enteric neurons. These findings led to the discovery of similar findings in humans with Hirschsprung disease.[48,49]

In adult life, the ET system is most important in the renal and cardiovascular system as shown by a chronically activated ET system, which results in a blood pressure-independent fibrosis of kidneys and lungs.

The creation and investigation of genetically modified animals has contributed significantly to the understanding of ET function in vivo.

Transgenic models	Phenotype	References
ET-$1^{+/+}$	Kidney: renal cysts, renal interstitial fibrosis, glomerulosclerosis lung: fibrosis and chronic inflammation	2,3,12
Endothelium specific $ET^{+/+}$	Vascular remodeling and endothelial dysfunction	13
Cardiac-specific $ET^{+/+}$	Inflammatory cardiomyopathy	14
ET-$2^{+/+}$	Glomerulosclerosis	15
ET-$3^{+/+}$	Hyperpigmentation	17
ET-$1^{+/+}$; $eNOS^{-/-}$	Elevated blood pressure	11

KO models	Phenotype	References
ET-$1^{-/-}$	Craniopharyngeal and cardiovascular malformations	18-20
ET-$1^{+/-}$	Elevated blood pressure	18,23-25
CD-specific ET-$1^{-/-}$	Elevated blood pressure	27,28

(continued)

(continued)

KO models	Phenotype	References
Cardiomyocyte-specific ET-1$^{-/-}$	Dilatative cardiomyopathy	30,31
ET-3$^{-/-}$	Aganglionic megacolon, pigmentary disorders	16,32
ETAR$^{-/-}$	Craniopharyngeal and cardiovascular malformations	22
ETBR$^{-/-}$	Aganglionic megacolon, pigmentary disorders	16,34
Rescued ETBR$^{-/-}$	Salt-sensitive hypertension	35,36,38
Diabetic ETBR$^{-/-}$	Low-renin hypertension, progressive renal failure	39
Endothelial cell-specific ETBR$^{-/-}$	Endothelial dysfunction, normotensive on high-salt diet	40
CD-specific ETBR$^{-/-}$	Elevated blood pressure, increasing blood pressure and impaired sodium excretion on high-salt diet	41
ECE-1$^{-/-}$	Severe cardiac developmental disorders, craniofacial abnormalities, aganglionic megacolon, pigmentary disorders	45
ECE-2$^{-/-}$	Healthy phenotype	46
ECE-1$^{-/-}$; ECE-2$^{-/-}$	Worsened ECE-1$^{-/-}$ embryonic phenotype	46

References

1. Kedzierski RM, Yanagisawa M. Endothelin system: the double-edged sword in health and disease. *Annu Rev Pharmacol Toxicol*. 2001;41:851-876.
2. Hocher B, ThoneReineke C, Rohmeiss P, et al. Endothelin-1 transgenic mice develop glomerulosclerosis, interstitial fibrosis, and renal cysts but not hypertension. *J Clin Invest*. 1997;99(6):1380-1389.
3. Shindo T, Kurihara H, Maemura K, et al. Renal damage and salt-dependent hypertension in aged transgenic mice overexpressing endothelin-1. *J Mol Med*. 2002;80(2):105-116.
4. Yanagisawa M, Kurihara H, Kimura S, et al. A novel potent vasoconstrictor peptide produced by vascular endothelial-cells. *Nature*. 1988;332(6163):411-415.
5. Hocher B, ThoneReineke C, Bauer C, Raschack M, Neumayer HH. The paracrine endothelin system: pathophysiology and implications in clinical medicine. *Eur J Clin Chem Clin Biochem*. 1997;35(3):175-189.
6. Miyauchi T, Masaki T. Pathophysiology of endothelin in the cardiovascular system. *Annu Rev Physiol*. 1999;61:391-415.
7. Rothermund L, Paul M. The role of endothelin in hypertension. *Curr Opin Nephrol Hypertens*. 1998;7(4):451-456.
8. Liefeldt L, Schonfelder G, Bocker W, et al. Transgenic rats expressing the human ET-2 gene: a model for the study of endothelin actions in vivo. *J Mol Med*. 1999;77(7):565-574.
9. Hocher B, Schwarz A, Slowinski T, et al. In-vivo interaction of nitric oxide and endothelin. *J Hypertens*. 2004;22(1):111-119.
10. Quaschning T, Voss F, Relle K, et al. Lack of endothelial nitric oxide synthase promotes endothelin-induced hypertension: lessons from endothelin-1 transgenic/endothelial nitric oxide synthase knockout mice. *J Am Soc Nephrol*. 2007;18(3):730-740.
11. Quaschning T, Voss F, Herzfeld S, et al. Lack of iNOS impairs endothelial function in endothelin-1 transgenic mice. *Kidney Blood Press Res*. 2008;31(2):127-134.
12. Hocher B, Schwarz A, Fagan KA, et al. Pulmonary fibrosis and chronic lung inflammation in ET-1 transgenic mice. *Am J Respir Cell Mol Biol*. 2000;23(1):19-26.

Part 3
Molecules and Mediators and Therapeutic Applications

Chapter 6
Endothelin-1 and Systemic Sclerosis

Jeremy D. Pearson, Xu Shi-wen, Sarah L. Howat, Christopher P. Denton, and David J. Abraham

6.1 Introduction

By painstakingly seeking and purifying a novel vasoconstrictor from the supernatants of large-scale cultures of endothelial cells, the 21 amino acid peptide endothelin-1 (ET-1) was discovered by Yanagisawa and co-workers[1] in Japan in 1988. They showed that it was synthesized in two stages by cleavage of a large precursor (prepro-ET-1), was structurally related to a snake venom toxin, and was a highly potent long-lasting vasoconstrictor (acting in the nanomolar range; still one of the most potent vasoconstrictors known). Within a few years, it had been shown that ET-1 secretion from endothelial cells can be enhanced by a series of cytokines, hypoxia, mechanical stress, and other stimuli, and that its action is mediated by two G-protein coupled receptors, ET(A) and ET(B).[2] These receptors are almost 50% identical with similar affinity for ET-1. Both are linked to elevation of intracellular calcium ions, and activation of several kinase signaling pathways, which are likely to differ in different cell types.

The pharmaceutical industry has produced a series of antagonists to these receptors and they have become very important both in understanding biology of ET-1 and for therapeutic use. The most widely used clinical antagonist is bosentan, a mixed antagonist that binds to both receptors with approximately equal affinity. There are also selective antagonists for the ET(A) or the ET(B) receptor.[2]

Primarily by the use of experimental animal models, ET-1 has been implicated in several disease processes, including pulmonary hypertension, cardiac hypertrophy and progression to chronic heart failure, progression of atherosclerosis and restenosis after angioplasty, development of chronic renal disease, and in the rare connective autoimmune connective tissue disease systemic sclerosis (scleroderma, SSc).[3] The underlying actions of ET-1 contributing to these diseases are cell contraction (particularly smooth muscle cells and fibroblasts), cellular hypertrophy, and promotion of extracellular matrix production and hence fibrosis.

J.D. Pearson (✉)
Cardiovascular Division, King's College London, Waterloo Campus, SE1 9NH, London, UK
e-mail: jeremy.pearson@kcl.ac.uk

D. Abraham et al. (eds.), *Advances in Vascular Medicine*,

DOI 10.1007/978-1-84882-637-3_6, © Springer-Verlag London Limited 2010

The first direct indication that overexpression of ET-1 is by itself sufficient to induce fibrosis came from experiments in which ET-1 overexpressing transgenic mice were made, and among other phenotypic changes these animals gradually develop significant organ fibrosis.[4] Before this, however, the benefit of ET receptor antagonists in experimental models, such as bleomycin-induced lung fibrosis, had demonstrated the involvement of ET-1 in inflammation-driven fibrotic reactions.[5]

6.2 Scleroderma

SSc is classically recognized by the fact that it causes hardened skin (hence the name scleroderma), though the disease in fact leads to multiorgan fibrosis and multiple consequent symptoms. Lifespan is reduced and death is usually due to compromised renal function or pulmonary hypertension. Significantly, over 90% of patients who go on to develop SSc have preexisting Raynaud's phenomenon — an inappropriate severe episodic vasoconstriction in peripheral blood vessels in response to cold temperature or other stresses. Current understanding of the pathology of SSc is far from complete, but it includes:

- Early vascular damage, directly perceived by Raynaud's phenomenon, involving endothelial cell perturbation
- Concurrent immune dysfunction with characteristic circulating autoantibodies present
- Microvascular loss and arteriolar occlusion by neointimal thickening
- Inflammatory cell traffic from small blood vessels into the local connective tissue
- Excessive deposition of matrix components and progressive fibrosis, including the differentiation of perivascular cells to a contractile myofibroblast phenotype[6]

The cause of vascular damage is not known, but altered paracrine signaling mechanisms between endothelial cells and fibroblasts may be key to the induction of fibrosis, and mediators for which there is evidence to support their involvement include transforming growth factor beta (TGFβ), connective tissue growth factor (CCN2), and ET-1.[6] Importantly, increased presence of perivascular ET-1 is a feature of early SSc lesions,[7] and raised levels of ET-1 are detected in patient plasma.[8]

6.3 Endothelin-1 and Scleroderma

In vitro experiments first demonstrated that ET-1 increases collagen production in human skin fibroblasts, acting via both ET(A) and ET(B) receptors on these cells,[9] and then showed that the phenotypic changes in a range of matrix-related gene products induced by ET-1 treatment were similar to the differences exhibited between SSc and normal skin fibroblasts in early passage in culture,[10] consistent with the view that ET-1 action contributes to the progression of SSc. More recent

Table 6.1 Endothelin-1 (ET-1) rapidly upregulates a series of matrix-related genes in fibroblasts

Gene	Fold increase at 4 h
COMP	11.9
FGF2	5.0
CCN2 (CTGF)	4.8
Integrin β3	4.4
ADAM12	4.4
MMP3	4.1
Collagen VII α1	3.9
TGFβ	3.5
ADAM19	3.2
MMP1	2.5
Thrombospondin	2.5
Integrin α2	2.3
TIMP3	2.1

Human lung fibroblasts in early passage culture were treated with 100 nM ET-1 for 4 h. mRNA levels were quantified on Affymetrix human U133A chips. Results are mean values from two independent experiments. Data from Xu et al[11]

gene array experiments indicate that expression of a large series of matrix-related gene mRNAs is enhanced by treating fibroblasts with ET-1, including CCN2 and TGFβ (Table 6.1), and that induction of several of these mRNAs requires activation of the mitogen-activated protein kinase ERK.[11] Induction of CCN2 and TGFβ indicates the initiation of an autocrine signaling loop that continually drives fibrosis, a hallmark of SSc. Gene array profiling of SSc vs. normal fibroblasts suggests strong similarities, though not identity, with the pattern of changes induced by ET-1.[12]

In addition to changes in matrix-related proteins, studies of this type also show that contractile proteins are increased in SSc fibroblasts. ET-1 directly induces contraction of skin fibroblasts in collagen gels, an action that seems to be mediated predominantly via ET(A) receptors and requires signaling via protein kinase C, ERK, and phosphoinositide-3 kinase (PI-3K) pathways,[10] and involves modification of phenotype to myofibroblasts expressing increased levels of the smooth muscle isoform of actin.[13] Further indication of autocrine signaling, directly involving ET-1, comes from the demonstration that SSc fibroblasts themselves in culture secrete increased levels of ET-1, and their enhanced contractile phenotype can be reduced by bosentan treatment (Fig. 6.1).[13] The increased contractility is also reduced by blocking TGFβ receptors, implicating endogenously produced TGFβ in addition to ET-1.[12]

Additional evidence of feedback interacting loops involving both mediators was found by treating normal fibroblasts with TGFβ, and demonstrating that the resulting increased smooth muscle actin production is inhibited by bosentan.[14] Gene array

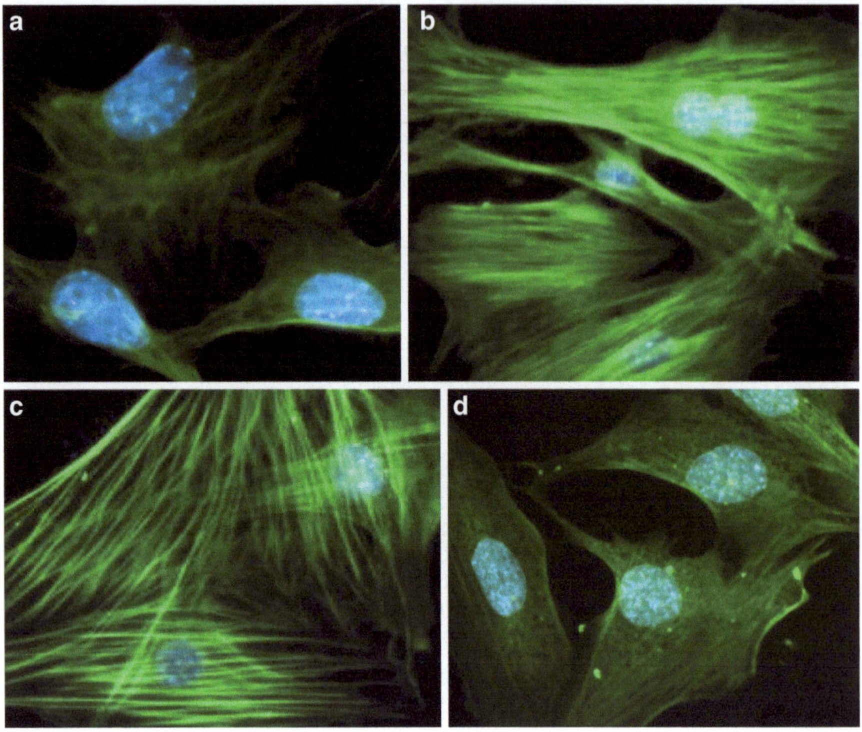

Fig. 6.1 Endothelin-1 (ET-1) induces a myofibroblast phenotype in normal fibroblasts and contributes to the maintenance of the phenotype in scleroderma (SSc) fibroblasts. (**a**) and (**b**): Normal human lung fibroblasts under control conditions (**a**) or after treatment with 100 nM ET-1 (**b**) for 24 h. (**c**) and (**d**): SSc fibroblasts under control conditions (**c**) or after treatment with 10 μM bosentan for 24 h (**d**). Cells are stained to detect smooth muscle alpha-actin. Figure derived from Shi-Wen et al[13]

data also show that the overall pattern of altered gene expression in SSc fibroblasts in culture is substantially shifted by bosentan treatment toward the pattern expressed by normal fibroblasts.[15]

If fibroblasts are transfected with an ET-1 reporter gene, basal expression is, as expected from the results above, higher in SSc fibroblasts. This elevated basal expression is not reduced by blocking TGFβ receptor activation (though it can be further increased by TGFβ treatment).[14]

Taken together, these results implicate ET-1, TGFβ, and CCN2 as important mediators of fibrosis in SSc, acting in a complex inter-related manner, summarized in Fig. 6.2. CCN2, first cloned as the secreted protein CTGF from endothelial cell cultures,[16] is one of the CCN family of proteins that have been implicated in cell–cell interactions regulating matrix biology and angiogenesis.[17] CCN2 is a heparin-binding protein, and specific receptors for CCN2 have not yet been defined, though integrins and low-density lipoprotein receptor-related protein (LRP, CD91) have been suggested.[17] CCN2 is upregulated by TGFβ and has often been regarded as a

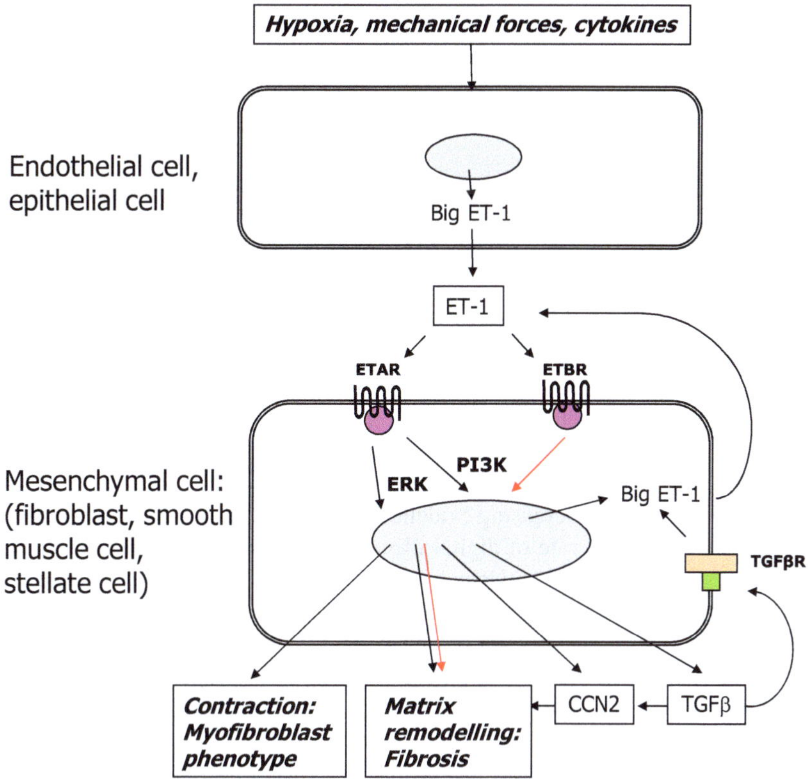

Fig. 6.2 Pathways leading to ET-1 induction of contractile and profibrotic phenotypes. ET-1 secretion from endothelial or epithelial cells is increased by stimuli including hypoxia, altered mechanical stress or shear, and some proinflammatory cytokines. It acts on target mesenchymal cells at ET(A) and ET(B) receptors via several signaling pathways to enhance contraction (primarily via ET(A) receptors) and matrix production (via both receptors). Autocrine loops involving mesenchymal cell production of ET-1, transforming growth factor beta (TGFβ), and CCN2 contribute. Not shown, for simplicity, are ET(B) and TGFβ receptors on endothelial cells that contribute to further paracrine feedback regulation

major downstream mediator of TGFβ action. Recent experiments with CCN2 knockout fibroblasts substantiate this, demonstrating that around a third of the several hundred genes upregulated by TGFβ treatment in normal fibroblasts fail to be upregulated in the CCN2 knockout cells; expression of some of these transcripts, including smooth muscle actin, can be restored by overexpression of AKT (the kinase acting downstream from PI-3K).[18] Similar feedback loops involving ET-1, TGFβ, and CCN2 are likely to be involved in fibrosis in other organs, including blood vessels, the heart, and the liver.

Several actions of exogenous ET-1 are inhibited in SSc fibroblasts, perhaps because of preexisting activation of pathways by autocrine actions of endogenous ET-1, or possibly relating to alterations in the relative levels of activation of downstream signaling pathways in these cells. Thus, ET-1 is a fibroblast mitogen, but much less effective on SSc cells.[9] In addition, whereas interleukin-1β is a more potent enhancer of ICAM-1 (a cell surface adhesion molecule) in SSc than normal fibroblasts, via a pathway requiring PKCδ, ET-1 (which acts via a pathway requiring PKCε) is much less effective at inducing ICAM-1 in SSc cells.[19,20]

6.4 Conclusions

The results described here strongly suggest that ET-1 contributes to the pathogenesis of SSc, in combination with other mediators. Further support comes from the finding that specific polymorphic variants of the genes coding for the ET(A) and ET(B) receptors appear to be over-represented in patients with diffuse SSc.[21] This has led to hopes that ET-1 receptor antagonists could be useful either in combating the early vasospastic symptoms of Raynaud's phenomenon in SSc, or the later progression of fibrosis. There is increasing evidence that bosentan can reduce the incidence and increase the healing rate of digital ulcers in SSc patients, a reflection of their Raynaud's phenomenon.[22] Similarly, bosentan has been shown to have detectable clinical benefit for pulmonary arterial hypertension in SSc patients,[23,24] though the patients being treated currently already have substantial deterioration in pulmonary function. Overall, effects of bosentan are thus positive, but modest. Since endothelial cells express ET(B) receptors, which are coupled to production of beneficial vasodilator agents including nitric oxide,[2] it is surmised that selective ET(A) receptor antagonists might have an improved profile of activity in SSc patients,[25] though only one study has to date been reported.[26] However, it seems likely that ET receptor antagonists will become increasingly part of multimodal therapy in SSc patients.

Acknowledgments Research from the authors' labs summarized here was funded by The Arthritis Research Campaign, The Wellcome Trust, and Raynaud's & Scleroderma Association.

References

1. Yanagisawa M, Kurihara H, Kimura S, et al. A novel potent vasoconstrictor peptide produced by vascular endothelial cells. *Nature*. 1988;332:411-415.
2. Gray GA, Webb DJ. The endothelin system and its potential as a therapeutic target in cardiovascular disease. *Pharmacol Ther*. 1996;72:109-148.
3. Kirkby NS, Hadoke PW, Bagnall AJ, Webb DJ. The endothelin system as a therapeutic target in cardiovascular disease: great expectations or bleak house? *Br J Pharmacol*. 2008;153:1105-1119.
4. Hocher B, Schwarz A, Fagan KA, et al. Pulmonary fibrosis and chronic lung inflammation in ET-1 transgenic mice. *Am J Respir Cell Mol Biol*. 2000;23:19-26.
5. Park SH, Saleh D, Giaid A, Michel RP. Increased endothelin-1 in bleomycin-induced pulmonary fibrosis and the effect of an endothelin receptor antagonist. *Am J Respir Crit Care Med*. 1997;156:600-608.
6. Denton CP, Black CM, Abraham DJ. Mechanisms and consequences of fibrosis in systemic sclerosis. *Nat Clin Pract Rheumatol*. 2006;2:134-144.

7. Abraham DJ, Vancheeswaran R, Dashwood MR, et al. Increased levels of endothelin-1 and differential endothelin type A and B receptor expression in scleroderma-associated fibrotic lung disease. *Am J Pathol.* 1997;151:831-841.
8. Vancheeswaran R, Magoulas T, Efrat G, et al. Circulating endothelin-1 levels in systemic sclerosis subsets–a marker of fibrosis or vascular dysfunction? *J Rheumatol.* 1994;21:1838-1844.
9. Xu S, Denton CP, Holmes A, et al. Endothelins: effect on matrix biosynthesis and proliferation in normal and scleroderma fibroblasts. *J Cardiovasc Pharmacol.* 1998:31(suppl 1);S360-S363.
10. Shi-Wen X, Denton CP, Dashwood MR, et al. Fibroblast matrix gene expression and connective tissue remodeling: role of endothelin-1. *J Invest Dermatol.* 2001;116:417-425.
11. Xu SW, Howat SL, Renzoni EA, et al. Endothelin-1 induces expression of matrix-associated genes in lung fibroblasts through MEK/ERK. *J Biol Chem.* 2004;279:23098-23103.
12. Chen Y, Shi-Wen X, van Beek J, et al. Matrix contraction by dermal fibroblasts requires transforming growth factor-beta/activin-linked kinase 5, heparan sulfate-containing proteoglycans, and MEK/ERK: insights into pathological scarring in chronic fibrotic disease. *Am J Pathol.* 2005;167:1699-1711.
13. Shi-Wen X, Chen Y, Denton CP, et al. Endothelin-1 promotes myofibroblast induction through the ETA receptor via a rac/phosphoinositide 3-kinase/Akt-dependent pathway and is essential for the enhanced contractile phenotype of fibrotic fibroblasts. *Mol Biol Cell.* 2004;15:2707-2719.
14. Shi-Wen X, Rodríguez-Pascual F, Lamas S, et al. Constitutive ALK5-independent c-Jun N-terminal kinase activation contributes to endothelin-1 overexpression in pulmonary fibrosis: evidence of an autocrine endothelin loop operating through the endothelin A and B receptors. *Mol Cell Biol.* 2006;26:5518-5527.
15. Shi-Wen X, Renzoni EA, Kennedy L, et al. Endogenous endothelin-1 signaling contributes to type I collagen and CCN2 overexpression in fibrotic fibroblasts. *Matrix Biol.* 2007;26:625-632.
16. Bradham DM, Igarashi A, Potter RL, Grotendorst GR. Connective tissue growth factor: a cysteine-rich mitogen secreted by human vascular endothelial cells is related to the SRC-induced immediate early gene product CEF-10. *J Cell Biol.* 1991;114:1285-1294.
17. Perbal B. CCN proteins: multifunctional signalling regulators. *Lancet.* 2004;363:62-64.
18. Shi-wen X, Stanton LA, Kennedy L, et al. CCN2 is necessary for adhesive responses to transforming growth factor-beta1 in embryonic fibroblasts. *J Biol Chem.* 2006;281:10715-10726.
19. Xu SW, Denton CP, Dashwood MR, et al. Endothelin-1 regulation of intercellular adhesion molecule-1 expression in normal and sclerodermal fibroblasts. *J Cardiovasc Pharmacol.* 1998:31(suppl 1);S545-S547.
20. Waters CE, Shi-Wen X, Denton CP, et al. Signaling pathways regulating intercellular adhesion molecule 1 expression by endothelin 1: comparison with interleukin-1beta in normal and scleroderma dermal fibroblasts. *Arthritis Rheum.* 2006;54:649-660.
21. Fonseca C, Renzoni E, Sestini P, et al. Endothelin axis polymorphisms in patients with scleroderma. *Arthritis Rheum.* 2006;54:3034-3042.
22. Korn JH, Mayes M, Matucci Cerinic M, et al. Digital ulcers in systemic sclerosis: prevention by treatment with bosentan, an oral endothelin receptor antagonist. *Arthritis Rheum.* 2004;50: 3985-3993.
23. Rubin LJ, Badesch DB, Barst RJ, et al. Bosentan therapy for pulmonary arterial hypertension. *N Engl J Med.* 2002;346:896-903.
24. Denton CP, Pope JE, Peter HH, et al. Long-term effects of bosentan on quality of life, survival, safety and tolerability in pulmonary arterial hypertension related to connective tissue diseases. *Ann Rheum Dis.* 2008;67:1222-1228.
25. Dhaun N, Pollock DM, Goddard J, Webb DJ. Selective and mixed endothelin receptor antagonism in cardiovascular disease. *Trends Pharmacol Sci.* 2007;28:573-579.
26. Girgis RE, Frost AE, Hill NS, et al. Selective endothelin A receptor antagonism with sitaxsentan for pulmonary arterial hypertension associated with connective tissue disease. *Ann Rheum Dis.* 2007;66:1467-1472.

Chapter 7
The Role of the Homeodomain Transcription Factor Nkx2-5 in the Cardiovascular System

Markella Ponticos

7.1 Introduction

Nkx2-5, also known as *C*ardiac *s*pecific homeobo*x* (CSX), belongs to the NK-2 family of homeobox DNA binding transcriptional activators that are structurally and functionally highly conserved in evolution (reviewed in ref.[1] Nkx2-5 is the mammalian homologue[2] of the Drosophila NK4 gene (*tinman*), which was originally identified as a gene essential in the development of the dorsal vessel, the Drosophila "heart"[3] and its mutation results in the complete absence of dorsal vessel formation.[4,5] *Tinman* is expressed in the trunk mesoderm early in development and is then restricted to the dorsal and visceral mesoderm.[6] Tinman homologues such as Nkx2-5 have been since described in many vertebrates from Xenopus to human and share a highly conserved protein structure of four distinct domains: an N-terminal TN-domain, a homeodomain, NK-2 domain, and a conserved C-terminal peptide.[7] The homeodomain that binds to a consensus DNA sequence (TNAAGEGG) through a helix–turn–helix motif[7,8] and interacts with other transcription factors[1,8-10] has been well characterized. The function(s) of the other domains is less well described although it is thought that the NK-2 domain has the ability to repress transcriptional activity through protein–protein interactions.

Expression of *Nkx2-5* is first observed in the precardiac mesoderm and the pharyngeal endoderm at 7.8 days postcoitum (d.p.c.) in the mouse embryo.[11] Expression during development is also observed in the tongue, spleen, liver, stomach, and anterior larynx. Expression is down regulated significantly after birth in all organs except the heart where it is expressed in cardiomyocytes of the atria and ventricles and it continues to be expressed into adulthood.

M. Ponticos (✉)
Division of Medicine, Research Department of Inflammation, Centre for Rheumatology and Connective Tissue Diseases, Royal Free and University College Medical School, University College London, Royal Free Campus, Rowland Hill Street, London, NW3 2PF, UK
e-mail: m.ponticos@medsch.ucl.ac.uk

D. Abraham et al. (eds.), *Advances in Vascular Medicine*,
DOI 10.1007/978-1-84882-637-3_7,

7.2 Role of Nkx2-5 in the Cardiovascular System

7.2.1 Nkx2-5 in Cardiac Development

The vertebrate heart is the first organ that is formed and its development involves a well-studied and precise sequence of events, requiring exquisite temporal and spatial regulations (reviewed in refs. [12,13]).

In brief, heart development initiates during the late gastrulation stage at the anterior, lateral mesoderm of the embryo. The cardiac mesoderm forms two distinct cardiac progenitor cell fields, the primary and secondary heart fields (SHFs), and the cardiac crescent. Cells from the primary heart field migrate and fold along the embryonic midline to form a linear heart tube of which the inflow–outflow region is aligned in a rostral–caudal direction, respectively. During further development of the heart tube, cells from the secondary heart field (SHF) contribute and the heart tube undergoes rightward looping. As the heart develops and matures, subsequent chamber septation occurs. The development of the valves and the division of the outflow track into two large vessels follow chamber septation. At the same time, the development of the cardiac conduction system (CCS) and the development of coronary arteries result in the mature organ.[14,15]

The role of Nkx2-5 in embryonic development has been studied extensively.[16,17] Expression of Nkx2-5 is one of the earliest markers of the cardiac lineage[18,19] and occurs in both primary and SHFs, observed in the precardiac mesoderm and the pharyngeal endoderm at 7.8 d.p.c. in the mouse embryo.[2,11] Nkx2-5 is continued to be expressed in postnatal hearts and is notably upregulated in hypertrophied hearts.[2,11,20]

Targeted disruption of Nkx2-5 in mice causes the arrest of heart development after the initial stage of looping, and is lethal by 9.0–11.0 d.p.c.[16,21,22] Nkx2-5 null mutant hearts exhibit growth retardation and do not form trabeculae and endocardial cushions. Commitment to the cardiac lineage is not compromised, although the expression of many essential cardiac-specific genes has been affected. For example, transcription factors such as eHAND and Iroquois homeobox gene 4 (Irx4), which are essential in the differentiation of embryonic ventricular myocardium are down-regulated in Nkx2-5 deficient mice.[23-25]

The role of Nkx2-5 in later stages of cardiac development has been investigated using Nkx2-5 conditional null mice where deletion of the gene was restricted to the ventricles using MLC2v-Cre-mediated recombination of the floxed Nkx2-5 allele.[26] The hearts from these mice showed hypertrabeculation and noncompaction of the myocardium; subsequently the mice exhibited chamber dilatation and progressive heart failure.

More recently, Prall et al.[27] showed that Nkx2-5 regulates SHF cell proliferation and outflow track morphology, and demonstrated using Nkx2-5 hypomorphic mice that Nkx2-5 orchestrates the transition between periods of heart induction, SHF progenitor cell proliferation, and outflow track morphogenesis via a Smad1-dependent negative feedback loop, which may be targeted in congenital heart disease (CHD).

7.2.1.1 Nkx2-5 in Congenital Heart Disease (CHD)

NKX2-5 is the most commonly mutated gene in CHD, accounting for 1–4% of specific malformations.[28-30] Heterozygous mutations in the human *NKX2.5* gene were found to cause familial CHD, secundum atrial septal defects (ASD), tetralogy of Fallot (a complex condition arising from stenosis of the Pulmonary artery), and other forms associated with atrioventricular (AV) conduction defects (see 7.2.2).[15,30] To date, 28 mutations of NKX2-5 have been reported; eight mutations have been identified in the homeodomain and are characterized with high disease penetrance (98% for AV block (see 7.2.2) and 83% for ASD).[31-33]

7.2.2 *Nkx2-5 in the CCS*

The CCS is a specialized structure responsible for the coordinated contraction of the heart by establishing and maintaining electrophysiological activities.[34] The development of the CCS is a highly complex process and closely associated with cardiac development (reviewed in ref. [14]). The CCS consists of atrial components such as the AV and sinoatrial nodes, and ventricular components such as atrioventricular bundle (AVB or His bundle) and the left and right bundle branches. The earliest morphologically distinct CCS structure is seen at the inner dorsal wall of the AV canal at 9–10 d.p.c. in the mouse embryo. By 11–12 d.p.c. (5–5.5 weeks in the human) the AV node, the sinoatrial node as well as the AV bundle have developed, and by 13 d.p.c. all components of the CCS, except the Purkinje fibers (PF), are distinguishable. PF are specialized cardiomyocytes whose role is to co-ordinate the rapid spread of action potential in the ventricular myocardium and their differentiation and maturation requires the precise regulation of Nkx2-5 expression.[35] Nkx2-5 expression is elevated and correlates with the recruitment of cells to the development of the ventricular conduction system when compared with the surrounding myocardium,[36] although the detailed understanding is lacking of the role of Nkx2-5 and the pathways in which it is involved. Recent studies have begun to elucidate some of these pathways and have revealed that Nkx2-5, along with the T-box transcription factor Tbx5, is required for the ventricular conduction system but not the AV node. Coordinated high-level activity of these two transcription factors on promoters of sensitive genes results in a restricted and specific program of ventricular conduction gene expression. One such target gene of Nkx2-5/Tbx5 transcriptional activation is the transcriptional repressor inhibitor of DNA 2 (Id2), which is required for normal structure and function of ventricular conduction system.[34,35,37,38]

7.2.2.1 Nkx2-5 and CCS disorders

Disorders of the CCS occur often and result in arrhythmias, which can be life-threatening. Dominant mutations in *Nkx2-5* result in electrophysiological abnormalities of the conduction system independently and in the absence of congenital heart defects.[26,30] Haploinsufficent mice for *Nkx7-5* and heterozygous Nkx2-5 null mutant mice (Nkx2-5$^{+/-}$) develop AV blocks and arrhythmias[39,40] and provide compelling evidence of the general role of Nkx2-5 in the formation of the CCS and in particular the development to the PF network.[41] In ventricular-restricted conditionally null Nkx2-5 mice, CCS defects have been attributed to hypoplasia of AV nodes and AV bundles and PF hypocellularity.[26,42]

Mutations identified in the human *NKX2-5* gene, such as the Gln170ter NKX2-5 truncation mutation, also result in AV blocks thereby highlighting the importance.[30]

Another disease that Nkx2-5 has been unexpectedly associated with is the inherited neuromuscular disorder Myotonic muscular dystrophy (DM1), a disease thought to be caused by RNA toxicity. Yadava et al. demonstrated using a reversible transgenic mouse model of RNA toxicity in DM1, that DM1 is associated with induced NKX2-5 expression, and that this resulted in cardiac conduction defects, increased expression *NKX2-5,* and profound disturbances in gap junction proteins connexin 40 and connexin 43. The effects on NKX2-5 and its downstream targets were reversed by silencing toxic RNA expression. This study was the first to show that *NKX*2-5 is the first genetic modifier of DM1-associated RNA toxicity in the heart.[43]

7.2.3 Nkx2-5 in the Vasculature

Expression of Nkx2-5 is not normally observed in adult vasculature, although a number of studies and observations now suggest that Nkx2-5 plays a significant role in vessels. The first piece of evidence came from the Nkx2-5 null mouse embryos which, in addition to the well-documented cardiac abnormalities, were shown to have poorly developed blood vessels such as the intersomitic and pharyngeal arch arteries and the dorsal aorta. Moreover, yolk sacs from null embryos at 9.5 d.p.c. had no defined vasculature as opposed to the wild-type embryos, which had a well-established network of large vitelline vessels and smaller vessels filled with blood.[22]

Indeed, Tanaka et al. showed that Nkx2-5 is required for a canalized outflow tract (OFT) conduit.[22] The cardiac OFT is a simple vascular conduit that connects the right ventricle to the aortic sac, which is the site of confluence of the arteries of the pharyngeal arch in the embryo. The embryonic OFT is lined with myocardial cells derived from the SHF caudal to the pharynx where *Nkx2-5* is expressed, and is subsequently divided into two channels, which direct blood flow toward either the aortic or the pulmonary circulation.[44,45] *Nkx2-5* was shown to be upstream of *Tbx1*, a gene important in OFT morphogenesis.[46] In a series of animal experiments using conditional deleted *Tbx1* in *Nkx2-5* expressing cells, these authors showed that the mutant mice had a defective aortic arch phenotype and this phenotype was attributed to the deletion of *Tbx1* in the *Nkx2-5* expressing cells.

In developing mice, *Nkx2-5* has also been shown to be expressed and regulate the formation of pulmonary myocardium.[47] The pulmonary myocardium is a myocardial layer, which forms a sheath around the pulmonary venous vessel. The origin of this vascular component is controversial, although it was shown that an

Nkx2-5-expressing population forms de novo arising in the pulmonary mesenchyme in the presence of transcription factor Pitx2c (see 7.4).[47] The pulmonary myocardium is an important source of electrical activity that initiates atrial fibrillation and expresses the gap-junction protein Connexin40 (Cx40), which is a target of Nkx2-5 regulation and is essential for fast atrial conduction (see 7.4).

The origin of cells involved in the morphogenesis of the cardiovascular system is a matter of intensive investigation with continuous progress being made in deciphering the origins of the myocardial and endocardial cell lineages from multipotential progenitor cells. The origins of vascular smooth muscle are less well defined. Vascular smooth muscle cells (vSMC) develop from diverse sources later than myocardial and endocardial cells.[48] Recently, Wu et al. isolated early Nkx2-5 expressing cells and demonstrated that these could differentiate into both cardiomyocytes and vSMC in vivo and suggested that these two cell lineages diverged from a common precursor.[48] The expression of *Nkx2-5* in de-differentiated collagen-producing vSMC provides additional support to this hypothesis.[49]

7.2.4 Nkx2-5 in Adult Tissues

The role of Nkx2-5 in adult tissues is not as yet well characterized, although with more information obtained on its downstream targets, it is very likely that novel functions will be assigned to it in the future. Nkx2-5 is expressed in adult hearts and it is overexpressed specifically in hypertrophied hearts.[11,20] During right ventricular pressure overload-induced as well as in agonist-induced hypertrophy, Nkx2-5 expression is upregulated suggesting a role of Nkx2-5 in cardiac hypertrophy in general.[50,51] Nkx2-5 overexpressing mice do not have cardiac hypertrophy, although the Nkx2-5 target genes are upregulated suggesting that Nkx2-5 alone is not sufficient to induce cardiac hypertrophy.[52]

Recently, studies carried out in transgenic mice overexpressing a dominant negative mutant of Nkx2-5 under the control of the α-myosin heavy chain (α-MHC) promoter showed impaired cardiac function and degeneration of cardiomyocytes.[53] In the same study, exposure of these mice to doxorubicin-induced myocardial damage showed severe cardiac dysfunction accompanied by cell apoptosis when compared with the mild effect on Nkx2-5 wild-type animals. The data suggest a protective role of Nkx2-5 in adult hearts against stress or cytotoxic damage.

7.3 Regulation of *NKX2-5*

7.3.1 Transcriptional Regulation of Nkx2-5

The human *NKX2.5* gene is localized to chromosome 5q35[54,55] and the murine *Nkx2-5* gene is localized to chromosome 17A3.3. Investigations of the murine gene structure have revealed a high degree of complexity. Various studies first determined that the regulatory region extends to 23 kb surrounding the gene and contains at least seven different activating regions (AR), and three possible inhibitory regions. These regulatory regions or enhancers are active in specific locations of the embryo, with more than one sometimes being active at a specific location and include specific regulatory elements for the heart (AR1-5), thyroid (AR6 and 7), spleen, pharynx, and stomach (AR7). However, there are expression patterns during development or in the adult that have yet to be mapped within this regulatory region (reviewed in ref. [56]). A more recent study has identified three distal enhancers over 20 kb upstream from the transcriptional start site, arranged in a modular manner and responsible for later cardiac chamber specification and expression in tongue.[57] The complexity of transcriptional regulation is compounded by the protein coding region that comprises of two exons, which can be spliced to either two alternative exons (1a and 1b) located 2–4 kb upstream of the transcriptional start site resulting in potential alternative isoforms.[15,58]

In the murine *Nkx2-5* gene, a regulatory element located between around −3 and −2.5 kb upstream of the transcriptional start site was identified as sufficient for initial expression during heart development and contains essential GATA binding sites.[59,60] Subsequently, it was shown that this element also contained a cluster of Smad binding sites, which have been suggested as targets of bone morphogenic protein (BMP) signaling (see 7.3.2).[61] In a separate study, Lien et al.[62] showed that the AR2 enhancer containing the GATA/Smad binding sites is a direct target of Smad4.[62] In an elegant study,[63] identified a novel enhancer region rich in GATA and Smad binding sites. They showed that BMP signaling activates Nkx2-5 directly through Smad1/4 and that there is co-dependence/interaction with GATA4 in vivo. They also speculated that TGF-β family members other than BMPs, such as activin, nodal, and TGF-β1 itself, may regulate *Nkx2-5* activity under certain conditions.

In a recent report, Monzen et al.[64] have demonstrated that HMGA2, a member of the high mobility group (HMG) of nuclear proteins, which are involved in the regulation of genes by chromatin remodeling and in synergy with Smad1/4, can trans-activate *Nkx2-5*.

Apart from GATA and Smad binding, myocyte enhancer factor 2C (MEF2C) has been shown to upregulate Nkx2-5 expression.[65] Moreover, it appears that Nkx2-5 and MEF2C can upregulate each other's expression. The MEF2 family of transcription factor plays a critical role in cell-type specific transcription of genes in cardiac, skeletal, and smooth muscle cells as well as some neuronal cells (reviewed in ref. [66]).

Recently, it was shown that the transcription factor HIF-1α, a regulator of the adaptive response to oxygen tension and of oxygen homeostasis in general, is an upstream activator of Nkx2-5 in Xenopus.[67] These authors suggested that a region between −2.5 kb and −1.2 kb was responsible for this activation, although they could not rule out an indirect regulation by HIF-1α.

7.3.2 Signaling Pathways Involved in Nkx2-5 Induction

The pathways that signal the induction of Nkx2-5 are complex. Indeed, signaling mechanisms that result in the activation or upregulation of Nkx2-5 are still under scrutiny.

BMPs are part of the transforming growth factor-β (TGF-β) superfamily of signaling molecule mediating many cellular processes[68] and have long been known to be involved in the commitment of cells to a cardiogenic fate. Studies in chicken embryos have shown that BMP-2, which is expressed in the anterior lateral endoderm, is sufficient to induce *Nkx2-5* expression in the anterior lateral mesoderm.[69] *Nkx2-5* expression is also regulated by BMP-4 in the chicken stomach and gizzard.[70,71] Transcriptional regulation in response to BMP signaling occurs through Smad transcription factor family. Smad1, Smad5, and Smad8 are specific activators of BMP signaling and Smad4 is a co-activating Smad required for Smad translocation into the cell nucleus. Smad6, the "inhibitory" Smad, is expressed in the developing heart and may repress signaling after the induction to the cardiogenic lineage.[68,72] Conserved target Smad binding sites have been identified in the murine *Nkx2-5* gene upstream regulatory region (see 7.3.1) through which BMPs can exert a positive or negative regulatory effect depending on the spatial and temporal conditions[61] and Smad4 has been shown to interact with this region.[62] Expression of the BMP inhibitor, noggin, was sufficient to suppress the induction of cardiomyogenesis by Nkx2-5.[73] Recently, Prall et al.[27] provided evidence of a Nkx2-5/BMP2/Smad1 negative feedback loop. They demonstrated that Nkx2-5 is required for cardiac progenitor cell proliferation in vivo through the repression of BMP2/Smad1 signaling.

Recent stem cell studies have proposed that TGF-β1 also regulates Nkx2-5 expression levels.

One study in bone marrow stem cells demonstrated that TGF-β1 treatment can induce the myogenic differentiation of CD117+ stem cells by upregulating *GATA4* and *Nkx2-5* expression.[74] Another report describes how TGF-β1 can upregulate *Nkx2-5* expression in skeletal muscle derived primitive cells.[75] Finally, in mouse embryonic teratocarcinoma stem cells, TGF-β1 neutralizing antibody inhibited induction of *Nkx2-5* and thus prevented cardiomyocyte differentiation.[76]

Another signaling pathway proposed to be involved in the regulation of *Nkx2-5* is the *Wnt* pathway. The *Wnt* family of secreted signaling molecules is involved in many developmental processes such as cell-fate specification, cell proliferation, migration, and adhesion (reviewed by ref.[77]) Canonical *Wnt* signaling involves the frizzled (Fzd) family of co-receptors and lipoprotein receptor related proteins (LRP) 5/6, β-catenin, and the LEF/TCF family of DNA-binding proteins. Activating canonical *Wnt* molecules (*Wnt1* and *Wnt3a*) in the anterior mesoderm inhibits *Nkx2-5* expression.[78,79] Cannonical *Wnt* signaling has a biphasic role in cardiac specification and blocking it in differentiating ES cells prior to cardiogenesis also inhibits the expression of early cardiogenic markers such as *Nkx2-5*.[80] In mouse embryonic stem cells, Pal et al.[81] have shown that treatment with soluble *Wnt* inhibitor prevented Nkx2-5 and GATA-4 activity. *Wnt* family members such as *Wnt5a* and *Wnt11* signal through an alternative pathway, the noncanonical *Wnt* pathway, which involves the Ca^{2+}/protein kinase C (PKC) cascade and/or the RhoA/JNK cascade.

Blocking *Wnt11* signaling in the anterior mesoderm *Xenopus* embryos blocks the expression of both *Nkx2-5* and *GATA4,* while expressing *Wnt11* in the posterior mesoderm induces ectopic expression of *Nkx2-5*.[79,82]

Endothelin-1 (ET-1) is a cytokine produced by endothelial cells (reviewed in refs.[83,84] and has been shown to be capable of regulating expression of PF-specific genes including *Nkx2-5*).[85,86] A report by Patel and Kos[87] demonstrated that ET-1 and Neuregulin-1, another endothelial-derived factor, upregulate *Nkx2-5* expression in murine embryonic cardiomyocytes. ET-1 requires activation by proteolytic cleavage by endothelin converting enzyme-1 (ECE-1) and recently it was reported that ECE-1 is a downstream target for *Nkx2-5* in H9c2 cardiomyoblasts.[88] These findings suggest a potential positive feedback loop involving Nkx2-5 and the endothelin gene axis.

7.3.3 Regulation of Nkx2-5 by Post-Translational Modification

Relatively little is known about posttranslational modification of Nkx2-5; however, it has been reported that phosphorylation on serine 163 by casein kinase II resulted in increased Nkx2-5 activity through increased DNA binding.[89]

More recently, it was reported that Nkx2-5 is a target of small ubiquitin-like modifiers (SUMOs).[90] The SUMO conjugation pathway is similar to the ubiquitination pathway although the proteins are not directed to the proteosome to be degraded; instead their function is modulated by nuclear–cytoplasmic shuttling.[91]

7.4 Regulation of Downstream Genes by NKX2-5

7.4.1 Direct Regulation of Genes

Nkx2-5 has been shown to regulate the transcription of many downstream target genes with a wide range of functions, which are summarized below (Table 7.1), although the list is by no means complete as the number of Nkx2-5 targets is rapidly increasing.

Initially, several genes have been reported as being in downregulated in the heart of Nkx2.5-null mice, including *atrial natriuretic factor* (ANF), *B-type natriuretic peptide*, *cardiac ankyrin repeat protein* (*CARP*), *eHAND*, *MEF2C*, *myosin light chain 2v* (MLC-2v), *SM22*, *N-myc*, *Msx2*, *Irx4*, and *Chisel* (Csl).[16,23,24,92-94]

Many transcription factors and transcriptional regulators are themselves transcriptional targets of Nkx2-5. Promoters of genes such as the MEF2C, homeodomain only protein (HOP), Xin, CARP, Id2 and Myocardin are directly regulated by Nkx2-5.

MEF2C is a member of the MADS-box transcription factor family that regulates the majority of muscle-specific genes and plays a role in myogenesis of all muscle types. Mef2c is a direct target of Foxh1, which physically and functionally interacts with Nkx2-5 to mediate strong Smad-dependent activation of a TGF-β response

Table 7.1 Downstream target genes of Nkx2-5

Gene	Function	Reference
Myocyte enhancer factor 2C (MEF2C)	Transcription factor involved in cardiac, smooth muscle, and skeletal muscle cell determination	65,95
Homeodomain only protein (HOP)	Non-DNA binding transcription factor involved in cardiac development	96,97
Xin	Transcriptional regulator involved in cardiac and skeletal muscle differentiation	98
Cardiac ankyrin repeat protein (CARP)	Transcriptional repressor and titin-binding protein involved in cardiogenesis	94
Id2	Transcriptional repressor in the ventricular conduction system	34
Tbx1	Transcription factor involved in he formation of the cardiac outflow tract (OFT)	46
Pitx2	Transcription factor involved in left/right symmetry during organogenesis and formation of the pulmonary myocardium	100
eHand/Hand1	Transcription factor involved in left ventrical development	23
Myocardin	Chromatin remodeling Transcription factor expressed in embryonic heart and vascular smooth muscle cells	104
Iroquois homeobox gene 4(Irx4)	Transcription factor involved in ventrical development	24
Cardiac α-Actin	Cardiomyocyte sarcomeric, contractile protein	18,102
Myosin light chain 2v (MLC-2v)	Ventricular differentiation	16,101
Atrial natriuretic peptide (ANP)	Chamber specification and postnatal chamber specificity	106,123
Chisel (Csl)	Muscle-specific regulatory protein	93
Sodium calcium exchanger I (NCX1)	Ion exchange protein involved in cardiogenesis	107
Calreticulin	Ca(2+) binding chaperone of the endoplasmic reticulum involved in cardiogenesis	108
Connexin 40	Gap junction protein involved in the cardiac conduction system (CCS)	110
Connexin 43	Gap junction protein involved in the CCS	109,110
Connexin 45	Gap junction protein involved in the CCS	109
Endothelin-converting enzyme-1 (ECE-1)	Metaloprotease involved in the endothelin cascade and in cardiac development	88
Collagen 1α2	Structural extracellular matrix protein involved in embryogenesis, adult wound repair, and fibrotic diseases	49
Procollagen lysyl hydroxylase 1 (PLOD1)	Enzyme that hydroxylates and stabilizes collagens	106
Csm	Cardiac-specific RNA Helicase	103
A1 adenosine receptor	G-protein coupled adenosine receptor involved in heart and brain development	112

element in the mouse *Mef2c* gene.[95] MEF2C can regulate NKx2-5 forming a positive feedback loop (see 7.3.1).[65] HOP is a non-DNA-binding transcriptional antagonist of serum response factor (SRF) and therefore modulates SRF activity during heart development. It is a direct target of Nkx2-5 activation.[96,97] Xin is a gene involved in heart tube formation and looping and is thought to be activated by BMP-induced Nkx2-5/MEF2C.[98] CARP is a nuclear factor with inhibitory activity and is involved in cardiac ventricular specification and is a downstream target of Nkx2-5.[94] Nkx2-5 synergizes with GATA-4 to achieve this regulation (see 7.4.2).[99]

Nkx2-5 and Tbx-5 co-activate the promoter of Inhibitor of DNA 2 (Id2), which is a helix–loop–helix transcriptional repressor and is required for the normal structure and function of the ventricular conduction system.[34] Pitx2 is transcription factor downstream of nodal signaling involved in left/right patterning and although its expression is not directly initiated by Nkx2-5, its maintenance requires Nkx2-5.[100] Nkx2-5 also is involved in synergistic regulation with Pit2x (see 7.4.2) of other promoters.

eHand/Hand1 is a member of the basic-helix–loop–helix transcription factor family involved in left/right asymmetry in embryonic development. In hearts of embryos lacking Nkx2-5, which do not loop, left-sided eHand expression was abolished.[23]

Irx4 is a transcription factor involved in ventricle formation and is transcriptionally regulated through the combinatorial activities of Nkx2-5 and dHand/Hand2.[25]

Cytoskeletal proteins Cardiac α-Actin[18] and MLC-2v[16,101] are both targets of Nkx2-5 transcriptional regulation. Cardiac α-Actin is transcriptionally activated by the synergistic effect of Nkx2-5 with SRF (see 7.4.2).[8,102]

Csl was identified through differential screening as target of Nkx2-5. Csl is expressed in the heart, skeletal and smooth muscle of the stomach and pulmonary veins, and is thought to have a regulatory role in myocyte structure and function.[93]

Csm is an RNA helicase specifically expressed in the mouse heart and is markedly downregulated in Nkx2.5 null animals and its promoter is directly activated by Nkx2-5. In addition, Csm potentiated phenylephrine-induced hypertrophic response in cardiacmyocytes.[103]

Myocardin A is a transcription factor, highly expressed in developing heart, which is transcriptionally regulated by Nkx2-5. In turn, trans-activates ANF through a serum response element through BMP-2 and transforming growth factor beta-activated kinase 1 (TAK-1) signaling.[104]

Atrial natriuretic factor or peptide (ANF or ANP) is directly activated by Nkx2-5 but in synergy with other transcription factors (see 7.4.2).[105,106] Sodium Calcium exchanger I (NCX1) is a transmembrane electrogenic protein that is the primary mechanism for calcium efflux in the heart. In neonatal cardiomyocytes, the NCX1 promoter is directly activated by Nkx2-5 and this activation is potentiated by SRF binding, although in adult cardiomyocytes this additive effect is lost.[107]

Calreticulin is a calcium-binding chaperone of the endoplasmic reticulum and essential in cardiac development. Nkx2.5 activates the calreticulin promoter directly and is antagonized by binding of chicken ovalbumin upstream promoter-transcription factor 1 (COUP-T1) to the Nkx2.5 binding site.[108]

Connexins are transmemberane proteins, which form intercellular channels known as gap junctions and in cardiomyocytes are responsible for co-ordinated contraction.

Connexins 40, 43, and 45 are expressed in heart and conduction system where they are all regulated by Nkx2-5.[109,110]

Procollagen lysyl hydroxylase 1 (PLOD1) is a member of an enzyme family that hydroxylate lysine residues in collagens. Mutations in PLOD1 result among other connective tissue defects with muscular and aortic defects.[111] In concert with PITX2C, Nkx2-5 regulates the PLOD1 promoter in embryonic heart development.

The structural extracellular matrix protein Collagen 1α2 is activated in vSMC by Nkx2-5 through a mechanism, which involves a response element located in the gene's far upstream enhancer region and the displacement of the repressor ZEB1(δEF1) from an overlapping binding site.[49]

ECE-1 is a zinc metalloprotease, which cleaves and activates the vasoconstrictor ET-1. It is involved in cardiovascular development and a downstream target of Nkx2-5.[88]

A1 adenosine receptor is a G–coupled receptor expressed in early cardiac embryogenesis and is activated by Nkx2-5/GATA-4 complex (see 7.4.2).[112]

7.4.2 *Combinatorial Regulation by Nkx2-5*

Nkx2-5 is known to interact with other proteins to achieve positive or negative regulation of downstream genes. One such well-characterized interaction is with the transcription factor GATA-4, which also regulates Nkx2-5 transcriptional activity (see 7.3.1). Physical interaction between the two proteins was shown to occur via the homeodomain of Nkx2-5 and the zinc-finger domain of GATA-4 in vitro and in vivo and the complex results in a synergistic transactivation of the ANP promoter. This co-operativity was independent of GATA-4 binding on the ANP promoter.[8,10,105,113] CARP is another gene, which is transcriptionally regulated by Nkx2-5/GATA-4 complex formation[99] as is the *A1 adenosine receptor*.[112] A further synergistic interaction of the Nkx2-5/GATA-4 complex was demonstrated with SRF on cardiac α-Actin promoter.[8,102]

Nkx2-5 also interacts with the T-box transcription factor Tbx-5, through the homeodomain of NKx2-5 and the N-terminal domain and T-box of Tbx-5, to synergistically activate the promoters of ANP, connexion 40, and Id2 (see 7.4.1).[34,114,115] A model of positive and negative transcriptional regulation was described by Habets et al., whereby another T-box factor Tbx-2 acts as a transcriptional repressor by displacing Tbx-5 from the complex with Nkx2-5 on the ANP promoter in areas such as the AV canal and the OFT where Tbx-2 is expressed and Tbx-5 is absent.[116]

Nkx2-5 is involved in a synergistic regulatory mechanism with PITX2C where together they can activate transcription of ANF and PLOD1 (see 7.4.1), although not through physical interaction.[106] It is also reported that Nkx2-5 acts to inhibit PITX2A activation of the ANF promoter in a cell-specific manner.

eHand/Hand1 is another transcription factor involved in cardiac development, which interacts with Nkx2-5 to synergistically co-activate the *ANP* promoter[117] and mice lacking both genes also lack ventricle specification. Another transcription factor that functionally interacts with Nkx2-5 is Foxh1, forkhead DNA-binding transcription factor in the TGF-β-Smad pathway. Together, they regulate the transcription of the MEF2C promoter in the anterior heart field.[95]

7.5 Future Perspectives

The role of Nkx2-5 in the development of the heart has been investigated at length and through these studies greater understanding of the origins of cells in different compartments of the cardiovascular system has been gained. However, greater understanding of the role of Nkx2-5 in adult tissues and how this is altered in disease is required if the full therapeutic potential of Nkx2-5 is to be realized.

Recently, Nkx2-5 has been recognized as a marker of a tripotent progenitor cell population,[118] which can give rise to endothelial, cardiac, and smooth muscle cell lineages. This embryonic stem cell population, which also expresses Isl1 and flk1, could be used in cardiovascular tissue regeneration by directing the differentiation into mature cardiac, pacemaker, smooth muscle, and endothelial cell types. Another cardiac progenitor cell population expressing Nkx2-5 and c-kit has also been reported with the ability to generate cardiac and smooth muscle cells but not endothelial cells.[48] More information about these embryonic stem cell populations has to be obtained and their potential relationship with adult heart progenitor cell populations has to be explored. However, as other types of progenitor cells found in adult hearts have not been very successful in generating cardiac muscle in vivo,[119] identification of these multipotent embryonic cell population is encouraging for future trial of cardiac repair using stem cells.[120]

Another future area of further study is the investigation of the role of Nkx2-5 in adult tissue repair and fibrosis in the cardiovascular system. Reports are already suggesting that cardiac fibrosis as a result of insult/damage coincides with Nkx2-5 induction in adult tissues.[50,121,122] Futhermore, induction by Nkx2-5 of genes related with the extracellular matrix remodeling, such as PLOD1, Collagen type I, and connexins, open up the possibility of Nkx2-5 being involved in repair and/or fibrosis in conditions were scar tissue is formed such as cardiac fibrosis, atrial fibrillation, and even in atherosclerosis.

7.6 Summary

- Nkx2-5 is a transcription factor involved in the early embryonic development and adult maintenance of the cardiovascular system and the CCS.
- Nkx2-5 induction is mediated by various signaling pathways including BMP, TGF-β, and Wnt.
- Targets of Nkx2-5 include cardiac and smooth muscle specific genes and ECM-related genes.

References

1. Harvey RP. NK-2 homeobox genes and heart development. *Dev Biol.* 1996;178:203-216.
2. Komuro I, Izumo S. Csx: a murine homeobox-containing gene specifically expressed in the developing heart. *Proc Natl Acad Sci U S A.* 1993;90:8145-8149.
3. Kim Y, Nirenberg M. Drosophila NK-homeobox genes. *Proc Natl Acad Sci U S A.* 1989;86:7716-7720.
4. Azpiazu N, Frasch M. tinman and bagpipe: two homeo box genes that determine cell fates in the dorsal mesoderm of Drosophila. *Genes Dev.* 1993;7:1325-1340.
5. Bodmer R. The gene tinman is required for specification of the heart and visceral muscles in Drosophila. *Development.* 1993;118:719-729.
6. Evans SM. Vertebrate tinman homologues and cardiac differentiation. *Semin Cell Dev Biol.* 1999;10:73-83.
7. Chen CY, Schwartz RJ. Identification of novel DNA binding targets and regulatory domains of a murine tinman homeodomain factor, Nkx-2.5. *J Biol Chem.* 1995;270:15628-15633.
8. Sepulveda JL, Belaguli N, Nigam V, Chen CY, Nemer M, Schwartz RJ. GATA-4 and Nkx-2.5 coactivate Nkx-2 DNA binding targets: role for regulating early cardiac gene expression. *Mol Cell Biol.* 1998;18:3405-3415.
9. Durocher D, Nemer M. Combinatorial interactions regulating cardiac transcription. *Dev Genet.* 1998;22:250-262.
10. Lee Y, Shioi T, Kasahara H, et al. The cardiac tissue-restricted homeobox protein Csx/Nkx2.5 physically associates with the zinc finger protein GATA4 and cooperatively activates atrial natriuretic factor gene expression. *Mol Cell Biol.* 1998;18:3120-3129.
11. Kasahara H, Bartunkova S, Schinke M, Tanaka M, Izumo S. Cardiac and extracardiac expression of Csx/Nkx2.5 homeodomain protein. *Circ Res.* 1998;82:936-946.
12. Brand T. Heart development: molecular insights into cardiac specification and early morphogenesis. *Dev Biol.* 2003;258:1-19.
13. Kelly RG, Buckingham ME. The anterior heart-forming field: voyage to the arterial pole of the heart. *Trends Genet.* 2002;18:210-216.
14. Jongbloed MR, Mahtab EA, Blom NA, Schalij MJ, Gittenberger-de Groot AC. Development of the cardiac conduction system and the possible relation to predilection sites of arrhythmogenesis. *ScientificWorldJournal.* 2008;8:239-269.
15. Tanaka M, Wechsler SB, Lee IW, Yamasaki N, Lawitts JA, Izumo S. Complex modular cis-acting elements regulate expression of the cardiac specifying homeobox gene Csx/Nkx2.5. *Development.* 1999;126:1439-1450.
16. Lyons I, Parsons LM, Hartley L, et al. Myogenic and morphogenetic defects in the heart tubes of murine embryos lacking the homeo box gene Nkx2-5. *Genes Dev.* 1995;9:1654-1666.
17. Tanaka M, Kasahara H, Bartunkova S, et al. Vertebrate homologs of tinman and bagpipe: roles of the homeobox genes in cardiovascular development. *Dev Genet.* 1998;22:239-249.
18. Chen CY, Schwartz RJ. Competition between negative acting YY1 versus positive acting serum response factor and tinman homologue Nkx-2.5 regulates cardiac alpha actin promoter activity. *Mol Endocrinol.* 1997;11:812-822.
19. Harvey RP, Lai D, Elliott D, et al. Homeodomain factor Nkx2–5 in heart development and disease. *Cold Spring Harb Symp Quant Biol.* 2002;67:107-114.
20. Shiojima I, Komuro I, Mizuno T, et al. Molecular cloning and characterization of human cardiac homeobox gene CSX1. *Circ Res.* 1996;79:920-929.
21. Biben C, Weber R, Kesteven S, et al. Cardiac septal and valvular dysmorphogenesis in mice heterozygous for mutations in the homeobox gene Nkx2–5. *Circ Res.* 2000;87:888-895.
22. Tanaka M, Chen Z, Bartunkova S, Yamasaki N, Izumo S. The cardiac homeobox gene Csx/Nkx2.5 lies genetically upstream of multiple genes essential for heart development. *Development.* 1999;126:1269-1280.
23. Biben C, Harvey RP. Homeodomain factor Nkx2–5 controls left/right asymmetric expression of bHLH gene eHand during murine heart development. *Genes Dev.* 1997;11:1357-1369.

24. Bruneau BG, Bao ZZ, Tanaka M, et al. Cardiac expression of the ventricle-specific homeobox gene Irx4 is modulated by Nkx2-5 and dHand. *Dev Biol.* 2000;217:266-277.
25. Yamagishi H, Yamagishi C, Nakagawa O, Harvey RP, Olson EN, Srivastava D. The combinatorial activities of Nkx2.5 and dHAND are essential for cardiac ventricle formation. *Dev Biol.* 2001;239:190-203.
26. Pashmforoush M, Lu JT, Chen H, et al. Nkx2-5 pathways and congenital heart disease; loss of ventricular myocyte lineage specification leads to progressive cardiomyopathy and complete heart block. *Cell.* 2004;117:373-386.
27. Prall OW, Menon MK, Solloway MJ, et al. An Nkx2-5/Bmp2/Smad1 negative feedback loop controls heart progenitor specification and proliferation. *Cell.* 2007;128:947-959.
28. Benson DW, Silberbach GM, Kavanaugh-McHugh A, et al. Mutations in the cardiac transcription factor NKX2.5 affect diverse cardiac developmental pathways. *J Clin Invest.* 1999;104: 1567-1573.
29. McElhinney DB, Geiger E, Blinder J, Benson DW, Goldmuntz E. NKX2.5 mutations in patients with congenital heart disease. *J Am Coll Cardiol.* 2003;42:1650-1655.
30. Schott JJ, Benson DW, Basson CT, et al. Congenital heart disease caused by mutations in the transcription factor NKX2-5. *Science.* 1998;281:108-111.
31. Akazawa H, Komuro I. Cardiac transcription factor Csx/Nkx2-5: Its role in cardiac development and diseases. *Pharmacol Ther.* 2005;107:252-268.
32. Kasahara H, Benson DW. Biochemical analyses of eight NKX2.5 homeodomain missense mutations causing atrioventricular block and cardiac anomalies. *Cardiovasc Res.* 2004;64: 40-51.
33. Pabst S, Wollnik B, Rohmann E, et al. A novel stop mutation truncating critical regions of the cardiac transcription factor NKX2-5 in a large family with autosomal-dominant inherited congenital heart disease. *Clin Res Cardiol.* 2008;97:39-42.
34. Moskowitz IP, Kim JB, Moore ML, et al. A molecular pathway including Id2, Tbx5, and Nkx2-5 required for cardiac conduction system development. *Cell.* 2007;129:1365-1376.
35. Harris BS, Spruill L, Edmonson AM, et al. Differentiation of cardiac Purkinje fibers requires precise spatiotemporal regulation of Nkx2-5 expression. *Dev Dyn.* 2006;235:38-49.
36. Thomas PS, Kasahara H, Edmonson AM, et al. Elevated expression of Nkx-2.5 in developing myocardial conduction cells. *Anat Rec.* 2001;263:307-313.
37. Moskowitz IP, Pizard A, Patel VV, et al. The T-Box transcription factor Tbx5 is required for the patterning and maturation of the murine cardiac conduction system. *Development.* 2004;131:4107-4116.
38. Wakimoto H, Kasahara H, Maguire CT, Moskowitz IP, Izumo S, Berul CI. Cardiac electrophysiological phenotypes in postnatal expression of Nkx2.5 transgenic mice. *Genesis.* 2003;37:144-150.
39. Jay PY, Izumo S. Elucidating the molecular and genetic interactions responsible for congenital heart disease. *Pediatr Res.* 2002;51:127.
40. Tanaka M, Berul CI, Ishii M, et al. A mouse model of congenital heart disease: cardiac arrhythmias and atrial septal defect caused by haploinsufficiency of the cardiac transcription factor Csx/Nkx2.5. *Cold Spring Harb Symp Quant Biol.* 2002;67:317-325.
41. Meysen S, Marger L, Hewett KW, et al. Nkx2.5 cell-autonomous gene function is required for the postnatal formation of the peripheral ventricular conduction system. *Dev Biol.* 2007;303:740-753.
42. Jay PY, Harris BS, Maguire CT, et al. Nkx2-5 mutation causes anatomic hypoplasia of the cardiac conduction system. *J Clin Invest.* 2004;113:1130-1137.
43. Yadava RS, Frenzel-McCardell CD, Yu Q, et al. RNA toxicity in myotonic muscular dystrophy induces NKX2-5 expression. *Nat Genet.* 2008;40:61-68.
44. Kelly RG, Brown NA, Buckingham ME. The arterial pole of the mouse heart forms from Fgf10-expressing cells in pharyngeal mesoderm. *Dev Cell.* 2001;1:435-440.
45. Waldo KL, Kumiski DH, Wallis KT, et al. Conotruncal myocardium arises from a secondary heart field. *Development.* 2001;128:3179-3188.
46. Xu H, Morishima M, Wylie JN, et al. Tbx1 has a dual role in the morphogenesis of the cardiac outflow tract. *Development.* 2004;131:3217-3227.

47. Mommersteeg MT, Brown NA, Prall OW, et al. Pitx2c and Nkx2-5 are required for the formation and identity of the pulmonary myocardium. *Circ Res*. 2007;101:902-909.
48. Wu SM, Fujiwara Y, Cibulsky SM, et al. Developmental origin of a bipotential myocardial and smooth muscle cell precursor in the mammalian heart. *Cell*. 2006;127:1137-1150.
49. Ponticos M, Partridge T, Black CM, Abraham DJ, Bou-Gharios G. Regulation of collagen type I in vascular smooth muscle cells by competition between Nkx2.5 and deltaEF1/ZEB1. *Mol Cell Biol*. 2004;24:6151-6161.
50. Saadane N, Alpert L, Chalifour LE. Expression of immediate early genes, GATA-4, and Nkx-2.5 in adrenergic-induced cardiac hypertrophy and during regression in adult mice. *Br J Pharmacol*. 1999;127:1165-1176.
51. Thompson JT, Rackley MS, O'Brien TX. Upregulation of the cardiac homeobox gene Nkx2-5 (CSX) in feline right ventricular pressure overload. *Am J Physiol*. 1998;274:H1569-H1573.
52. Takimoto E, Mizuno T, Terasaki F, et al. Up-regulation of natriuretic peptides in the ventricle of Csx/Nkx2–5 transgenic mice. *Biochem Biophys Res Commun*. 2000;270:1074-1079.
53. Toko H, Zhu W, Takimoto E, et al. Csx/Nkx2-5 is required for homeostasis and survival of cardiac myocytes in the adult heart. *J Biol Chem*. 2002;277:24735-24743.
54. Shiojima I, Komuro I, Inazawa J, et al. Assignment of cardiac homeobox gene CSX to human chromosome 5q34. *Genomics*. 1995;27:204-206.
55. Turbay D, Wechsler SB, Blanchard KM, Izumo S. Molecular cloning, chromosomal mapping, and characterization of the human cardiac-specific homeobox gene hCsx. *Mol Med*. 1996;2:86-96.
56. Schwartz RJ, Olson EN. Building the heart piece by piece: modularity of cis-elements regulating Nkx2-5 transcription. *Development*. 1999;126:4187-4192.
57. Chi X, Chatterjee PK, Wilson W III, Zhang SX, DeMayo FJ, Schwartz RJ. Complex cardiac Nkx2-5 gene expression activated by noggin-sensitive enhancers followed by chamber-specific modules. *Proc Natl Acad Sci U S A*. 2005;102:13490-13495.
58. Reecy JM, Li X, Yamada M, et al. Identification of upstream regulatory regions in the heart-expressed homeobox gene Nkx2-5. *Development*. 1999;126:839-849.
59. Lien CL, Wu C, Mercer B, Webb R, Richardson JA, Olson EN. Control of early cardiac-specific transcription of Nkx2-5 by a GATA-dependent enhancer. *Development*. 1999;126:75-84.
60. Searcy RD, Vincent EB, Liberatore CM, Yutzey KE. A GATA-dependent nkx-2.5 regulatory element activates early cardiac gene expression in transgenic mice. *Development*. 1998;125:4461-4470.
61. Liberatore CM, Searcy-Schrick RD, Vincent EB, Yutzey KE. Nkx-2.5 gene induction in mice is mediated by a Smad consensus regulatory region. *Dev Biol*. 2002;244:243-256.
62. Lien CL, McAnally J, Richardson JA, Olson EN. Cardiac-specific activity of an Nkx2-5 enhancer requires an evolutionarily conserved Smad binding site. *Dev Biol*. 2002;244:257-266.
63. Brown CO III, Chi X, Garcia-Gras E, Shirai M, Feng XH, Schwartz RJ. The cardiac determination factor, Nkx2-5, is activated by mutual cofactors GATA-4 and Smad1/4 via a novel upstream enhancer. *J Biol Chem*. 2004;279:10659-10669.
64. Monzen K, Ito Y, Naito AT, et al. A crucial role of a high mobility group protein HMGA2 in cardiogenesis. *Nat Cell Biol*. 2008;10:567-574.
65. Skerjanc IS, Petropoulos H, Ridgeway AG, Wilton S. Myocyte enhancer factor 2C and Nkx2-5 up-regulate each other's expression and initiate cardiomyogenesis in P19 cells. *J Biol Chem*. 1998;273:34904-34910.
66. Black BL, Olson EN. Transcriptional control of muscle development by myocyte enhancer factor-2 (MEF2) proteins. *Annu Rev Cell Dev Biol*. 1998;14:167-196.
67. Nagao K, Taniyama Y, Kietzmann T, Doi T, Komuro I, Morishita R. HIF-1alpha signaling upstream of NKX2.5 is required for cardiac development in Xenopus. *J Biol Chem*. 2008;283:11841-11849.
68. Massague J, Chen YG. Controlling TGF-beta signaling. *Genes Dev*. 2000;14:627-644.
69. Schultheiss TM, Burch JB, Lassar AB. A role for bone morphogenetic proteins in the induction of cardiac myogenesis. *Genes Dev*. 1997;11:451-462.
70. Smith DM, Nielsen C, Tabin CJ, Roberts DJ. Roles of BMP signaling and Nkx2.5 in patterning at the chick midgut-foregut boundary. *Development*. 2000;127:3671-3681.

71. Smith DM, Tabin CJ. BMP signalling specifies the pyloric sphincter. *Nature*. 1999;402: 748-749.
72. Yamada M, Szendro PI, Prokscha A, Schwartz RJ, Eichele G. Evidence for a role of Smad6 in chick cardiac development. *Dev Biol*. 1999;215:48-61.
73. Jamali M, Karamboulas C, Rogerson PJ, Skerjanc IS. BMP signaling regulates Nkx2-5 activity during cardiomyogenesis. *FEBS Lett*. 2001;509:126-130.
74. Li TS, Suzuki R, Ueda K, Murata T, Hamano K. Analysis of the origin and population dynamics of cardiac progenitor cells in a donor heart model. *Stem Cells*. 2007;25:911-917.
75. bdel-Latif A, Zuba-Surma EK, Case J, Tiwari S, Hunt G, Ranjan S, Vincent RJ, Srour EF, Bolli R, Dawn B. TGF-beta1 enhances cardiomyogenic differentiation of skeletal muscle-derived adult primitive cells. *Basic Res Cardiol*. 2008;103(6):514-524.
76. Lim JY, Kim WH, Kim J, Park SI. Involvement of TGF-beta1 signaling in cardiomyocyte differentiation from P19CL6 cells. *Mol Cells*. 2007;24:431-436.
77. Cohen ED, Tian Y, Morrisey EE. Wnt signaling: an essential regulator of cardiovascular differentiation, morphogenesis and progenitor self-renewal. *Development*. 2008;135:789-798.
78. Marvin MJ, Di RG, Gardiner A, Bush SM, Lassar AB. Inhibition of Wnt activity induces heart formation from posterior mesoderm. *Genes Dev*. 2001;15:316-327.
79. Schneider VA, Mercola M. Wnt antagonism initiates cardiogenesis in Xenopus laevis. *Genes Dev*. 2001;15:304-315.
80. Naito AT, Shiojima I, Akazawa H, et al. Developmental stage-specific biphasic roles of Wnt/beta-catenin signaling in cardiomyogenesis and hematopoiesis. *Proc Natl Acad Sci U S A*. 2006;103:19812-19817.
81. Pal R, Khanna A. Role of smad- and wnt-dependent pathways in embryonic cardiac development. *Stem Cells Dev*. 2006;15:29-39.
82. Eisenberg CA, Eisenberg LM. WNT11 promotes cardiac tissue formation of early mesoderm. *Dev Dyn*. 1999;216:45-58.
83. Abraham D, Distler O. How does endothelial cell injury start? The role of endothelin in systemic sclerosis. *Arthritis Res Ther*. 2007;9(Suppl 2):S2.
84. Abraham D, Ponticos M, Nagase H. Connective tissue remodeling: cross-talk between endothelins and matrix metalloproteinases. *Curr Vasc Pharmacol*. 2005;3:369-379.
85. Gourdie RG, Wei Y, Kim D, Klatt SC, Mikawa T. Endothelin-induced conversion of embryonic heart muscle cells into impulse-conducting Purkinje fibers. *Proc Natl Acad Sci U S A*. 1998;95:6815-6818.
86. Takebayashi-Suzuki K, Pauliks LB, Eltsefon Y, Mikawa T. Purkinje fibers of the avian heart express a myogenic transcription factor program distinct from cardiac and skeletal muscle. *Dev Biol*. 2001;234:390-401.
87. Patel R, Kos L. Endothelin-1 and Neuregulin-1 convert embryonic cardiomyocytes into cells of the conduction system in the mouse. *Dev Dyn*. 2005;233:20-28.
88. Funke-Kaiser H, Lemmer J, Langsdorff CV, et al. Endothelin-converting enzyme-1 (ECE-1) is a downstream target of the homeobox transcription factor Nkx2-5. *FASEB J*. 2003;17:1487-1489.
89. Kasahara H, Izumo S. Identification of the in vivo casein kinase II phosphorylation site within the homeodomain of the cardiac tisue-specifying homeobox gene product Csx/Nkx2.5. *Mol Cell Biol*. 1999;19:526-536.
90. Wang J, Zhang H, Iyer D, Feng XH, Schwartz RJ. Regulation of cardiac specific Nkx2.5 gene activity by sumo modificaiton. *J Biol Chem*. 2008;283(34):23235-23243.
91. Lin X, Sun B, Liang M, et al. Opposed regulation of corepressor CtBP by SUMOylation and PDZ binding. *Mol Cell*. 2003;11:1389-1396.
92. Christoffels VM, Habets PE, Franco D, et al. Chamber formation and morphogenesis in the developing mammalian heart. *Dev Biol*. 2000;223:266-278.
93. Palmer S, Groves N, Schindeler A, et al. The small muscle-specific protein Csl modifies cell shape and promotes myocyte fusion in an insulin-like growth factor 1-dependent manner. *J Cell Biol*. 2001;153:985-998.
94. Zou Y, Evans S, Chen J, Kuo HC, Harvey RP, Chien KR. CARP, a cardiac ankyrin repeat protein, is downstream in the Nkx2-5 homeobox gene pathway. *Development*. 1997;124:793-804.

95. Von Both I, Silvestri C, Erdemir T, et al. Foxh1 is essential for development of the anterior heart field. *Dev Cell*. 2004;7:331-345.
96. Chen F, Kook H, Milewski R, et al. Hop is an unusual homeobox gene that modulates cardiac development. *Cell*. 2002;110:713-723.
97. Shin CH, Liu ZP, Passier R, et al. Modulation of cardiac growth and development by HOP, an unusual homeodomain protein. *Cell*. 2002;110:725-735.
98. Wang DZ, Reiter RS, Lin JL, et al. Requirement of a novel gene, Xin, in cardiac morphogenesis. *Development*. 1999;126:1281-1294.
99. Kuo H, Chen J, Ruiz-Lozano P, Zou Y, Nemer M, Chien KR. Control of segmental expression of the cardiac-restricted ankyrin repeat protein gene by distinct regulatory pathways in murine cardiogenesis. *Development*. 1999;126:4223-4234.
100. Shiratori H, Sakuma R, Watanabe M, et al. Two-step regulation of left-right asymmetric expression of Pitx2: initiation by nodal signaling and maintenance by Nkx2. *Mol Cell*. 2001;7:137-149.
101. Ross RS, Navankasattusas S, Harvey RP, Chien KR. An HF-1a/HF-1b/MEF-2 combinatorial element confers cardiac ventricular specificity and established an anterior-posterior gradient of expression. *Development*. 1996;122:1799-1809.
102. Sepulveda JL, Vlahopoulos S, Iyer D, Belaguli N, Schwartz RJ. Combinatorial expression of GATA4, Nkx2-5, and serum response factor directs early cardiac gene activity. *J Biol Chem*. 2002;277:25775-25782.
103. Ueyama T, Kasahara H, Ishiwata T, Yamasaki N, Izumo S. Csm, a cardiac-specific isoform of the RNA helicase Mov10l1, is regulated by Nkx2.5 in embryonic heart. *J Biol Chem*. 2003;278:28750-28757.
104. Ueyama T, Kasahara H, Ishiwata T, Nie Q, Izumo S. Myocardin expression is regulated by Nkx2.5, and its function is required for cardiomyogenesis. *Mol Cell Biol*. 2003;23:9222-9232.
105. Durocher D, Charron F, Warren R, Schwartz RJ, Nemer M. The cardiac transcription factors Nkx2-5 and GATA-4 are mutual cofactors. *EMBO J*. 1997;16:5687-5696.
106. Ganga M, Espinoza HM, Cox CJ, et al. PITX2 isoform-specific regulation of atrial natriuretic factor expression: synergism and repression with Nkx2.5. *J Biol Chem*. 2003;278:22437-22445.
107. Muller JG, Thompson JT, Edmonson AM, et al. Differential regulation of the cardiac sodium calcium exchanger promoter in adult and neonatal cardiomyocytes by Nkx2.5 and serum response factor. *J Mol Cell Cardiol*. 2002;34:807-821.
108. Guo L, Lynch J, Nakamura K, et al. COUP-TF1 antagonizes Nkx2.5-mediated activation of the calreticulin gene during cardiac development. *J Biol Chem*. 2001;276:2797-2801.
109. Dupays L, Jarry-Guichard T, Mazurais D, et al. Dysregulation of connexins and inactivation of NFATc1 in the cardiovascular system of Nkx2-5 null mutants. *J Mol Cell Cardiol*. 2005;38:787-798.
110. Kasahara H, Ueyama T, Wakimoto H, et al. Nkx2.5 homeoprotein regulates expression of gap junction protein connexin 43 and sarcomere organization in postnatal cardiomyocytes. *J Mol Cell Cardiol*. 2003;35:243-256.
111. Hyland J, la-Kokko L, Royce P, Steinmann B, Kivirikko KI, Myllyla R. A homozygous stop codon in the lysyl hydroxylase gene in two siblings with Ehlers-Danlos syndrome type VI. *Nat Genet*. 1992;2:228-231.
112. Rivkees SA, Chen M, Kulkarni J, Browne J, Zhao Z. Characterization of the murine A1 adenosine receptor promoter, potent regulation by GATA-4 and Nkx2.5. *J Biol Chem*. 1999;274:14204-14209.
113. Shiojima I, Komuro I, Oka T, et al. Context-dependent transcriptional cooperation mediated by cardiac transcription factors Csx/Nkx-2.5 and GATA-4. *J Biol Chem*. 1999;274:8231-8239.
114. Bruneau BG, Nemer G, Schmitt JP, et al. A murine model of Holt-Oram syndrome defines roles of the T-box transcription factor Tbx5 in cardiogenesis and disease. *Cell*. 2001;106:709-721.
115. Hiroi Y, Kudoh S, Monzen K, et al. Tbx5 associates with Nkx2-5 and synergistically promotes cardiomyocyte differentiation. *Nat Genet*. 2001;28:276-280.
116. Habets PE, Moorman AF, Clout DE, van Roon MA, Lingbeek M, van LM, Campione M, Christoffels VM. Cooperative action of Tbx2 and Nkx2.5 inhibits ANF expression in the atrioventricular canal: implications for cardiac chamber formation. *Genes Dev* 2002;16:1234-1246.

117. Thattaliyath BD, Firulli BA, Firulli AB. The basic-helix-loop-helix transcription factor HAND2 directly regulates transcription of the atrial naturetic peptide gene. *J Mol Cell Cardiol.* 2002;34:1335-1344.
118. Moretti A, Caron L, Nakano A, et al. Multipotent embryonic isl1+ progenitor cells lead to cardiac, smooth muscle, and endothelial cell diversification. *Cell.* 2006;127:1151-1165.
119. Beltrami AP, Barlucchi L, Torella D, et al. Adult cardiac stem cells are multipotent and support myocardial regeneration. *Cell.* 2003;114:763-776.
120. Basson M. Singling out heart cells. *Nat Med.* 2007;13:33.
121. Luo MH, Li YS, Yang KP. Fibrosis of collagen I and remodeling of connexin 43 in atrial myocardium of patients with atrial fibrillation. *Cardiology.* 2007;107:248-253.
122. Saadane N, Alpert L, Chalifour LE. Altered molecular response to adrenoreceptor-induced cardiac hypertrophy in Egr-1-deficient mice. *Am J Physiol Heart Circ Physiol.* 2000;278:H796-H805.
123. Durocher D, Chen CY, Ardati A, Schwartz RJ, Nemer M. The atrial natriuretic factor promoter is a downstream target for Nkx-2.5 in the myocardium. *Mol Cell Biol.* 1996;16:4648-4655.

Chapter 8
Cell Therapy for Cardiovascular Disease

Jonathan R. Murrow, Saurabh S. Dhawan, and Arshed A. Quyyumi

8.1 Introduction

Advances in diagnosis and treatment have dramatically impacted morbidity and mortality from cardiovascular disease over the past several decades.[1] The discovery in 1960 of stem cells capable of regeneration and repair sparked interest in a new mode of therapy for heart disease beyond pharmaceuticals and cardiac devices.[2] Over the past 10 years, work has focused on five key cell types – the endothelial mononuclear progenitor cell, the autologous skeletal myoblast, the allogeneic mesenchymal stem cell, the resident cardiac stem cell, and the human embryonic stem cell – as potential therapeutic agents, which may further contribute to gains in treating cardiovascular disease. This chapter aims to review these cell types, their preclinical underpinnings, the nascent clinical studies, and limitations observed in their use.

8.2 Endothelial Progenitor Cells/Bone Marrow-Derived Mononuclear Cells

8.2.1 Background

The pathobiology of atherosclerosis has been largely attributed to processes of repeated vascular injury leading to plaque development and expansion, subsequent tissue ischemia, and ultimately infarction.[3] More recently, the discovery of a population of endogenous mononuclear cells that reside in the bone marrow mobilize in response to tissue injury, and repair injured vascular tissue has challenged the

A.A. Quyyumi (✉)
Division of Cardiology, Emory University,
1364 Clifton Road, Suite D403C, Atlanta, GA, 30322, USA
e-mail: aquyyum@emory.edu

D. Abraham et al. (eds.), *Advances in Vascular Medicine*,
DOI 10.1007/978-1-84882-637-3_8, © Springer-Verlag London Limited 2010

classic Ross hypothesis, offering the possibility of another mode of treating atherosclerotic vascular disease. The precise identity of these so-called endothelial progenitor cells (EPCs) remains elusive. EPCs have thus necessarily been defined by their functions – that they originate in the bone marrow, circulate in peripheral blood, home to sites of vascular injury, and participate in new blood vessel formation.[4] Asahara and colleagues in 1997 provided the strongest evidence that new blood vessel formation is partly attributable to a population of bone marrow-derived monocytes. CD34-positive mononuclear cells isolated from humans demonstrated an endothelial phenotype in vitro. Moreover, these cells participated in neo-angiogenesis in a mouse hindlimb ischemia model,[5] thus suggesting a possible therapeutic role for regeneration after ischemic vascular injury.

Preclinical work has shown that exogenously administered EPCs home to areas of ischemia or infarct and participate in tissue regeneration or repair. Kalka and co-workers described the first use of EPCs in therapeutic neovascularization when human EPCs were transplanted into athymic nude mice in a model of hindlimb ischemia.[6] Blood flow recovery and capillary density was reported to markedly improve. Human EPC transplanted into ischemic regions of a rat model of myocardial infarction resulted in improvements in left ventricular function.[7] In a swine model of chronic myocardial ischemia, coronary collateral development and left ventricular ejection fraction improved after transplantation of autologous CD34-positive mononuclear cells.[8] Moreover, other work has shown that transplanted EPCs injected after myocardial infarct may home to infarct border zones.[9] When athymic nude rats were injected with radiolabeled human EPCs from peripheral blood, radioactivity was measured in cardiac tissues in greater proportion in infarcted hearts rather than in controls (although the total percentage was dwarfed by the percentage localizing to spleen and liver [1% vs. >70%]).

Although investigators initially hypothesized that EPCs directly participate in the formation of new blood vessels at sites of injury, the true mechanism of repair remains to be fully described. Three potential mechanisms by which EPCs exert influence on injured tissue include differentiation, cell fusion, and paracrine effects.[10] Pathologic correlations show that areas of infarcted myocardium injected with autologous bone marrow derived mononuclear cells have a higher capillary density with cellular changes including hyperplasia of pericytes, mural cells, and adventitia.[11] The presence of contractile proteins in these cells suggests an active process of angiogenesis. Still, the role that these EPCs play at sites of tissue injury remains controversial, with some work suggesting that they may perform tasks other than angiogenesis.[12] Badorff and co-workers reported that peripheral blood mononuclear cells from healthy adults transdifferentiated in vitro into functionally active cardiomyocytes (expressing alpha-sarcomeric actinin, cardiac troponin I, atrial natriuretic peptide, and formation of gap junctions) when co-cultured with rat cardiomyocytes, via a cell-to-cell contact mechanism rather than cellular fusion.[13] Other possible beneficial mechanisms of EPC repair via cell–cell signaling mechanisms are evident in work wherein co-incubation of bone marrow cells with cardiomyocytes in an ex vivo model of ischemia resulted in mitigation of cell death in a protein kinase C and p24 mitogen-activated kinase dependent process.[14]

Other work suggests that bone marrow-derived mononuclear cells contribute to new blood vessel formation in mouse hindlimb ischemia by a paracrine effect rather than directly incorporating into new vessel walls.[15,16]

8.2.2 *Clinical Studies*

While these repair mechanisms of EPCs continue to be elaborated, clinical studies of EPCs or of a broader population of bone marrow-derived mononuclear cells have been performed in settings of acute myocardial infarction, refractory angina, and chronic ischemic cardiomyopathy[17-33] (Table 8.1). Although double-blinded placebo-controlled studies are limited, available data from over 350 subjects enrolled to date in various trials suggests that therapy with EPCs or bone marrow-derived mononuclear cells in these settings poses no excess hazard, and may result in improvements in ventricular function and remodeling. No single study has convincingly demonstrated convincing efficacy of cell therapy, yet generally favorable results have encouraged investigators to continue tackling key questions such as cell type, source of cells, delivery mode, dose, and timing of therapy.

Several studies have pursued a strategy of infusing autologous EPCs or bone marrow cells down the infarct-related artery of subjects several days after acute myocardial infarction, with the expectation that amplification of the existing repair mechanisms by delivery of a concentrated dose of cells to areas of recent injury might have demonstrable clinical benefit. Strauer and coworkers found that infusion of intracoronary autologous mononuclear bone marrow cells in the infarct-related arteries of 10 patients 5–7 days after acute transmural myocardial infarction resulted in improvements in regional contractility, reduction in infarct size, and improved perfusion.[17] Cells were obtained by bone marrow aspiration with isolation of mononuclear bone marrow mononuclear cells via Ficoll and subsequent washing. CD133-positive cells (a more refined putative marker of EPCs) accounted for $0.65 \pm 0.4\%$ and CD34-positive cells $2.1 \pm 0.28\%$ of an average yield of $2.8(\pm 2.2) \pm 10^7$ cells. Notably, no patient in the study underwent revascularization of the infarct-related artery within 4 h of onset of symptoms, suggesting that improvements may be attributable to cell therapy effects rather than recovery of hibernating myocardium. Subsequently, Fernandez-Aviles and co-workers found improvement in end-systolic volume and improvement in LV function after infusing autologous bone marrow mononuclear cells down the infarct-related artery of patients suffering from acute ST-elevation myocardial infarction in a nonrandomized controlled trial.[22] Cells were infused around 2 weeks after initial presentation, and no excess adverse events were noted in the study group. Though of similar trial design, the TOPCARE-MI trial examined whether adequate and effective doses of cells could be obtained from the peripheral circulation, thus avoiding the need for bone-marrow harvest. Fifty-nine subjects with acute myocardial infarction underwent intracoronary infusion of either circulating progenitor cells or bone marrow-derived mononuclear cells 4.9 ± 1.5 days after presentation.[18,34] Though in the immediate follow–up two

Table 8.1 Clinical trials of cell therapy in patients with acute myocardial infarction

					Outcome			
Trial, reference, and year	Design and follow-up	Cell number and type, route and, timing	Treated patients and controls	Imaging modality	*Hemodynamic* LVEF, SVI	*Geometry* LVEDV, LVESV LVEDVI, LVESVI, wall thickness	*Contractility* IWM, WMSI, shortening	*Perfusion defect* Size, viability
Strauer et al[17] (2002)	Non-RCT, 3 month	$9–28 \times 10^6$ BMC, IC, 5–9 days s/p AMI	10 BMC 10 controls	LV angio, stress echo, radionuclide ventriculography	↔ LVEF ↑ SVI	↔ LVEDV ↓ LVESV	↑ IWM ↔ Shortening	↓ Size
Fernandez-Avilez et al[22] (2004)	Non-RCT, 6–21 month	$11–175 \times 10^6$ BMC, IC, 5–29 days s/p AMI	20 BMC 13 controls	LV angio, stress echo, cardiac MR	↑ LVEF	↔ LVEDV ↓ LVESV ↑ Wall thickness	↑ WMSI	
TOPCARE-AMI Assmus et al[34] (2002)	Non-RCT, 4 month	$213 \pm 75 \times 10^6$ BMC, $16 \pm 12 \times 10^6$ CPC IC, 4.3 ± 1.5 days s/p AMI	9 BMC 11 CPC 11 controls	LV angio, echo, FDG-PET	↑ LVEF	↔ LVEDV ↓ LVESV	↑ WMSI	↓ Size ↑ Viability
TOPCARE-AMI Schachinger et al[18] (2004)	Non-RCT, 1 year	$213 \pm 75 \times 10^6$ BMC, $16 \pm 12 \times 10^6$ CPC IC, 4.9 ± 1.5 days s/p AMI	29 BMC 30 CPC 59 controls	LV angio, echo, FDG-PET, cardiac MR	↑ LVEF	↔ LVEDV ↓ LVESV ↔ Reactive hypertrophy	↑ WMSI	↓ Size ↑ Viability
Chen et al[23] (2004)	RCT, 6 month	$8–10 \times 10^9$ BMC, IC, 18 days s/p AMI	34 BMC 35 controls	EPS, echo, FDG-PET	↑ LVEF ↑ SVI	↓ LVEDV ↓ LVESV	↑ IWM	↓ Size
Bartunek et al[24] (2005)	Non-RCT, 4 month	$1.5–33.6 \times 10^6$ BMC, IC, 11.6 ± 1.4 days s/p AMI	19 BMC 16 controls	LV angio, EPS, SPECT, FDG-PET	↑ LVEF	↔ LVEDVI ↑ LVSP to LVESVI ratio	↑ Chordae shortening	↓ Size
BOOST Wollert et al[25] (2004)	RCT, 6 month	$24.6 \pm 9.4 \times 10^8$ BMC, IC, 4.8 ± 1.3 days s/p AMI	30 BMC 30 controls	Cardiac MR	↑ LVEF	↔ LVEDVI ↔ LVESVI ↔ Wall thickness	↑ IWM	
BOOST Meyer et al[26] (2006)	RCT, 18 month	$24.6 \pm 9.4 \times 10^8$ BMC, IC, 4.8 ± 1.3 days s/p AMI	30 BMC 30 controls	Cardiac MR	↔ LVEF ↑ Speed of LVEF recovery	↔ LVEDVI ↔ LVESVI ↔ Wall thickness	↔ IWM	
REPAIR-AMI Schachinger et al[27] (2006)	RCT, 4 month	$236 \pm 174 \times 10^6$ BMC, IC, 3–7 days s/p AMI	101 BMC 103 controls	LV angio	↑ LVEF	↔ LVEDV ↓ LVESV	↑ IWM	NR

ASTAMI Lunde et al[28] (2006)	RCT, 6 month	54–130 × 10^6 BMC, IC, 5–6 days s/p AMI	50 BMC 50 controls	Echo, SPECT, cardiac MR	↔ LVEF	↔ LVEDV	NR	↓ Size
Janssens et al[29] (2006)	RCT, 4 month	172 ± 72 × 10^6 BMC, IC, 1 day s/p AMI	33 BMC 34 controls	Echo, cardiac MR, acetate-PET	↔ LVEF	↔ LVEDVI ↔ LVESVI ↔ Wall thickness	↑ IWM	↓ Size
TCT-STAMI Ge et al[30] (2006)	RCT, 6 month	4 × 10^7 BMC, IC, 12 h s/p AMI	10 BMC 10 controls	Echo, SPECT	↑ LVEF	↔ LVDd	NR	↓ Size
Meluzin et al[31] (2006)	RCT, 3 month	0.9–2 × 10^7 BMC, 0.9–2 × 10^8 BMC	22 high-dose BMC 22 low-dose BMC 22 controls	Echo, SPECT, FDG-PET	↑ LVEF	↔ LVEDV ↓ LVESV	↑ IWM	↔ Size
MAGIC CELL-3-DES (AMI) Kang et al[32] (2006)	RCT, 6 month	1–2 × 10^9 G-CSF-mobilized BMC (G-CSF × 3 days), IC, 10–11 days s/p AMI	25 G-CSF 25 controls	Cardiac MR	↑ LVEF	↔ LVEDV ↔ LVESV	NR	↓ Size
Li ZQ et al[33] (2006)	Non-RCT, 5 month	7.25 ± 7.33 × 10^7 BMC, (G-CSF × 5 days), IC, 1–7 days s/p AMI	35 G-CSF 35 controls	Echo	↑ LVEF	↔ LVEDV ↓ LVESV	↑ WMSI	NR
AMR-001 PI: Quyyumi AA NCT00313339	RCT	BMC, IC	Under analysis					
MESENDO PI: Lasasla GP NCT00548613	RCT	BMC, IC and IM	Still recruiting					
PI: Traverse J NCT00268307	RCT	BMC, IC	Still recruiting					
PI: Hare J NCT00114452	RCT	MSC IV	Stopped recruiting					
PI: Perin E NCT00555828	RCT	MPC IM	Still recruiting					

RCT randomized control trial; *AMI* acute myocardial infarction; *BMC* bone marrow mononuclear cells; *CPC* circulating progenitor cells; *MSC* mesenchymal stem cells; *MPC* mesenchymal progenitor cells; *G-CSF* granulocyte colony stimulating factor; *IC* intra-coronary; *IM* intra-myocardial; *LV* left ventricle; *LV angio* left ventricular angiography; *Echo* echocardiography; *SPECT* single photon emission computed tomography; *MR* magnetic resonance; *FDG-PET* F18-fluoxydeoxyglucose positron emission tomography; *LVEF* left ventricular ejection fraction; *SVI* stroke volume index; *LVEDV* left ventricular end diastolic volume; *LVEDVI* left ventricular end diastolic volume index; *LVESV* left ventricular end systolic volume; *LVESVI* left ventricular end systolic volume index; *LVDd* LV diastolic diameter; *IWM* infarct wall motion; *WMSI* wall motion score index; *Psyst/ESV* ration of peak systolic pressure to end systolic volume; *LVSP* left ventricular systolic pressure; *NR* not reported subjects developed recurrent infarction with one resultant death, in the 1 year follow-up period no significant ventricular arrhythmias or other adverse effects were observed. Both groups demonstrated favorable effects on LV remodeling as assessed by ejection fraction ($50 \pm 10\%$–$58 \pm 10\%$; $p < 0.001$) and end-systolic LV volume.

Randomized controlled trials of progenitor cell therapy have shown conflicting results with regard to efficacy. Improvements in left ventricular performance were observed by Wollert and co-workers when 30 subjects were randomized to receive intracoronary autologous BMC or placebo 4.8 days after presenting with acute ST elevation myocardial infarction.[25] In comparison with 30 control subjects, cell therapy was associated with enhanced systolic function in segments adjacent to infarcted areas, without an excess risk of adverse clinical events. Long-term follow-up at 18 months, however, failed to show a persistent benefit of cell therapy over controls.[26] In a multicenter study in which 204 subjects were randomized to intracoronary autologous bone marrow mononuclear cells or placebo, intracoronary injection of cells 3–7 days after acute myocardial infarction resulted in improvement in left ventricular ejection fraction measured by left ventricular angiography ($5.5 \pm 7.3\%$ vs. $3.0 \pm 6.5\%$; $p = 0.01$) in the treatment group.[27] At 1-year follow-up, cell therapy was associated with a reduced risk of major adverse cardiovascular events (death, recurrence of myocardial infarction, and revascularization). In contrast, Lunde and co-workers failed to observe any significant difference in left ventricular ejection fraction by myocardial perfusion study or cardiac magnetic resonance imaging at 6 months in a randomized study of 100 patients receiving either intracoronary autologous bone marrow mononuclear cells or placebo 6 days after acute myocardial infarction.[28]

Likewise, studies questioning whether early delivery of cell therapy would affect efficacy and safety have found conflicting results. Evaluating the effects of emergent intracoronary infusion of bone marrow mononuclear cells on clinical outcomes, Ge and colleagues randomized 20 subjects to cell therapy infusion down the infarct-related artery vs. controls within 24 h of presentation with acute ST elevation myocardial infarction. As measured by echocardiography and myocardial perfusion study, both ventricular systolic function and perfusion improved in the cell therapy group at 6 months.[30] In contrast, early cell therapy (within 24 h of reperfusion after acute myocardial infarction) using intracoronary autologous bone marrow mononuclear cells was not associated with significant improvement in left ventricular function or perfusion relative to standard medical therapy in 67 subjects randomized by Janssens and co-workers.[29]

Most progenitor cell therapy trials have employed a broad population of mononuclear cells, but more recent work has focused on whether a more narrowly defined population of progenitor cells might be isolated and used in therapy. Based on preliminary data suggesting that selected hematopoietic cell have a higher engraftment potential when compared with bone marrow-derived mononuclear cells, Bartunek and co-workers infused CD133-positive bone marrow mononuclear cells down the infarct-related artery of patients 11.6 ± 1.4 days after acute myocardial infarction.[24] In comparison with controls, treated subjects evinced improvements in left ventricular performance as measured by ventriculography and myocardial perfusion studies. Ongoing studies examine whether isolated CD34-positive mononuclear cells benefit patients with acute myocardial infarction as well as chronic ischemic cardiomyopathy.

While most studies have obtained mononuclear progenitor cells from bone marrow niches, several studies have mobilized mononuclear progenitor cells from the bone marrow into the peripheral circulation using granulocyte colony-stimulating factor (G-CSF). Kang and co-workers found improvement in left ventricular structure and function at 6 months among patients treated with intracoronary peripheral blood mononuclear cells infused 10–11 days after acute myocardial infarction.[32] No effect was noted in subjects with old (>14 days) myocardial infarction or in control groups. Similarly, Li and colleagues observed improvements in left ventricular ejection fraction and remodeling in a cohort of 35 nonrandomized subjects treated with G-CSF-mobilized autologous peripheral mononuclear cells infused down the infarct-related artery 7 days after presentation with acute myocardial infarction.[33] Notably, there was a high incidence of complications with regard to mobilization, separation, and infusion. Ten other studies have employed G-CSF injections alone as a means of mobilizing bone marrow mononuclear cells in the setting of acute myocardial infarction with the aim of improving left ventricular function. According to a meta-analysis by Zohlnhofer and colleagues, among the 445 patients treated with this approach there was no significant improvement in left ventricular function over placebo.[35]

In 13 studies of progenitor cell therapy for acute myocardial infarction, doses of infused cells range from 10 million to 1 billion – in large part affected by available reservoir in the bone marrow and the degree of selection of progenitor cell subpopulations. Whether these doses are inadequate for clinical effect – thereby accounting for the modest clinical results seen across studies – remains unanswered. Meluzin and co-workers addressed the effect of dose of bone marrow mononuclear cells on clinical response in randomizing 66 patients to intracoronary infusion of high dose (10^8 cells), low dose (10^7 cells), or control at 7 days after acute myocardial infarction.[31] Regional myocardial function improved in a dose-dependent manner as measured by tissue Doppler echocardiography and by gated myocardial perfusion imaging, thus suggesting that an adequate amount of delivered cells is important.

The potential role of progenitor cell therapy in chronic atherosclerotic coronary artery disease has been addressed in several studies (Table 8.2). Here, the route of delivery of cells varies from intracoronary infusion to direct intramyocardial injection. Strauer and co-workers found that intracoronary transplantation of autologous bone marrow mononuclear cells was associated with improvements in infarct size and global contractility (although in this cohort ejection fraction was relatively preserved).[36] Versus controls, treatment with bone marrow mononuclear cells was associated with improvement in maximum oxygen uptake and regional fluoro deoxyglucose uptake. Perin and co-workers reported a durable improvement in left ventricular perfusion defects and exercise capacity in subjects treated with endomyocardial injection of autologous bone marrow mononuclear cells for ischemic cardiomyopathy.[37] Treatment of chronic ischemia with intracoronary peripheral mononuclear cells was associated

Table 8.2 Clinical trials of cell therapy in patients with ischemic cardiomyopathy

					Outcome				
Trial, year and, reference	Design and follow-up	Cell Number and type, route and, timing	Treated patients and controls	Imaging modality	*Hemodynamic* LVEF	*Geometry* LVEDV, LVESV, LVEDVI, LVESVI	*Contractility* IWM, Systolic thickening WMSI	*Perfusion defect* Size, Viability Hibernating myocardium	Other
Ischemic cardiomyopathy (ICM)									
Perin et al[37] (2004)	Non-RCT, 12 month	$25.5 \pm 6.3 \times 10^6$ BMC IM, timing NR	11 BMC 9 controls	Spiroergometry, echo, SPECT	↔ LVEF	NR	NR	↓ Size	↑ VO_2 max
IACT Strauer et al[36] (2005)	Non-RCT, 3 month	$60–132 \times 10^6$ BMC, IC, 823 ± 945 days s/p MI	18 BMC 18 controls	Spiroergometry, LV angio, SPECT, FDG-PET	↑ LVEF	NR	↑ IWM	↓ Size ↑ Viability	↑ VO_2 max
Erbs et al[38] (2005)	RCT, 3 month	$69 \pm 14 \times 10^6$ CPC IC, 225 ± 87 days s/p MI	13 CPC 13 controls	Cor. angiography, cardiac MR, FDG-PET	↑ LVEF	↑ LVEDV ↓ LVESV	↑ IWM ↑ Systolic thickening of infarct area	↓ Size ↑ Viability ↓ Hibernating segments of myocardium	↓ ED ↑CFR
Katritsis et al[40] (2005)	Non-RCT, 4 month	$2–4 \times 10^6$ MSC+EPC, IC, 242 ± 464 days s/p MI	11 MSC+EPC 11 controls	Echo, stress echo, radionuclide ventriculography, SPECT	↔ LVEF	↔ LVEDV ↔ LVESV ↔ LVEDD ↔ LVESD	↑ WMSI	↓ Non-viable segments	
Mocini et al[39] (2006)	Non-RCT, 3 month	$292 \pm 232 \times 10^6$ BMC IM, timing NR	17 BMC 18 controls	Echo, cardiac MR	↑ LVEF	NR	↑ WMSI	NR	

MAGIC CELL-3-DES (OMI) Kang et al.[32] (2006)	RCT, 6 month	1–2 × 10^9 G-CSF-mobilized BMC (G-CSF × 3 days), IC, 517 ± 525 days s/p MI	16 BMC 16 controls	Cardiac MR	↔ LVEF	↔ LVEDV ↔ LVESV	NR	↔ Size	↑CFR
PI: Perin E NCT00203203	RCT	BMC IM	Still recruiting						
PI: Bolli R NCT00474461	Non-RCT	CSC IM	Not yet recruiting						
PI: Adler D and Lazarus H NCT00365326	Non-RCT	BMC IC	Still recruiting						
PI: Zeiher AM and Walter DH NCT00326989	RCT	BMC mobilized by shock wave, IC	Still recruiting						
PI: Asahara T NCT00221182	Non-RCT	EPC IC	Still recruiting						
PI: Hare JM et al NCT00587990	RCT	MSC IM	Still recruiting						

RCT randomized control trial; *OMI* old myocardial infarction; *BMC* bone marrow mononuclear cells; *CPC* circulating progenitor cells; *MSC* mesenchymal stem cells; *MPC* mesenchymal progenitor cells; *CSC* cardiac stem cells; *G-CSF* granulocyte colony stimulating factor; *IC* intra-coronary; *IM* intra-myocardial; *Trans-endo* trans-endocardial; *LV* left ventricle; *LV Angio* left ventricular angiography; *Echo* echocardiography; *SPECT* single photon emission computed tomography; *Tc-SPECT* technetium single photon emission computed tomography; *Thal-SPECT* thallium single photon emission computed tomography; *MR* magnetic resonance; *FDG-PET* F18-fluoxydeoxyglucose positron emission tomography; *LVEF* left ventricular ejection fraction; *SVI* stroke volume index; *LVEDV* left ventricular end diastolic volume; *LVEDD* left ventricular end diastolic diameter; *LVESV* left ventricular end systolic volume; *LVESD* left ventricular end systolic diameter; *IWM* infarct wall motion; *WMSI* wall motion score index; *WMSI* wall motion score index; *ED* endothelial dysfunction; *CFR* coronary flow reserve; VO_2 *max* maximum peak oxygen uptake; *NR* not reported with improvement in hibernating myocardium and infarct size at 3 months vs. controls.[38] A study of similar subjects with refractory angina found improved perfusion and left ventricular function with endomyocardial injection of autologous bone-marrow-derived mononuclear cells.[19] Similar results were reported in an unblinded series of subjects with endomyocardial injection of autologous bone marrow mononuclear cells.[20] Mocini and colleagues describe a series of subjects with chronic ischemic cardiomyopathy finding improvement in left ventricular ejection fraction by echocardiography with direct injection of autologous bone marrow mononuclear cells into infarct border zones at the time of coronary artery bypass grafting.[39] Endomyocardial delivery of CD133-positive bone marrow derived mononuclear cells at the time of coronary artery bypass grafting proved safe with modest improvements in ventricular performance at 6 months.[21] Katritsis and co-workers infused a combination of so-called progenitor cells and mesenchymal cells from bone marrow in the coronary arteries of subjects suffering with chronic ischemic cardiomyopathy, finding improvements in left ventricular function measured by echocardiography and myocardial perfusion studies.[40]

8.2.3 Limitations and Future Directions

As mentioned, variables affecting the success of trials of progenitor cell therapy include the type and dose of cell, the mode of harvest, as well as the method and timing of delivery. Other possible limitations with the use of autologous bone marrow mononuclear cells rise from potential dysfunction in those subjects' EPCs. Bone marrow derived mononuclear cells from subjects with chronic ischemic cardiomyopathy have less migratory and proliferative capacity in vitro, with reduced angiogenic capacity in vivo.[41] While past studies have employed cellular admixtures obtained from bone marrow niches as reservoirs of cells with regenerative potential, ongoing studies have sought to refine the cell type by selecting for cell populations (such as CD34-positive cells) thought to be enriched for EPCs. Finally, the mechanism of potential benefit of progenitor cell therapy remains to be fully elucidated. Whether cells directly transdifferentiate into mature endothelial cells, fuse with existing vascular or muscle cells, or contribute angiogenic growth factors remains controversial.[10]

8.3 Mesenchymal Stem Cells – Autologous Skeletal Myoblasts

8.3.1 Background

The concept that skeletal myoblasts might serve as a source of myogenic cells was advanced by Taylor and coworkers in demonstrating that transplanted regenerative skeletal myoblasts incorporated into cardiac tissue and improve cardiac performance in a model of cryoinfarct in rabbits.[42,43] Here, autologous skeletal myoblasts appeared to regenerate viable striated tissue among immature cardiocytes and was associated with improvement in ventricular compliance and diastolic performance; systolic changes were less robust.[44-47] Benefits appear to extend to models of nonischemic cardiomyopathy as well.[48] Ex vivo measures of skeletal myoblasts co-cultured with cardiomyocytes demonstrated transdifferentiation with formation of direct cell-to-cell signaling.[49] The histologic and functional changes are similar to transplantation of bone marrow-derived progenitor cells.[50]

Preclinical data from rat models of autologous skeletal myoblast transplant into infarcted myocardium suggests that culture technique measurably impacts ventricular performance after transplant.[51] Cryopreservation of harvested cells in a hamster model was not associated with decreased cell viability.[52] Transplanted myoblasts, however, demonstrate formation of myofibers and markers of cell division that suggest viability after transfer in rat models.[53] In canine models, implanted skeletal myoblasts were associated with induction of cellular hypertrophy of host myocytes at infarct sites.[54] In other measures of ventricular remodeling, improvements in ventricular performance are additive to those observed with angiotensin converting enzyme inhibitor treatment.[25] These effects remain durable for up to a year, with observed mechanisms of colonization of scar fibrosis by skeletal muscle cells thought to augment compliance.[55] Other models of delivery include intracoronary infusion, which in preclinical canine models demonstrated a phenotype of striated muscle in infarct zones.[56]

8.3.2 *Clinical Studies*

From this basis, clinical studies have examined the efficacy of this cell type in improving ventricular function in chronic cardiomyopathy. In contrast to progenitor cell studies, here the cell identity and mode of harvest is better defined. Moreover, issues of timing are less pertinent when exploring cell therapy of chronic heart failure rather than acute myocardial infarction. As in therapeutic progenitor cell studies, investigators have sought to address issues of therapeutic dose and of cell delivery in trials of autologous skeletal myoblast therapy. Clinical benefits have likewise been modest, but observed salutary effects of skeletal myoblast therapy have encouraged continued study (though mindful of safety concerns from tachyarrhythmias).

Early clinical experience has provided anecdotes of successful cell engraftment in several case series using direct intramyocardial injection at the time of cardiac surgery. One subject transplanted with skeletal myoblasts at the time of coronary artery bypass grafting evinced long-term viability (17.5 months) and an interesting phenotypic change in the grafted cells to slow-twitch fibers.[57] Another series of five subjects with end-stage ischemic cardiomyopathy transplanted with autologous skeletal myoblasts at the time of left ventricular assist device (LVAD) implantation found myoblast cell survival (mean 124 days), with alignment of myofibers with host myocardial fibers.[58]

Yield from muscle biopsy averages around 870 million cells arising at 2–3 weeks after quadriceps muscle biopsy. Among these, more than 70% are CD56-positive myoblasts with greater than 90% viability.[59] Long-term survival, however, has been measured at less than 1% from a delivered dose of 300×10^6 cells.[58]

Larger studies have found improvements in left ventricular function associated with direct injection of autologous skeletal myoblasts. Menasche and co-workers implanted 10 subjects with advanced left ventricular dysfunction with 871×10^6 skeletal myoblasts in myocardial scar at the time of coronary artery bypass grafting.[60] In this unblinded phase I study, investigators reported improvements in New York Heart Association functional class, improvement in global left ventricular ejection fraction, and improvement in regional ejection fraction in the majority of scar sites injected. Meanwhile, adverse events included sustained ventricular tachycardia in four subjects. A larger phase 2 study (MAGIC) randomized 97 subjects to autologous skeletal myoblast injection at the time of bypass surgery vs. placebo, finding no significant improvement in regional or global ventricular performance by echocardiography; however, left ventricular volumes improved relative to placebo therapy.[61,62] A similar study of 10 subjects referred for coronary artery bypass grafting reported improvement in left ventricular ejection fraction at 4 months and 1 year after implant.[63] Notably, culture yields (and thus doses) ranged from 400,000 to 50 million cells (65.4% myoblasts). Also, ventricular tachycardia was observed in five subjects.[61] Dib and co-workers observed evidence of increase areas of myocardial viability by positron emission tomography scanning in 30 subjects treated with direct injection of autologous skeletal myoblasts at the time of coronary artery bypass surgery ($n=24$) or LVAD implantation ($n=6$).[64,65]

Alternative models of cell delivery include percutaneous endomyocardial injection with electromechanical mapping of the ventricle to identify areas of scar.[66] A pilot study of five subjects with ischemic cardiomyopathy injected with a 296 ± 199 million skeletal myoblasts found improvement in global left ventricular ejection fraction and in regional wall motion by magnetic resonance imaging. One subject had an implantable cardioverter defibrillator placed for non unsustained ventricular tachycardia. At 1 year, functional class, contractile reserve, and end-systolic volumes improved.[67] A subsequent case-controlled study of transcatheter transplant of autologous skeletal myoblasts demonstrated improvements in left ventricular ejection fraction, walking distance, and functional class.[68] Cellular yield was 210 ± 150 million cells, implanted over an average of 19 ± 10 injection sites. Finally, skeletal myoblasts have been implanted via injection across the coronary sinus in 10 subjects with ischemic cardiomyopathy, resulting in up to 100 million cells injected and improvements in functional class.[69]

8.3.3 Limitations and Future Directions

In addition to modest improvements to date in blinded, placebo-controlled trials of autologous skeletal myoblasts for ischemic cardiomyopathy, the incidence of ventricular tachycardia with this therapy has raised safety concerns. Mechanisms of ventricular tachycardia in this setting has been attributed to several factors. The absence of gap junctions on skeletal muscle cells is thought to promote spiral waves conducive for reentrant ventricular tachycardia.[70,71] Genetic modification of cells to increase expression of gap junction (i.e., connexion 43) has been a suggested method of decreasing arrhythmic risk.[70,72-74] Presently, phase II/III multicenter studies are underway to evaluate the efficacy of skeletal myoblast therapy, with an implanted cardioverter-defibrillator as a prerequisite to study entry.

8.4 Mesenchymal Stem Cells – Allogeneic Mesenchymal Stem Cells

A population of nonhematopoietic pluripotent cells resident in the bone marrow that give rise to cardiomyocytes (as well as osteocytes, chondrocytes, and adipocytes) are known as mesenchymal stem cells (MSC).[75,76] Given its propensity to form cardiomyocytes, this cell line has been the focus of work targeting left ventricular dysfunction. Interestingly, though expressing human leukocyte antigen major histocompatibility complex class I molecules, MSCs do not express co-stimulatory molecules and thus escape recognition by alloreactive T cells.[77] As a result, they enjoy a relatively immune-privileged state, which allows for allogenic cell transplant as a mode of therapy.

Preclinical work has supported the concept that these immune-privileged bone marrow derived allogeneic cells could contribute to regeneration.[78] Rat models of cryoinjured myocardium exhibited improvement in left ventricular function after direct injection of bone marrow-derived MSCs.[78] Allogeneic MSC injected into the scar of a model of infarction in rats demonstrated transient improvement in ventricular performance, while expressing muscle-specific markers and long-term survival at 6 months.[79] Another work in rat models has shown formation of clusters of cells on infarct border zones in addition to microvessel formation.[80] Direct injection of infarcted myocardium in a swine model in the peri-infarct period with allogenic MSC resulted in persistent engraftment at 8 weeks, improvement in infarct size and tissue perfusion by magnetic resonance imaging, and improvements in hemodynamics.[81-84] Short-term survival may improve in this model and cell type with adjunctive transmyocardial laser revascularization.[85] The mechanism of these improvements, in addition to supposed direct differentiation into cardiomyocytes, contribution of paracrine signaling, and fusion with existing cardiac cells, has been supposed to include stimulation of stem cell niches resident in cardiac tissue.[86] A phase I study of allogeneic MSC therapy administered intravenously after acute myocardial infarction has been completed with results forthcoming.

8.5 Resident Cardiac Stem Cells

A population of undifferentiated cells has been isolated from subcultures of postnatal atrial or ventricular human biopsy specimens and from mouse hearts, lending to the support of the existence of stem cells that reside within the adult heart.[87-90]

Growing in self-adherent clusters ex vivo, these "cardiospheres" are clonogenic, are capable of self-renewal, and can differentiate in vitro and in vivo into cells with myocardial and vascular phenotypes.

Preclinical work has uncovered that these cardiospheres in culture express endothelial markers (KDR or flk-1, CD31) and stem cell markers (CD34, c-kit, and Sca-1), as well as other proteins essential for contractile and electrical function.[89] Cardiomyocyte differentiation transcription factors are upregulated in these cell colonies.[91] These cells appear to migrate in vitro in response to cytokines elaborated by circulating EPCs.[92] Functionally, cells derived from cardiospheres demonstrate electrical coupling in vitro.[89] Moreover, when transplanted in a mouse model of myocardial infarction, they appear to yield the phenotype of cardiomyocytes, endothelial cells, and smooth muscle cells with improvements in ventricular performance and reduction in scar size.[87,89] Notably, these effects were not observed with implanted fibroblasts, suggesting a possible role for these cells as an autologous source of regenerative tissue.

Presently, no human studies have been performed using resident cardiac stem cells in treating cardiovascular disease.

8.6 Human Embryonic Stem Cells

Ethical and practical concerns have limited the study of stem cells derived from human embryos in both preclinical and clinical settings.[93] In vitro studies of human embryonic stem cells have demonstrated that cultivated aggregates termed embryoid bodies (EB) exhibit potential for developing into cardiomyocytes.[73] Spontaneously contracting regions of EBs stain for elements of cardiac tissue (cardiac myosin heavy chain, alpha-actinin, desmin, troponin I, and atrial natriuretic peptide) show myofibrillar organization under electron microscopy, and electrical properties. Spontaneous contraction is reported in a percentage of EBs.[73] Another model of cardiomyocyte differentiation from human embryonic stem cells involves co-culture with visceral-endoderm (VE)-like cells from mice, highlighting paracrine interactions between the endoderm in the development of cardiac cells.[94,95] These cells in culture demonstrate coupling via gap junctions and functional calcium ion channels.[94]

Xue and co-workers demonstrated a model of human embryonic stem cell-derived cardiomyocytes that integrate into "recipient" cardiac tissue and demonstrate electrical and mechanical coupling in vitro and in vivo, supporting the possibility of cell-based pacemakers.[96] Other investigators have shown in canine models the durability of human MSCs for cardiac pacing as long as 6 weeks after implantation, when dosed at >700,000 cells per injection.[97] Laflamme and co-workers demonstrated that injection of differentiated cardiac-enriched human embryonic stem cells into athymic rats resulted in proliferation of cardiomyocytes with time, typified by angiogenesis and expression of cardiac markers.[98] Understanding of the molecular mechanisms of differentiation of human embryonic stem cells remains rudimentary, but work by Singh and colleagues suggests that the nuclear protein Chibby facilitates cardiac cell development.[99] Notably, embryonic stem cell-derived cardiomyocytes implanted in infarcted cardiac tissue resulted in attenuation of scar thinning, improvements in left ventricular dilatation, wall motion score index, and left ventricular diastolic dimensions.[100,101]

In addition to ethical and logistical issues related to the use of human embryonic stem cells in studies or therapies of cardiovascular disease, further concerns about aberrant development have been raised when undifferentiated human embryonic stem cells and human EB injected into normal rat myocardium resulted in teratoma formation.[100,101]

8.7 Technical Considerations

As with pharmaceutical therapy for common cardiovascular conditions, the clinical outcome in cell therapy is in part informed by the dose, the potency, and the mechanism of effect.[2,102] Preparation of cell product can reduce the number and robustness of available cells. Several studies have sought to establish a dose–response relationship to therapy, but the adequate dosing of cell-based therapies remains largely empiric. It is estimated that a typical myocardial infarction results in loss of a billion cardiomyocytes, while most trials have isolated and transplanted a dose on the order of 10 to 100 million cells. On delivery, cells are subjected to migration away from the target site, acute oxidative stress, ischemia, and inflammation.[102] Moreover, determining an effect attributable to injected cells must be separated from improvements due to the passive effects of injected biomaterials.[103] Addressing the concern that the extracellular matrix may be perturbed and thus contribute to the pathophysiology of chronic ischemic cardiomyopathy, recent work has employed a cell-seeded collagen matrix implanted in subjects undergoing coronary artery bypass surgery – finding increases in infarct scar and improvements in global ejection fraction.[104] Additionally, application of a fibrin glue has been shown to be beneficial with regard to cell transplant survival, infarct size reduction, and blood flow restoration.[105]

8.8 Conclusions

Advances in stem cell biology over the past decade have fuelled interest in new therapies for acute and chronic cardiovascular diseases. Preclinical work with a variety of cell types has suggested efficacy in improving ischemia and ventricular function, although mechanisms of effect remain to be fully explained. Human studies using certain cell types have shown modest clinical efficacy, while the safety of these experimental therapies supports continuing patient-oriented research.

References

1. Tunstall-Pedoe H, Vanuzzo D, Hobbs M, et al. Estimation of contribution of changes in coronary care to improving survival, event rates, and coronary heart disease mortality across the WHO MONICA project populations. *Lancet*. 2000;355:688–700.
2. Welt FG, Losordo DW. Cell therapy for acute myocardial infarction: curb your enthusiasm? *Circulation*. 2006;113:1272–1274.
3. Ross R. The pathogenesis of atherosclerosis: a perspective for the 1990s. *Nature*. 1993;362: 801–809.
4. Ingram DA, Caplice NM, Yoder MC. Unresolved questions, changing definitions, and novel paradigms for defining endothelial progenitor cells. *Blood*. 2005;106:1525–1531.
5. Asahara T, Murohara T, Sullivan A, et al. Isolation of putative progenitor endothelial cells for angiogenesis. *Science*. 1997;275:964–966.
6. Kalka C, Masuda H, Takahashi T, et al. Transplantation of ex vivo expanded endothelial progenitor cells for therapeutic neovascularization. *Proc Natl Acad Sci U S A*. 2000;97: 3422–3427.
7. Kawamoto A, Gwon HC, Iwaguro H, et al. Therapeutic potential of ex vivo expanded endothelial progenitor cells for myocardial ischemia. *Circulation*. 2001;103:634–637.
8. Kawamoto A, Tkebuchava T, Yamaguchi J, et al. Intramyocardial transplantation of autologous endothelial progenitor cells for therapeutic neovascularization of myocardial ischemia. *Circulation*. 2003;107:461–468.
9. Aicher A, Brenner W, Zuhayra M, et al. Assessment of the tissue distribution of transplanted human endothelial progenitor cells by radioactive labeling. *Circulation*. 2003;107:2134-2139.
10. Urbich C, Dimmeler S. Endothelial progenitor cells: characterization and role in vascular biology. *Circ Res*. 2004;95:343–353.
11. Dohmann HF, Perin EC, Takiya CM, et al. Transendocardial autologous bone marrow mononuclear cell injection in ischemic heart failure: postmortem anatomicopathologic and immunohistochemical findings. *Circulation*. 2005;112:521–526.
12. Murry CE, Soonpaa MH, Reinecke H, et al. Haematopoietic stem cells do not transdifferentiate into cardiac myocytes in myocardial infarcts. *Nature*. 2004;428:664–668.
13. Badorff C, Brandes RP, Popp R, et al. Transdifferentiation of blood-derived human adult endothelial progenitor cells into functionally active cardiomyocytes. *Circulation*. 2003;107: 1024–1032.
14. Kubal C, Sheth K, Nadal-Ginard B, Galinanes M. Bone marrow cells have a potent anti-ischemic effect against myocardial cell death in humans. *J Thorac Cardiovasc Surg*. 2006;132:1112–1118.
15. Ziegelhoeffer T, Fernandez B, Kostin S, et al. Bone marrow-derived cells do not incorporate into the adult growing vasculature. *Circ Res*. 2004;94:230–238.
16. Rehman J, Li J, Orschell CM, March KL. Peripheral blood "endothelial progenitor cells" are derived from monocyte/macrophages and secrete angiogenic growth factors. *Circulation*. 2003;107:1164–1169.
17. Strauer BE, Brehm M, Zeus T, et al. Repair of infarcted myocardium by autologous intracoronary mononuclear bone marrow cell transplantation in humans. *Circulation*. 2002;106:1913–1918.
18. Schachinger V, Assmus B, Britten MB, et al. Transplantation of progenitor cells and regeneration enhancement in acute myocardial infarction: final one-year results of the TOPCARE-AMI trial. *J Am Coll Cardiol*. 2004;44:1690–1699.
19. Beeres SLMA, Bax JJ, Dibbets-Schneider P, et al. Sustained effect of autologous bone marrow mononuclear cell injection in patients with refractory angina pectoris and chronic myocardial ischemia: twelve-month follow-up results. *Am Heart J*. 2006;152:684.e11–684.e16.
20. de la Fuente LM, Stertzer SH, Argentieri J, et al. Transendocardial autologous bone marrow in chronic myocardial infarction using a helical needle catheter: 1-year follow-up in an open-label, nonrandomized, single-center pilot study (the TABMMI study). *Am Heart J*. 2007; 154:79.e1–79.e7.

21. Stamm C, Kleine HD, Choi YH, et al. Intramyocardial delivery of CD133+ bone marrow cells and coronary artery bypass grafting for chronic ischemic heart disease: safety and efficacy studies. *J Thorac Cardiovasc Surg*. 2007;133:717–725.
22. Fernandez-Aviles F, San Roman JA, Garcia-Frade J, et al. Experimental and clinical regenerative capability of human bone marrow cells after myocardial infarction. *Circ Res*. 2004;95:742-748.
23. Chen S-L, Fang W-W, Ye F, et al. Effect on left ventricular function of intracoronary transplantation of autologous bone marrow mesenchymal stem cell in patients with acute myocardial infarction. *Am J Cardiol*. 2004;94:92–95.
24. Bartunek J, Vanderheyden M, Vandekerckhove B, et al. Intracoronary injection of CD133-positive enriched bone marrow progenitor cells promotes cardiac recovery after recent myocardial infarction: feasibility and safety. *Circulation*. 2005;112:I178–I183.
25. Wollert KC, Meyer GP, Lotz J, et al. Intracoronary autologous bone-marrow cell transfer after myocardial infarction: the BOOST randomised controlled clinical trial. *Lancet*. 2004;364:141–148.
26. Meyer GP, Wollert KC, Lotz J, et al. Intracoronary bone marrow cell transfer after myocardial infarction: eighteen months' follow-up data from the randomized, controlled BOOST (BOne marrOw transfer to enhance ST-elevation infarct regeneration) trial. *Circulation*. 2006;113:1287–1294.
27. Schachinger V, Erbs S, Elsasser A, et al. Intracoronary bone marrow-derived progenitor cells in acute myocardial infarction. *N Engl J Med*. 2006;355:1210–1221.
28. Lunde K, Solheim S, Aakhus S, et al. Intracoronary injection of mononuclear bone marrow cells in acute myocardial infarction. *N Engl J Med*. 2006;355:1199–1209.
29. Janssens S, Dubois C, Bogaert J, et al. Autologous bone marrow-derived stem-cell transfer in patients with ST-segment elevation myocardial infarction: double-blind, randomised controlled trial. *Lancet*. 2006;367:113–121.
30. Ge J, Li Y, Qian J, et al. Efficacy of emergent transcatheter transplantation of stem cells for treatment of acute myocardial infarction (TCT-STAMI). *Heart*. 2006;92:1764–1767.
31. Meluzin J, Mayer J, Groch L, et al. Autologous transplantation of mononuclear bone marrow cells in patients with acute myocardial infarction: The effect of the dose of transplanted cells on myocardial function. *Am Heart J*. 2006;152:975.e9–975.e15.
32. Kang H-J, Lee H-Y, Na S-H, et al. Differential effect of intracoronary infusion of mobilized peripheral blood stem cells by granulocyte colony-stimulating factor on left ventricular function and remodeling in patients with acute myocardial infarction versus old myocardial infarction: the MAGIC Cell-3-DES randomized, controlled trial. *Circulation*. 2006;114: I145–I151.
33. Li Z-Q, Zhang M, Jing Y-Z, et al. The clinical study of autologous peripheral blood stem cell transplantation by intracoronary infusion in patients with acute myocardial infarction (AMI). *Int J Cardiol*. 2007;115:52–56.
34. Assmus B, Schachinger V, Teupe C, et al. Transplantation of progenitor cells and regeneration enhancement in acute myocardial infarction (TOPCARE-AMI). *Circulation*. 2002;106: 3009–3017.
35. Zohlnhofer D, Dibra A, Koppara T, et al. Stem cell mobilization by granulocyte colony-stimulating factor for myocardial recovery after acute myocardial infarction: a meta-analysis. *J Am Coll Cardiol*. 2008;51:1429–1437.
36. Strauer BE, Brehm M, Zeus T, et al. Regeneration of human infarcted heart muscle by intracoronary autologous bone marrow cell transplantation in chronic coronary artery disease: the IACT Study. *J Am Coll Cardiol*. 2005;46:1651–1658.
37. Perin EC, Dohmann HFR, Borojevic R, et al. Improved exercise capacity and ischemia 6 and 12 months after transendocardial injection of autologous bone marrow mononuclear cells for ischemic cardiomyopathy. *Circulation*. 2004;110:II213–II218.
38. Erbs S, Linke A, Adams V, et al. Transplantation of blood-derived progenitor cells after recanalization of chronic coronary artery occlusion: first randomized and placebo-controlled study. *Circ Res*. 2005;97:756–762.
39. Mocini D, Staibano M, Mele L, et al. Autologous bone marrow mononuclear cell transplantation in patients undergoing coronary artery bypass grafting. *Am Heart J*. 2006;151:192–197.

40. Katritsis DG, Sotiropoulou PA, Karvouni E, et al. Transcoronary transplantation of autologous mesenchymal stem cells and endothelial progenitors into infarcted human myocardium. *Catheter Cardiovasc Interv*. 2005;65:321–329.
41. Heeschen C, Lehmann R, Honold J, et al. Profoundly reduced neovascularization capacity of bone marrow mononuclear cells derived from patients with chronic ischemic heart disease. *Circulation*. 2004;109:1615–1622.
42. Taylor DA, Atkins BZ, Hungspreugs P, et al. Regenerating functional myocardium: improved performance after skeletal myoblast transplantation. *Nat Med*. 1998;4:929–933.
43. Taylor DA. Cellular cardiomyoplasty with autologous skeletal myoblasts for ischemic heart disease and heart failure. *Curr Control Trials Cardiovasc Med*. 2001;2:208–210.
44. Atkins BZ, Lewis CW, Kraus WE, Hutcheson KA, Glower DD, Taylor DA. Intracardiac transplantation of skeletal myoblasts yields two populations of striated cells in situ. *Ann Thorac Surg*. 1999;67:124–129.
45. Atkins BZ, Hueman MT, Meuchel J, Hutcheson KA, Glower DD, Taylor DA. Cellular cardiomyoplasty improves diastolic properties of injured heart. *J Surg Res*. 1999;85:234–242.
46. Atkins BZ, Hueman MT, Meuchel JM, Cottman MJ, Hutcheson KA, Taylor DA. Myogenic cell transplantation improves in vivo regional performance in infarcted rabbit myocardium. *J Heart Lung Transplant*. 1999;18:1173–1180.
47. Hutcheson KA, Atkins BZ, Hueman MT, Hopkins MB, Glower DD, Taylor DA. Comparison of benefits on myocardial performance of cellular cardiomyoplasty with skeletal myoblasts and fibroblasts. *Cell Transplant*. 2000;9:359–368.
48. Pouly J, Hagege AA, Vilquin JT, et al. Does the functional efficacy of skeletal myoblast transplantation extend to nonischemic cardiomyopathy? *Circulation*. 2004;110:1626–1631.
49. Iijima Y, Nagai T, Mizukami M, et al. Beating is necessary for transdifferentiation of skeletal muscle-derived cells into cardiomyocytes. *Faseb J*. 2003;17:1361–1363.
50. Thompson RB, Emani SM, Davis BH, et al. Comparison of intracardiac cell transplantation: autologous skeletal myoblasts versus bone marrow cells. *Circulation*. 2003;108(1): II264–II271.
51. Chazaud B, Hittinger L, Sonnet C, et al. Endoventricular porcine autologous myoblast transplantation can be successfully achieved with minor mechanical cell damage. *Cardiovasc Res*. 2003;58:444–450.
52. Ohno N, Fedak PW, Weisel RD, Mickle DA, Fujii T, Li RK. Transplantation of cryopreserved muscle cells in dilated cardiomyopathy: effects on left ventricular geometry and function. *J Thorac Cardiovasc Surg*. 2003;126:1537–1548.
53. Gulbins H, Schrepfer S, Uhlig A, et al. Myoblasts survive intracardiac transfer and divide further after transplantation. *Heart Surg Forum*. 2002;5:340–344.
54. Zhong H, Zhu H, Wei H, Zhang Z. Influence of skeletal muscle satellite cells implanted into infarcted myocardium on remnant myocyte volumes. *Chin Med J (Engl)*. 2003;116:1088–1091.
55. Ghostine S, Carrion C, Souza LC, et al. Long-term efficacy of myoblast transplantation on regional structure and function after myocardial infarction. *Circulation*. 2002;106:I131–I136.
56. Zhong H, Zhu H, Zhang Z. Affects of different access routes on autologous satellite cell implantation stimulating myocardial regeneration. *Chin Med J (Engl)*. 2002;115:1521–1524.
57. Hagege AA, Carrion C, Menasche P, et al. Viability and differentiation of autologous skeletal myoblast grafts in ischaemic cardiomyopathy. *Lancet*. 2003;361:491–492.
58. Pagani FD, DerSimonian H, Zawadzka A, et al. Autologous skeletal myoblasts transplanted to ischemia-damaged myocardium in humans. Histological analysis of cell survival and differentiation. *J Am Coll Cardiol*. 2003;41:879–888.
59. Menasche P. Myoblast transplantation: feasibility, safety and efficacy. *Ann Med*. 2002;34: 314–315.
60. Menasche P, Hagege AA, Vilquin JT, et al. Autologous skeletal myoblast transplantation for severe postinfarction left ventricular dysfunction. *J Am Coll Cardiol*. 2003;41:1078–1083.
61. Hagege AA, Marolleau JP, Vilquin JT, et al. Skeletal myoblast transplantation in ischemic heart failure: long-term follow-up of the first phase I cohort of patients. *Circulation*. 2006;114: I108–I113.

62. Menasche P, Alfieri O, Janssens S, et al. The myoblast autologous grafting in ischemic cardiomyopathy (MAGIC) trial: first randomized placebo-controlled study of myoblast transplantation. *Circulation*. 2008;117:1189–1200.
63. Siminiak T, Kalawski R, Fiszer D, et al. Autologous skeletal myoblast transplantation for the treatment of postinfarction myocardial injury: phase I clinical study with 12 months of follow-up. *Am Heart J*. 2004;148:531–537.
64. Dib N, Michler RE, Pagani FD, et al. Safety and feasibility of autologous myoblast transplantation in patients with ischemic cardiomyopathy: four-year follow-up. *Circulation*. 2005;112:1748–1755.
65. Dib N, McCarthy P, Campbell A, et al. Feasibility and safety of autologous myoblast transplantation in patients with ischemic cardiomyopathy. *Cell Transplant*. 2005;14:11–19.
66. Smits PC, van Geuns RJ, Poldermans D, et al. Catheter-based intramyocardial injection of autologous skeletal myoblasts as a primary treatment of ischemic heart failure: clinical experience with six-month follow-up. *J Am Coll Cardiol*. 2003;42:2063–2069.
67. Biagini E, Valgimigli M, Smits PC, et al. Stress and tissue Doppler echocardiographic evidence of effectiveness of myoblast transplantation in patients with ischaemic heart failure. *Eur J Heart Fail*. 2006;8:641–648.
68. Ince H, Petzsch M, Rehders TC, Chatterjee T, Nienaber CA. Transcatheter transplantation of autologous skeletal myoblasts in postinfarction patients with severe left ventricular dysfunction. *J Endovasc Ther*. 2004;11:695–704.
69. Brasselet C, Morichetti MC, Messas E, et al. Skeletal myoblast transplantation through a catheter-based coronary sinus approach: an effective means of improving function of infarcted myocardium. *Eur Heart J*. 2005;26:1551–1556.
70. Abraham MR, Henrikson CA, Tung L, et al. Antiarrhythmic engineering of skeletal myoblasts for cardiac transplantation. *Circ Res*. 2005;97:159–167.
71. Chen M, Fan ZC, Liu XJ, et al. Effects of autologous stem cell transplantation on ventricular electrophysiology in doxorubicin-induced heart failure. *Cell Biol Int*. 2006;30:576–582.
72. Roell W, Lewalter T, Sasse P, et al. Engraftment of connexin 43-expressing cells prevents post-infarct arrhythmia. *Nature*. 2007;450:819–824.
73. Kehat I, Kenyagin-Karsenti D, Snir M, et al. Human embryonic stem cells can differentiate into myocytes with structural and functional properties of cardiomyocytes. *J Clin Invest*. 2001;108:407–414.
74. Tolmachov O, Ma YL, Themis M, et al. Overexpression of connexin 43 using a retroviral vector improves electrical coupling of skeletal myoblasts with cardiac myocytes in vitro. *BMC Cardiovasc Disord*. 2006;6:25.
75. Grinnemo KH, Mansson A, Dellgren G, et al. Xenoreactivity and engraftment of human mesenchymal stem cells transplanted into infarcted rat myocardium. *J Thorac Cardiovasc Surg*. 2004;127:1293–1300.
76. Di Nicola M, Carlo-Stella C, Magni M, et al. Human bone marrow stromal cells suppress T-lymphocyte proliferation induced by cellular or nonspecific mitogenic stimuli. *Blood*. 2002;99:3838–3843.
77. Tse WT, Pendleton JD, Beyer WM, Egalka MC, Guinan EC. Suppression of allogeneic T-cell proliferation by human marrow stromal cells: implications in transplantation. *Transplantation*. 2003;75:389–397.
78. Onishchenko NA, Potapov IV, Bashkina LV, Krasheninnikov ME, Zaidenov VA, Avramov PV. Recovery of contractile function of cryodamaged rat myocardium after transplantation of fetal cardiomyocytes and predifferentiated bone marrow stromal stem cells. *Bull Exp Biol Med*. 2004;138:357–360.
79. Dai W, Hale SL, Martin BJ, et al. Allogeneic mesenchymal stem cell transplantation in postinfarcted rat myocardium: short- and long-term effects. *Circulation*. 2005;112:214–223.
80. Piao H, Youn TJ, Kwon JS, et al. Effects of bone marrow derived mesenchymal stem cells transplantation in acutely infarcting myocardium. *Eur J Heart Fail*. 2005;7:730–738.
81. Amado LC, Saliaris AP, Schuleri KH, et al. Cardiac repair with intramyocardial injection of allogeneic mesenchymal stem cells after myocardial infarction. *Proc Natl Acad Sci U S A*. 2005;102:11474–11479.

82. Amado LC, Schuleri KH, Saliaris AP, et al. Multimodality noninvasive imaging demonstrates in vivo cardiac regeneration after mesenchymal stem cell therapy. *J Am Coll Cardiol.* 2006;48:2116–2124.
83. Poh KK, Sperry E, Young RG, Freyman T, Barringhaus KG, Thompson CA. Repeated direct endomyocardial transplantation of allogeneic mesenchymal stem cells: safety of a high dose, "off-the-shelf," cellular cardiomyoplasty strategy. *Int J Cardiol.* 2007;117:360–364.
84. Schuleri KH, Amado LC, Boyle AJ, et al. Early improvement in cardiac tissue perfusion due to mesenchymal stem cells. *Am J Physiol Heart Circ Physiol.* 2008;294(5):H2002–H2011.
85. Patel AN, Spadaccio C, Kuzman M, et al. Improved cell survival in infarcted myocardium using a novel combination transmyocardial laser and cell delivery system. *Cell Transplant.* 2007;16:899–905.
86. Mazhari R, Hare JM. Mechanisms of action of mesenchymal stem cells in cardiac repair: potential influences on the cardiac stem cell niche. *Nat Clin Pract Cardiovasc Med.* 2007;4(1):S21–S26.
87. Messina E, De Angelis L, Frati G, et al. Isolation and expansion of adult cardiac stem cells from human and murine heart. *Circ Res.* 2004;95:911–921.
88. Beltrami AP, Barlucchi L, Torella D, et al. Adult cardiac stem cells are multipotent and support myocardial regeneration. *Cell.* 2003;114:763–776.
89. Smith RR, Barile L, Cho HC, et al. Regenerative potential of cardiosphere-derived cells expanded from percutaneous endomyocardial biopsy specimens. *Circulation.* 2007;115:896–908.
90. Oh H, Bradfute SB, Gallardo TD, et al. Cardiac progenitor cells from adult myocardium: homing, differentiation, and fusion after infarction. *Proc Natl Acad Sci U S A.* 2003;100: 12313–12318.
91. Di Meglio F, Nurzynska D, Castaldo C, et al. In vitro cultured progenitors and precursors of cardiac cell lineages from human normal and post-ischemic hearts. *Eur J Histochem.* 2007;51:275–282.
92. Urbich C, Aicher A, Heeschen C, et al. Soluble factors released by endothelial progenitor cells promote migration of endothelial cells and cardiac resident progenitor cells. *J Mol Cell Cardiol.* 2005;39:733–742.
93. Condorelli G, Catalucci D. Human stem cells for heart failure treatment ready for prime time? *J Am Coll Cardiol.* 2007;50:1894–1895.
94. Mummery C, Ward-van Oostwaard D, Doevendans P, et al. Differentiation of human embryonic stem cells to cardiomyocytes: role of coculture with visceral endoderm-like cells. *Circulation.* 2003;107:2733–2740.
95. Nakamura T, Schneider MD. The way to a human's heart is through the stomach: visceral endoderm-like cells drive human embryonic stem cells to a cardiac fate. *Circulation.* 2003;107:2638–2639.
96. Xue T, Cho HC, Akar FG, et al. Functional integration of electrically active cardiac derivatives from genetically engineered human embryonic stem cells with quiescent recipient ventricular cardiomyocytes: insights into the development of cell-based pacemakers. *Circulation.* 2005;111:11–20.
97. Plotnikov AN, Shlapakova I, Szabolcs MJ, et al. Xenografted adult human mesenchymal stem cells provide a platform for sustained biological pacemaker function in canine heart. *Circulation.* 2007;116:706–713.
98. Laflamme MA, Gold J, Xu C, et al. Formation of human myocardium in the rat heart from human embryonic stem cells. *Am J Pathol.* 2005;167:663–671.
99. Singh AM, Li FQ, Hamazaki T, Kasahara H, Takemaru K, Terada N. Chibby, an antagonist of the Wnt/beta-catenin pathway, facilitates cardiomyocyte differentiation of murine embryonic stem cells. *Circulation.* 2007;115:617–626.
100. Leor J, Gerecht S, Cohen S, et al. Human embryonic stem cell transplantation to repair the infarcted myocardium. *Heart.* 2007;93:1278–1284.
101. Caspi O, Huber I, Kehat I, et al. Transplantation of human embryonic stem cell-derived cardiomyocytes improves myocardial performance in infarcted rat hearts. *J Am Coll Cardiol.* 2007;50:1884–1893.

102. Tang GH, Fazel S, Weisel RD, Verma S, Li RK. Optimizing cardiac cell therapy: from processing to delivery. *J Thorac Cardiovasc Surg*. 2005;130:966–968.
103. Gaudette GR, Cohen IS. Cardiac regeneration: materials can improve the passive properties of myocardium, but cell therapy must do more. *Circulation*. 2006;114:2575–2577.
104. Chachques JC, Trainini JC, Lago N, Cortes-Morichetti M, Schussler O, Carpentier A. Myocardial assistance by grafting a new bioartificial upgraded myocardium (MAGNUM trial): clinical feasibility study. *Ann Thorac Surg*. 2008;85:901–908.
105. Christman KL, Vardanian AJ, Fang Q, Sievers RE, Fok HH, Lee RJ. Injectable fibrin scaffold improves cell transplant survival, reduces infarct expansion, and induces neovasculature formation in ischemic myocardium. *J Am Coll Cardiol*. 2004;44:654–660.

Chapter 9
Tissue-Engineered Vascular Substitutes: New Models Toward Successful Small Diameter Grafts

Pedro D'Orléans-Juste, Dan Lacroix, Lucie Germain, and François A. Auger

Abbreviations

EC	Endothelial cells
EPC	Endothelial progenitor cells
eNOS	Endothelial nitric oxide synthase
ePTFE	Expanded poly(tetrafluoroethylene)
HIF-1α	Hypoxia inducible factor-1 alpha
htPA	Human tissue-type plasminogen activator
NO	Nitric oxide
PU	Polyurethane
PVR	Peripheral vascular resistance
SMC	Smooth muscle cells
TEVS	Tissue-engineered vascular substitutes
VS	Vascular substitutes

9.1 Introduction

The first successful Vinyon-N-synthetic vascular grafts was developed by Voorhees in 1952. New materials quickly followed, and grafts started being made with expanded poly(tetrafluoroethylene) (ePTFE) and Dacron®. Those materials provided successful large diameter grafts but proved to be a clinical failure when applied to small diameter grafts.[1] Since then, research has been spurred to investigate new materials, preimplantation modifications involving the use of therapeutic agents or living cells, new approaches such as tissue engineering and nanotechnology or any combination of the above to achieve a clinically reliable small-diameter vascular

P. D'Orléans-Juste (✉), D. Lacroix, L. Germain, and F.A. Auger
Department of Pharmacology, Université de Sherbrooke, 3001,12th Avenue North, Sherbrooke, QC J1H 5N4, Canada
e-mail: pedro.dorleans-juste@usherbrooke.ca

D. Abraham et al. (eds.), *Advances in Vascular Medicine*,
DOI 10.1007/978-1-84882-637-3_9,

substitute (VS). Although progress has been made, the ultimate goal has yet to be achieved. As experimental tissue-engineered vascular substitutes (TEVS) have evolved to become physiologically closer to native blood vessels, they have only started to be used as in vitro models for research in vascular physiology since approximately 2001. The advent of biologically relevant TEVS indeed, is fairly recent, necessitates specific expertise, and currently still involves lengthy procedures. For those reasons, most of the published research is still based on cell culture and animal experimentation. Nonetheless, the very few available reports on this topic have demonstrated that biological TEVS can provide interesting models to study normal and pathological vascular physiology and will probably gain in visibility and popularity in the coming years. It should be noted that to simplify the text, the expression TEVS used by the authors of this chapter covers a broad range of constructs with very different compositions and properties.

9.2 Vascular Substitutes Overview

9.2.1 Large vs. Small Diameter Vascular Prostheses

The first VS developed for grafts were based on synthetic materials made of a synthetic fluoropolymer, ePTFE (also known as Teflon® or Gore-Tex®) and a thermoplastic polymer resin of the polyester family called polyethylene terephthalate commonly abbreviated to PET or PETE and well known under the name of Dacron®. The microstructure of ePTFE is a composite of irregular-shaped solid membranes called nodes that are linked by a scaffold of fine fibrils stretched between them. The porosity of the material is defined by the internodal distance (IND) and, in vascular prostheses applications, it became customary to refer to ePTFE grafts with IND equal to or greater than 45 μm as "high-porosity ePTFE grafts" and those with IND equal or lower to 30 μM as "low-porosity ePTFE grafts." Grafts made of ePTFE can also be made impenetrable with the addition of an external wrap. In contrast to the irregular structure of ePTFE, Dacron® basically comes as fibers of polyester and the major difference in VS made from this material is based on whether woven Dacron® or knitted Dacron® was used to fabricate the graft. Another category of materials that came to be used later in vascular grafts were those made of polyurethane (PU). Various methods of production can be used to generate VS with PU, leading to distinctions such as fibrillar or foamy PU grafts.

It is notable that such materials have provided very acceptable clinical results for almost half a century with large diameter vascular prostheses used in high blood flow and low resistance conditions, without the need for further modifications.[1] However, the use of similar materials for small diameter VS results in a high rate of failure due to surface thrombogenicity, poor healing, lack of compliance and anastomotic intimal hyperplasia that lead to occlusion of the graft.[1] Because of this important functional difference between large and small diameter VS, research mostly focused on this last application while commercial grafts for the first have continued to be manufactured with the same basic materials and methods for the last decades.[1] For more details regarding vascular prosthetic materials currently used in commercially available prostheses for vascular surgery (ePTFE, Dacron®, PU), the readers are invited to consult the recent review by Kapadia et al[2]

9.2.2 Small Diameter Vascular Prostheses

Since the first synthetic VS started to be used successfully for large diameter grafts and then met with clinical failure when applied to small diameter ones, a tremendous amount of research has been devoted to small diameter VS. In fact, so much has been published that it would be a daunting task to present an exhaustive list of all relevant publications. However, the authors felt that there was a need to provide the readers with some classification and for a systematic presentation of the relevant categories of VS that have been developed over the years. Table 9.1 strives to provide such an overview. Although it does not follow, strictly speaking, a chronological order in presenting the various categories and examples of VS, there is a general top-to-bottom flow of going from the older to the more recent developments. The examples were classified in stages of development going from in vitro to animal experimentation and then to clinical data, whenever applicable.

9.2.3 Genetic Engineering and TEVS

Enhancement of the antithrombotic and antihyperplasic properties of endothelial cells (EC) by genetic modification was also investigated.[3-5] Human EC overexpressing endothelial nitric oxide synthase (eNOS), for example, reduce human platelet aggregation and bovine aorta smooth muscle cells (SMC) proliferation.[3] Furthermore, adenoviral vectors infected human EC enhances the secretion of human tissue-type plasminogen activator (htPA) in those cells seeded in small diameter ePTFE grafts for in vitro studies.[4] The same authors demonstrated that EC secreted higher levels of htPA after transduction and did not affect cell retention on the luminal surface of the graft even under perfusion. These results indicate that with proper screening of the genetically modified EC to provide safeguards against aberrant phenotypes, the success rate of endothelialized grafts could eventually be improved by such modifications of the cells prior to seeding. However, although the approach shows promise, work remains to be done to develop viral vectors that show sufficient safety and efficacy to be clinically acceptable in that context. Also, the fact that optimal application of the technique would involve autologous EC means that genetic modification of those cells would impose a further burden of time to the approach.

Table 9.1 Various materials and modifications investigated for small diameter vascular grafts

Composition		Modifications		Developmental stages	Selected references
Bio materials		ePTFE[a]	None	Animal[b]	63–69
		–	–	Clinical	70
		–	Treated[c]	In vitro	71
		–	–	Animal[b]	46,72–80
		–	–	Clinical	81-84
		–	Seeded[d]	In vitro	85
	Nonresorbable	–	–	Animal[b]	86–93
		–	–	Clinical	21,94–97
		Dacron[e]	None	Animal[b]	98–101
		–	–	Clinical	102,103
		–	Treated[f]	In vitro	104
		–	–	Canine	105
		–	–	Clinical	82,83,106
		–	Seeded[d]	Animal[b]	20,23,24,88, 107,108
		PU[g,h]	None	Animal[b]	109-112
		–	Treated[i]	In vitro	113,114
		–	–	Animal[b]	115–117
		–	–	Clinical	118
		–	Seeded[j]	In vitro	85,119
		–	–	Animal[b]	28,119–121
		Silk fibroin	None	In vitro	122,123
		–	–	Animal (sheep)	124
		PGA, PLA, etc.	Seeded[j]	In vitro	32,125,126
		–	–	Animal[b]	29,127-131
		–	–	Clinical	132,133
	Biodegradable	Chitosan	None	In vitro	134
		Hyaluronan	Seeded with SMC	In vitro	135
		Collagen	Seeded cells only[k]	In vitro	136
		–	Mechanical	In vitro	137,138
		–	Reinforcement[l]	In vitro	13,139–143
		–		Animal (dogs)	144,145
	Gels	Fibrin	Seeded cells only[k]	In vitro	137
		–	Fibrinolysis inhibitors[m]	In vitro	146
		–	–	Animal (sheep)	147
Biological membranes		Peritoneal seeding	–	Rodents and dogs	148,149
		Decellularized scaffolds[n]	–	In vitro	150–152
	–		–	Animal[o]	153–159
		Self-assembly[p]	–	In vitro	36,37,45
		–	–	Animal[q]	160
		–	–	Clinical	161

(continued)

Table 9.1 (continued)

Composition	Modifications	Developmental stages	Selected references
New assembly technologies	–	In vitro	6,7,162

[a]Without distinction between sealed, low-porosity, or high-porosity expanded poly(tetrafluoroethylene) (ePTFE)
[b]Animal models vary. canine, porcine, ovine, caprine, simian, rats, and rabbits
[c]ePTFE treated with additional factors to improve endothelialization and/or reduce thrombogenicity: fibronectin, plasmin-treated fibrin, heparin, VEGF, anti-CD34, rapamycin, PEG-hirudin/iloprost, forskolin. Exception: one group used untreated prostheses, but animals were injected or not with G-CSF post-graft[76]
[d]Many experiments used autologous endothelial cells (EC), but some resorted to immortalized murine fibroblasts or human umbilical vein endothelial cells (HUVEC)[85] bone marrow CD34+ cells induced into EC prior to use[92] or undifferentiated mesenchymal stem cells (MSC)[28]
[e]Without distinction between Dacron membranes generated by knitting[104] or weaving[163]
[f]Treaments cited were ionic polyurethane (PU), tissue factor pathway inhibitor, and heparin+collagen
[g]Without distinction between fibrillar PU generated by weaving, knitting, electrostatic spinning[164] or winding,[165] or foamy PU[109,166]
[h]For more details, a recent review of various modifications and results has been published[166]
[i]Treatments included dipyridamole, NO-donor substances, collagen with hyaluronan, heparin, heparin and bFGF, heparin with VEGF and sirolimus, and PEG+diazeniumdiolate, and YIGDSR peptide
[j]Vascular smooth muscle cells (SMC), EC, human umbilical vein endothelial cells (HUVEC), bone marrow MSC differentiated into vascular cells. Other treatments such as fibronectin coatings have been used to improve EC attachment to the luminal surface. With nonresorbable materials, seeding usually aims at endothelialization to reduce thrombogenicity, but biodegradable porous or fibrillar materials can also be seeded with various cellular types to promote remodeling
[k]All vascular substitutes (VS) based on collagen or fibrin gels incorporate vascular cells imbedded during gel formation, regardless of any further treatment or reinforcement that may be applied. Cells embedded in the gels are usually SMC to generate a pseudo-media, but Weinberg and Bell[13] used an additional gel seeded with fibroblasts to add a pseudo-adventitia. Some gel-based VS experiments added a step of luminal seeding with either EC or EC progenitor cells
[l]Reinforcement approaches used either addition of a sheath (e.g., PU, Dacron, nylon, PTFE), incorporation of monomers such as elastin or silk fibroin, or treatment such as glycation
[m]Fibrinolysis inhibitors used were aprotinin[147] or epsilon-aminocaproic acid (ACA)[146]
[n]Decellularized material originated from bovine or porcine small intestine submucosa (SIS), from canine or porcine vascular explants or from bovine ureter
[o]Animal models were canine, porcine, or rabbits
[p]Also called "sheet-based tissue engineering"
[q]Animal models were dogs, nude rats, and nonhuman primates

9.2.4 *New Assembly Technologies*

New technologies are emerging that can be used to investigate ways to achieve a better control over the manipulation of various material molecules and cells in the production of VS. Examples of those new technologies have been published that made use of magnetic micro-beads and nanoparticles. One approach has been to use magnetic forces to position thrombin-coated magnetic micro-beads in a defined spatial array to guide the self-assembly of fibrin fibrils through catalytic cleavage of soluble fibrinogen substrate.[6] This led to nanoscale definition of the scaffold's architecture along a geodesic pattern. Since it is becoming clear that cellular function is regulated in part by subtle environmental clues such as cell–cell and cell–scaffold contacts, this work is an interesting proof-of-concept of a new way to manipulate the environmental cues that will be provided to cells in tissue-engineered scaffolds. Another approach that has been described to impose an initial seeding pattern to cells in TEVS assembly was to label SMC with CD44 Dynabeads and subject them to radial magnetic force to drive the cells onto the luminal surface of a tubular scaffold and finally to immobilize them on the substrate's surface.[7,8] Another technology that has been drawing attention covers a number of different approaches that could be categorized as "bioprinting." Specific techniques vary from actually moving cells in "inkjet printing" fashion to directing where cells will adhere and grow by soft lithographic patterning of proteins and cells.[9-12] Those approaches are still experimental but provide indications of future tools that could be used to produce the next generation of TEVS.

9.3 Use of Tissue-Engineered Substitutes in Vascular Research

As described in Sect. 9.2, the first VS developed for surgery were focused on the use of inert biocompatible materials such as ePTFE and Dacron®. It is thus understandable that most published experiments related to VS were aimed at small diameter products and that one of the major goals remained the improvement of their antithrombotic properties. Those experiments often involved trials in animal models and yielded interesting data on the differences in vascular cell physiology between different species. One example of those data is that human EC show a very limited capacity for de novo endothelialization of surgically implanted VS. In patients who receive vascular prostheses, it is a well-documented fact that only the perianastomotic region will display transanastomotic endothelialization, leaving the internal section bare of EC.[1] It can thus be seen that experiments aimed solely at improving the existing VS indirectly yielded cellular and physiological information on human vascular cell types. However, very few publications can be found that have specifically used TEVS as models to investigate vascular pathogenesis.

9.3.1 Tissue Engineered Vascular Substitutes

Until very recently, most scientific research specifically aimed at normal and pathological vascular physiology has been performed either with more conventional tools and models without resorting to the use of TEVS. Part of the reason for the relative scarcity of such publications based on the use of VS could be that in a notable number of cases involving the use of tissue-engineered constructs, they were not used as vascular models per se, but as the main object of improved clinical applications in vascular surgery. They were thus used more as an end in itself than as the means for other research objectives. Another possible reason for the small number of such publications is that scientific efforts toward better understanding of vascular pathogenesis or normal vascular physiology naturally adopt well-known and proven models such as conventional cell culture and animal models. Chronologically speaking, the simpler VS that have been known since the 1970s might not be seen as valid models for the complexity of human blood vessels, while the more sophisticated TEVS that are closer to physiological reality have only started to be the subject of scientific interests in the last decade. The earliest report of a method to generate TEVS that involved the use of living cells throughout the construct was probably the one by Weinberg and Bell in 1986.[13] Since then, many approaches have been reported (see Table 9.1). Although they might be seen to be more experimentally demanding to produce than simpler models, TEVS that offer models closer to physiological reality have many advantages that can be considered to make it worth their investment in time and resources. Examples of those advantages are the availability and uniformity of TEVS, compared with native human blood vessels, and the capacity to manipulate experimental features such as the use of specific vascular layers (Fig. 9.1). Moreover, such TEVS can be made of human cells that are kept in a more natural three-dimensional environment with pulse flow conditions that are much closer to the physiological reality of human blood vessels. The endothelium of TEVS generated with the self-assembly technique can be subjected to pulse flow in vitro for many days and EC display a morphology similar to native human vessels, including alignment along the axis of the flow and expression of Vascular Endothelial Cadherin (Fig. 9.2). The merits of using experimental models that are relevant to human vascular physiology have been elaborately discussed in a recent review by Dr. Peter Zilla.[1]

9.3.2 *In Vitro Cell Culture Approaches*

In vitro cell culture models have been immensely useful to elucidate many of the cellular and molecular mysteries of the vascular cells and tissues. However, vascular cells cultured as monolayers in static conditions on plastic also have limitations. For example, it has been clearly demonstrated that EC behave in a completely different fashion in the absence of shear stress, which is a normal physiological condition in native blood vessels.[14] Close contact with SMC of the media is another normal physiological condition that is absent from monolayer cultures of EC.[14] These observations explain why a growing number of scientific teams have used experimental models in which EC are subjected to shear stress. Some of those systems were based on simple laminar flow chambers made with glass plates or other planar materials.[15-18] Other experimental systems involved more elaborate conditions such as one that used cocultures of EC and SMC separated by collagen

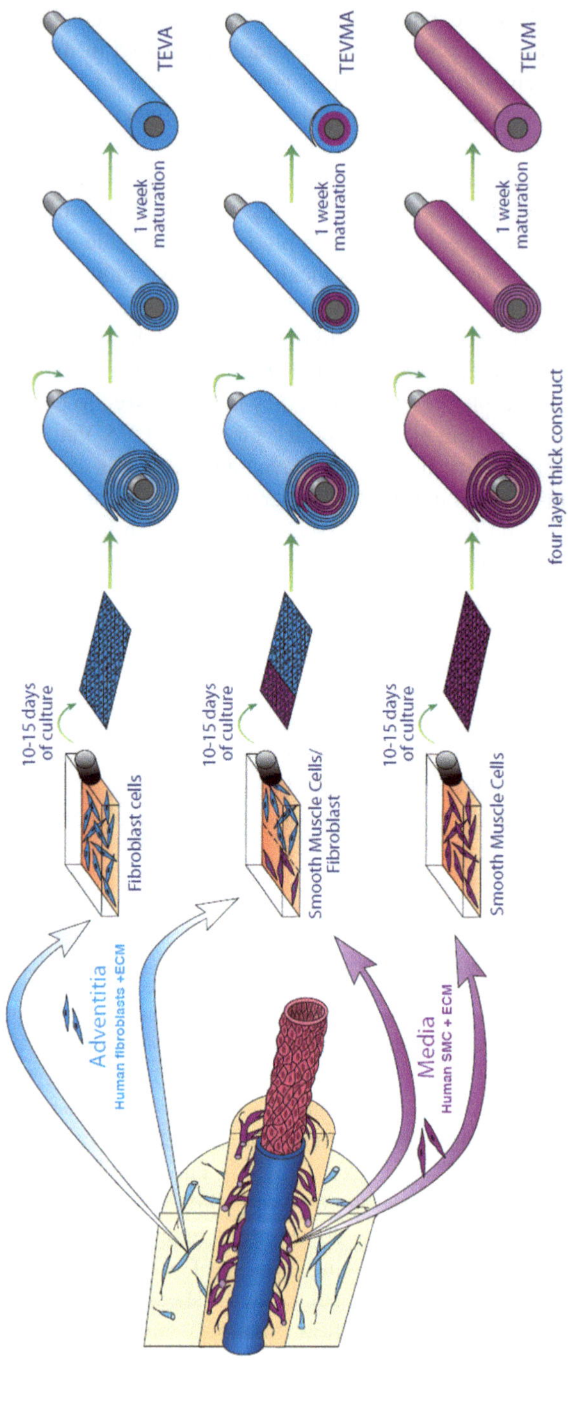

Fig. 9.1 Schematic representation of the experimental design in which human cells are used to generate various tissue-engineered vascular substitutes (TEVS) that incorporate only an adventitia (TEVA), a media (TEVM), or both (TEVMA). From ref.[36]

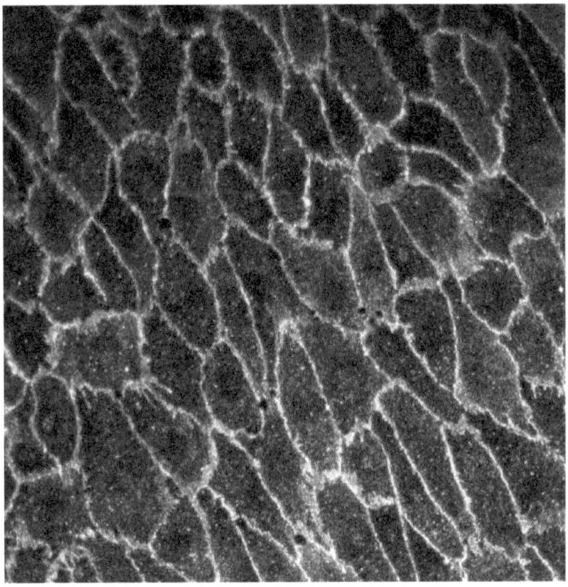

Fig. 9.2 Confocal immunofluorescence picture of the endothelium in a TEVS generated by the LOEX self-assembly technique, taken after 5 days under pulse flow conditions. Primary antibody was directed against VE-cadherin. Picture taken at 60× magnification

gels and supported by porous membranes.[19] There are thus alternative models that are less complex to produce than TEVS that are quite valid to probe the effects of certain physiological conditions on vascular cells. However, especially when it comes to preclinical studies, there are questions that are better addressed with models that are closer to physiological reality than rigid laminar flow chambers. In the following sections, we will strive to provide examples that will demonstrate the use of VS in vascular research.

9.3.3 Use of TEVS in Vascular Physiology

The effort spent on improving antithrombotic properties of small diameter VS naturally led to numerous publications on endothelialization.[1,20-25] These publications clearly demonstrated the superiority of endothelialized grafts, prior to implantation, to achieve long-term patency in patients. They also spurred the search for a reliable source of cells for seeding, which led to experiments on the use of endothelial progenitor cells (EPC) from peripheral blood[26,27] or from bone marrow mesenchymal stem cells (MSC).[28,29] Although experiments of seeding grafts with EPC yielded in vitro results that seemed promising, questions have been raised regarding clinically relevant aspects such as graft patency.[30] Another aspect of endothelialized grafts that has drawn attention is the tendency of EC to detach from the luminal surface when exposed to physiological conditions of shear stress or perturbed laminar flow. Consequently, with the general goal of improving EC attachment and functions in VS, more fundamental analyzes were also performed. One study investigated the effect of shear stress on EC–SMC interaction and showed that coculture with SMC under static condition induced EC gene expression of two growth-related factors that was abolished under shear stress conditions.[14] Other groups examined the interaction between EC basal surface and components of the extracellular matrix.[31] It was demonstrated, for example, that EC used different integrin complexes to attach to quiescent SMC, fibronectin adsorbed to plastic or polyacrylamide gels with similar elastic moduli as SMC and that changes in the type of adhesion resulted in modified focal adhesion formation and the rate of cell spreading.[31] In the examples provided above, research was not always done with VS, but sometimes with simpler models. It can be argued that experiments involving TEVS models will gain popularity once they become better known.

9.3.4 Use of TEVS in Vascular Pharmacology

Pharmacological reactivity in TEVS has been reported as early as 1999, in a publication where contraction was observed in response to prostaglandin $F_{2\alpha}$.[32] Since then, there have been a number of publications in which TEVS were used in vitro to investigate vascular tone and vascular pharmacological responses.[33-39] Two publications examined the capacity of TEVS generated with human SMC to respond to vasoconstrictor agonists such as histamine, bradykinin, ATP, and UTP.[38,39] These publications provided a proof-of-concept that such models can be of interest in pharmacological research to replace native human blood vessels, which are difficult to obtain and provide heterogeneous samples of varying quality. The same models that were also used to investigate the response of TEVS to endothelin demonstrated the presence of functional endothelin receptors and yielded data supporting the important concept that the adventitia plays a role in the control of vascular ton.[33,36,37,40] Similar TEVS were also used to investigate the mechanisms and the signaling pathways through which polyphenols (provinols) modulate vascular contraction in human blood vessels.[34,35]

9.3.5 Vascular Pathology

We already mentioned in Sect. 9.3.2 that the effect of shear stress on the migration of SMC in a coculture system where SMC and EC, separated by a collagen layer, was explored with cells seeded on a membrane filter.[19] This later study was aimed at a better understanding of the effect of shear stress conditions on SMC migration in hyperplasia. Another study used a parallel-plate flow system to investigate whether vascular calcification affects plaque stability under various conditions of shear stress.[41] Those two studies exemplify the importance attributed to dynamic flow conditions to investigate vascular phenomena under more physiological conditions.

Some experiments need in vitro models that are closer to native vascular physiology. To that end, some studies have been performed with tissue explants in ex vivo experiments. In one example of such studies, porcine carotid arteries were cultured for 7 days in a simplified ex vivo artery organ culture system with pulse flow under hypertensive or normotensive pressure conditions.[42] According to the authors, this system could maintain the arteries viable for 7 days. The aim of this study was to evaluate if early stage changes in hypertensive arteries could have an impact on long-term adaptation. Another study demonstrated that TEVS based on collagen gels can be kept for 6 days in a bioreactor.[43] It is not clear if that represents a limit to the viability of the TEVS or merely an experimental choice made by the authors since another publication mentions TEVS that were maintained 7 weeks in a system that did not keep luminal flow but, rather, applied pulse radial distension.[44] It would be interesting to see if biological TEVS made with human cells could be kept viable and responsive for longer periods in bioreactors that maintain pulse flow under normotensive and hypertensive conditions. Incubation periods of at least a few weeks could be more suitable for the study of certain pathologies, such as atherosclerosis. The long-term mechanical stability of TEVS made by the self-assembly method is discussed in the first publication that presented this model, in which the authors report that some tissue-engineered adventitias were matured for 24 weeks without significant change in strength.[45] This same publication also demonstrated that such TEVS could maintain patency in vivo at least 7 days in a xenografts model (Fig. 9.3).

Considering the importance of neointimal hyperplasia in the clinical use of small diameter vascular grafts, it is not surprising to find that experiments targeted this adaptive mechanism. The aforementioned study on EC genetic modification to reduce thrombosis and hyperplasia is an example of such work.[3] More conventional approaches were also used, such as experiments to investigate the use of time-released rapamycin in a 6 mm ePTFE graft.[46] In that study, experimental grafts were performed in a porcine model and the results showed a reduction in neointimal hyperplasia.[46] However, as recently revisited, the problem of small diameter grafts has not yet been quite satisfactorily solved either by various therapeutic agents,[47] induced vascular atrophy,[48] gene therapy,[49] or modifications of prosthetic conduits.[2] One publication was based on work done with TEVS to investigate the underlying causes of stenosis.[50] The results indicate that loss of mechanical loading in tissue-engineered sheets of SMC can induce a dedifferentiation of the cells toward a proliferative phenotype, which raises the question of whether a similar phenomenon could contribute to stenosis of vascular grafts with native vessels.

The term TEVS is often associated with large or small diameter vascular prostheses of macroscopic dimensions. However, the generation of microvascular networks in tissue engineering applications can also be considered a form of TEVS. As such, it is relevant to mention that the concept of artificially generating

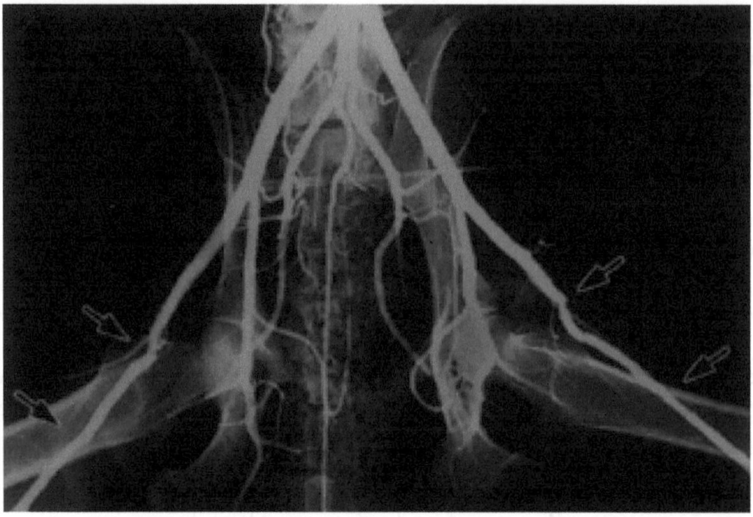

Fig. 9.3 Angiogram of the lower limbs of a canine model 7 days after implantation of a TEVS made entirely of human cells for all three vascular layers: intima, media, and adventitia. From ref. [45]

microvascular networks has evoked much interest in angiogenic mechanisms. One study described in vivo experiments with TEVS to explore the role of nitric oxide (NO) on angiogenesis through mural cells recruitment and vascular morphogenesis.[51] These in vivo experiments were performed in addition to in vitro experiments and provided additional data showing that EC-derived NO played an important role in the recruitment of mural cells and subsequent stabilization of angiogenic vessels.[51] Other publications have demonstrated the interest of the tissue engineering community for microvascular networks.[52-55] Such technology could also have applications in the treatment of ischemic tissue.[56-58] One study presents a biomaterials approach that provides an interesting structure based on polyhedral oligomeric silsesquioxane-PU nanocomposites to generate microvascular network.[59]

9.3.6 TEVS and Aging Processes

While aging is not pathological per se, it does have an impact on cellular characteristics such as viability and proliferative capacity. Considering that vascular cells harvested for the production of autologous TEVS will come from patients of different age, including the elderly, a team of scientists chose to use porcine cells in a TEVS model to investigate the effects of vascular cell age on extracellular matrix deposition, cellular mitosis, and protein synthesis.[44] Their results showed differences in the ECM of TEVS produced with SMC from infant rather than adult animals and demonstrated that such a model could be used to investigate various aspects of aging and vascular remodeling. This indicates that TEVS made with human cells from donors of different age could be useful tools to study fundamental cellular mechanisms in the effect of age on different vascular pathologies.

9.4 Future Perspectives

9.4.1 Bioengineering

The use of TEVS in the research on vascular physiology will become a question of selecting the appropriate tool for the aim of a specific research project. The use of human cells to produce tissue-engineered constructs with specific cell types being present or absent is one way to create models that are adapted for a given experimental objective. Bioreactors are another tool that will become more sophisticated and better adapted to vascular research. New bioreactors will be developed to allow fine control over pulse flow conditions to reliably mimic normal or pathological conditions. They will also be designed to allow TEVS made of living cells to be kept viable and physiologically normal for periods extending as long as several weeks to provide better research tools into conditions that need more than a few days to develop detectable changes. Finally, there are the tools of genetic engineering that allow controlled upregulation or downregulation of specific genes, such as viral vectors to provide efficient stable integration of inducible constructs in a specific cell type prior to their use in a TEVS and the use of RNA inhibition tools such as siRNA, shRNA, and morpholinos. Taken together, it will become possible to generate human vascular models in which many parameters can be experimentally controlled while keeping a cellular and tissular organization close to the physiological reality.

9.4.2 New Clinical Indications

Another yet to be completely perfected application is the use of bio-engineered prosthesis in hemodialysis. Indeed, in end-stage renal failure patients, arterio-venous shunting is the preferred approach for hemodialysis, albeit vascular prostheses are also currently used. In a significant percentage of those patients, however, currently used prostheses promote a marked increased of peripheral vascular resistance (PVR).[60,61] The increased PVR is positively correlated with increased local vascular laminar flow disturbance, type of biomaterial used, as well as with an imbalanced production of vascular endothelial derived factors.[62] For example, a recent report even suggests that ePTFE-containing vascular prostheses trigger the systemic release of inflammatory mediators such as Hypoxia Inducible Factor-1 alpha (HIF-1α)[61] in a more marked fashion than in patients subjected to arterio-venous shunting. Developing more compliant and biocompatible vascular prostheses may therefore significantly reduce the complications associated with their use in dialysis. Finally, rather than implanting large vessels, the advent of a sieve-like network of small diameter vessels may ultimately lead to more compatible and compliant types of prostheses with additional extra-corporeal ionic exchange capacities for dialysis patients.

9.5 Conclusion

Recent progress in the research for a clinically successful tissue-engineered small diameter vascular graft gives us reason to be optimistic about reaching the stage of conclusive clinical trials within the next decade. The real challenge will be to develop cost-effective solutions that will allow such grafts to become readily available to all categories of patients. Meanwhile, various models of TEVS could provide physiologically relevant alternatives to human blood vessels and become increasingly used in fundamental and clinical vascular research.

Acknowledgments We thank Dr. Caroline Basoni and Dr. Valérie Cattan for their help with the literature review, and Miss Marie-Christine Fiola for Fig. 9.2.

References

1. Zilla P, Bezuidenhout D, Human P. Prosthetic vascular grafts: wrong models, wrong questions and no healing. *Biomaterials*. 2007;28(34):5009-5027.
2. Kapadia MR, Popowich DA, Kibbe MR. Modified prosthetic vascular conduits. *Circulation*. 2008;117(14):1873-1882.
3. Kader KN, Akella R, Ziats NP, et al. eNOS-overexpressing endothelial cells inhibit platelet aggregation and smooth muscle cell proliferation in vitro. *Tissue Eng*. 2000;6(3):241-251.
4. Sugawara Y, Sakata Y, Minowada S, et al. Adenovirus-mediated transfer of tissue-type plasminogen activator gene to human endothelial cells. *Surgery*. 1997;122(1):91-100.
5. Kong D, Melo LG, Mangi AA, et al. Enhanced inhibition of neointimal hyperplasia by genetically engineered endothelial progenitor cells. *Circulation*. 2004;109(14):1769-1775.
6. Alsberg E, Feinstein E, Joy MP, Prentiss M, Ingber DE. Magnetically-guided self-assembly of fibrin matrices with ordered nano-scale structure for tissue engineering. *Tissue Eng*. 2006;12(11):3247-3256.
7. Perea H, Aigner J, Heverhagen JT, Hopfner U, Wintermantel E. Vascular tissue engineering with magnetic nanoparticles: seeing deeper. *J Tissue Eng Regen Med*. 2007;1(4):318-321.
8. Perea H, Aigner J, Hopfner U, Wintermantel E. Direct magnetic tubular cell seeding: a novel approach for vascular tissue engineering. *Cells Tissues Organs*. 2006;183(3):156-165.
9. Campbell PG, Weiss LE. Tissue engineering with the aid of inkjet printers. *Expert Opin Biol Ther*. 2007;7(8):1123-1127.
10. Gao D, Kumar G, Co C, Ho CC. Formation of capillary tube-like structures on micropatterned biomaterials. *Adv Exp Med Biol*. 2008;614:199-205.

11. Kane RS, Takayama S, Ostuni E, Ingber DE, Whitesides GM. Patterning proteins and cells using soft lithography. *Biomaterials*. 1999;20(23-24):2363-2376.
12. Xia N, Thodeti CK, Hunt TP, et al. Directional control of cell motility through focal adhesion positioning and spatial control of Rac activation. *Faseb J*. 2008;22(6):1649-1659.
13. Weinberg CB, Bell E. A blood vessel model constructed from collagen and cultured vascular cells. *Science*. 1986;231(4736):397-400.
14. Chiu JJ, Chen LJ, Chen CN, Lee PL, Lee CI. A model for studying the effect of shear stress on interactions between vascular endothelial cells and smooth muscle cells. *J Biomech*. 2004;37(4):531-539.
15. Fernandez P, Bourget C, Bareille R, Daculsi R, Bordenave L. Gene response in endothelial cells cultured on engineered surfaces is regulated by shear stress. *Tissue Eng*. 2007;13(7):1607-1614.
16. Ozawa N, Shichiri M, Iwashina M, Fukai N, Yoshimoto T, Hirata Y. Laminar shear stress up-regulates inducible nitric oxide synthase in the endothelium. *Hypertens Res*. 2004;27(2):93-99.
17. Reinhardt PH, Kubes P. Differential leukocyte recruitment from whole blood via endothelial adhesion molecules under shear conditions. *Blood*. 1998;92(12):4691-4699.
18. Sharefkin JB, Diamond SL, Eskin SG, McIntire LV, Dieffenbach CW. Fluid flow decreases preproendothelin mRNA levels and suppresses endothelin-1 peptide release in cultured human endothelial cells. *J Vasc Surg*. 1991;14(1):1-9.
19. Sakamoto N, Ohashi T, Sato M. Effect of fluid shear stress on migration of vascular smooth muscle cells in co-cultured model. *Ann Biomed Eng*. 2006;34(3):408-415.
20. Baitella-Eberle G, Groscurth P, Zilla P, et al. Long-term results of tissue development and cell differentiation on Dacron prostheses seeded with microvascular cells in dogs. *J Vasc Surg*. 1993;18(6):1019-1028.
21. Bordenave L, Fernandez P, Remy-Zolghadri M, Villars S, Daculsi R, Midy D. In vitro endothelialized ePTFE prostheses: clinical update 20 years after the first realization. *Clin Hemorheol Microcirc*. 2005;33(3):227-234.
22. Meinhart J, Deutsch M, Zilla P. Eight years of clinical endothelial cell transplantation. Closing the gap between prosthetic grafts and vein grafts. *ASAIO J*. 1997;43(5):M515-M521.
23. Pasic M, Muller-Glauser W, von Segesser L, Odermatt B, Lachat M, Turina M. Endothelial cell seeding improves patency of synthetic vascular grafts: manual versus automatized method. *Eur J Cardiothorac Surg*. 1996;10(5):372-379.
24. Stanley JC, Burkel WE, Ford JW, et al. Enhanced patency of small-diameter, externally supported Dacron iliofemoral grafts seeded with endothelial cells. *Surgery*. 1982;92(6):994-1005.
25. Seifalian AM, Tiwari A, Hamilton G, Salacinski HJ. Improving the clinical patency of prosthetic vascular and coronary bypass grafts: the role of seeding and tissue engineering. *Artif Organs*. 2002;26(4):307-320.
26. Kaushal S, Amiel GE, Guleserian KJ, et al. Functional small-diameter neovessels created using endothelial progenitor cells expanded ex vivo. *Nat Med*. 2001;7(9):1035-1040.
27. Shirota T, He H, Yasui H, Matsuda T. Human endothelial progenitor cell-seeded hybrid graft: proliferative and antithrombotic potentials in vitro and fabrication processing. *Tissue Eng*. 2003;9(1):127-136.
28. Mirza A, Hyvelin JM, Rochefort GY, et al. Undifferentiated mesenchymal stem cells seeded on a vascular prosthesis contribute to the restoration of a physiologic vascular wall. *J Vasc Surg*. 2008;47(6):1313-1321.
29. Zhang L, Zhou J, Lu Q, Wei Y, Hu S. A novel small-diameter vascular graft: in vivo behavior of biodegradable three-layered tubular scaffolds. *Biotechnol Bioeng*. 2008;99(4):1007-1015.
30. Rotmans JI, Heyligers JM, Stroes ES, Pasterkamp G. Endothelial progenitor cell-seeded grafts: rash and risky. *Can J Cardiol*. 2006;22(11):929-932.
31. Wallace CS, Strike SA, Truskey GA. Smooth muscle cell rigidity and extracellular matrix organization influence endothelial cell spreading and adhesion formation in co-culture. *Am J Physiol Heart Circ Physiol*. 2007;293(3):H1978-H1986.
32. Niklason LE, Gao J, Abbott WM, et al. Functional arteries grown in vitro. *Science*. 1999;284(5413):489-493.

33. Auger FA, D'Orleans-Juste P, Germain L. Adventitia contribution to vascular contraction: hints provided by tissue-engineered substitutes. *Cardiovasc Res*. 2007;75(4):669-678.
34. Diebolt M, Germain L, Auger FA, Andriantsitohaina R. Mechanism of potentiation by polyphenols of contraction in human vein-engineered media. *Am J Physiol Heart Circ Physiol*. 2005;288(6):H2918-H2924.
35. Diebolt M, Laflamme K, Labbe R, Auger FA, Germain L, Andriantsitohaina R. Polyphenols modulate calcium-independent mechanisms in human arterial tissue-engineered vascular media. *J Vasc Surg*. 2007;46(4):764-772.
36. Laflamme K, Roberge CJ, Grenier G, et al. Adventitia contribution in vascular tone: insights from adventitia-derived cells in a tissue-engineered human blood vessel. *Faseb J*. 2006;20(8):1245-1247.
37. Laflamme K, Roberge CJ, Labonte J, et al. Tissue-engineered human vascular media with a functional endothelin system. *Circulation*. 2005;111(4):459-464.
38. L'Heureux N, Stoclet JC, Auger FA, Lagaud GJ, Germain L, Andriantsitohaina R. A human tissue-engineered vascular media: a new model for pharmacological studies of contractile responses. *Faseb J*. 2001;15(2):515-524.
39. Stoclet JC, Andriantsitohaina R, L'Heureux N, Martinez C, Germain L, Auger F. Use of human vessels and human vascular smooth muscle cells in pharmacology. *Cell Biol Toxicol*. 1996;12(4-6):223-225.
40. Laflamme K, Roberge CJ, Pouliot S, D'Orleans-Juste P, Auger FA, Germain L. Tissue-engineered human vascular media produced in vitro by the self-assembly approach present functional properties similar to those of their native blood vessels. *Tissue Eng*. 2006;12(8):2275-2281.
41. Lin TC, Tintut Y, Lyman A, Mack W, Demer LL, Hsiai TK. Mechanical response of a calcified plaque model to fluid shear force. *Ann Biomed Eng*. 2006;34(10):1535-1541.
42. Han HC, Ku DN. Contractile responses in arteries subjected to hypertensive pressure in seven-day organ culture. *Ann Biomed Eng*. 2001;29(6):467-475.
43. Stegemann JP, Nerem RM. Phenotype modulation in vascular tissue engineering using biochemical and mechanical stimulation. *Ann Biomed Eng*. 2003;31(4):391-402.
44. Solan A, Niklason L. Age effects on vascular smooth muscle: an engineered tissue approach. *Cell Transplant*. 2005;14(7):481-488.
45. L'Heureux N, Paquet S, Labbe R, Germain L, Auger FA. A completely biological tissue-engineered human blood vessel. *Faseb J*. 1998;12(1):47-56.
46. Cagiannos C, Abul-Khoudoud OR, DeRijk W, et al. Rapamycin-coated expanded polytetrafluoroethylene bypass grafts exhibit decreased anastomotic neointimal hyperplasia in a porcine model. *J Vasc Surg*. 2005;42(5):980-988.
47. Wallitt EJ, Jevon M, Hornick PI. Therapeutics of vein graft intimal hyperplasia: 100 years on. *Ann Thorac Surg*. 2007;84(1):317-323.
48. Min SK, Kenagy RD, Clowes AW. Induction of vascular atrophy as a novel approach to treating restenosis. A review. *J Vasc Surg*. 2008;47(3):662-670.
49. Bhardwaj S, Roy H, Yla-Herttuala S. Gene therapy to prevent occlusion of venous bypass grafts. *Expert Rev Cardiovasc Ther*. 2008;6(5):641-652.
50. Grenier G, Remy-Zolghadri M, Bergeron F, et al. Mechanical loading modulates the differentiation state of vascular smooth muscle cells. *Tissue Eng*. 2006;12(11):3159-3170.
51. Kashiwagi S, Izumi Y, Gohongi T, et al. NO mediates mural cell recruitment and vessel morphogenesis in murine melanomas and tissue-engineered blood vessels. *J Clin Invest*. 2005;115(7):1816-1827.
52. Ford MC, Bertram JP, Hynes SR, et al. A macroporous hydrogel for the co-culture of neural progenitor and endothelial cells to form functional vascular networks in vivo. *Proc Natl Acad Sci U S A*. 2006;103(8):2512-2517.
53. Hudon V, Berthod F, Black AF, Damour O, Germain L, Auger FA. A tissue-engineered endothelialized dermis to study the modulation of angiogenic and angiostatic molecules on capillary-like tube formation in vitro. *Br J Dermatol*. 2003;148(6):1094-1104.
54. Melero-Martin JM, Khan ZA, Picard A, Wu X, Paruchuri S, Bischoff J. In vivo vasculogenic potential of human blood-derived endothelial progenitor cells. *Blood*. 2007;109(11):4761-4768.

55. Black AF, Berthod F, L'Heureux N, Germain L, Auger FA. In vitro reconstruction of a human capillary-like network in a tissue-engineered skin equivalent. *FASEB J.* 1998;12:1331-1340.
56. Suuronen EJ, Muzakare L, Doillon CJ, et al. Promotion of angiogenesis in tissue engineering: developing multicellular matrices with multiple capacities. *Int J Artif Organs*. 2006;29(12): 1148-1157.
57. Suuronen EJ, Veinot JP, Wong S, et al. Tissue-engineered injectable collagen-based matrices for improved cell delivery and vascularization of ischemic tissue using CD133+ progenitors expanded from the peripheral blood. *Circulation*. 2006;114(suppl 1):I138-I144.
58. Weisz A, Koren B, Fischer L, Lewis BS, Flugelman MY. Therapeutic angiogenesis for ischemic syndromes. *Isr Med Assoc J.* 2000;suppl 2:52-57.
59. Kannan RY, Salacinski HJ, Edirisinghe MJ, Hamilton G, Seifalian AM. Polyhedral oligomeric silsequioxane-polyurethane nanocomposite microvessels for an artificial capillary bed. *Biomaterials*. 2006;27(26):4618-4626.
60. Karakitsos D, Patrianakos AP, Parthenakis FI, et al. Altered proximal aortic stiffness and endothelin plasma levels in diabetic patients with end-stage renal disease. *ASAIO J.* 2007;53(3):343-350.
61. Misra S, Fu AA, Rajan DK, et al. Expression of hypoxia inducible factor-1 alpha, macrophage migration inhibition factor, matrix metalloproteinase-2 and -9, and their inhibitors in hemodialysis grafts and arteriovenous fistulas. *J Vasc Interv Radiol*. 2008;19(2 pt 1):252-259.
62. Chou KJ, Lee PT, Chen CL, et al. Physiological changes during hemodialysis in patients with intradialysis hypertension. *Kidney Int*. 2006;69(10):1833-1838.
63. Clowes AW, Kirkman TR, Reidy MA. Mechanisms of arterial graft healing. Rapid transmural capillary ingrowth provides a source of intimal endothelium and smooth muscle in porous PTFE prostheses. *Am J Pathol*. 1986;123(2):220-230.
64. Geary RL, Kohler TR, Vergel S, Kirkman TR, Clowes AW. Time course of flow-induced smooth muscle cell proliferation and intimal thickening in endothelialized baboon vascular grafts. *Circ Res*. 1994;74(1):14-23.
65. Kenagy RD, Fischer JW, Lara S, Sandy JD, Clowes AW, Wight TN. Accumulation and loss of extracellular matrix during shear stress-mediated intimal growth and regression in baboon vascular grafts. *J Histochem Cytochem*. 2005;53(1):131-140.
66. Kraiss LW, Geary RL, Mattsson EJ, Vergel S, Au YP, Clowes AW. Acute reductions in blood flow and shear stress induce platelet-derived growth factor-A expression in baboon prosthetic grafts. *Circ Res*. 1996;79(1):45-53.
67. Miura H, Nishibe T, Yasuda K, et al. The influence of node-fibril morphology on healing of high-porosity expanded polytetrafluoroethylene grafts. *Eur Surg Res*. 2002;34(3):224-231.
68. Shi Q, Bhattacharya V, Hong-De Wu M, Sauvage LR. Utilizing granulocyte colony-stimulating factor to enhance vascular graft endothelialisation from circulating blood cells. *Ann Vasc Surg*. 2002;16(3):314-320.
69. Sterpetti AV, Hunter WJ, Schultz RD, Farina C. Healing of high-porosity polytetrafluoroethylene arterial grafts is influenced by the nature of the surrounding tissue. *Surgery*. 1992;111(6):677-682.
70. Davids L, Dower T, Zilla P. The lack of healing in conventional vascular grafts. In: Zilla P, Greisler H, eds. *Tissue Engineering of Prosthetic Vascular Grafts*. Austin: Landes; 1999:3-44.
71. Crombez M, Chevallier P, Gaudreault RC, Petitclerc E, Mantovani D, Laroche G. Improving arterial prosthesis neo-endothelialisation: application of a proactive VEGF construct onto PTFE surfaces. *Biomaterials*. 2005;26(35):7402-7409.
72. Begovac PC, Thomson RC, Fisher JL, Hughson A, Gallhagen A. Improvements in GORE-TEX vascular graft performance by Carmeda BioActive surface heparin immobilization. *Eur J Vasc Endovasc Surg*. 2003;25(5):432-437.
73. Christenson JT, Thulesius O, Owunwanne A, Nazzal M. Forskolin impregnation of small calibre PTFE grafts lowers early platelet graft sequestration and improves patency in a sheep model. *Eur J Vasc Surg*. 1991;5(3):271-275.
74. Heise M, Schmidmaier G, Husmann I, et al. PEG-hirudin/iloprost coating of small diameter ePTFE grafts effectively prevents pseudointima and intimal hyperplasia development. *Eur J Vasc Endovasc Surg*. 2006;32(4):418-424.

75. Lin PH, Chen C, Bush RL, Yao Q, Lumsden AB, Hanson SR. Small-caliber heparin-coated ePTFE grafts reduce platelet deposition and neointimal hyperplasia in a baboon model. *J Vasc Surg*. 2004;39(6):1322-1328.
76. Murray-Wijelath J, Lyman DJ, Wijelath ES. Vascular graft healing. III. FTIR analysis of ePTFE graft samples from implanted bigrafts. *J Biomed Mater Res B Appl Biomater*. 2004;70(2):223-232.
77. Nishibe T, Okuda Y, Kumada T, Tanabe T, Yasuda K. Enhanced graft healing of high-porosity expanded polytetrafluoroethylene grafts by covalent bonding of fibronectin. *Surg Today*. 2000;30(5):426-431.
78. Pfeiffer T, Kever M, Grabitz K, et al. Healing characteristics of small-calibre vascular prostheses coated with plasmin-treated fibrin – an experimental study. *Vasa*. 2000;29(2):117-124.
79. Shimada T, Nishibe T, Miura H, et al. Improved healing of small-caliber, long-fibril expanded polytetrafluoroethylene vascular grafts by covalent bonding of fibronectin. *Surg Today*. 2004;34(12):1025-1030.
80. Rotmans JI, Heyligers JM, Verhagen HJ, et al. In vivo cell seeding with anti-CD34 antibodies successfully accelerates endothelialisation but stimulates intimal hyperplasia in porcine arteriovenous expanded polytetrafluoroethylene grafts. *Circulation*. 2005;112(1):12-18.
81. Bosiers M, Deloose K, Verbist J, et al. Heparin-bonded expanded polytetrafluoroethylene vascular graft for femoropopliteal and femorocrural bypass grafting: 1-year results. *J Vasc Surg*. 2006;43(2):313-318; discussion 318-319.
82. Devine C, McCollum C. Heparin-bonded Dacron or polytetrafluorethylene for femoropopliteal bypass: five-year results of a prospective randomized multicenter clinical trial. *J Vasc Surg*. 2004;40(5):924-931.
83. Robinson BI, Fletcher JP. Fluoropolymer coated Dacron or polytetrafluoroethylene for femoropopliteal bypass grafting: a multicentre trial. *ANZ J Surg*. 2003;73(3):95-99.
84. Walluscheck KP, Bierkandt S, Brandt M, Cremer J. Infrainguinal ePTFE vascular graft with bioactive surface heparin bonding. First clinical results. *J Cardiovasc Surg (Torino)*. 2005;46(4):425-430.
85. Andrews KD, Feugier P, Black RA, Hunt JA. Vascular prostheses: performance related to cell-shear responses. *J Surg Res*. 2007;149(1):39-46.
86. Fields C, Cassano A, Makhoul RG, et al. Evaluation of electrostatically endothelial cell seeded expanded polytetrafluoroethylene grafts in a canine femoral artery model. *J Biomater Appl*. 2002;17(2):135-152.
87. Graham LM, Burkel WE, Ford JW, Vinter DW, Kahn RH, Stanley JC. Expanded polytetrafluoroethylene vascular prostheses seeded with enzymatically derived and cultured canine endothelial cells. *Surgery*. 1982;91(5):550-559.
88. Herring M, Baughman S, Glover J, et al. Endothelial seeding of Dacron and polytetrafluoroethylene grafts: the cellular events of healing. *Surgery*. 1984;96(4):745-755.
89. Hussain SM, Long GW, Juleff RS, et al. Comparison of immediate seeding of endothelial cells with culture lining of small diameter ePTFE carotid interposition grafts. *J Surg Res*. 1991;51(1):33-39.
90. Poole-Warren LA, Schindhelm K, Graham AR, Slowiaczek PR, Noble KR. Performance of small diameter synthetic vascular prostheses with confluent autologous endothelial cell linings. *J Biomed Mater Res*. 1996;30(2):221-229.
91. Zamora JL, Navarro LT, Ives CL, Weilbaecher DG, Gao ZR, Noon GP. Seeding of arteriovenous prostheses with homologous endothelium. A preliminary report. *J Vasc Surg*. 1986;3(6):860-866.
92. Zhang N, Zhang H, Han S, et al. Induction of bone marrow CD34+ cells and endothelialisation of polytetrafluoroethylene prostheses. *ANZ J Surg*. 2007;77(6):469-473.
93. Zilla P, Preiss P, Groscurth P, et al. In vitro-lined endothelium: initial integrity and ultrastructural events. *Surgery*. 1994;116(3):524-534.
94. Deutsch M, Meinhart J, Fischlein T, Preiss P, Zilla P. Clinical autologous in vitro endothelialisation of infrainguinal ePTFE grafts in 100 patients: a 9-year experience. *Surgery*. 1999;126(5):847-855.
95. Deutsch M, Meinhart J, Vesely M, et al. In vitro endothelialisation of expanded polytetrafluoroethylene grafts: a clinical case report after 41 months of implantation. *J Vasc Surg*. 1997;25(4):757-763.

96. Meinhart JG, Deutsch M, Fischlein T, Howanietz N, Froschl A, Zilla P. Clinical autologous in vitro endothelialisation of 153 infrainguinal ePTFE grafts. *Ann Thorac Surg*. 2001;71 (suppl 5):S327-S331.
97. Leseche G, Ohan J, Bouttier S, Palombi T, Bertrand P, Andreassian B. Above-knee femoropopliteal bypass grafting using endothelial cell seeded PTFE grafts: five-year clinical experience. *Ann Vasc Surg*. 1995;suppl 9:S15-S23.
98. Keough EM, Callow AD, Connolly RJ, Weinberg KS, Aalberg JJ, O'Donnell TF Jr. Healing pattern of small caliber dacron grafts in the baboon: an animal model for the study of vascular prostheses. *J Biomed Mater Res*. 1984;18(3):281-292.
99. Shi Q, Wu MH, Hayashida N, Wechezak AR, Clowes AW, Sauvage LR. Proof of fallout endothelialisation of impervious Dacron grafts in the aorta and inferior vena cava of the dog. *J Vasc Surg*. 1994;20(4):546-556; discussion 556-557.
100. Stewart GJ, Essa N, Chang KH, Reichle FA. A scanning and transmission electron microscope study of the luminal coating on Dacron prostheses in the canine thoracic aorta. *J Lab Clin Med*. 1975;85(2):208-226.
101. Szilagyi DE, Smith RF, Elliott JP, Allen HM. Long-term behavior of a dacron arterial substitute: clinical, roentgenologic and histologic correlations. *Ann Surg*. 1965;162(3):453-477.
102. Berger K, Sauvage LR, Rao AM, Wood SJ. Healing of arterial prostheses in man: its incompleteness. *Ann Surg*. 1972;175(1):118-127.
103. Schultz GA, Sauvage LR, Mathisen SR, et al. A five- to seven-year experience with externally-supported Dacron prostheses in axillofemoral and femoropopliteal bypass. *Ann Vasc Surg*. 1986;1(2):214-224.
104. Phaneuf MD, Dempsey DJ, Bide MJ, Quist WC, LoGerfo FW. Coating of Dacron vascular grafts with an ionic polyurethane: a novel sealant with protein binding properties. *Biomaterials*. 2001;22(5):463-469.
105. Sun LB, Utoh J, Moriyama S, Tagami H, Okamoto K, Kitamura N. Pretreatment of a Dacron graft with tissue factor pathway inhibitor decreases thrombogenicity and neointimal thickness: a preliminary animal study. *ASAIO J*. 2001;47(4):325-328.
106. Lambert AW, Fox AD, Williams DJ, Horrocks M, Budd JS. Experience with heparin-bonded collagen-coated grafts for infrainguinal bypass. *Cardiovasc Surg*. 1999;7(5):491-494.
107. Burkel WE, Ford JW, Vinter DW, Kahn RH, Graham LM, Stanley JC. Fate of knitted dacron velour vascular grafts seeded with enzymatically derived autologous canine endothelium. *Trans Am Soc Artif Intern Organs*. 1982;28:178-184.
108. Shepard AD, Eldrup-Jorgensen J, Keough EM, et al. Endothelial cell seeding of small-caliber synthetic grafts in the baboon. *Surgery*. 1986;99(3):318-326.
109. Bezuidenhout D, Davies N, Zilla P. Effect of well defined dodecahedral porosity on inflammation and angiogenesis. *ASAIO J*. 2002;48(5):465-471.
110. Jeschke MG, Hermanutz V, Wolf SE, Koveker GB. Polyurethane vascular prostheses decreases neointimal formation compared with expanded polytetrafluoroethylene. *J Vasc Surg*. 1999;29(1):168-176.
111. Therrien M, Guidoin R, Adnot A, Paynter R. Hydrophobic and fibrillar microporous polyetherurethane urea prosthesis: an ESCA study on the internal and external surfaces of explanted grafts. *Biomaterials*. 1989;10(8):517-520.
112. Zhang Z, Wang Z, Liu S, Kodama M. Pore size, tissue ingrowth, and endothelialisation of small-diameter microporous polyurethane vascular prostheses. *Biomaterials*. 2004;25(1):177-187.
113. Aldenhoff YB, van Der Veen FH, ter Woorst J, Habets J, Poole-Warren LA, Koole LH. Performance of a polyurethane vascular prosthesis carrying a dipyridamole (Persantin) coating on its lumenal surface. *J Biomed Mater Res*. 2001;54(2):224-233.
114. Taite LJ, Yang P, Jun HW, West JL. Nitric oxide-releasing polyurethane-PEG copolymer containing the YIGSR peptide promotes endothelialisation with decreased platelet adhesion. *J Biomed Mater Res B Appl Biomater*. 2008;84(1):108-116.
115. Fleser PS, Nuthakki VK, Malinzak LE, et al. Nitric oxide-releasing biopolymers inhibit thrombus formation in a sheep model of arteriovenous bridge grafts. *J Vasc Surg*. 2004;40(4):803-811.

116. Ishii Y, Kronengold RT, Virmani R, et al. Novel bioengineered small caliber vascular graft with excellent one-month patency. *Ann Thorac Surg*. 2007;83(2):517-525.
117. Doi K, Matsuda T. Enhanced vascularization in a microporous polyurethane graft impregnated with basic fibroblast growth factor and heparin. *J Biomed Mater Res*. 1997;34(3):361-370.
118. Pierce CM, Wade A, Mok Q. Heparin-bonded central venous lines reduce thrombotic and infective complications in critically ill children. *Intensive Care Med*. 2000;26(7):967-972.
119. Yue X, van der Lei B, Schakenraad JM, et al. Smooth muscle cell seeding in biodegradable grafts in rats: a new method to enhance the process of arterial wall regeneration. *Surgery*. 1988;103(2):206-212.
120. Turner NJ, Murphy MO, Kielty CM, et al. Alpha2(VIII) collagen substrata enhance endothelial cell retention under acute shear stress flow via an alpha2beta1 integrin-dependent mechanism: an in vitro and in vivo study. *Circulation*. 2006;114(8):820-829.
121. Miwa H, Matsuda T. An integrated approach to the design and engineering of hybrid arterial prostheses. *J Vasc Surg*. 1994;19(4):658-667.
122. Lovett M, Cannizzaro C, Daheron L, Messmer B, Vunjak-Novakovic G, Kaplan DL. Silk fibroin microtubes for blood vessel engineering. *Biomaterials*. 2007;28(35):5271-5279.
123. Zhang X, Baughman CB, Kaplan DL. In vitro evaluation of electrospun silk fibroin scaffolds for vascular cell growth. *Biomaterials*. 2008;29(14):2217-2227.
124. Christenson JT, Owunwanne A, Nazzal M. Thrombogenicity and long-term patency in autologous vein, polytetrafluoroethylene (PTFE) and silk grafts in a sheep model: evaluated through the use of indium-III-labeled platelets. *Am J Physiol Imaging*. 1990;5(2):80-83.
125. Chu CF, Lu A, Liszkowski M, Sipehia R. Enhanced growth of animal and human endothelial cells on biodegradable polymers. *Biochim Biophys Acta*. 1999;1472(3):479-485.
126. Niklason LE, Abbott W, Gao J, et al. Morphologic and mechanical characteristics of engineered bovine arteries. *J Vasc Surg*. 2001;33(3):628-638.
127. Lim SH, Cho SW, Park JC, et al. Tissue-engineered blood vessels with endothelial nitric oxide synthase activity. *J Biomed Mater Res B Appl Biomater*. 2008;85(2):537-546.
128. Roh JD, Brennan MP, Lopez-Soler RI, et al. Construction of an autologous tissue-engineered venous conduit from bone marrow-derived vascular cells: optimization of cell harvest and seeding techniques. *J Pediatr Surg*. 2007;42(1):198-202.
129. Shinoka T, Shum-Tim D, Ma PX, et al. Creation of viable pulmonary artery autografts through tissue engineering. *J Thorac Cardiovasc Surg*. 1998;115(3):536-545; discussion 545-546.
130. Shum-Tim D, Stock U, Hrkach J, et al. Tissue engineering of autologous aorta using a new biodegradable polymer. *Ann Thorac Surg*. 1999;68(6):2298-2304; discussion 2305.
131. Hoerstrup SP, Cummings Mrcs I, Lachat M, et al. Functional growth in tissue-engineered living, vascular grafts: follow-up at 100 weeks in a large animal model. *Circulation*. 2006;114(suppl I159-I166.
132. Shin'oka T, Imai Y, Ikada Y. Transplantation of a tissue-engineered pulmonary artery. *N Engl J Med*. 2001;344(7):532-533.
133. Shin'oka T, Matsumura G, Hibino N, et al. Midterm clinical result of tissue-engineered vascular autografts seeded with autologous bone marrow cells. *J Thorac Cardiovasc Surg*. 2005;129(6):1330-1338.
134. Zhang L, Ao Q, Wang A, et al. A sandwich tubular scaffold derived from chitosan for blood vessel tissue engineering. *J Biomed Mater Res A*. 2006;77(2):277-284.
135. Arrigoni C, Camozzi D, Imberti B, Mantero S, Remuzzi A. The effect of sodium ascorbate on the mechanical properties of hyaluronan-based vascular constructs. *Biomaterials*. 2006;27(4):623-630.
136. Hirai J, Kanda K, Oka T, Matsuda T. Highly oriented, tubular hybrid vascular tissue for a low pressure circulatory system. *ASAIO J*. 1994;40(3):M383-M388.
137. Cummings CL, Gawlitta D, Nerem RM, Stegemann JP. Properties of engineered vascular constructs made from collagen, fibrin, and collagen-fibrin mixtures. *Biomaterials*. 2004;25(17):3699-3706.
138. Seliktar D, Nerem RM, Galis ZS. Mechanical strain-stimulated remodeling of tissue-engineered blood vessel constructs. *Tissue Eng*. 2003;9(4):657-666.

139. Couet F, Mantovani D. Experimental validation of a new approach for the development of mechano-compatible composite scaffolds for vascular tissue engineering. *J Mater Sci Mater Med.* 2008;19(7):2551-2554.
140. Kobashi T, Matsuda T. Fabrication of compliant hybrid grafts supported with elastomeric meshes. *Cell Transplant.* 1999;8(5):477-488.
141. Lv Q, Hu K, Feng Q, Cui F. Fibroin/collagen hybrid hydrogels with crosslinking method: preparation, properties, and cytocompatibility. *J Biomed Mater Res A.* 2008;84(1):198-207.
142. Berglund JD, Nerem RM, Sambanis A. Incorporation of intact elastin scaffolds in tissue-engineered collagen-based vascular grafts. *Tissue Eng.* 2004;10(9-10):1526-1535.
143. Girton TS, Oegema TR, Grassl ED, Isenberg BC, Tranquillo RT. Mechanisms of stiffening and strengthening in media-equivalents fabricated using glycation. *J Biomech Eng.* 2000;122(3):216-223.
144. He H, Shirota T, Yasui H, Matsuda T. Canine endothelial progenitor cell-lined hybrid vascular graft with nonthrombogenic potential. *J Thorac Cardiovasc Surg.* 2003;126(2):455-464.
145. Hirai J, Matsuda T. Venous reconstruction using hybrid vascular tissue composed of vascular cells and collagen: tissue regeneration process. *Cell Transplant.* 1996;5(1):93-105.
146. Grassl ED, Oegema TR, Tranquillo RT. A fibrin-based arterial media equivalent. *J Biomed Mater Res A.* 2003;66(3):550-561.
147. Swartz DD, Russell JA, Andreadis ST. Engineering of fibrin-based functional and implantable small-diameter blood vessels. *Am J Physiol Heart Circ Physiol.* 2005;288(3):H1451-H1460.
148. Campbell JH, Efendy JL, Campbell GR. Novel vascular graft grown within recipient's own peritoneal cavity. *Circ Res.* 1999;85(12):1173-1178.
149. Chue WL, Campbell GR, Caplice N, et al. Dog peritoneal and pleural cavities as bioreactors to grow autologous vascular grafts. *J Vasc Surg.* 2004;39(4):859-867.
150. Amiel GE, Komura M, Shapira O, et al. Engineering of blood vessels from acellular collagen matrices coated with human endothelial cells. *Tissue Eng.* 2006;12(8):2355-2365.
151. Dahl SL, Koh J, Prabhakar V, Niklason LE. Decellularized native and engineered arterial scaffolds for transplantation. *Cell Transplant.* 2003;12(6):659-666.
152. McFetridge PS, Daniel JW, Bodamyali T, Horrocks M, Chaudhuri JB. Preparation of porcine carotid arteries for vascular tissue engineering applications. *J Biomed Mater Res A.* 2004;70(2):224-234.
153. Bader A, Steinhoff G, Strobl K, et al. Engineering of human vascular aortic tissue based on a xenogeneic starter matrix. *Transplantation.* 2000;70(1):7-14.
154. Huynh T, Abraham G, Murray J, Brockbank K, Hagen PO, Sullivan S. Remodeling of an acellular collagen graft into a physiologically responsive neovessel. *Nat Biotechnol.* 1999;17(11):1083-1086.
155. Martin ND, Schaner PJ, Tulenko TN, et al. In vivo behavior of decellularized vein allograft. *J Surg Res.* 2005;129(1):17-23.
156. Matsuura JH, Black KS, Levitt AB, et al. Cellular remodeling of depopulated bovine ureter used as an arteriovenous graft in the canine model. *J Am Coll Surg.* 2004;198(5):778-783.
157. Nemcova S, Noel AA, Jost CJ, Gloviczki P, Miller VM, Brockbank KG. Evaluation of a xenogeneic acellular collagen matrix as a small-diameter vascular graft in dogs – preliminary observations. *J Invest Surg.* 2001;14(6):321-330.
158. Roeder RA, Lantz GC, Geddes LA. Mechanical remodeling of small-intestine submucosa small-diameter vascular grafts – a preliminary report. *Biomed Instrum Technol.* 2001;35(2):110-120.
159. Teebken OE, Pichlmaier AM, Haverich A. Cell seeded decellularized allogeneic matrix grafts and biodegradable polydioxanone-prostheses compared with arterial autografts in a porcine model. *Eur J Vasc Endovasc Surg.* 2001;22(2):139-145.
160. L'Heureux N, Dusserre N, Konig G, et al. Human tissue-engineered blood vessels for adult arterial revascularization. *Nat Med.* 2006;12(3):361-365.
161. L'Heureux N, McAllister TN, de la Fuente LM. Tissue-engineered blood vessel for adult arterial revascularization. *N Engl J Med.* 2007;357(14):1451-1453.
162. Nieponice A, Soletti L, Guan J, et al. Development of a tissue-engineered vascular graft combining a biodegradable scaffold, muscle-derived stem cells and a rotational vacuum seeding technique. *Biomaterials.* 2008;29(7):825-833.

163. Etz CD, Homann T, Silovitz D, et al. Vascular graft replacement of the ascending and descending aorta: do Dacron grafts grow? *Ann Thorac Surg.* 2007;84(4):1206-1212; discussion 1212-1213.
164. Vascular prostheses: performance related to cell-shear responses. <http://www.ncbi.nlm.nih.gov/entrez/query.fcgi?cmd=Retrieve&db=PubMed&dopt=Citation&list_uids=18395748/>; 2007. Accessed 27.05.08.
165. Wilson GJ, MacGregor DC, Klement P, et al. Anisotropic polyurethane nonwoven conduits: a new approach to the design of a vascular prosthesis. *Trans Am Soc Artif Intern Organs.* 1983;29:260-268.
166. Zdrahala RJ. Small caliber vascular grafts. Part II: polyurethanes revisited. *J Biomater Appl.* 1996;11(1):37-61.

Chapter 10
Role of Perivascular Adipose Tissue in Vascular Function

Maria S. Fernández-Alfonso, Marta Gil-Ortega, and Beatriz Somoza

10.1 Introduction

The study of vascular tone regulation has been traditionally focused on layer-specific mechanisms and has changed through the years. Most early functional studies characterized vasoconstrictor and vasodilator agents and their receptor types and subtypes. At the same time, a bulk of investigation focused on the neural regulation of medial function, characterized perivascular innervation in the adventitia and adventitial–medial border, and described both vasoconstrictor and vasodilator neurotransmitters. The identification in the 1980s of nitric oxide (NO) as an endothelium-derived relaxing factor (EDRF) reoriented vascular function studies of the next two decades. As a consequence, the endothelial layer is now considered a paracrine tissue, which produces and releases a variety of contractile and relaxant factors that directly and indirectly regulate medial function through modulation of neurotransmitter release. During this time, the adventitia was only regarded as a structural support for the media and its functional role was ignored. However, in recent years, there is increasing evidence of a direct modulation of the adventitia on blood vessel function in a variety of situations (for review see ref.).[1-3]

It has to be kept in mind that many blood vessels are surrounded by adipose tissue. Similar to the adventitia, perivascular adipose tissue (PVAT) was considered only as a passive structural support for the blood vessel and was routinely removed for isolated blood vessel studies. Soltis and Cassis[4] demonstrated for the first time that PVAT reduced contractions to noradrenaline in rat aorta. Since then, the interest in analyzing the potential involvement of perivascular fat in the paracrine regulation of vascular tone has been continuously increasing.

Adipose tissue is a source of substances, generically called adipokines. These include leptin, adiponectin, the unidentified adipose-derived relaxing factor (ADRF), as well as an increasing number of macrophage-derived factors (cytokines). In addition, the complete renin–angiotensin system (RAS) is expressed in the

M.S. Fernández-Alfonso (✉)
Instituto Pluridisciplinar, Universidad Complutense, Juan XXIII, 1, E-28040, Madrid, Spain
e-mail: marisolf@farm.ucm.es

D. Abraham et al. (eds.), *Advances in Vascular Medicine*,
DOI 10.1007/978-1-84882-637-3_10,

adipose tissue. These adipose-derived compounds allow the adipose tissue to establish a cross-talk with other organs and tissues by means of a true autocrine, paracrine, and endocrine communication. Some of the substances secreted by adipocytes play a significant role in cardiovascular physiology, either by a direct modulation of vascular tone or by the stimulation of inflammation. Thus, there is increasing evidence suggesting that dysregulation of adipokines contributes to obesity-related cardiovascular diseases.[5,6] The aim of this review is to focus on the adipokines synthesized and released by PVAT and their paracrine role in the direct regulation of vascular function.

10.2 Perivascular Adipose Tissue

Blood vessels are surrounded by adipose tissue, termed as periadventitial or PVAT. This includes large vessels (aorta, superior mesenteric artery, coronary arteries, etc.), as well as resistance vessels (small mesenteric arteries, intramuscular arteries, the stromal–vascular fraction of adipose tissue, etc.). The type of PVAT differs between vascular beds. Some studies have suggested that periaortic adipose tissue is brown (BAT), whereas mesenteric adipose tissue is white (WAT).[7-9] This has been confirmed recently by the differential expression of the uncoupling protein 1 (UCP-1), a specific marker for BAT, in rat periaortic but not in rat mesenteric adipose tissue (MAT).[10]

According to the different types of adipose tissues, there are also some histological differences between periaortic and mesenteric adipocytes from the rat (Fig. 10.1). Perivascular mesenteric adipocytes are bigger in size with a mean diameter of approximately 40 μm when compared with periaortic adipocytes, which are four times smaller (mean diameter of approximately 9 μm).[10]

An important issue regarding PVAT is whether it can be considered as visceral adipose tissue. In its classical concept, visceral adipose tissue is intraperitoneal, and thus mainly composed of omental, mesenteric, and retroperitoneal fat masses.[11] In view of this definition, adipose tissue surrounding mesenteric arteries can be considered visceral fat, whereas periaortic adipose tissue and fat surrounding other blood vessels outside the intraperitoneal cavity are not included in this definition. However, if visceral adipose tissue is considered as the fat deposited around internal organs, i.e. periorganic adipose tissue, then PVAT can also be included in the definition of visceral adipose tissue. This change in concept is crucial to include PVAT in all the ulterior implications of visceral adipose tissue for cardiovascular regulation and cardiovascular risk.

10.3 Vasoactive Factors Released by PVAT

During the last decade, an important number of adipocyte-derived peptides with vasoactive effects have been identified.

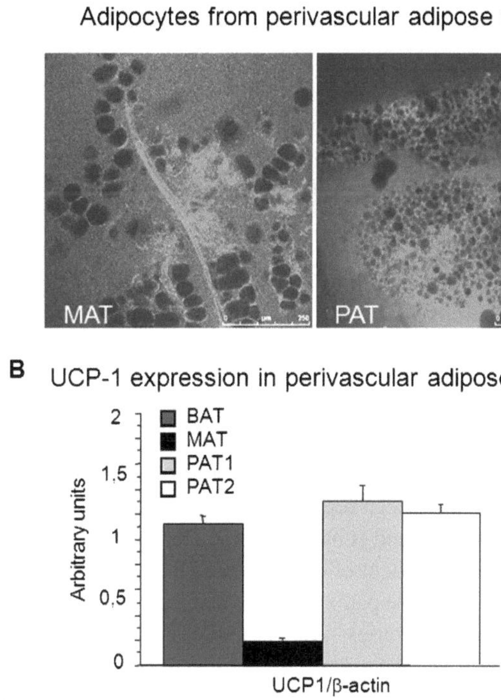

Fig. 10.1 (**a**) Adipocytes from rat perivascular adipose tissue (PVAT). *Left* panel shows mesenteric adipose tissue (MAT) and *right* panel shows pariaortic adipose tissue (PAT). (**b**) Expression of uncoupling protein-1 (UCP-1) in PAT but not in MAT, demonstrating that PAT is brown adipose tissue (BAT)

10.3.1 Leptin

Leptin is a hormone mainly synthesized by adipocytes.[12] It participates in the regulation of vascular tone through two different mechanisms: (1) an indirect vasoconstriction through stimulation of sympathetic activity at hypothalamic level,[13,14] and (2) a direct vasodilatation that depends on an intact and functional endothelium through mechanisms that vary between different vascular beds. In vivo, experiments have revealed that leptin infusion reduces arterial pressure by increasing NO release.[15] In aorta, such an effect involves a mechanism linked to the activation of endothelial nitric oxide synthase (eNOS) in the endothelium.[16,17] In mesenteric arteries, leptin induces the release of endothelium-derived hyperpolarizing factor.[16] Moreover, an endothelium-independent anticontractile effect of leptin on angiotensin II-induced contractions has recently been described.[18]

Leptin is produced by PVAT, both in the aorta and in mesenteric arteries at active concentrations.[19] This leads to the suggestion of a paracrine role of leptin from PVAT in the regulation of vascular tone. Unfortunately, the lack until now of selective leptin antagonists does not allow to discriminate between the effects elicited by either circulating or PVAT-derived leptin. Further studies will be necessary to address this question.

10.3.2 Adipocyte-Derived Relaxing Factor (ADRF)

In 2002, Löhn et al described the inhibitory action of PVAT on aortic contractions to a variety of vasoconstrictors. This anticontractile action is induced by a transferable protein factor released by adipocytes, which the authors called ADRF, in analogy to EDRF.

The anticontractile effect of ADRF has been described in rat,[20-24] mouse,[25] and human arteries.[26,27] It seems to be mediated by different mechanisms depending on the vascular bed. In rat aorta, both an endothelium-dependent[22] and an endothelium-independent relaxation[22,23] have been reported. The endothelium-dependent relaxation is mediated through NO release and subsequent calcium-dependent K^+ channel activation.[22] The endothelium-independent effect is mediated either by H_2O_2 formation[22] or by the activation of tyrosine kinase pathways and opening of ATP-dependent K^+ (K_{ATP}) channels.[23] In rat mesenteric arteries, ADRF induces an endothelium-independent relaxation through the activation of voltage-dependent K^+ channels (K_v).[21,24]

ADRF has been also found in the human thoracic artery.[26,27] In this vessel, the anticontractile effect of PVAT is also due to K^+ channels opening[26] and independent on the release of NO or prostacyclin.[27] Since the human thoracic artery is the gold standard graft for coronary artery surgery, the finding of an anticontractile effect of PVAT has led to the suggestion that removal of PVAT, i.e. skeletonization of the vessel, might be deleterious for the graft patency.[27-29]

ADRF is released by perivascular adipocytes. The mechanism of ADRF release from rat aortic PVAT is dependent on calcium and cAMP.[20] It does not involve Na^+ channels, neuronal N-type Ca^{2+}-channels, vanilloid/cannabinoid or calcitonin gene-related peptide receptors excluding the participation of perivascular nerves in ADRF release.[20]

Interestingly, the anti-contractile effect in these vessels is directly dependent on the amount of periadventitial fat.[21,24] The resting membrane potential of mesenteric vascular smooth muscle cells is more hyperpolarized in intact mesenteric rings surrounded by fat than in rings without fat,[24] suggesting that PVAT contributes to the maintenance of basal mesenteric artery tone.

An essential question that remains to be answered concerns the chemical structure of ADRF. The identity of ADRF with leptin has been discarded, since the lack of functional leptin receptors in the Zucker fa/fa rats did not modify the effect of perivascular fat.[23] Moreover, the identity of ADRF with adiponectin (see below) has also been discarded, since the anticontractile effect of PVAT is still present in adiponectin knock-out mice ($APN^{-/-}$).[25] Some candidates of different masses (74–13.8 kDa) have been proposed,[30] and further studies will be necessary to finally identify ADRF.

10.3.3 Adiponectin

Adiponectin is an adipokine mainly synthesized by the adipose tissue. This hormone has been shown to induce vasodilatation in rat aorta and in mouse mesenteric arteries through an endothelium-independent mechanism involving the activation of K_v channels.[25] Other authors have shown that adiponectin increases NO release from vascular endothelial cells in culture.[31,32] Moreover, adiponectin seems to preserve endothelial function through inhibition of endothelial cell activation[33] and synthesis of inflammatory markers.[34] To which degree circulating or PVAT-derived adiponectin contributes to regulation of vascular tone is a subject that deserves future research.

10.3.4 Reactive Oxygen Species: H_2O_2 and O_2^-

Hydrogen peroxide (H_2O_2), a nonradical form of reactive oxygen species, can be produced in perivascular adipocytes either by the membrane-bound NAD(P)H oxidase system[22] and also by the activity of the superoxide dismutase (SOD). H_2O_2 has been shown to be a vasoactive substance that induces both contractile and relaxant responses on blood vessels by different mechanisms depending on vessels type, contractile status, concentrations, and animal species.[35] On the one hand, it seems that the contractile effect mediated by H_2O_2 is probably due to direct activation of cyclooxygenase (COX), and it has also been attributed to an increase in intracellular Ca^{2+}.[36,37] On the other hand, H_2O_2 can also induce relaxation and several mechanisms have been proposed. The best known is endothelium-independent relaxation mediated by activation of smooth muscle K^+ channels. H_2O_2 directly opens K^+ channels by oxidation of their cystein residues. In addition, H_2O_2 is related to endothelium-dependent relaxation as a result of an increased NO production by NOS activation secondary to the activation of endothelial K^+ channels.[36] Moreover, it is well known that H_2O_2 might also induce relaxation by direct activation of smooth muscle soluble guanylate cyclase (sGC).[22,38] Whereas the contractile effects seem to occur at lower micromolar concentrations, relaxation appears at concentrations higher than 0.1 mM.[36]

On the other hand, superoxide anions can be also produced in PVAT by the NAD(P)H oxidase system.[22] Recent studies[39] demonstrated that superoxide anion potentiated vasoconstriction to norepinephrine in rat mesenteric arteries, but the mechanism for this effect has not been elucidated yet. Moreover, O_2^- indirectly promotes vasoconstriction through the inactivation of endothelial NO.[10]

10.3.5 The Renin–Angiotensin System (RAS) in PVAT

During the last decade, the existence of a local RAS in adipose tissue and its functional importance has attracted closer attention.[41-43] White adipose tissue is the most abundant source of angiotensinogen, the substrate of the system, after the liver.[44] Angiotensinogen is cleaved by renin producing angiotensin I (Ang I), which is then converted to angiotensin II (Ang II) by either angiotensin-converting enzyme (ACE) and/or chymase. The decapeptide Ang I is also a substrate of ACE2, which catalyzes Ang 1–7 production. Ang II is considered to be the major effector peptide of the RAS by binding two different receptors, AT_1 and AT_2. AT_1 receptors are widely distributed and mediate most of the biological responses that contribute to the known pressor, trophic, and proinflammatory effects of Ang II, whereas AT_2 receptors antagonize several of the AT_1 receptor-mediated responses (for review see ref. [43]).

Moreover, almost all components of the system have been described in both white and brown adipose tissue.[43,45,46] However, while most of the studies investigating the RAS in adipose tissue have focused on subcutaneous and visceral adipose tissues, little attention has been paid to PVAT. In this context, our group has recently shown expression of all components of the RAS, except renin, in PVAT of the aorta and mesenteric arteries from Wistar Kyoto rats (WKR) (Fig. 10.2). An important new finding is the expression of the (pro)renin receptor, ACE2 and of three AT_{1a} receptor isoforms in PVAT.

An intriguing and important question that deserves future investigation is the physiological role of the perivascular adipose RAS. Speculating, the most obvious implication of perivascular adipose RAS is to contribute to the regulation of vascular tone. Local adipose Ang II could induce direct contractile responses and modulate the response to other adipokines. Interestingly, the inhibitory effect of both ADRF[23] and leptin[47] on Ang II-induced contractions is more potent than their anti-contractile effect on other vasoconstrictors. This suggests that there might be a balance between adipose tissue-derived vasodilator and vasoconstrictor factors essential for the maintenance of vascular resistance. Moreover, owing to the trophic

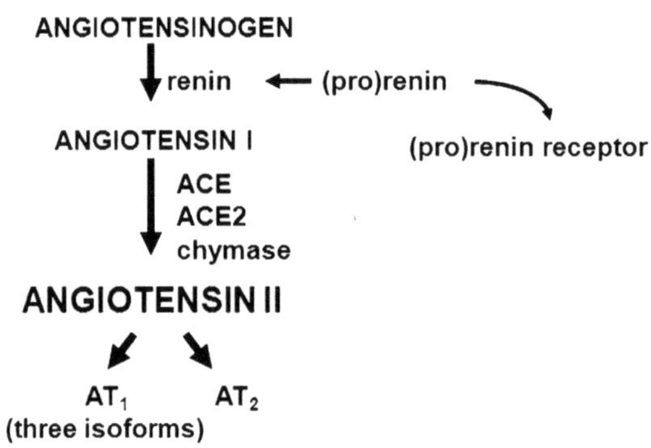

Fig. 10.2 Components of the renin–angiotensin system (RAS) expressed in PVAT of normotensive Wistar Kyoto rats (WKY). ACE>angiotensin-converting enzyme; ACE2>angiotensin-converting enzyme 2; AT_1>angiotensin receptor type 1; AT_2>angiotensin receptor type 2. Levels of ACE, chymase, angiotensin II, AT_1 (Exon 1+2+6bp+3), and AT_2 receptors are higher in mesenteric PVAT when compared with periaortic PVAT[10]

effects of Ang II, the local production of this peptide in PVAT would also have a substantial impact on vascular structure.

10.3.6 Inflammatory Cytokines

Adipose tissue produces a high variety of inflammatory cytokines such as interleukins (IL-6, IL-1β), TNF-α, monocyte chemotactic protein (MCP-1), etc. These cytokines have been suggested to participate in inflammation of adipose tissue and to contribute to both obesity-associated insulin resistance and vascular dysfunction. Recently, several excellent reviews have been published specifically focused on this issue.[48,49]

10.4 Perivascular Adipose Tissue and Cardiovascular Pathophysiology

Visceral adipose tissue is recognized as an important indicator of high cardiovascular and metabolic risk. However, there are very few studies analyzing the role of PVAT on vascular function in a number of pathophysiological situations, particularly in obesity and metabolic syndrome.

10.4.1 Hypertension

The relevance of PVAT to hypertension has been demonstrated in several recent publications. The spontaneously hypertensive rat (SHR) is a well-known and widely used model of hypertension. These animals are lean and exhibit lighter adipose pads when compared with normotensive WKY.[21] The reduction in adipose tissue is due to the presence of smaller adipocytes in SHR when compared with WKY with no changes in the adipocyte number between strains. Reductions in PVAT mass correlate with a reduced anti-contractile effect of PVAT in mesenteric artery rings from SHR and with a lower release of vasodilatory adipocytokines, such as ADRF and leptin.[21] Interestingly, these changes in periadventitial fat observed in SHR precede hypertension since alterations in PVAT mass and function are already present in 4-week-old SHR, which are prehypertensive.[50]

Leptin deficiency in PVAT is also related to hypertension. Plasma leptin concentration is lower in SHR when compared with WKY (7.6 ± 1.0 ng/ml vs. 11.5 ± 2.2 ng/ml; $p < 0.05$). This positively correlates with the less amount of PVAT observed in SHR ($r > 0.84$; $p < 0.05$) as well as with lower leptin levels in MAT. Moreover, SHR show an impairment of leptin-induced vasodilatation in SHR. These findings support the novel concept that a reduction in the amount of PVAT contributes to the increased vascular resistance observed in lean SHR.[21]

10.4.2 Obesity

In light of the findings reviewed above, there seems to be a contradiction between the anticontractile effect of PVAT depending on the amount of fat and obesity-related hypertension. The volume of PVAT increases throughout the vasculature in obesity both in animal models[51] and in humans.[52] Thus, it would be conceivable to think that in these circumstances, the anticontractile effect of PVAT would be increased. However, there is evidence that in obese patients and animal models of obesity, the production of adipokines might be unbalanced in favor of vasoconstrictor and proinflammatory substances.

Gao et al[26] demonstrated that the anticontractile effect of PVAT is lost in an animal model of obesity despite higher amounts of perivascular fat. Similarly, NZO mice, which have a severe metabolic syndrome associated with hypertension,[53] show a reduced anticontractile effect of PVAT.[25] Furthermore, a pathophysiological relevance of Ang II production in PVAT for blood pressure regulation cannot be excluded.[41,54] Thus, an increased local formation of angiotensinogen and Ang II in rat adipocytes due to overfeeding might represent a link between increased adipose tissue mass and hypertension in rodents.[55,56] Similarly, local Ang II formation in adipose tissue is increased in obese hypertensive subjects.[42,57]

10.5 Regional Differences Between PVAT and Vascular Beds

According to the different types of PVAT, several findings suggest that there might be also regional differences in the paracrine effects of PVAT depending on the vascular bed.

Regarding ADRF, the channels activated by PVAT in mesenteric arteries (K_v) differ from the channels proposed to be activated in the aorta (K_{ATP}), suggesting that there might be vascular regional differences in the effects of ADRF. Based on these findings, the existence of different ADRFs has been suggested.[24]

Moreover, there are also differences between normotensive and SHR (see above) correlating with differences in the amount of PVAT. We have shown that mesenteric PVAT has similar mRNA levels between the strains, but almost tenfold lower leptin protein content in SHR. This correlates with the lower amount of MAT found in the hypertensive strain. In contrast, periaortic adipose tissue amount is similar between strains, correlating with similar leptin protein content. Therefore, in SHR there is a reduction in perivascular leptin restricted to white PVAT and resistance arteries supporting, again, the concept that a reduction in the amount of PVAT contributes to the increased vascular resistance observed in lean SHR.

The RAS in PVAT also shows regional differences between vascular beds. Gene expression levels of some RAS components are different between periaortic (brown adipose tissue) and perimesenteric fat (white adipose tissue). The finding that Ang II levels and both AT_{1a} and AT_2 receptor expression levels are higher in mesenteric fat suggests that RAS-mediated effects might differ depending on the type of surrounding adipose tissue and the type of vascular bed.[10]

10.6 Hypothesis of the Role of PVAT in Health and Disease

According to the data reviewed here, PVAT seems to have a dual function in health and disease (Fig. 10.3). Under physiological conditions, PVAT releases a number of adipokines (leptin, ADRF, adiponectin, H_2O_2) that are essential for the maintenance of vascular resistance and elicit a beneficial effect on vascular function. Consequently, in situations of normal weight, the protective and beneficial role of perivascular adipokines increase parallel to the amount of PVAT. However, in situations of overweight and obesity, there seems to be a shift in the balance between adipose tissue-derived vasodilator and vasoconstrictor factors. This unbalance toward a predominance of vasoconstrictor factors in obesity could provide the link between obesity, cardiovascular functional and structural alterations, and cardiovascular diseases.

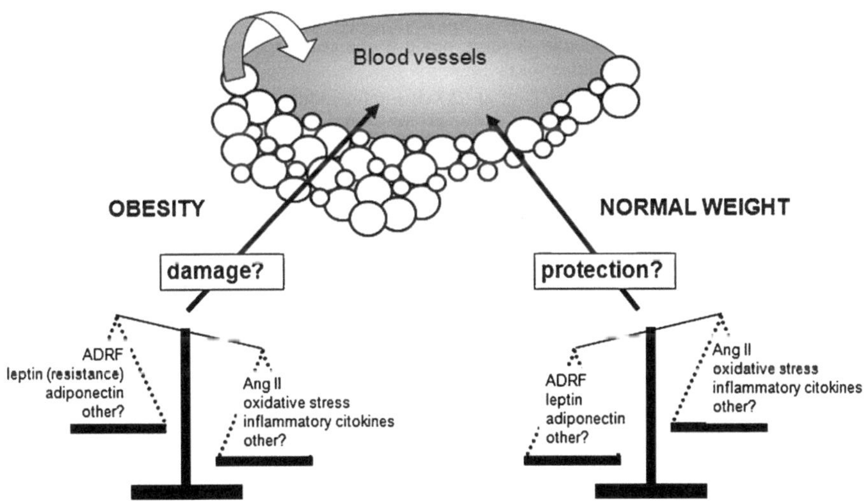

Fig. 10.3 Hypothesis of the role of PVAT and PVAT-derived adipokines in health and disease

References

1. Gutterman DD (1999) Adventitia-dependent influences on vascular function. Am J Physiol 277:H1265–H1272
2. Pagano PJ, Gutterman DD (2007) The adventitia: the outs and ins of vascular disease. Cardiovasc Res 75:636–639
3. Rey FE, Pagano PJ (2002) The reactive adventitia. Arterioscler Thromb Vasc Biol 22:1962–1971
4. Soltis EE, Cassis LA (1991) Influence of perivascular adipose tissue on rat aortic smooth muscle responsiveness. Clin Exp Hypertens 275:681–692
5. Goldstein BJ, Scalia R (2004) Adiponectin: a novel adipokine linking adipocytes and vascular function. J Clin Endocrinol Metab 89:2563–2568
6. Trayhurn P, Beattie JH (2001) Physiological role of adipose tissue: white adipose tissue as an endocrine and secretory organ. Proc Nutr Soc 60:329–339
7. Hausman DB, DiGirolamo M, Bartness TJ et al (2002) The biology of white adipocyte proliferation. Obesity Rev 2:239–254
8. Kortelainen ML, Pelletier G, Ricquier D et al (1993) Immunohistochemical detection of human brown adipose tissue uncoupling protein in an autopsy series. J Histochem Cytochem 41:759–764
9. Matthias A, Richards SM, Dora KA et al (1994) Characterization of perfused periaortic brown adipose tissue from the rat. Can J Physiol Pharmacol 72:344–352
10. Gálvez-Prieto B, Bolbrinker J, Stucchi P et al (2008) Comparative expression analysis of the renin-angiotensin system components between white and brown perivascular adipose tissue. J Endocrinol 197:55–64
11. Märin P, Andersson B, Ottosson M et al (1992) The morphology and metabolism of intraabdominal adipose tissue in men. Metabolism 41:1242–1248
12. Zhang Y, Ji H, Fabucci ME et al (2004) Translational control of the rat angiotensin type 1a receptor by alternative splicing. Gene 341:93–100
13. Elias CF, Lee C, Kelly J et al (1998) Leptin activates hypothalamic CART neurons projecting to the spinal cord. Neuron 21:1375–1385
14. Hall JE, Brands MW, Hildebrandt DA et al (2000) Role of sympathetic nervous system and neuropeptides in obesity hypertension. Braz J Med Biol Res 33:605–618
15. Frühbeck G (1999) Pivotal role of nitric oxide in the control of blood pressure after leptin administration. Diabetes 48:903–908
16. Lembo G, Vecchione C, Fratta L et al (2000) Leptin induces direct vasodilatation through distinct endothelial mechanisms. Diabetes 49:293–297
17. Vecchione C, Maffei A, Colella S et al (2002) Leptin effect on endothelial nitric oxide is mediated through Akt-endothelial nitric oxide synthase phosphorylation pathway. Diabetes 51:168–173
18. Rodriguez A, Fortuno A, Gomez-Ambrosi J et al (2007) The inhibitory effect of leptin on angiotensin II-induced vasoconstriction in vascular smooth muscle cells is mediated via a nitric oxide-dependent mechanism. Endocrinology 148:324–331
19. Gonzalez MC, Abderrahim F, Galvez-Prieto B et al (2005) Leptin-induced vasodilatation is impaired in spontaneously hypertensive rats. Hypertension 46:893
20. Dubrovska G, Verlohren S, Luft FC et al (2004) Mechanisms of ADRF release from rat aortic adventitial adipose tissue. Am J Physiol 286:1107–1113
21. Gálvez B, de Castro J, Herold D et al (2006) Perivascular adipose tissue and mesenteric vascular function in spontaneously hypertensive rats. Arterioscler Thromb Vasc Biol 58:989–994
22. Gao YJ, Takemori K, Su LY, An WS, Lu C, Sharma AM, Lee RM (2006) Perivascular adipose tissue promotes vasoconstriction: the role of superoxide anion. Cardiovasc Res 71:363–373
23. Löhn M, Dubrovska G, Lauterbach B et al (2002) Periadventitial fat releases a vascular relaxing factor. FASEB J 16:1057–1063

24. Verlohren S, Dubrovska G, Tsang S-Y et al (2004) Visceral periadventitial adipose tissue regulates arterial tone of mesenteric arteries. Hypertension 44:271–276
25. Fesus G, Dubrovska G, Gorzelniak K et al (2007) Adiponectin is a novel humoral vasodilator. Ardiovasc Res 75:719–727
26. Gao YL, Zeng Z, Teoh K et al (2005) Perivascular adipose tissue modulates vascular function in the human internal thoracic artery. J Thorac Cardiovasc Surg 130:1130–1136
27. Malinowsky M, Deja MA, Golba KS et al (2008) Perivascular tissue of internal thoracic atery releases a potent nitric oxide and prostacyclin-independent anticontractile factor. Eur J Cardio-Thorac Surg 33:225–231
28. Dashwood M, Souza D, Fernandez-Alfonso MS (2008) RE: Perivascular tissue of internal thoracic artery releases potent nitric oxide and prostacyclin-independent anticontractile factor. Europ J Cardio-Thorac Surg . doi:10.1016/j.ejcts.2008.03.021
29. Souza DRS, Johansson B, Böjö L et al (2006) Harvesting the saphenous vein with surrounding tissue for CABG provides long-term graft patency comparable to the left internal thoracic artery: results of a randomized longitudinal trial. J Thorac Cardiovasc Surg 132:373–378
30. Yang L, Hu BR, Xiang JZ et al (2005) Adventitium-derived relaxing factor may be a protein factor secreted by adipocytes with non-species-specificity and not limited periadventitial fat. Chin J Pharmacol Toxiciol 19:401–406
31. Chen H, Montagnani M, Funahashi T et al (2003) Adiponectin stimulates production of nitric oxide in vascular endothelial cells. J Biol Chem 278:45021–45026
32. Hattori Y, Suzuki M, Hattori S et al (2003) Globular adiponectin upregulates nitric oxide production in vascular endothelial cells. Diabetologia 46:1543–1549
33. Ouchi N, Kihara S, Funahashi T et al (2003) Obesity, adiponectin and vascular inflammatory disease. Curr Opin Lipidol 14:561–566
34. Kobashi C, Urakaze M, Kishida M et al (2005) Adiponectin inhibits endothelial synthesis of interleukin-8. Circ Res 97:1245–1252
35. Ardanaz P (2006) Hydrogen peroxide as a paracrine vascular mediator: regulation and signaling leading to dysfunction. Exp Biol Med (Maywood) 231:237–251
36. Gil-Longo J, González-Vázquez C (2005) Characterization of four different effects elicited by H2O2 in rat aorta. Vascul Pharmacol 43:128–138
37. Suvorava T, Lauer N, Kumpf S et al (2005) Endogenous vascular hydrogen peroxide regulates arteriolar tension in vivo. Circulation 112:2487–2495
38. Rogers PA, Dick GM, Knudson JD et al (2006) H2O2-induced redox-sensitive coronary vasodilation is mediated by 4-aminopyridine-sensitive K+ channels. Am J Physiol Heart Circ Physiol 291:H2473–H2482
39. Girouard H, de Champlain J (2005) Acute and chronic effects of free radicals on alpha1-adrenergic-induced vasoconstriction in mesenteric beds of spontaneously hypertensive rats. J Hypertens 23:807–814
40. Virdis A, Colucci R, Fornai M et al (2005) Cyclooxygenase-2 inhibition imporves vascular endothelial dysfunction in a rat model of endotoxic shock: role of inducible nitric-oxide synthase and oxidative stress. J Pharmacol Exp Ther 312:945–953
41. Engeli S, Schling P, Gorzelniak K et al (2003) The adipose-tissue renin-angiotensin-aldosterone system: role in the metabolic syndrome? Int J Biochem Cell Biol 35:807–825
42. Gorzelniak K, Engeli S, Janke J et al (2002) Hormonal regulation of the human adipose-tissue renin-angiotensin system: relationship to obesity and hypertension. J Hypertens 20:965–973
43. Paul M, Mehr AP, Kreutz R (2006) Physiology of the renin angiotensin systems. Physiol Rev 86:747–803
44. Phillips MI, Speakman EA, Kimura B (1993) Levels of angiotensin and molecular biology of the tissue renin angiotensin systems. Regul Pept 43:1–20
45. Cassis LA, Lynch KR, Peach MJ (1988) Localization of angiotensinogen messenger RNA in rat aorta. Circ Res 62:1259–1262
46. Engeli S, Gorzelniak K, Kreutz R et al (1999) Co-expression of renin-angiotensin system genes in human adipose tissue. J Hypertens 17:555–560

47. Fortuño A, Rodriguez A, Gomez-Ambrosi J et al (2002) Leptin inhibits angiotensin II-induced intracellular calcium increase and vasoconstriction in rat aorta. Endocrinology 143:3555–3560
48. Eringa EC, Bakker W, Smulders YM et al (2007) Regulation of vascular function and insulin sensitiviy by adipose tissue: focus on perivascular adipose tissue. Microcirculation 14:1–14
49. Thalmann S, Meier CA (2007) Local adipose tissue depots as cardiovascular risk factors. Cardiovasc Res 75:690–701
50. Gálvez-Prieto B, Cano MV, Delgado M et al (2008) A reduction in the amount and anti-contractile effect of periadventitial mesenteric adipose tissue precedes hypertension development in SHR. Hypertens Res 31:1415–1423
51. Somoza B, Guzmán R, Cano V et al (2007) Induction of cardiac uncoupling protein-2 expression and adenosine 5'-monophosphate-activated protein kinase phosphorylation during early states of diet-induced obesity in mice. Endocrinology 148:924–931
52. Iacobellis G, Ribaudo MC, Assael F et al (2003) Echocardiographic epicardial adipose tissue is related to anthropometric and clinical parameters of metabolic syndrome: a new indicator of cardiovascular risk. J Clin Endocrinol Metab 88:5163–5168
53. Ortlepp JR, Kluge R, Giesen K et al (2000) A metabolic syndrome of hypertension, hyperinsulinaemia and hypercholesterolemia in the New Zealand obese mouse. Eur J Clin Invest 30:195–202
54. Massiera F, Bloch-Faure M, Ceiler D et al (2001) Adipose angiotensinogen is involved in adipose tissue growth and blood pressure regulation. FASEB J 15:2727–2729
55. Boustany CM, Bharadwaj K, Daugherty A et al (2004) Activation of the systemic and adipose renin-angiotensin system in rats with diet-induced obesity and hypertension. Am J Physiol 287:943–949
56. Frederich RC Jr, Kahn BB, Peach MJ et al (1992) Tissue-specific nutritional regulation of angiotensinogen in adipose tissue. Hypertension 19:339–344
57. Giacchetti G, Faloia E, Mariniello B et al (2002) Overexpression of the renin-angiotensin system in human visceral adipose tissue in normal and overweight subjects. Am J Hypertens 15:381–388

Chapter 11
Endothelium: Dysfunction and Repair

Muhiddin A. Ozkor, Jonathan R. Murrow, and Arshed A. Quyyumi

11.1 Introduction

Endothelial function refers to a multitude of physiological processes of the vascular endothelium that maintain healthy homeostasis of the vascular wall and may be used as a barometer of the injury/repair inflicted by multiple environmental and genetic factors.[1-3]

This chapter reviews the structure and function of the normal endothelium, the factors that lead to abnormal endothelial function, the methods for measuring vascular dysfunction, and the prognostic value of these measures.

11.2 Endothelium – Structure and Function

The endothelium is a monolayer of cells covering the inner surface of blood vessels in arterial, venous, and capillary beds. Though initially seen as a simple barrier providing impermeability to blood vessels, demonstration of an endothelium-derived relaxing factor by Furchgott and Zawadzki and later the description of nitric oxide (NO) and the L-arginine-NO-cyclic guanosine monophosphate (cGMP) pathway by Ignarro and Murad instigated an avalanche of interest in this.[4]

Current understanding indicates that the endothelium is a dynamic organ, regulating the circulation in response to physical and chemical signals by the production of a wide range of factors. In response to shear stress, temperature change, bradykinin, and acetylcholine, the endothelium modulates flow by controlling vasodilator tone. Effector molecules elaborated by the healthy endothelium include NO, prostacycline, endothelium-derived hyperpolarizing

A.A. Quyyumi (✉)
Division of Cardiology, Emory University, 1364 Clifton Road, Suite D403C,
Atlanta, GA, 30322, USA
e-mail: aquyyum@emory.edu

D. Abraham et al. (eds.), *Advances in Vascular Medicine*,
DOI 10.1007/978-1-84882-637-3_11, © Springer-Verlag London Limited 2010

factors (EDHF), carbon monoxide (CO), and endothelin. These substances act in a paracrine fashion on the vascular smooth muscle and thus modulate vasomotor tone. In addition, the healthy endothelium promotes flow by its rheologic effects on blood components that include inhibition of clotting factor activation, platelet aggregation, and adhesion of inflammatory cells. Finally, recent work suggests that damage or denudation of the endothelium is repaired and regenerated by the action of local and circulating endothelial progenitor cells (EPCs).

11.3 Endothelial Dysfunction – Features and Mechanisms

In response to damaging environmental exposures and possibly exacerbated by genetic factors, protective features of the healthy endothelium may be diminished – predisposing individuals to the development of overt atherosclerosis and its untoward consequences. If endothelial health is typified by appropriate vasodilation, inhibition of platelet aggregation and clotting factors, maintenance of a barrier to inflammatory infiltrate, and by adequate measures of repair, then endothelial dysfunction is characterized by vasoconstriction, promotion of thrombosis, inflammation and smooth muscle proliferation, and reduced number and function of endothelial progenitors.[5]

Traditional cardiovascular risk factors that precipitate atherosclerosis are also associated with endothelial dysfunction. These include sedentary lifestyle, obesity, hypercholesterolemia, hypertension, diabetes mellitus, insulin resistance, tobacco smoking, aging, and others[3,6,7] (Table 11.1). The extent of endothelial dysfunction appears to correlate with the traditional risk factor "burden," implying that combined or repeated injury to the vascular endothelium results in greater dysfunction. However, there is considerable heterogeneity in the magnitude of dysfunction observed in individuals with similar risk factor profiles. Novel risk factors such as infections, hyperhomocysteinemia, genetic heterogeneity, and the variable duration of exposure to individual risk factors presumably account for some of this observed variability[8-12] (Table 11.1).

These exposures promote oxidative processes in the vascular wall that result in the production of reactive oxygen species, which ultimately lead to activation of the vascular renin–angiotensin system, transcriptional factors, growth factors, proinflammatory cytokines, chemoattractant substances, and adhesion molecules.[6,7] This complex cascade of events reduces the bioavailability of NO and underlies the transition that occurs from normal endothelial function to endothelial dysfunction. In this way, endothelial dysfunction can be construed as a precursor or *forme fruste* of atherosclerosis, as these features precede lesion development.[13] Thus, the severity and progression of atherosclerotic disease are critically determined by the functional integrity of the endothelium.[14-26]

Table 11.1 Risk factors associated with endothelial dysfunction in the absence of clinical atherosclerosis

Sedentary lifestyle
Hypertension
Hypercholesterolemia
Diabetes
Heart failure
Estrogen withdrawal
Age
Smoking
Multiple infections
Homocystinuria
Prinzmetal's angina
Insulin resistance
Inflammation
Rheumatoid arthritis, systemic lupus erythemotosis
Chronic renal disease
Multiple infections
Periodontitis
Sleep apnea
Erectile dysfunction
Polycystic ovary syndrome
Nitrate tolerance
Alzheimer's disease

11.4 Nitric Oxide

NO is the predominant mediator of normal vascular function, and its reduced bioavailability is associated with endothelial dysfunction, oxidative stress, and inflammation.[27,28] NO is a soluble gas synthesized by endothelial NO synthase (eNOS) from L-arginine, a process that requires the cofactor tetrahydrobiopterin for catalytic activity[29] (Fig. 11.1). Endothelial NOS is normally present as a dimer that acts in a "coupled" state to generate NO in the presence of adequate levels of tetrahydrobiopterin.[30] In the initiating step of L-arginine oxidation, tetrahydrobiopterin donates an electron to the ferrous–dioxygen complex in the oxygenase domain, leading to separation of the dioxygen and formation of an iron–oxy species. On generation of NO, tetrahydrobiopterin is converted into 7,8-dihydrobiopterin and is recycled back to tetrahydrobiopterin in a two-step process. Lack of tetrahydrobiopterin impairs this process and leads to generation of superoxide from the ferrous–dioxygen complex instead of NO.[30] The superoxide anion is an unstable free radical that rapidly reacts with free NO to form peroxynitrite that can in turn react with other compounds to generate several reactive oxygen species. These reactive oxygen species are intermediate products of the electron transport chain, are electrochemically unstable, and lead to a state of increased oxidative stress. This phenomenon, often precipitated by tetrahydrobiopterin deficiency, has been referred to

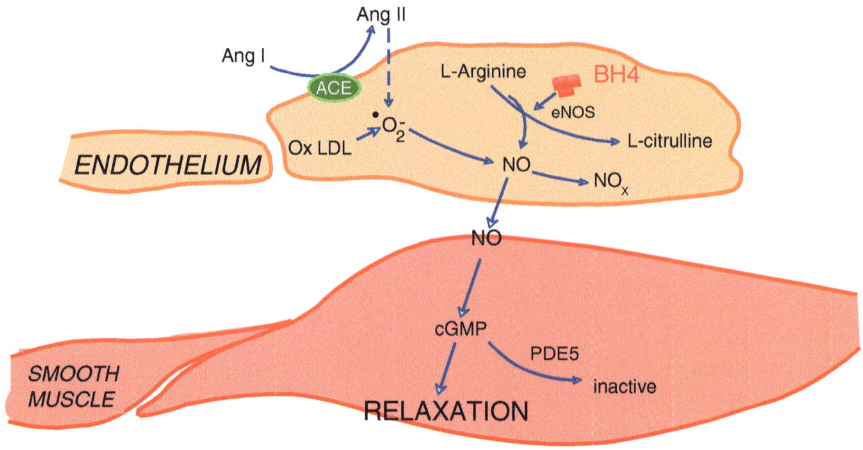

Fig. 11.1 Nitric oxide stimulated smooth muscle relaxation. NO is a soluble gas synthesized by endothelial NO synthase from L-arginine, a process that requires the cofactor tetrahydrobiopterin for catalytic activity. Endothelial NOS is normally present as a dimer that acts in a "coupled" fashion to generate NO. Lack of tetrahydrobiopterin leads to the generation of superoxide anions and peroxynitrite, resulting in a state of increased oxidative stress. This phenomenon is referred to as eNOS "uncoupling." *Ang I* Angiotensin I; *Ang II* angiotensin II; *ACE* angiotensin converting enzyme; *oxLDL* oxidized low-density lipoprotein; (O_2^-) superoxide anions; *BH4* tetrahydrobiopterin; *NO* nitric oxide; *eNOS* endothelial nitric oxide synthase; NO_x peroxynitrite; *cGMP* cyclic guanosine monophosphate; *PDE5* phosphodiesterase type 5

as eNOS "uncoupling," and characterizes an abnormal state of eNOS function that underlies endothelial dysfunction, where there is diminution of NO and increased production of reactive oxygen species.

NO release can be stimulated by endogenous factors such as bradykinin, catecholamines, and ischemia, and by physical or mechanical stimuli, including temperature changes and increases in shear stress. Release of NO leads to smooth muscle vasodilation by a cGMP-dependent mechanism activating intracellular guanylate cyclase, and hence is a major determinant of resting vasomotor tone.

Absent or reduced activity of NO predisposes the vessel wall not only to increased vasoconstrictor tone that could lead to spasm, hypertension, and exacerbation of ischemia, but also to a variety of long-term deleterious effects on the vasculature. Reduced NO bioavailability is responsible for nuclear factor kappa beta-(NF-κβ-) dependent activation of selectins and other adhesion molecules that promote vascular inflammation and ultimate development of atherosclerosis. It increases thrombogenicity of platelets and coagulability of blood, increasing the likelihood of thrombosis.[31] Finally, reduced NO bioavailability is associated with reduced counts and functional measures of EPCs – in part since the release of EPCs from the bone marrow is an NO-dependent process.[32-34] Given the central role of NO in mediating endothelial dysfunction, impairment of vasodilation due to decreased NO availability has often been used as a measure of endothelial function.[35-39]

11.5 Prostaglandins

Factors that stimulate endothelial release of NO also activate the cyclooxygenase pathway to generate vasoactive metabolites of arachidonic acid. Prostaglandin generation differs by species and circulatory beds; however, foremost among these active biological products are the vasodilator, prostacyclin, and vasoconstrictor, thromboxane A2. Prostacyclin elicits smooth muscle relaxation by activating specific cell surface receptors (IP) that are G-protein-coupled to adenylyl cyclase (AC), thereby elevating cyclic adenosine monophosphate levels (Fig. 11.2). It acts independently of NO as another endothelium-derived vasodilator by desensitization of the smooth muscle contractility to calcium.[40] Prostacyclin also possesses nonvasoactive properties that include promotion of fibrinolysis and inhibition of platelet adhesion and aggregation.[41]

The contribution of prostacyclin to endothelium-dependent relaxation is often evident only after inhibition of eNOS, that is believed to be due to a tonic inhibitory effect of NO. NO may either enhance or inhibit cyclooxygenase activity and expression.[42] Thromboxane also promotes platelet aggregation, and this balance between prostacyclin and thromboxane can vary critically in healthy and diseased states. In human blood vessels, it appears that NO and not prostacyclin, is the major mediator of vascular tone.

11.6 Endothelium-Derived Hyperpolarizing Factor (EDHF)

Even after complete inhibition of NO and prostaglandin synthesis, endothelium-dependent vasodilation persists via hyperpolarization of vascular smooth muscle cells. This NO- and prostaglandin-independent component of vasodilator tone has been attributed to EDHF. EDHF can compensate for loss of NO-mediated vasodilator tone, particularly in the microcirculation.[43]

Potential EDHFs differ by species and vascular bed. They act by increasing potassium conductance resulting in the subsequent propagation of depolarization of vascular smooth muscle cells and relaxation[44-46] (Fig. 11.2). In the human vasculature, endothelium-dependent hyperpolarization is at least partly caused by the release of epoxyeicosatrienoic acids from the cytochrome P450-dependent metabolism of arachidonic acid. These products promote vasodilation by stimulating small and large calcium-dependent potassium channels on endothelial cells and subsequent hyperpolarization[47] (Fig. 11.2). The EDHF phenomena may be further explained by the transmission of endothelial cell hyperpolarization to the vascular smooth muscle via gap junctions.[37,48,49] These couple endothelial cells to other endothelial cells and to smooth muscle cells, providing a low-resistance electrical pathway between the cell layers. Gap junctions are formed by the docking of two connexons present in adjacent cells that creates an aqueous pore permitting the transfer of ions and electrical continuity that establishes a uniform membrane potential across cells.[50,51] Their number increases with diminution in

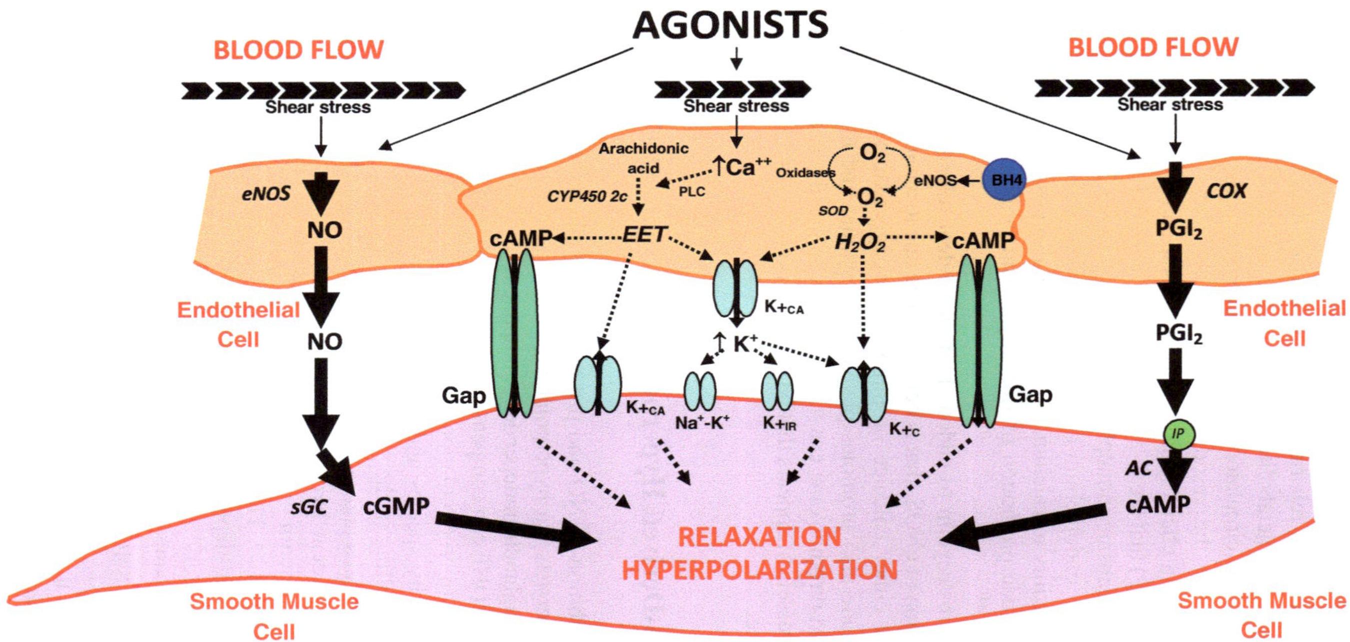

Fig. 11.2 Mechanisms for endothelial cell mediated relaxation. Agonist (bradykinin/acetylcholine/substance P) or shear stress increases the activity of eNOS and cyclooxygenase (COX), providing NO and prostacyclin (PGI_2)-mediated dilation. There are multiple potential EDHF pathways. Increases in intracellular calcium activates phospholipase A2 (PLC) to produce arachadonic acid. Its metabolism by cytochrome P450 2C (CYP4502c) generates eicosatrienoic acids (EETs) that can stimulate calcium-dependent potassium (K_{Ca}) channels in endothelial and smooth muscle cells. EETs may also directly activate gap junctions (Gap). The increase in potassium in the interstitium may activate K_{Ca} channels, inwardly rectifying potassium channels (K_{IR}^+), or the Na^+-K^+ pump on smooth muscle cells and cause hyperpolarization. The action of eNOS (with cofactor tetrahydrobiopterin [BH4]) and oxidases on oxygen (O_2) produces the reactive oxygen species superoxide ($\cdot O_2^-$). Hydrogen peroxide (H_2O_2) generated by dismutation of superoxide anions by superoxide dismutase (SOD) can also cause hyperpolarization by activating endothelial and smooth muscle K_{Ca} channels or by gap junctions. *AC* Adenylyl cyclase; *cAMP* cyclic adenosine monophosphate; *cGMP* cyclic guanosine monophosphate; *sGC* soluble guanylyl cyclase; *IP* prostacyclin receptor

the size of the artery, paralleling the importance of EDHF to vessel size with a greater influence in the resistance than in the conductance vessels. Various endothelial oxidases, including NO synthase, lipoxygenases, cytochrome P450 epoxygenases, NAD(P)H oxidases, and xanthine oxidase generate superoxide anions that are degraded to hydrogen peroxide spontaneously or through superoxide dismutase – dependent dismutation.[52] Hydrogen peroxide also has the ability to activate calcium-dependent potassium channels and remains a contender as an EDHF[53] (Fig. 11.2).

The three main mediators of endothelial vasodilator function, NO, prostacyclin, and EDHF, appear not to be mutually exclusive and act synergistically in a complex manner to maintain the health of the vasculature (Fig. 11.2). These interactions have been demonstrated in the physiological maintenance of vascular tone where the contribution of each vasodilator is not equally distributed, and further, as a compensatory mechanism in response to the deficiency of an alternate mediator. In conduit arteries, NO is the predominant endothelium-derived vasodilator but has relatively less prominent contribution in the resistance vessels of the microcirculation where EDHF appears to predominate.[54] NO may tonically inhibit EDHF responses as some studies could only demonstrate EDHF responses once NO production had been inhibited.[55]

Alteration of EDHF-mediated responses have been reported with aging, hypertension, atherosclerosis, hypercholesterolemia, heart failure, angioplasty, eclampsia, diabetes, and sepsis. Depending on the vascular bed, this may either contribute to endothelial dysfunction, or compensate for the loss of NO bioavailability.[56-58] In the human forearm of hypertensive subjects, endothelium-dependent vasodilation is maintained despite decreased NO bioavailability because of the compensatory increased activity of EDHF.[57,58] Hypercholesterolemia is generally associated with preserved EDHF responses where its enhanced activity may compensate for the decrease in NO-mediated relaxation.[59-61] Endothelium-dependent hyperpolarization appears to be inhibited in isolated gastroepiploic arteries from atherosclerotic patients, an effect that may be secondary to the duration of hypercholesterolemic injury.[62] In contrast, EDHF-mediated responses are depressed in some models of type I and type II diabetes with the exception of murine models.[63] Thus, whether EDHF plays a causal or compensatory role in the endothelial dysfunction in the human circulation remains to be elucidated.

11.7 Carbon Monoxide

CO is a potent vasodilator in most vascular beds. It is an endogenously derived gas formed predominantly from the breakdown of heme by the enzyme heme oxygenase. This oxidative reaction serves as the first and rate-limiting step in heme catabolism by one of two distinct isoforms of heme oxygenase that are expressed in the vascular endothelium and smooth muscle.[64] Mechanisms underlying CO-induced vasodilation include the stimulation of soluble guanylate cyclase, inhibition of cytochrome P450, or activation of potassium channels.[64]

Evidence suggests that the heme oxygenase-CO system plays a beneficial role against atherosclerosis progression with its expression in the endothelium and foam cells of atherosclerotic lesions.[65] Oxidized low-density lipoprotein is a potent inducer of heme oxygenase in human vascular cells,[66] and its deficiency is associated with early atherosclerotic changes, and a polymorphism in the promoter region that is linked to reduced expression of heme oxygenase appears to be associated with susceptibility to coronary artery disease.[67-70] Though it is a vasodilator, CO is also a tonic inhibitor of eNOS as a result of binding to its prosthetic heme.

11.8 Endothelin

Shortly after the identification of the endothelial-derived relaxing factor, an opposing and potent vasoconstrictor agent with a novel 21-amino acid peptide sequence, called endothelin-1, was discovered.[71] Two additional endothelin-like peptides termed endothelin-2 and endothelin-3 were subsequently identified. Endothelin-1 is the predominant isoform that is produced and released primarily by endothelial cells and is found to a lesser extent in cardiomyocytes, macrophages, leukocytes, and mesangial cells. By stimulating specific receptors, endothelin-1 manifests a profound vasoconstrictor effect, stimulates proliferation and hypertrophy of vascular smooth muscle cells, and promotes cell adhesion and thrombosis.[71,72] It is synthesized by cleavage of a 203-amino acid, pre-pro-endothelin by a family of metalloproteases located in the vascular endothelial and smooth muscle cells, in a protein kinase C-dependent manner. Endothelin-1 is released in response to physical and chemical stimuli such as stretch, shear stress, and pH, and is upregulated with exercise, hypoxia, elevated levels of oxidized low-density lipoprotein and glucose, estrogen deficiency, obesity, cocaine use, aging, and thrombin.

Circulating levels of endothelin-1 appear to be lower in adults when compared with children and differ by ethnicity. Elevated levels have been found in both tissue and plasma of those with coronary artery disease, myocardial infarction, heart failure, pulmonary hypertension, renal failure, and after mechanical vascular injury from percutaneous coronary intervention. In experimental hypercholesterolemia, chronic endothelin receptor antagonism preserves coronary endothelial function and increases NO activity.[73] Moreover, selective or nonselective endothelin receptor blockade normalized NO-mediated endothelial dysfunction and reduces atheroma formation in models of atherosclerosis.[74-76] Thus, vasomotor tone in vivo is modulated by a balance between endothelially derived vasoconstrictors and vasodilators, with increased constrictor tone being mediated by endothelin-1 and vasoconstrictor prostanoids.[77,78]

11.9 Measures of Endothelial Dysfunction

In humans, endothelial dysfunction can be assessed by measuring vasodilation in response to specific endothelium-dependent pharmacologic or physiologic stimuli either in vitro (isolated arteries), or in vivo in the forearm, coronary, or peripheral circulations.

11.9.1 Noninvasive Measures of Endothelium-Dependent Vasodilation

Flow-mediated dilation (FMD) involves the measurement of the change in diameter of the brachial or radial artery using high-resolution ultrasound where vasodilation is induced by increased shear stress, a stimulus for NO release by the normal endothelium. Reactive hyperemia following a 5-min period of forearm ischemia causes arterial NO-dependent vasodilation and has been used extensively and reproducibly in clinical studies, but is highly operator-dependent.[38,79,80] FMD is predominantly NO-mediated, it correlates with coronary endothelial function,[39] and is depressed in subjects with atherosclerosis and those with cardiovascular risk factors.[35,37] Vascular responses to the endothelium-independent dilator, nitroglycerine, is used to ensure that any depression in measured FMD is not secondary to smooth muscle dysfunction.

Alternative approaches for measurement of endothelial function have involved modification of arterial compliance tests. Systolic pulse contour analysis by radial artery tonometry provides an indirect assessment of central aortic waveforms, and the reduction in arterial stiffness with the β2 agonist salbutamol appears to be an NO-dependent effect.[81] Changes in augmentation index in response to salbutamol can be measured from the peripheral arterial waveform and correlate with blood flow responses to acetylcholine, as a measure of NO bioavailability, which have been validated in adults and shown to change with exposure to risk factors and atherosclerotic disease.[81-85]

Pulsatile arterial tonometry has also been introduced as a reproducible technique for assessment of endothelial function.[86] The ratio of the finger-tip hyperemic response to the basal flow appears to correlate with coronary endothelial function,[87] and with risk factors.[38] This technique is now being extensively evaluated in large epidemiologic and interventional studies.

11.9.2 Invasive Measures of Endothelium-Dependent Vasodilation

Endothelial function in the microcirculation (resistance vessels) can be assessed by measurement of blood flow changes using strain gauge venous plethysmography of the forearm in response to pharmacologic stimulation.[88-90] Intra-arterial infusions of endothelium-dependent vasodilators such as acetylcholine, bradykinin, and substance P evaluates the capacity for release of endothelium-derived vasodilators, and use of specific antagonists such as L-N^G monomethyl arginine allows assessment of NO activity, and the change in response after vitamin C indicates the contribution of oxidative stress to the reduced endothelial function. Importantly, endothelial dysfunction detected as reduced microvascular vasodilator response to acetylcholine was associated with a worse long-term outcome.[24]

11.9.3 Coronary Endothelial Function

Endothelial vasomotion measured in the coronary circulation may be considered to be the "gold standard" method for assessment of endothelial function. It can be measured as the vasodilator response to intracoronary infusion of acetylcholine or other endothelium-dependent probes, by the use of quantitative coronary angiography for assessment of changes in conduit vessel diameter, and intracoronary Doppler for the direct calculation of changes in coronary blood flow and coronary vascular resistance.[41,91] The normal endothelial response to acetylcholine is dilation of both epicardial vessels and microcirculation, resulting in an increase in coronary blood flow. With endothelial dysfunction, the epicardial vasodilator responses become attenuated or paradoxical vasoconstriction occurs, and a less robust increase in blood flow will be seen indicating epicardial and microcirculatory endothelial dysfunction, respectively. Endothelium-independent function can be assessed as the vasodilator response to sodium nitroprusside and flow reserve as the vasodilator response to adenosine.

11.10 Endothelial Dysfunction and Prognosis

Endothelial dysfunction, both in the coronary and peripheral vasculature, is a predictor of adverse long-term cardiovascular events in patients with coronary disease, hypertension, heart failure, or atherosclerosis.[14,15,25,92-97] In patients with and without significant coronary disease undergoing coronary angiography, the presence of epicardial or microvascular endothelial dysfunction appears to be an independent predictor of cardiovascular events, including death, myocardial infarction, stroke, and hospitalization for unstable angina.[14-16,92,98,99] Endothelial dysfunction is also a predictor of future development of hypertension, diabetes, progression of atherosclerosis, and of adverse cardiovascular events, independent of the risk factor burden.[16,100]

11.11 Biomarkers of Vasodilator Capacity

Abnormalities of endothelial function may also be detected by the measurement of circulating biomarkers reflective of endothelial health. These include measures of NO biology, inflammatory cytokines, adhesion molecules, regulators of thrombosis, as well as markers of endothelial damage and repair.[101]

Direct measurement of NO or its oxidized products, nitrites, and nitrates in circulating blood is fraught with difficulties because of the diet-related and non-vascular sources of nitrates. Asymmetric dimethylarginine (ADMA) is an endogenously derived competitive antagonist of eNOS that may reflect the functional status of endothelial health.[102] Intra-arterial infusion of ADMA reduced forearm blood flow as measured by venous occlusion plethysmography in the initial investigations reporting its role as an endogenous inhibitor of NO.[103,104] The enzymatic breakdown of ADMA is sensitive to altered redox conditions associated with inflammation and cardiovascular risk factors.[105] ADMA levels are elevated in cardiovascular risk states and with risk factor burden, including hypercholesterolemia,[106] hypertension,[107] diabetes,[108,109] hyperhomocystinemia,[110] and in individuals with overt atherosclerotic disease.[111,112] Levels are increased in heart failure,[113-115] renal failure,[116] and pulmonary hypertension.[117] A fall in ADMA level occurs during normal pregnancy, but is increased in women with preeclampsia.[118-120]

Inflammatory cytokines and adhesion molecules may also provide an indication of endothelial function with levels increasing with endothelial cell activation. Such molecules include E-selectin, vascular cell adhesion molecule 1, intercellular adhesion molecule 1, and P-selectin.[38,121] Elevations of C-reactive protein[122] levels have an inverse correlation with forearm blood flow responses to acetylcholine in patients with coronary artery disease.[123] In the community-based Framingham Heart Study of 2,113 participants, FMD positively correlated with plasma N-terminal proatrial natriuretic peptide and renin after multivariable analyses (adjusting for known risk factors) but did not correlate with CRP.[124]

Novel markers of oxidative stress may be important in providing an indication of endothelial dysfunction. The quantification of oxidative stress in terms of the balance of the endogenous glutathione and cysteine antioxidant systems provides a measure of the balance of prooxidants and antioxidants.[125] We have recently performed a cross-sectional study in 124 healthy nonsmokers, without risk factors or known cardiovascular disease, between 35 and 60 years of age.[126] Several markers of oxidative stress including the plasma thiols, reduced and oxidized glutathione and glutathione redox, plasma cysteine and its oxidized form, cystine and cysteine redox and their mixed disulfide, CySSG, were measured. The lipid hydroperoxides were estimated using the "determination of reactive oxygen metabolites" or d-ROMs assay, allowing quantification of peroxide and alcohol-like species in the plasma.[126-128] Among the biomarkers of oxidative stress, significant correlations between endothelial function and cystine, cysteine redox, the mixed disulfide, CySSG, and d-ROMS were observed, indicating that a higher level of oxidative stress was associated with lower flow-mediated vasodilation, even after adjustment for high sensitivity C-reactive protein (hsCRP) and other risk factors. Thus, serum markers of oxidative stress appear to be predictors of endothelial function.[129]

11.12 Endothelial Repair and Regeneration

The classic paradigm attributes repeated endothelial cell injury, from cardiovascular risk factors, as the stimulus for atherosclerotic plaque development.[130] This repetitive or prolonged exposure to risk factors leads to dysfunction and anatomical changes in the integrity of endothelial cells that may culminate in detachment of endothelial cells and their presence in the circulation. Recently, the concept of unchecked risk factor-mediated injury to the vascular wall has been revised in light of the discovery of a population of endogenous mononuclear cells that reside in the bone marrow, mobilize to areas of tissue injury, and repair injured vascular tissue. These have been termed EPCs. A second potential mechanism for vascular repair after injury is the local replication of neighboring mature endothelial cells. Under normal circumstances, endothelial cells are quiescent and replicate at a very slow rate with a turnover time greater than hundreds of days. It has been suggested that this can accelerate to less than 5 days in acute stress or disease states, which results in endothelial damage.[131,132] However, the capacity of endothelial cells to divide is limited and the protective role of this mechanism may not be sufficient in settings of increased risk factor burden.[133] Ultimately, the balance between injury and repair is believed to be the major determinant of vascular disease progression, with EPC mobilization and function a key factor in the process.[134-137]

The precise identity of EPCs remains elusive. As a result, they have been defined by their functions – originating largely in the bone marrow,[137,138] circulating in peripheral blood, homing to sites of vascular injury, and participating in new blood vessel formation,[139] in addition to having the capabilities of differentiating into endothelial cells both in vitro and in vivo.[140] Ever since the notion was first entertained that blood vessels could be formed de novo from circulating blood components, work has sought to define the true identity of these EPCs, to understand their role in vascular homeostasis and pathology, and to relate their functions to common clinical syndromes.

Asahara and colleagues in 1997 provided the strongest evidence that new blood vessel formation is partly attributable to a population of bone marrow-derived monocytes. CD34-positive mononuclear cells isolated from humans demonstrated an endothelial phenotype in vitro. Moreover, these cells participated in neo-angiogenesis in a mouse hindlimb ischemia model.[135] Thus, this work defined a cell population that exhibited the key features of an endothelial progenitor – a bone marrow heritage, an ability to mobilize and to home to sites of injury, and a role in the making of new blood vessels.

11.12.1 Identifying Endothelial Progenitor Cells – Cell Surface Markers

Human hematopoietic stem cells have been isolated primarily through their expression of the marker CD34,[141,142] and by the lack of lineage markers.[143] EPCs (and thus endothelial cells) are believed to arise from the differentiation of this common precursor. While this marker is lost by hematopoietic cells as they differentiate into blood components, it is retained by a population of cells that have phenotypic characteristics of an endothelial progenitor.[144-146] As these EPCs differentiate into mature endothelial cells, the CD34 surface marker is lost. Thus, the CD34 marker serves to identify in the peripheral circulation a population of cells enriched for EPCs but not mature endothelial cells.

AC133 (CD133) is a more primitive hematopoietic stem cell marker that is expressed on the majority of CD34-positive cells. It is a 5-transmembrane antigen identifying a population of cells that demonstrate an EPC phenotype when cultured with vascular endothelial growth factor (VEGF) and stem cell growth factor.[147,148] Unlike CD34, though, the expression of AC133 is lost during maturation of EPCs and thus allows an earlier and perhaps more precise identification of EPCs. The ability of injected CD34-positive cells to localize and colonize an implanted Dacron aortic graft and the presence of AC133-positive cells in the neo-intima that lines implanted ventricular assist devices in heart failure patients provides evidence for the contribution of EPCs to re-endothelialization of tissues.[136,149]

Vascular endothelial growth factor receptor (VEGFR-2, kinase-domain receptor) is another endothelial-specific marker expressed on mature endothelial cells, as well as, on cell populations enriched for endothelial progenitors.[149,150] The VEGFR-2 marker defines peripheral mononuclear cells that contribute to neo-endothelialization, and its coexpression with AC133 has been used to identify a population enriched for endothelial progenitors.[149,151] Of note, these double positive cells are quite rare, representing between 0.01 and 0.0001% of the circulating mononuclear population.[152]

However, difficulties in the precise characterization of EPCs arise from the presence of nonhematopoietic stem cells that have multipotent potential and the observation in peripheral blood of CD34-negative and AC133-negative cells capable of proliferating to express mature endothelial cell markers and forming endothelial colonies.[153-155]

11.12.2 Mechanisms: Mobilization, Homing, Pathogenesis

In the setting of tissue hypoxemia, EPCs are released into the blood stream from the bone marrow, circulate in the blood stream, and home to areas of cell injury and ischemia.[137,156-158] Here, they contribute to new blood vessel formation during tissue repair, by a variety of paracrine and proliferative mechanisms, including the production of proangiogenic growth factors,[140] in a process that closely resembles embryonic vasculogenesis.[140,158-161]

11.12.2.1 Mobilization

Quantitative measures of EPCs by flow cytometry and by culture demonstrates an ability of this cell population to mobilize in response to a variety of homeostatic and pathologic stimuli. This mobilization from the bone marrow to sites of vascular injury is thought to be mediated by matrix metalloproteinases in an NO-dependent process.[162] Matrix metalloproteinases-2 and matrix metalloproteinases-9 are proteolytic enzymes that contribute to hematopoietic cell migration and tissue localization in both health and in response to injury.

Circulating endogenous cytokines also contribute to EPC mobilization. VEGF activates specific tyrosine-kinase receptors resulting in EPC mobilization from the bone marrow. Increased levels of VEGF are associated with angiogenesis, and exogenously administered VEGF increases EPC counts in peripheral blood as well as postnatal angiogenesis in vivo.[163] Further, after coronary bypass, an early sharp rise in circulating EPCs (up to 50-fold increase) coincided with a VEGF peak which returned to baseline after 48–72 h.[164]

Mobilization of EPCs is in part NO-dependent and as demonstrated by its importance in neovascularization in eNOS-deficient mice. VEGF-mediated mobilization of EPCs is reduced in these mice. Infusion of wild-type EPCs, but not bone marrow transplantation, rescues mice and suggests a defect in progenitor cell mobilization (mediated by matrix metalloproteinases-9) as the mechanism of impaired neovascularization to hindlimb ischemia in this mouse model.[165] Mobilization may also be impaired in patients with cardiovascular risk factors due to there effect on NO.

Erythropoetin when administered exogenously can also enhance EPC mobilization from the bone marrow. Moreover, serum erythropoeitin and VEGF levels are associated with bone marrow derived progenitor cells as well as with the number and function of circulating EPCs.[166] Other more potent mobilizing agents include granulocyte colony stimulating factor and granulocyte macrophage colony stimulating factor.

11.12.2.2 Homing

Once mobilized from the bone marrow, circulating EPCs home to sites of vascular injury via binding of locally produced factors to cell surface receptors.[167] At the site of injury, an endogenous integrin-linked kinase (ILK) responds to hypoxia in endothelial cells by upregulating intracellular adhesion molecules including stromal-derived factor-1 that is able to bind receptors expressed on circulating CD34-positive mononuclear cells such as CXCR4.[168-171]

11.12.2.3 Pathogenesis

Circulating levels of EPCs are depressed in subjects with atherosclerosis or cardiac risk factors.[134,172] Preclinical work has identified several mechanisms by which EPC function is impaired. Whether this represents a primary deficiency in vascular repair potential or a suppression of circulating reparative cells in response to exposure to risk factors (i.e., tobacco smoke, diabetes, etc.) is still undetermined. C-reactive protein has been identified as a biomarker associated with increased cardiovascular risk.[173] While hsCRP is associated with increased colony forming units within a healthy cohort, when isolated EPCs from peripheral blood were cultured in the presence of CRP (>15 μg/mL), the number of colony forming units and expression of endothelial markers were reduced.[174,175] Further, exposure to CRP was associated in a decrease in eNOS mRNA expression by EPCs.[175] Additionally, plasma concentrations of ADMA in subjects with stable angina inversely correlates with the number of circulating CD34-positive/AC133-positive progenitor cells and EPC colony forming units. Ex vivo studies show that incubation with ADMA results in decreased EPC colony forming unit counts, decreased differentiation, and attenuated tubule formation. Endothelial NOS activity was decreased in EPCs incubated in ADMA, suggesting a possible mechanism of effect. Notably, coincubation of EPCs with rosuvastatin abolished these deleterious effects of ADMA.[176]

11.13 EPCs, Risk Factors, and Cardiovascular Events

Recent studies including ours have shown a reduction in the number and migratory activity of EPCs in patients with cardiovascular disease, and a negative correlation between EPC function and cardiovascular risk factors or the Framingham risk factor score.[33,134,172,177-181] Endothelial dysfunction in our studies and others correlated with these functional measures of EPCs.[134,177] We have also found that vascular injury, mediated by coronary stenting, causes a robust increase in circulating levels of EPCs for at least a week. Moreover, this increase was attenuated in older subjects, diabetics, and those with a history of stroke. It is intriguing that low EPC counts were an independent risk factor of poor outcome in two separate disease states characterized by endothelial injury, (a) in patients with CVD,[182,183] and (b) in our study of patients with acute lung injury.[184]

Clinical translation of these observations reveals that circulating EPCs levels and their function are (1) modulated by age and risk factors for atherosclerosis, (2) stimulated by vascular and tissue injury and medical therapy such as statins, and (3) may predict cardiovascular outcome in patients with coronary artery disease.[33,134,172,177-183]

11.14 Endothelial Dysfunction – Potential Therapy

Current strategies associated with improving endothelial function to reduce cardiovascular events in those at risk include life style changes such as weight loss and increased physical activity, as well as pharmacologic interventions with statins, angiotensin converting enzyme (ACE) inhibitors, angiotensin-1 receptor antagonists, or peroxisome proliferator-activated receptors – γ agonists.[34,172,185] Novel strategies to improve endothelial function may therefore have the potential to improve prognosis in vascular disease.

References

1. Quyyumi AA, Dakak N, Mulcahy D, et al. Nitric oxide activity in the atherosclerotic human coronary circulation. *J Am Coll Cardiol*. 1997;29:308–317.
2. Ashfaq S, Beinart SC, Abramson JL, et al. Plasma glutathione redox state: a novel marker of oxidative stress, correlates with early atherosclerosis in humans. *J Am Coll Cardiol*. 2003;41(suppl A):293A–294A.
3. Quyyumi AA. Endothelial function in health and disease: new insights into the genesis of cardiovascular disease. *Am J Med*. 1998;105:32S–39S.
4. Furchgott RF, Zawadzki JV. The obligatory role of endothelial cells in the relaxation of arterial smooth muscle by acetylcholine. *Nature*. 1980;288:373–376.
5. Quyyumi AA. Prognostic value of endothelial function. *Am J Cardiol*. 2003;91:19H–24H.
6. Cai H, Harrison DG. Endothelial dysfunction in cardiovascular diseases. The role of oxidant stress. *Circ Res*. 2000;87:840–844.
7. Dzau VJ. Theodore cooper lecture: tissue angiotensin and pathobiology of vascular disease: a unifying hypothesis. *Hypertension*. 2001;37:1047–1052.
8. Prasad A, Zhu J, Halcox JP, Waclawiw MA, Epstein SE, Quyyumi AA. Predisposition to atherosclerosis by infections: role of endothelial dysfunction. *Circulation*. 2002;106:184–190.
9. Quyyumi AA. Does acute improvement of endothelial dysfunction in coronary artery disease improve myocardial ischemia? A double–blind comparison of parenteral D– and L–arginine. *J Am Coll Cardiol*. 1998;32:904–911.
10. McDermott DH, Halcox JP, Schenke WH, et al. Association between polymorphism in the chemokine receptor CX3CR1 and coronary vascular endothelial dysfunction and atherosclerosis. *Circ Res*. 2001;89:401–407.
11. Zhu J, Quyyumi AA, Wu H, et al. Increased serum levels of heat shock protein 70 are associated with low risk of coronary artery disease. *Arterioscler Thromb Vasc Biol*. 2003;23:1055–1059.
12. Zhu J, Quyyumi AA, Rott D, et al. Antibodies to human heat-shock protein 60 are associated with the presence and severity of coronary artery disease: evidence for an autoimmune component of atherogenesis. *Circulation*. 2001;103:1071–1075.
13. Landmesser U, Hornig B, Drexler H. Endothelial function: a critical determinant in atherosclerosis? *Circulation*. 2004;109:II27–II33.
14. Schachinger V, Britten MB, Zeiher AM. Prognostic impact of coronary vasodilator dysfunction on adverse long-term outcome of coronary heart disease. *Circulation*. 2000;101:1899–1906.
15. Suwaidi JA, Hamasaki S, Higano ST, Nishimura RA, Holmes DR Jr, Lerman A. Long-term follow-up of patients with mild coronary artery disease and endothelial dysfunction. *Circulation*. 2000;101:948–954.
16. Halcox JP, Schenke WH, Zalos G, et al. Prognostic value of coronary vascular endothelial dysfunction. *Circulation*. 2002;106:653–658.
17. Sorensen KE, Kristensen IB, Celermajer DS. Atherosclerosis in the human brachial artery. *J Am Coll Cardiol*. 1997;29:318–322.
18. Luscher TF, Barton M. Biology of the endothelium. *Clin Cardiol*. 1997;20:II–3–II–10.
19. Pepine CJ. Why vascular biology matters. *Am J Cardiol*. 2001;88:5K–9K.
20. Anderson TJ, Uehata A, Gerhard MD, et al. Close relation to endothelial function in the human coronary and peripheral circulation. *J Am Coll Cardiol*. 1995;26:1235–1241.

21. Quyyumi AA, Dakak N, Andrews NP, et al. Nitric oxide activity in the human coronary circulation. Impact of risk factors for coronary atherosclerosis. *J Clin Invest.* 1995;95:1747–1755.
22. Hirooka Y, Egashira K, Imaizumi T, et al. Effect of L-arginine on acetylcholine-induced endothelium-dependent vasodilation differs between the coronary and forearm vasculatures in humans. *J Am Coll Cardiol.* 1994;24:948–955.
23. Neunteufl T, Heher S, Katzenschlager R, et al. Late prognostic value of flow-mediated dilation in the brachial artery of patients with chest pain. *Am J Cardiol.* 2000;86:207–210.
24. Heitzer T, Schlinzig T, Krohn K, Meinertz T, Munzel T. Endothelial dysfunction, oxidative stress, and risk of cardiovascular events in patients with coronary artery disease. *Circulation.* 2001;104:2673–2678.
25. Perticone F, Ceravolo R, Pujia A, et al. Prognostic significance of endothelial dysfunction in hypertensive patients. *Circulation.* 2001;104:191–196.
26. Gokce N, Keaney JF Jr, Hunter LM, Watkins MT, Menzoian JO, Vita JA. Risk stratification for postoperative cardiovascular events via noninvasive assessment of endothelial function: a prospective study. *Circulation.* 2002;105:1567–1572.
27. Vita JA. Endothelial function and clinical outcome. *Heart.* 2005;91:1278–1279.
28. Davignon J, Ganz P. Role of endothelial dysfunction in atherosclerosis. *Circulation.* 2004;109:III27–III32.
29. Channon KM. Tetrahydrobiopterin: regulator of endothelial nitric oxide synthase in vascular disease. *Trends Cardiovasc Med.* 2004;14:323–327.
30. Schmidt TS, Alp NJ. Mechanisms for the role of tetrahydrobiopterin in endothelial function and vascular disease. *Clin Sci (Lond).* 2007;113:47–63.
31. Dudzinski DM, Igarashi J, Greif D, Michel T. The regulation and pharmacology of endothelial nitric oxide synthase. *Annu Rev Pharmacol Toxicol.* 2006;46:235–276.
32. Aicher A, Heeschen C, Mildner-Rihm C, et al. Essential role of endothelial nitric oxide synthase for mobilization of stem and progenitor cells. *Nat Med.* 2003;9:1370–1376.
33. Vasa M, Fichtlscherer S, Adler K, et al. Increase in circulating endothelial progenitor cells by statin therapy in patients with stable coronary artery disease. *Circulation.* 2001;103:2885–2890.
34. Walter DH, Rittig K, Bahlmann FH, et al. Statin therapy accelerates reendothelialization: a novel effect involving mobilization and incorporation of bone marrow-derived endothelial progenitor cells. *Circulation.* 2002;105:3017–3024.
35. Celermajer DS, Sorensen KE, Gooch VM, et al. Non-invasive detection of endothelial dysfunction in children and adults at risk of atherosclerosis. *Lancet.* 1992;340:1111–1115.
36. Corretti MC, Anderson TJ, Benjamin EJ, et al. Guidelines for the ultrasound assessment of endothelial-dependent flow-mediated vasodilation of the brachial artery: a report of the International Brachial Artery Reactivity Task Force. *J Am Coll Cardiol.* 2002;39:257–265.
37. Joannides R, Haefeli WE, Linder L, et al. Nitric oxide is responsible for flow-dependent dilatation of human peripheral conduit arteries in vivo. *Circulation.* 1995;91:1314–1319.
38. Deanfield J, Donald A, Ferri C, et al. Endothelial function and dysfunction. Part I: methodological issues for assessment in the different vascular beds: a statement by the working group on endothelin and endothelial factors of the European society of hypertension. *J Hypertens.* 2005;23:7–17.
39. Anderson TJ, Uehata A, Gerhard MD, et al. Close relation of endothelial function in the human coronary and peripheral circulations. *J Am Coll Cardiol.* 1995;26:1235–1241.
40. Moncada S, Higgs EA, Vane JR. Human arterial and venous tissues generate prostacyclin (prostaglandin x), a potent inhibitor of platelet aggregation. *Lancet.* 1977;1:18–20.
41. Bonetti PO, Lerman LO, Lerman A. Endothelial dysfunction: a marker of atherosclerotic risk. *Arterioscler Thromb Vasc Biol.* 2003;23:168–175.
42. Mollace V, Muscoli C, Masini E, Cuzzocrea S, Salvemini D. Modulation of prostaglandin biosynthesis by nitric oxide and nitric oxide donors. *Pharmacol Rev.* 2005;57:217–252.
43. Halcox JP, Narayanan S, Cramer-Joyce L, Mincemoyer R, Quyyumi AA. Characterization of endothelium-derived hyperpolarizing factor in the human forearm microcirculation. *Am J Physiol Heart Circ Physiol.* 2001;280:H2470–H2477.

44. Feletou M, Vanhoutte PM. Endothelium-derived hyperpolarizing factor: where are we now? *Arterioscler Thromb Vasc Biol*. 2006;26:1215–1225.
45. Fleming I, Busse R. Endothelium–derived epoxyeicosatrienoic acids and vascular function. *Hypertension*. 2006;47:629–633.
46. Busse R, Edwards G, Feletou M, Fleming I, Vanhoutte PM, Weston AH. EDHF: bringing the concepts together. *Trends Pharmacol Sci*. 2002;23:374–380.
47. Miura H, Wachtel RE, Liu Y, et al. Flow-induced dilation of human coronary arterioles: important role of Ca(2+)-activated K(+) channels. *Circulation*. 2001;103:1992–1998.
48. Gilligan DM, Panza JA, Kilcoyne CM, Waclawiw MA, Casino PR, Quyyumi AA. Contribution of endothelium-derived nitric oxide to exercise-induced vasodilation. *Circulation*. 1994;90:2853–2858.
49. Quyyumi AA, Dakak N, Andrews NP, Gilligan DM, Panza JA, Cannon RO 3rd. Contribution of nitric oxide to metabolic coronary vasodilation in the human heart. *Circulation*. 1995;92:320–326.
50. Christ GJ, Spray DC, el-Sabban M, Moore LK, Brink PR. Gap junctions in vascular tissues. Evaluating the role of intercellular communication in the modulation of vasomotor tone. *Circ Res*. 1996;79:631–646.
51. Griffith TM. Endothelium-dependent smooth muscle hyperpolarization: do gap junctions provide a unifying hypothesis? *Br J Pharmacol*. 2004;141:881–903.
52. Fleming I, Michaelis UR, Bredenkotter D, et al. Endothelium-derived hyperpolarizing factor synthase (cytochrome P450 2C9) is a functionally significant source of reactive oxygen species in coronary arteries. *Circ Res*. 2001;88:44–51.
53. Matoba T, Shimokawa H, Kubota H, et al. Hydrogen peroxide is an endothelium-derived hyperpolarizing factor in human mesenteric arteries. *Biochem Biophys Res Commun*. 2002;290:909–913.
54. Shimokawa H, Yasutake H, Fujii K, et al. The importance of the hyperpolarizing mechanism increases as the vessel size decreases in endothelium-dependent relaxations in rat mesenteric circulation. *J Cardiovasc Pharmacol*. 1996;28:703–711.
55. Nishikawa Y, Stepp DW, Chilian WM. Nitric oxide exerts feedback inhibition on EDHF-induced coronary arteriolar dilation in vivo. *Am J Physiol Heart Circ Physiol*. 2000;279:H459–H465.
56. Feletou M, Vanhoutte PM. Endothelium-dependent hyperpolarizations: past beliefs and present facts. *Ann Med*. 2007;39:495–516.
57. Taddei S, Versari D, Cipriano A, et al. Identification of a cytochrome P450 2C9-derived endothelium-derived hyperpolarizing factor in essential hypertensive patients. *J Am Coll Cardiol*. 2006;48:508–515.
58. Taddei S, Virdis A, Ghiadoni L, Versari D, Salvetti A. Endothelium, aging, and hypertension. *Curr Hypertens Rep*. 2006;8:84–89.
59. Brandes RP, Behra A, Lebherz C, et al. N(G)-nitro-L-arginine– and indomethacin-resistant endothelium-dependent relaxation in the rabbit renal artery: effect of hypercholesterolemia. *Atherosclerosis*. 1997;135:49–55.
60. Morikawa K, Matoba T, Kubota H, et al. Influence of diabetes mellitus, hypercholesterolemia, and their combination on EDHF-mediated responses in mice. *J Cardiovasc Pharmacol*. 2005;45:485–490.
61. Wolfle SE, de Wit C. Intact endothelium-dependent dilation and conducted responses in resistance vessels of hypercholesterolemic mice in vivo. *J Vasc Res*. 2005;42:475–482.
62. Urakami-Harasawa L, Shimokawa H, Nakashima M, Egashira K, Takeshita A. Importance of endothelium-derived hyperpolarizing factor in human arteries. *J Clin Invest*. 1997;100:2793–2799.
63. Feletou M, Vanhoutte PM. EDHF: new therapeutic targets? *Pharmacol Res*. 2004;49:565–580.
64. Wu L, Wang R. Carbon monoxide: endogenous production, physiological functions, and pharmacological applications. *Pharmacol Rev*. 2005;57:585–630.
65. Durante W, Johnson FK, Johnson RA. Role of carbon monoxide in cardiovascular function. *J Cell Mol Med*. 2006;10:672–686.

66. Agarwal A, Balla J, Balla G, Croatt AJ, Vercellotti GM, Nath KA. Renal tubular epithelial cells mimic endothelial cells upon exposure to oxidized LDL. *Am J Physiol.* 1996;271:F814–F823.
67. Yachie A, Niida Y, Wada T, et al. Oxidative stress causes enhanced endothelial cell injury in human heme oxygenase-1 deficiency. *J Clin Invest.* 1999;103:129–135.
68. Kawashima A, Oda Y, Yachie A, Koizumi S, Nakanishi I. Heme oxygenase-1 deficiency: the first autopsy case. *Hum Pathol.* 2002;33:125–130.
69. Chen YH, Lin SJ, Lin MW, et al. Microsatellite polymorphism in promoter of heme oxygenase-1 gene is associated with susceptibility to coronary artery disease in type 2 diabetic patients. *Hum Genet.* 2002;111:1–8.
70. Kaneda H, Ohno M, Taguchi J, et al. Heme oxygenase-1 gene promoter polymorphism is associated with coronary artery disease in Japanese patients with coronary risk factors. *Arterioscler Thromb Vasc Biol.* 2002;22:1680–1685.
71. Yanagisawa M, Kurihara H, Kimura S, et al. A novel potent vasoconstrictor peptide produced by vascular endothelial cells. *Nature.* 1988;332:411–415.
72. Schiffrin EL. State-of-the-art lecture. Role of endothelin-1 in hypertension. *Hypertension.* 1999;34:876–881.
73. Best PJ, McKenna CJ, Hasdai D, Holmes DR Jr, Lerman A. Chronic endothelin receptor antagonism preserves coronary endothelial function in experimental hypercholesterolemia. *Circulation.* 1999;99:1747–1752.
74. Barton M, Haudenschild CC, D'Uscio LV, Shaw S, Munter K, Luscher TF. Endothelin ETA receptor blockade restores NO-mediated endothelial function and inhibits atherosclerosis in apolipoprotein E-deficient mice. *Proc Natl Acad Sci U S A.* 1998;95:14367–14372.
75. McKenna CJ, Burke SE, Opgenorth TJ, et al. Selective ET(A) receptor antagonism reduces neointimal hyperplasia in a porcine coronary stent model. *Circulation.* 1998;97:2551–2556.
76. Reel B, Ozkal S, Islekel H, et al. The role of endothelin receptor antagonism in collar-induced intimal thickening and vascular reactivity changes in rabbits. *J Pharm Pharmacol.* 2005;57:1599–1608.
77. Kinlay S, Behrendt D, Wainstein M, et al. Role of endothelin-1 in the active constriction of human atherosclerotic coronary arteries. *Circulation.* 2001;104:1114–1118.
78. Saye JA, Singer HA, Peach MJ. Role of endothelium in conversion of angiotensin I to angiotensin II in rabbit aorta. *Hypertension.* 1984;6:216–221.
79. Brunner H, Cockcroft JR, Deanfield J, et al. Endothelial function and dysfunction. Part II: association with cardiovascular risk factors and diseases. A statement by the working group on endothelins and endothelial factors of the European society of hypertension. *J Hypertens.* 2005;23:233–246.
80. Endemann DH, Schiffrin EL. Endothelial dysfunction. *J Am Soc Nephrol.* 2004;15:1983–1992.
81. Hayward CS, Kraidly M, Webb CM, Collins P. Assessment of endothelial function using peripheral waveform analysis: a clinical application. *J Am Coll Cardiol.* 2002;40:521–528.
82. Wilkinson IB, Fuchs SA, Jansen IM, et al. Reproducibility of pulse wave velocity and augmentation index measured by pulse wave analysis. *J Hypertens.* 1998;16:2079–2084.
83. Oliver JJ, Webb DJ. Noninvasive assessment of arterial stiffness and risk of atherosclerotic events. *Arterioscler Thromb Vasc Biol.* 2003;23:554–566.
84. Wilkinson IB, Hall IR, MacCallum H, et al. Pulse-wave analysis: clinical evaluation of a noninvasive, widely applicable method for assessing endothelial function. *Arterioscler Thromb Vasc Biol.* 2002;22:147-152.
85. Dawes M, Chowienczyk PJ, Ritter JM. Effects of inhibition of the L-arginine/nitric oxide pathway on vasodilation caused by beta-adrenergic agonists in human forearm. *Circulation.* 1997;95:2293-2297.
86. Kuvin JT, Patel AR, Sliney KA, et al. Assessment of peripheral vascular endothelial function with finger arterial pulse wave amplitude. *Am Heart J.* 2003;146:168-174.
87. Bonetti PO, Pumper GM, Higano ST, Holmes DR Jr, Kuvin JT, Lerman A. Noninvasive identification of patients with early coronary atherosclerosis by assessment of digital reactive hyperemia. *J Am Coll Cardiol.* 2004;44:2137–2141.

88. Panza JA, Quyyumi AA, Brush JE Jr, Epstein SE. Abnormal endothelium-dependent vascular relaxation in patients with essential hypertension. *N Engl J Med.* 1990;323:22-27.
89. John S, Schmieder RE. Impaired endothelial function in arterial hypertension and hypercholesterolemia: potential mechanisms and differences. *J Hypertens.* 2000;18:363-374.
90. Linder L, Kiowski W, Buhler FR, Luscher TF. Indirect evidence for release of endothelium-derived relaxing factor in human forearm circulation in vivo. Blunted response in essential hypertension. *Circulation.* 1990;81:1762-1767.
91. Hasdai D, Lerman A. The assessment of endothelial function in the cardiac catheterization laboratory in patients with risk factors for atherosclerotic coronary artery disease. *Herz.* 1999;24:544-547.
92. Chan SY, Mancini GB, Kuramoto L, Schulzer M, Frohlich J, Ignaszewski A. The prognostic importance of endothelial dysfunction and carotid atheroma burden in patients with coronary artery disease. *J Am Coll Cardiol.* 2003;42:1037-1043.
93. Fichtlscherer S, Breuer S, Zeiher AM. Prognostic value of systemic endothelial dysfunction in patients with acute coronary syndromes: further evidence for the existence of the "vulnerable" patient. *Circulation.* 2004;110:1926-1932.
94. Gokce N, Keaney JF Jr, Hunter LM, et al. Predictive value of noninvasively determined endothelial dysfunction for long-term cardiovascular events in patients with peripheral vascular disease. *J Am Coll Cardiol.* 2003;41:1769-1775.
95. Heitzer T, Baldus S, von Kodolitsch Y, Rudolph V, Meinertz T. Systemic endothelial dysfunction as an early predictor of adverse outcome in heart failure. *Arterioscler Thromb Vasc Biol.* 2005;25:1174-1179.
96. Bonetti PO, Barsness GW, Keelan PC, et al. Enhanced external counterpulsation improves endothelial function in patients with symptomatic coronary artery disease. *J Am Coll Cardiol.* 2003;41:1761-1768.
97. Szmitko PE, Fedak PW, Weisel RD, Stewart DJ, Kutryk MJ, Verma S. Endothelial progenitor cells: new hope for a broken heart. *Circulation.* 2003;107:3093-3100.
98. Murakami T, Mizuno S, Ohsato K, Moriuchi I, Arai Y, Nio Y, Kaku B, Takahashi Y, Ohnaka M. Effects of troglitazone on frequency of coronary vasospastic-induced angina pectoris in patients with diabetes mellitus. *Am J Cardiol* 1999;84:92-94, A8.
99. Brevetti G, Silvestro A, Schiano V, Chiariello M. Endothelial dysfunction and cardiovascular risk prediction in peripheral arterial disease: additive value of flow-mediated dilation to ankle-brachial pressure index. *Circulation.* 2003;108:2093-2098.
100. Celermajer DS, Sorensen KE, Bull C, Robinson J, Deanfield JE. Endothelium-dependent dilation in the systemic arteries of asymptomatic subjects relates to coronary risk factors and their interaction. *J Am Coll Cardiol.* 1994;24:1468-1474.
101. Smith SC Jr, Anderson JL, Cannon RO 3rd, et al. CDC/AHA workshop on markers of inflammation and cardiovascular disease: application to clinical and public health practice: report from the clinical practice discussion group. *Circulation.* 2004;110:e550-e553.
102. Boger RH, Maas R, Schulze F, Schwedhelm E. Elevated levels of asymmetric dimethylarginine (ADMA) as a marker of cardiovascular disease and mortality. *Clin Chem Lab Med.* 2005;43:1124-1129.
103. Vallance P, Leone A, Calver A, Collier J, Moncada S. Accumulation of an endogenous inhibitor of nitric oxide synthesis in chronic renal failure. *Lancet.* 1992;339:572-575.
104. Vallance P, Leiper J. Cardiovascular biology of the asymmetric dimethylarginine:dimethylarginine dimethylaminohydrolase pathway. *Arterioscler Thromb Vasc Biol.* 2004;24:1023-1030.
105. Boger RH, Sydow K, Borlak J, et al. LDL cholesterol upregulates synthesis of asymmetrical dimethylarginine in human endothelial cells: involvement of S-adenosylmethionine-dependent methyltransferases. *Circ Res.* 2000;87:99-105.
106. Boger RH, Bode-Boger SM, Sydow K, Heistad DD, Lentz SR. Plasma concentration of asymmetric dimethylarginine, an endogenous inhibitor of nitric oxide synthase, is elevated in monkeys with hyperhomocyst(e)inemia or hypercholesterolemia. *Arterioscler Thromb Vasc Biol.* 2000;20:1557-1564.

107. Goonasekera CD, Rees DD, Woolard P, Frend A, Shah V, Dillon MJ. Nitric oxide synthase inhibitors and hypertension in children and adolescents. *J Hypertens*. 1997;15:901-909.
108. Xiong Y, Fu YF, Fu SH, Zhou HH. Elevated levels of the serum endogenous inhibitor of nitric oxide synthase and metabolic control in rats with streptozotocin-induced diabetes. *J Cardiovasc Pharmacol*. 2003;42:191-196.
109. Paiva H, Lehtimaki T, Laakso J, et al. Plasma concentrations of asymmetric-dimethyl-arginine in type 2 diabetes associate with glycemic control and glomerular filtration rate but not with risk factors of vasculopathy. *Metabolism*. 2003;52:303-307.
110. Boger RH, Lentz SR, Bode-Boger SM, Knapp HR, Haynes WG. Elevation of asymmetrical dimethylarginine may mediate endothelial dysfunction during experimental hyperhomocyst(e) inaemia in humans. *Clin Sci (Lond)*. 2001;100:161-167.
111. Miyazaki H, Matsuoka H, Cooke JP, et al. Endogenous nitric oxide synthase inhibitor: a novel marker of atherosclerosis. *Circulation*. 1999;99:1141-1146.
112. Schulze F, Lenzen H, Hanefeld C, et al. Asymmetric dimethylarginine is an independent risk factor for coronary heart disease: results from the multicenter coronary artery risk determination investigating the influence of ADMA concentration (CARDIAC) study. *Am Heart J*. 2006;152:493 e491-e498.
113. Feng Q, Lu X, Fortin AJ, et al. Elevation of an endogenous inhibitor of nitric oxide synthesis in experimental congestive heart failure. *Cardiovasc Res*. 1998;37:667-675.
114. Saitoh M, Osanai T, Kamada T, et al. High plasma level of asymmetric dimethylarginine in patients with acutely exacerbated congestive heart failure: role in reduction of plasma nitric oxide level. *Heart Vessels*. 2003;18:177-182.
115. Usui M, Matsuoka H, Miyazaki H, Ueda S, Okuda S, Imaizumi T. Increased endogenous nitric oxide synthase inhibitor in patients with congestive heart failure. *Life Sci*. 1998;62:2425-2430.
116. Kielstein JT, Boger RH, Bode-Boger SM, et al. Asymmetric dimethylarginine plasma concentrations differ in patients with end-stage renal disease: relationship to treatment method and atherosclerotic disease. *J Am Soc Nephrol*. 1999;10:594-600.
117. Gorenflo M, Zheng C, Werle E, Fiehn W, Ulmer HE. Plasma levels of asymmetrical dimethyl-L-arginine in patients with congenital heart disease and pulmonary hypertension. *J Cardiovasc Pharmacol*. 2001;37:489-492.
118. Savvidou MD, Hingorani AD, Tsikas D, Frolich JC, Vallance P, Nicolaides KH. Endothelial dysfunction and raised plasma concentrations of asymmetric dimethylarginine in pregnant women who subsequently develop pre-eclampsia. *Lancet*. 2003;361:1511-1517.
119. Fickling SA, Williams D, Vallance P, Nussey SS, Whitley GS. Plasma concentrations of endogenous inhibitor of nitric oxide synthesis in normal pregnancy and pre-eclampsia. *Lancet*. 1993;342:242-243.
120. Holden DP, Fickling SA, Whitley GS, Nussey SS. Plasma concentrations of asymmetric dimethylarginine, a natural inhibitor of nitric oxide synthase, in normal pregnancy and preeclampsia. *Am J Obstet Gynecol*. 1998;178:551-556.
121. Ridker PM, Brown NJ, Vaughan DE, Harrison DG, Mehta JL. Established and emerging plasma biomarkers in the prediction of first atherothrombotic events. *Circulation*. 2004;109:IV6-IV19.
122. Danesh J, Wheeler JG, Hirschfield GM, et al. C-reactive protein and other circulating markers of inflammation in the prediction of coronary heart disease. *N Engl J Med*. 2004;350:1387-1397.
123. Fichtlscherer S, Rosenberger G, Walter DH, Breuer S, Dimmeler S, Zeiher AM. Elevated C-reactive protein levels and impaired endothelial vasoreactivity in patients with coronary artery disease. *Circulation*. 2000;102:1000-1006.
124. Kathiresan S, Gona P, Larson MG, et al. Cross-sectional relations of multiple biomarkers from distinct biological pathways to brachial artery endothelial function. *Circulation*. 2006;113:938-945.
125. Jones DP. Extracellular redox state: refining the definition of oxidative stress in aging. *Rejuvenation Res*. 2006;9:169-181.
126. Ashfaq S, Abramson JL, Jones DP, et al. Endothelial function and aminothiol biomarkers of oxidative stress in healthy adults. *Hypertension*. 2008;52(1):80-85.

127. Cesarone MR, Belcaro G, Carratelli M, et al. A simple test to monitor oxidative stress. *Int Angiol.* 1999;18:127-130.
128. Alberti ABL, Macciantelli D, Caratelli M. The radical cation of N,N-diethyl-para-phenylendiamine: a possible indicator of oxidative stress in biological samples. *Res Chem Intermed.* 2000;26:253-267.
129. Ashfaq SJD, Kolm P, Rhodes SD, et al. Relationship of novel oxidative stress markers to vascular endothelial function in healthy adults. *J Am Coll Cardiol.* 2005;45:415A.
130. Ross R. The pathogenesis of atherosclerosis: a perspective for the 1990s. *Nature.* 1993;362: 801-809.
131. Schwartz SM, Benditt EP. Aortic endothelial cell replication. I. Effects of age and hypertension in the rat. *Circ Res.* 1977;41:248-255.
132. Feletou M, Vanhoutte PM. Endothelial dysfunction: a multifaceted disorder (The Wiggers Award Lecture). *Am J Physiol Heart Circ Physiol.* 2006;291:H985-H1002.
133. Op den Buijs J, Musters M, Verrips T, Post JA, Braam B, van Riel N. Mathematical modeling of vascular endothelial layer maintenance: the role of endothelial cell division, progenitor cell homing, and telomere shortening. *Am J Physiol Heart Circ Physiol.* 2004;287:H2651-H2658.
134. Hill JM, Zalos G, Halcox JP, et al. Circulating endothelial progenitor cells, vascular function, and cardiovascular risk. *N Engl J Med.* 2003;348:593-600.
135. Asahara T, Murohara T, Sullivan A, et al. Isolation of putative progenitor endothelial cells for angiogenesis. *Science.* 1997;275:964-967.
136. Shi Q, Rafii S, Wu MH, et al. Evidence for circulating bone marrow-derived endothelial cells. *Blood.* 1998;92:362-367.
137. Lin Y, Weisdorf DJ, Solovey A, Hebbel RP. Origins of circulating endothelial cells and endothelial outgrowth from blood. *J Clin Invest.* 2000;105(1):71-77.
138. Asahara T, Masuda H, Takahashi T, et al. Bone marrow origin of endothelial progenitor cells responsible for postnatal vasculogenesis in physiological and pathological neovascularization. *Circ Res.* 1999;85:221-228.
139. Ingram DA, Caplice NM, Yoder MC. Unresolved questions, changing definitions, and novel paradigms for defining endothelial progenitor cells. *Blood.* 2005;106:1525-1531.
140. Urbich C, Dimmeler S. Endothelial progenitor cells functional characterization. *Trends Cardiovasc Med.* 2004;14:318-322.
141. Krause DS, Fackler MJ, Civin CI, May WS. CD34: structure, biology, and clinical utility. *Blood.* 1996;87:1-13.
142. Berenson RJ. Transplantation of CD34+ hematopoietic precursors: clinical rationale. *Transplant Proc.* 1992;24:3032-3034.
143. Baum CM, Weissman IL, Tsukamoto AS, Buckle AM, Peault B. Isolation of a candidate human hematopoietic stem-cell population. *Proc Natl Acad Sci U S A.* 1992;89:2804-2808.
144. Civin CI, Strauss LC, Brovall C, Fackler MJ, Schwartz JF, Shaper JH. Antigenic analysis of hematopoiesis. III. A hematopoietic progenitor cell surface antigen defined by a monoclonal antibody raised against KG-1a cells. *J Immunol.* 1984;133:157-165.
145. Katz FE, Davis L, Myers CD, Greaves MF. Selective expression of class-II MHC antigens during hemopoietic differentiation. *Exp Hematol.* 1985;13:1182-1187.
146. Andrews RG, Singer JW, Bernstein ID. Monoclonal antibody 12-8 recognizes a 115-kd molecule present on both unipotent and multipotent hematopoietic colony-forming cells and their precursors. *Blood.* 1986;67:842-845.
147. Yin AH, Miraglia S, Zanjani ED, et al. AC133, a novel marker for human hematopoietic stem and progenitor cells. *Blood.* 1997;90:5002-5012.
148. Gehling UM, Ergun S, Schumacher U, et al. In vitro differentiation of endothelial cells from AC133-positive progenitor cells. *Blood.* 2000;95:3106-3112.
149. Peichev M, Naiyer AJ, Pereira D, et al. Expression of VEGFR-2 and AC133 by circulating human CD34(+) cells identifies a population of functional endothelial precursors. *Blood.* 2000;95:952-958.

150. Terman BI, Dougher-Vermazen M, Carrion ME, et al. Identification of the KDR tyrosine kinase as a receptor for vascular endothelial cell growth factor. *Biochem Biophys Res Commun.* 1992;187:1579-1586.
151. Nowak G, Karrar A, Holmen C, et al. Expression of vascular endothelial growth factor receptor-2 or Tie-2 on peripheral blood cells defines functionally competent cell populations capable of reendothelialization. *Circulation.* 2004;110:3699-3707.
152. Khan SS, Solomon MA, McCoy JP Jr. Detection of circulating endothelial cells and endothelial progenitor cells by flow cytometry. *Cytometry B Clin Cytom.* 2005;64:1-8.
153. Bhatia M, Bonnet D, Murdoch B, Gan OI, Dick JE. A newly discovered class of human hematopoietic cells with SCID-repopulating activity. *Nat Med.* 1998;4:1038-1045.
154. Harraz M, Jiao C, Hanlon HD, Hartley RS, Schatteman GC. CD34- blood-derived human endothelial cell progenitors. *Stem Cells.* 2001;19:304-312.
155. Goodell MA, Rosenzweig M, Kim H, et al. Dye efflux studies suggest that hematopoietic stem cells expressing low or undetectable levels of CD34 antigen exist in multiple species. *Nat Med.* 1997;3:1337-1345.
156. Kocher AA, Schuster MD, Szabolcs MJ, et al. Neovascularization of ischemic myocardium by human bone-marrow-derived angioblasts prevents cardiomyocyte apoptosis, reduces remodeling and improves cardiac function. *Nat Med.* 2001;7:430-436.
157. Jackson KA, Majka SM, Wang H, et al. Regeneration of ischemic cardiac muscle and vascular endothelium by adult stem cells. *J Clin Invest.* 2001;107:1395-1402.
158. Murohara T, Ikeda H, Duan J, et al. Transplanted cord blood-derived endothelial precursor cells augment postnatal neovascularization. *J Clin Invest.* 2000;105(6):1527-1536.
159. Asahara T, Kalka C, Isner JM. Stem cell therapy and gene transfer for regeneration. *Gene Ther.* 2000;7:451-457.
160. Takahashi T, Kalka C, Masuda H, et al. Ischemia- and cytokine-induced mobilization of bone marrow-derived endothelial progenitor cells for neovascularization. *Nat Med.* 1999;5:434-438.
161. Springer ML, Chen AS, Kraft PE, Bednarski M, Blau HM. VEGF gene delivery to muscle: potential role for vasculogenesis in adults. *Mol Cell.* 1998;2:549-558.
162. Lapidot T, Dar A, Kollet O. How do stem cells find their way home? *Blood.* 2005;106:1901-1910.
163. Asahara T, Takahashi T, Masuda H, et al. VEGF contributes to postnatal neovascularization by mobilizing bone marrow-derived endothelial progenitor cells. *EMBO J.* 1999;18:3964-3972.
164. Gill M, Dias S, Hattori K, et al. Vascular trauma induces rapid but transient mobilization of VEGFR2(+)AC133(+) endothelial precursor cells. *Circ Res.* 2001;88:167-174.
165. Aicher A, Zeiher AM, Dimmeler S. Mobilizing endothelial progenitor cells. *Hypertension.* 2005;45:321-325.
166. Heeschen C, Aicher A, Lehmann R, et al. Erythropoietin is a potent physiologic stimulus for endothelial progenitor cell mobilization. *Blood.* 2003;102:1340-1346.
167. Askari AT, Unzek S, Popovic ZB, et al. Effect of stromal-cell-derived factor 1 on stem-cell homing and tissue regeneration in ischaemic cardiomyopathy. *Lancet.* 2003;362:697-703.
168. Lee SP, Youn SW, Cho HJ, et al. Integrin-linked kinase, a hypoxia-responsive molecule, controls postnatal vasculogenesis by recruitment of endothelial progenitor cells to ischemic tissue. *Circulation.* 2006;114:150-159.
169. Yamaguchi J, Kusano KF, Masuo O, et al. Stromal cell-derived factor-1 effects on ex vivo expanded endothelial progenitor cell recruitment for ischemic neovascularization. *Circulation.* 2003;107:1322-1328.
170. Ceradini DJ, Kulkarni AR, Callaghan MJ, et al. Progenitor cell trafficking is regulated by hypoxic gradients through HIF-1 induction of SDF-1. *Nat Med.* 2004;10:858-864.
171. Mohle R, Bautz F, Rafii S, Moore MAS, Brugger W, Kanz L. The chemokine receptor CXCR-4 is expressed on CD34+ hematopoietic progenitors and leukemic cells and mediates transendothelial migration induced by stromal cell-derived factor-1. *Blood.* 1998;91:4523-4530.
172. Vasa M, Fichtlscherer S, Aicher A, et al. Number and migratory activity of circulating endothelial progenitor cells inversely correlate with risk factors for coronary artery disease. *Circ Res.* 2001;89:1e-7e.

173. Pearson TA, Mensah GA, Alexander RW, et al. Markers of inflammation and cardiovascular disease: application to clinical and public health practice: a statement for healthcare professionals from the centers for disease control and prevention and the American heart association. *Circulation* 2003;107:499-511.
174. Ciulla MM, Giorgetti A, Silvestris I, et al. Endothelial colony forming capacity is related to C-reactive protein levels in healthy subjects. *Curr Neurovasc Res*. 2006;3:99-106.
175. Verma S, Kuliszewski MA, Li SH, et al. C-reactive protein attenuates endothelial progenitor cell survival, differentiation, and function: further evidence of a mechanistic link between C-reactive protein and cardiovascular disease. *Circulation*. 2004;109:2058-2067.
176. Thum T, Tsikas D, Stein S, et al. Suppression of endothelial progenitor cells in human coronary artery disease by the endogenous nitric oxide synthase inhibitor asymmetric dimethylarginine. *J Am Coll Cardiol*. 2005;46:1693-1701.
177. Heiss C, Keymel S, Niesler U, Ziemann J, Kelm M, Kalka C. Impaired progenitor cell activity in age-related endothelial dysfunction. *J Am Coll Cardiol*. 2005;45:1441-1448.
178. Scheubel RJ, Zorn H, Silber RE, et al. Age-dependent depression in circulating endothelial progenitor cells in patients undergoing coronary artery bypass grafting. *J Am Coll Cardiol*. 2003;42:2073-2080.
179. Britten MB, Abolmaali ND, Assmus B, et al. Infarct remodeling after intracoronary progenitor cell treatment in patients with acute myocardial infarction (TOPCARE-AMI): mechanistic insights from serial contrast-enhanced magnetic resonance imaging. *Circulation*. 2003;108:2212-2218.
180. Tepper OM, Galiano RD, Capla JM, et al. Human endothelial progenitor cells from type II diabetics exhibit impaired proliferation, adhesion, and incorporation into vascular structures. *Circulation*. 2002;106:2781-2786.
181. Ghani U, Shuaib A, Salam A, et al. Endothelial progenitor cells during cerebrovascular disease. *Stroke*. 2005;36:151-153.
182. Schmidt-Lucke C, Rossig L, Fichtlscherer S, et al. Reduced number of circulating endothelial progenitor cells predicts future cardiovascular events: proof of concept for the clinical importance of endogenous vascular repair. *Circulation*. 2005;111:2981-2987.
183. Werner N, Kosiol S, Schiegl T, et al. Circulating endothelial progenitor cells and cardiovascular outcomes. *N Engl J Med*. 2005;353:999-1007.
184. Burnham EL, Taylor WR, Quyyumi AA, Rojas M, Brigham KL, Moss M. Increased circulating endothelial progenitor cells are associated with survival in acute lung injury. *Am J Respir Crit Care Med*. 2005;172:854-860.
185. Dimmeler S, Aicher A, Vasa M, et al. HMG-CoA reductase inhibitors (statins) increase endothelial progenitor cells via the PI 3-kinase/Akt pathway. *J Clin Invest*. 2001;108:391-397.

Section 2

Part 1
Clinical Aspects of Vascular Biology

Chapter 12
Chronic Thromboembolic Pulmonary Hypertension

Joanna Pepke-Zaba

12.1 Introduction

Chronic thromboembolic pulmonary hypertension (CTEPH) results from obstruction of the pulmonary vascular bed by nonresolving thromboemboli. Concomitant small vessel arteriopathy is present at various degrees in patients with CTEPH[1,2]; it is thought to account for a large number of pulmonary hypertension diagnoses.[3] In the absence of treatment, it has historically had a poor prognosis with 3-year survival as low as 10%, although in the majority of cases, it can now be treated effectively with surgery in the form of pulmonary endarterectomy (PEA).[4]

12.2 Epidemiology

The prevalence and incidence of CTEPH following acute pulmonary embolism (PE) are difficult to ascertain. However, recent follow-up studies in patients presenting with acute PE give an indication of the scale of the problem. Pengo et al[5] showed a cumulative incidence of 3.8% at 2 years, while even more recent studies by Becattini et al and Miniati et al have described cumulative incidences of 4/320 (1.3%) and 2/259 (0.8%), respectively.[6,7] Given that acute PE is as common as 1 per 1,000 of the population per year,[8] it suggests that the annual incidence of CTEPH following acute venous thromboembolism is of the order of 8–51 cases per 1 million population. Moreover, given that CTEPH is classically described as presenting some time following an acute event,[9] the results from these relatively short studies may represent an underestimate.

A further consideration is that up to 60% of patients with CTEPH do not have a previously documented VTE event.[3] In these patients, it is presumed that clinically

J. Pepke-Zaba (✉)
Director of Pulmonary Vascular Disease Unit, Papworth Hospital,
Papworth Everard, Cambridge, UK
e-mail: joanna.pepkezaba@papworth.nhs.uk

D. Abraham et al. (eds.), *Advances in Vascular Medicine*,
DOI 10.1007/978-1-84882-637-3_12, © Springer-Verlag London Limited 2010

"silent" emboli in the past have triggered the disease process. It has been estimated that there are 2,500 new cases of CTEPH each year in the United States.[10]

In the recent paper of Condliffe et al, 420 patients were diagnosed prospectively with CTEPH in the UK PH Centers, during the period between year 2001 and 2005.[11] Calculated incidence in 2005 was 1.75 cases/year/million. This is the first to report an incidence of CTEPH based on the number of diagnosed and treated cases within a whole population. The observed incidence increased by 75% over a period of 5 years, which is most likely to be due to increased awareness of the condition over that period. The figure of 1.75 cases/year/million is conservative, as it does not include patients who had been diagnosed but not referred to a PH center, or had been undiagnosed.

12.3 Current Pathogenetic Concepts and Genetics

This is discussed elsewhere (Prof. Morrell)

12.3.1 Associated Medical Conditions (AMC)

A history of deep venous thrombosis, PE, a known thrombophilia, a history VA-shunt, splenectomy, inflammatory bowel disease, or osteomyelitis as well as chronic central intravenous lines[12,13] increases the likelihood of CTEPH in a patient presenting with symptoms suggestive of pulmonary hypertension. However, the absence of such history does not rule out CTEPH. Approximately 50% of CTEPH patients have no clear history of diagnosed acute PE.[3] CTEPH may result from one embolic event or multiple recurrent events.

12.3.2 Predictors of Outcome in CTEPH

It has been demonstrated that the presence of AMCs predicts increased operative risk and worse long-term outcome in CTEPH.[14] Patients with AMC represent a distinct subentity of CTEPH, potentially requiring more aggressive presurgical as well as perioperative and postsurgical vasodilator treatment.

Overall, there was no correlation between CTEPH type and the presence of AMC, and no decreased likelihood of patients with AMC to undergo PEA. However, despite technically successful surgery, patients with AMC had worse outcomes with higher perioperative mortality (24 vs. 9%) and an increased incidence of postoperative pulmonary hypertension.

12.4 Diagnostic Assessments in CTEPH: Making a Diagnosis and Determining Candidacy for Surgery

The diagnosis of CTEPH involves process starting from clinical suspicion to definitive testing. Once a diagnosis of CTEPH is made, determining whether a patient is a candidate for PEA involves additional analysis.

12.4.1 Physical Examination

The physical examination in a patient with CTEPH is similar to that of any patient with pulmonary hypertension, except for one distinctive physical examination finding: pulmonary flow murmurs. These are soft "bruits" heard over the lung fields in areas corresponding to partially occluded pulmonary arteries. Pulmonary flow murmurs are best heard during a breath-hold at mid-inspiration.[15] The presence of unilateral leg edema, suggestive of deep venous thrombosis, should also raise the suspicion for chronic thromboembolic disease.

12.4.2 Imaging

The chest radiograph in a patient with chronic thromboembolic disease is often unremarkable.

The V/Q scan is the examination of choice for ruling out CTEPH. However, the complete absence of perfusion to one lung should raise suspicion for other processes, such as malignancy, mediastinal fibrosis, or vasculitis.

There are characteristic CT and MRA findings in CTEPH, but pulmonary angiography is still considered the "gold standard" diagnostic tool in the work-up of CTEPH.

The imaging of CTEPH is described in details in a separate chapter.

12.4.3 Hemodynamic Assessment

The right heart catheterization is mandatory for full hemodynamic assessment in the diagnosis and management of CTEPH. The presence of pulmonary hypertension is defined by either a resting mean pulmonary artery pressure (mPAP) of at least 25 mmHg or exercise PAP of at least 30 mmHg with wedge pressure less than 15 mmHg (or normal left atrium on ECHO). In addition, assessment of the degree of right-sided heart failure by measuring right atrial pressure, cardiac output, and mixed venous O_2 saturation is important in determining the severity of the disease and in calculating pulmonary vascular resistance for assessment of risk from surgical intervention.

Routine coronary angiography is generally recommended for patients over age 40–45 years considered for surgical treatment, as concomitant coronary artery bypass grafting is an option at the time of PEA.

12.4.4 Specialized Techniques

Pulmonary angioscopy – may help in assessment of surgically accessible disease.

Pulmonary artery occlusion pressure waveform analysis[16] – can help "partition" the upstream resistance (proximal thromboemboli) from the downstream resistance (secondary arteriopathy).

Accepted criteria for surgical intervention and factors that increase surgical risk are discussed elsewhere (surgical treatment).

12.5 The Current Role of Medical Therapy

12.5.1 Anticoagulation

CTEPH patients should receive lifelong anticoagulation adjusted to a target international ratio between 2.0 and 3.0. This is to prevent from recurrent thromboembolic events. When the disease is fully established, one should not expect significant regression of pulmonary hypertension from anticoagulation.[2]

12.5.2 When is Medical Therapy for CTEPH Appropriate?

Data from clinical trials of medical therapy in CTEPH are currently limited.

There is one multicenter RCT in nonoperable CTEPH – BENEFiT,[17] which showed improver PVR, CI, and pro-BNP, but no significant change in 6MWD after 16 weeks treatment with Bosentan and one small single-center RCT showing similar results with sildenafil.[18] The majority of findings are from small uncontrolled studies with prostacyclin,[19-25] endothelin receptor antagonists,[26-29] Phosphodiesterase-5 inhibition.[18,30-32]

Pharmacotherapy may be beneficial: (1) where PEA is not suitable due to significant distal disease; (2) when surgery is contraindicated due to significant co-morbidity that increases the risk of peri- and post-operative mortality; (3) in patients who are "high-risk" due to extremely poor hemodynamics prior to PEA (i.e., bridging to surgery); (4) in patients with persistent or residual PH after PEA.[33]

12.5.3 Patients with Inoperable Disease

Elevation of PVR out of proportion to what is attributable to mechanical thrombus obstruction is frequently seen, and signals significant and, in many cases, inoperable levels of arteriopathy. High PVR is indeed associated with bad outcome; however, these patients (if they survive) represent the persistent and residual PH after PEA. Management of these patients was in the past supportive. As the histopathological changes of pulmonary arterial hypertension are seen in patients with CTEPH, the disease-modifying therapies used in other forms of pulmonary arterial hypertension have been prescribed on a compassionate basis in the UK for patients with nonsurgical CTEPH. They have also been used as a bridge to surgery in selected patients with surgically accessible disease.[11]

12.5.4 Pre-PEA "Bridging" Therapy

The concept of introducing medical treatment as a "therapeutic bridge" between CTEPH diagnosis and PEA was initially proposed for continuous IV epoprostenol.[19,20] There are two main scenarios where preoperative "bridging" with medical therapy may provide benefits. A significant proportion of CTEPH patients undergoing PEA are hemodynamically unstable in the preoperative period to the point where risks from surgery in general are significantly raised. "High-risk" patients can be defined by various criteria: NYHA class IV disease; mPAP >50 mm Hg; cardiac index (CI) <2.0; PVR>1,000 dynes. $S.cm^{-5}$. It can be hypothesized that medical therapies demonstrating significant effects on pulmonary hemodynamics in patients with CTEPH may improve hemodynamic stability pre-surgery and post-operative outcome by providing control of pre-operative PVR.

A second application of bridging therapy is in cases where the delay to surgery is hazardously prolonged. Treatment to control pulmonary hemodynamics during the waiting period may prove valuable by precluding hemodynamic deterioration.

12.5.5 Post-PEA Therapy

Approximately 10–15%[34] of patients show persistent or residual PH after PEA surgery. Medical treatment for patients with persistent postoperative PH and consequent hemodynamic instability requires the development of criteria and guidance on how and when medical therapy should be initiated, for how long it should be continued, and what stopping rules should be applied.

12.5.6 *Safety Aspects*

Disease co-morbidity (e.g., COPD, cardiac disease) is an important factor in the choice of appropriate management for patients with CTEPH because patients are generally older than those with PAH, and have a greater frequency and severity of co-morbidity. It is therefore important to identify any safety issues related to the use of targeted therapy in CTEPH, either alone or in conjunction with PEA.

12.5.7 *Combination Therapy*

Further studies are required to establish if the benefits of combination therapy seen in IPAH extend also to CTEPH.

12.6 Survival

Survival from before the modern treatment era with targeted therapies was poor with 3-year survival as low as 10% in patients with an mPAP of >30 mmHg.[35]

In a study of 26 patients who were not treated with surgery or disease-modifying therapies, 2-year survival was only 10% in those with an mPAP of over 50 mmHg.[36] Despite another study involving 48 Japanese patients who had a mean mPAP of 50 mmHg reporting median survival of almost 6 years,[37] it is generally recognized that most patients with CTEPH treated with anticoagulation alone have progressive disease.[2] In the above studies, no clear distinction between the survival in surgical and nonsurgical was made.

In recently published paper by Condliffe et al, authors present the first study involving long-term objective follow-up of a national cohort of patients with CTEPH, showing improved survival for those with nonsurgical distribution, with 1- and 3-year survival of 83 and 76% in the period 2003–2006, when treated with disease-modifying drugs (Fig. 12.1).[11] The improved outcomes to those published by Suntharalingam et al with reported 1- and 3-year survival rates were 77% and 53%[38] on a group of 35 patients with distal CTEPH who were diagnosed between 1994 and 2005. In contrast, the 1-year survival was 96% in a study of 47 patients with nonsurgical disease or persistent PH postsurgery who were all treated with bosentan.[29]

The survival in patients with persistent PH 3-months following PEA, who were treated in selected cases with disease-modifying therapy, was excellent.[11]

Survival of patients with CTEPH treated with PEA surgery is discussed elsewhere (surgical treatment).

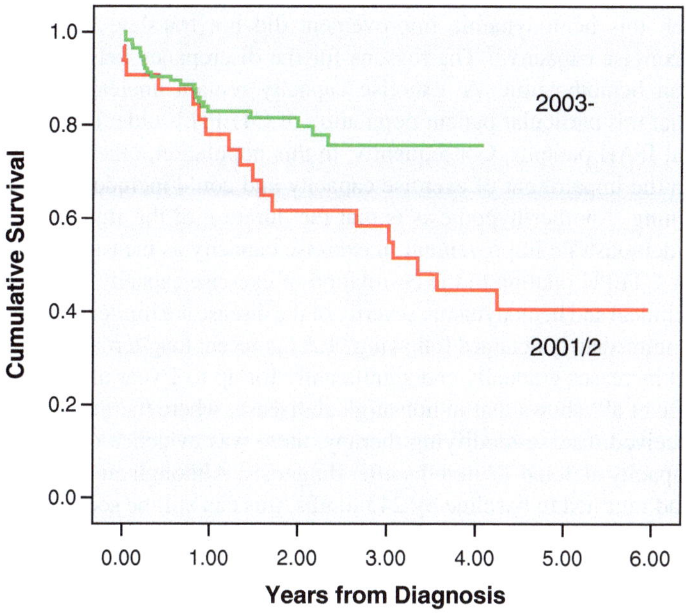

115	81	45	18	1		2003-
33	24	18	17	13	4	2001/2

Patients at risk

Fig. 12.1 Cumulative survival from date of diagnosis. Patients with nonsurgical disease diagnosed during 2001/2002 and 2003 onwards. Log-rank test, $p=0.023$

12.7 Functional Status

In open-label studies,[18,29,32] there was statistically significant improvement in functional status measured with 6 min walking test (6MWT) after 1 year of treatment with targeted therapies. This is in contrast to the findings of RCT BENEFiT, wherein there was no improvement in 6MWT after 4 months treatment with bosentan.[17] The main finding of this study was a statistically significant improvement in PVR. As pulmonary hemodynamics are believed to be important prognostic factors,[36,39] and the assessment of PVR is considered a prerequisite for optimal management of CTEPH, this finding carries clinical relevance. The reduction of PVR in the bosentan-treated group of over 22% from baseline values is consistent with the findings from previous open-label studies of bosentan treatment in CTEPH.[27-29] This was also paralleled by a reduction in NT-pro-BNP levels, which is of interest as NT-pro-BNP concentrations have been shown to correspond with pulmonary hemodynamics and functional parameters in PAH patients[40] and correlate with the severity of disease in CTEPH patients.[41]

However, this hemodynamic improvement did not translate into a favorable effect on exercise capacity.[17] The reasons for the discrepancy between the results observed on hemodynamic vs. exercise capacity remain unclear. One potential reason is that this particular patient population of CTEPH is older (mean age of 63) than typical IPAH patients. Consequently, in this population, other factors may be involved in the impairment of exercise capacity and could include skeletal muscle deconditioning. Another hypothesis is that the duration of the study was not long enough to demonstrate improvement in exercise capacity as measured by 6MWD.

Data for CTEPH relating to the correlation of exercise capacity with parameters reflecting clinical and hemodynamic severity of the disease are limited. Hemodynamic recovery is immediately obtained following PEA but recent long-term data suggest that the 6MWD increases gradually and significantly for up to 1 year after surgery.[41-44]

Condliffe et al[11] shows that in nonsurgical disease, where the majority (85%) of patients received disease-modifying therapy, there was evidence of an increase in exercise capacity at 3 and 12 months after diagnosis. Although an average exercise capacity had returned to baseline by 24 months, this can still be seen as a favorable outcome in a condition where the outcome was historically so poor.

12.8 Conclusions

While PEA is accepted as the first choice of treatment for CTEPH, evidence is accumulating on the potential use of medical therapies alongside surgery. Pharmacotherapy may be particularly useful in treating patients with predominant small-vessel disease who are poor candidates for surgery, or as bridging therapy in those where there is significant preoperative risk. Postsurgery use in controlling persistent or recurrent PH also appears a distinct possibility.

During initial studies, the prostanoids, dual endothelin receptor antagonist, bosentan, and the phosphodiesterase-5 inhibitor, sildenafil, have all shown potential in the treatment of inoperable CTEPH. Evidence from only one (up to date) multicenter prospective, randomized, controlled clinical trial BENEFiT suggests significantly improved hemodynamics at week 16 of treatment with bosentan but 6MWD remained unchanged.

The improvements in outcome during the modern treatment era highlight the importance of identifying patients with this increasingly treatable condition. There is a further need for the prospective RCT to clarify the role of targeted therapies in treatment of CTEPH.

References

1. Moser KM, Bloor CM. Pulmonary vascular lesions occurring in patients with chronic major vessel thromboembolic pulmonary hypertension. *Chest*. 1993;103:685-692.
2. Hoeper MM, Mayer E, Simonneau G, Rubin LJ. Chronic thromboembolic pulmonary hypertension. *Circulation*. 2006;113(16):2011-2020.

3. Lang IM. Chronic thromboembolic pulmonary hypertension – not so rare after all. *New Engl J Med.* 2004;350:2236-2238.
4. Klepetko W, Mayer E, Sandoval J, et al. Interventional and surgical modalities of treatment for pulmonary arterial hypertension. *J Am Coll Cardiol.* 2004;43(12 suppl S):73S-80S.
5. Pengo V, Lensing A, Prins M, et al. Incidence of chronic thromboembolic pulmonary hypertension after pulmonary embolism. *New Engl J Med.* 2004;350:2257-2264.
6. Becattini C, Agnelli G, Pesavento R, et al. Incidence of chronic thromboembolic pulmonary hypertension after a first episode of pulmonary embolism. *Chest.* 2006;130(1):172-175.
7. Miniati M, Monti S, Bottai M, et al. Survival and restoration of pulmonary perfusion in a long-term follow-up of patients after acute pulmonary embolism. *Medicine (Baltimore).* 2006;85(5):253-262.
8. Guidelines on diagnosis and management of acute pulmonary embolism. Task force on pulmonary embolism, European society of cardiology. *Eur Heart J.* 2000;21(16):1301-1336.
9. Moser KM, Auger WR, Fedullo PF. Chronic major-vessel thromboembolic pulmonary hypertension. *Circulation.* 1990;81(6):1735-1743.
10. Fedullo PF, Auger WR, Kerr KM, Rubin LJ. Chronic thromboembolic pulmonary hypertension. *N Engl J Med.* 2001;345:1465-1472.
11. Condliffe R, Kiely DG, Gibbs JS, et al. Improved outcomes in medically and surgically treated chronic thromboembolic pulmonary hypertension. *Am J Respir Crit Care Med.* 2008;177:1122-1127.
12. Bonderman D, Jakowitsch J, Aldbrecht C, et al. Medical conditions increasing the risk of chronic thromboembolic pulmonary hypertension. *Thromb Haemost.* 2005;93:512-516.
13. Jais X, Ioos V, Jardim C, et al. Splenectomy and chronic thromboembolic pulmonary hypertension. *Thorax.* 2005;60:1031-1034.
14. Bonderman D, Skoro-Sajer N, Jakowitsch J, et al. Predictors of outcome in chronic thromboembolic pulmonary hypertension. *Circulation.* 2007;115:2153-2158.
15. Auger WR, Channick RN, Kerr KM, Fedullo PF. Evaluation of patients with suspected chronic thromboembolic pulmonary hypertension. *Semin Thorac Cardiovasc Surg.* 1999;11:179-190.
16. Kim NH, Fesler P, Channick RN, et al. Preoperative partitioning of pulmonary vascular resistance correlates with early outcome after thromboendarterectomy for chronic thromboembolic pulmonary hypertension. *Circulation.* 2004;109:18-22.
17. Jais X, Ghofrani A, Hoeper M, et al. Bosentan for Inoperable chronic thromboembolic pulmonary hypertension (CTEPH): a randomized, placebo-controlled trial – BENEFIT. *Am J Respir Crit Care Med.* 2007;175:A896.
18. Suntharalingam J, Treacy CM, Doughty NJ, et al. Long term use of sildenafil in inoperable chronic thromboembolic pulmonary hypertension. *Chest.* 2008;134(2):229-236.
19. Kerr KM, Rubin LJ. Epoprostenol therapy as a bridge to pulmonary thromboendarterectomy for chronic thromboembolic pulmonary hypertension. *Chest.* 2003;123:319-320.
20. Nagaya N, Sasaki N, Ando M, et al. Prostacyclin therapy before pulmonary thromboendarterectomy in patients with chronic thromboembolic pulmonary hypertension. *Chest.* 2003;123:338-343.
21. Ono F, Nagaya N, Kyotani S, Oya H, Nakanishi N, Miyatake K. Hemodynamic and hormonal effects of beraprost sodium, an orally active prostacyclin analogue, in patients with secondary precapillary pulmonary hypertension. *Circulation.* 2003;67:375-378.
22. Ono F, Nagaya N, Okumura H, et al. Effect of orally active prostacyclin analogue on survival in patients with chronic thromboembolic pulmonary hypertension without major vessel obstruction. *Chest.* 2003;123:1583-1588.
23. Bresser P, Fedullo PF, Auger W, et al. Continuous intravenous epoprostenol for chronic thromboembolic pulmonary hypertension. *Eur Resp J.* 2004;23:595-600.
24. Scelsi L, Ghio S, Campana C, et al. Epoprostenol in chronic thromboembolic pulmonary hypertension with distal lesions. *Ital Heart J.* 2004;5:618-623.
25. Kramm T, Eberle B, Guth S, Mayer E. Inhaled iloprost to control residual pulmonary hypertension following pulmonary endarterectomy. *Eur J Cardiothorac Surg.* 2005;28:882-888.
26. Bonderman D, Nowotny R, Skoro-Sajer N, et al. Bosentan therapy for inoperable chronic thromboembolic pulmonary hypertension. *Chest.* 2005;128:2599-2603.

27. Hoeper MM, Kramm T, Wilkens H, et al. Bosentan therapy for inoperable chronic thromboembolic pulmonary hypertension. *Chest*. 2005;128:2363-2367.
28. Hughes R, George P, Parameshwar J, et al. Bosentan in inoperable chronic thromboembolic pulmonary hypertension. *Thorax*. 2005;60:707.
29. Hughes R, Jais X, Bonderman D, et al. Bosentan in inoperable chronic thromboembolic pulmonary hypertension: efficacy at 1 year. *Eur Resp J*. 2006;28:138-143.
30. Ghofrani HA, Schermuly RT, Rose F, et al. Sildenafil for long-term treatment of nonoperable chronic thromboembolic pulmonary hypertension. *Am J Respir Crit Care Med*. 2003;167: 1139-1141.
31. Sheth A, Park JES, Ong YE, Ho TB, Madden BP. Early haemodynamic benefit of sildenafil in patients with coexisting chronic thromboembolic pulmonary hypertension and left ventricular dysfunction. *Vasc Pharmacol*. 2005;42:41-45.
32. Reichenberger F, Voswinckel R, Enke B, et al. Long-term treatment with sildenafil in chronic thromboembolic pulmonary hypertension. *Eur Respir J*. 2007;30(5):922-927.
33. Bresser P, Pepke-Zaba J, Jais X, Humbert M, Hoeper MM. Medical therapies for chronic thromboembolic pulmonary hypertension: an evolving treatment paradigm. *Proc Am Thorac Soc*. 2006;3:594-600.
34. Auger WR, Kerr KM, Kim NH, et al. Chronic thromboembolic pulmonary hypertension. *Cardiol Clin*. 2004;22(3):453-466, vii.
35. Lewczuk J, Piszko P, Jagas J, et al. Prognostic factors in medically treated patients with chronic pulmonary embolism. *Chest*. 2001;119:818-823.
36. Riedel M, Stanek V, Widimisky J, Perovsky I. Longterm follow-up of patients with pulmonary thromboembolism: late prognosis and evolution of hemodynamic and respiratory data. *Chest*. 1982;81:151-158.
37. Kunieda T, Nakanishi N, Satoh T, Kyotani S, Okano Y, Nagaya N. Prognoses of primary pulmonary hypertension and chronic majorvessel thromboembolic pulmonary hypertension determined from cumulative survival curves. *Intern Med*. 1999;38:543-546.
38. Suntharalingam J, Machado RD, Sharples LD, et al. Demographic features, BMPR2 status and outcomes in distal chronic thromboembolic pulmonary hypertension. *Thorax*. 2007;62:617-622.
39. Fijalkowska A, Kurzyna M, Torbicki A, et al. Serum N-terminal brain natriuretic peptide as prognostic parameter in patients with pulmonary hypertension. *Chest*. 2006;19:1313-1321.
40. Suntharalingam J, Goldsmith K, Toshner M, et al. Role of NT-proBNP and 6MWD in chronic thromboembolic pulmonary hypertension. *Respir Med*. 2007;101:2254-2262.
41. Matsuda H, Ogino H, Minatoya K, et al. Long-term recovery of exercise ability after pulmonary endarterectomy for chronic thromboembolic pulmonary hypertension. *Ann Thorac Surg*. 2006;82(4):1338-1343.
42. Reesink HJ, van der Plas MN, Verhey NE, et al. Six-minute walk distance as parameter of functional outcome after pulmonary endarterectomy for chronic thromboembolic pulmonary hypertension. *J Thorac Cardiovasc Surg*. 2007;133(2):510-516.
43. Iwase T, Nagaya N, Ando M, et al. Acute and chronic effects of surgical thromboendarterectomy on exercise capacity and ventilatory efficiency in patients with chronic thromboembolic pulmonary hypertension. *Heart*. 2001;86(2):188-192.
44. D'Armini AM, Cattadori B, Monterosso C, et al. Pulmonary thromboendarterectomy in patients with chronic thromboembolic pulmonary hypertension: hemodynamic characteristics and changes. *Eur J Cardiothorac Surg*. 2000;18(6):696-701.

Chapter 13
Pathophysiology of Chronic Thromboembolic Pulmonary Hypertension (CTEPH)

Jay Suntharalingam and Nicholas W. Morrell

13.1 Epidemiology of CTEPH

The prevalence and incidence of chronic thromboembolic pulmonary hypertension (CTEPH) following acute pulmonary embolism (PE) is difficult to ascertain. However, recent follow-up studies in patients presenting with acute PE give an indication of the scale of the problem. An early study followed an unselected series of PE patients and found evidence of CTEPH in four patients (5.1%).[1] Pengo et al. subsequently showed a cumulative incidence of 3.8% at 2 years, while even more recent studies by Becattini et al. and Miniati et al. have described cumulative incidences of 4/320 (1.3%) and 2/259 (0.8%), respectively.[2-4] Given that acute PE is as common as 1 per 1,000 of the population per year,[5] suggesting the annual incidence of CTEPH following acute venous thromboembolism is of the order of 8 to 51 cases per 1 million population. Moreover, given that CTEPH may be classically described as presenting some time following an acute event,[6] the results from these relatively short studies may represent an underestimate. A further consideration is that up to 60% of patients with CTEPH do not have a previously documented VTE event.[7] In these patients, it is presumed that clinically "silent" emboli in the past have triggered the disease process.

13.2 Etiology of CTEPH

Although CTEPH was first described over 50 years ago,[8] the etiology of this condition remains poorly understood. A number of studies have sought evidence for the same defects in the coagulation pathways that are frequently found in patients with acute VTE. Generally speaking, these studies have been unable to find similar associations in CTEPH.[6,9,10] However, abnormalities within other prothrombotic pathways

J. Suntharalingam (✉) and N.W. Morrell
Consultant Respiratory Physician, Respiratory Department, Royal United Hospital Bath, Combe Park, Bath, BA1 3NG, UK
e-mail: jay.suntharalingam@ruh-bath.swest.nhs.uk

D. Abraham et al. (eds.), *Advances in Vascular Medicine*,
DOI 10.1007/978-1-84882-637-3_13,

have been described, including antiphospholipid antibodies (aPL),[10] lipoprotein (a),[11] and elevated factor VIII levels.[12]

The lack of the classical VTE genetic risk factors in CTEPH has led to the proposal that some cases of CTEPH, particularly distal disease, may represent in situ thrombosis as distinct from thromboembolism. Although the current favored opinion is that CTEPH does represent a primary thromboembolic disease complicated by secondary arteriopathic changes, there remains some uncertainty.[13] In a recent study from our center, we compared the demographics and genetics of proximal and distal CTEPH with idiopathic pulmonary arterial hypertension. This study demonstrated that proximal and distal CTEPH share a common presentation, demographic and genetic profile when compared with IPAH.[14]

13.3 Pathogenesis of CTEPH

CTEPH can be considered as a failure of the resolution of the clot burden following acute PE. Treatment of patients with heparin and warfarin effectively prevents the generation of further thrombus. The process of fibrinolysis is then achieved by endogenous pathways. In the initial "early" phase, thrombus resolution probably results from a combination of thrombus fragmentation and endogenous fibrinolysis. Imaging and hemodynamic studies suggest that this early phase takes place in the first 30–40 days following presentation, and may in some cases lead to complete disease resolution.[15] However, a significant proportion of patients have continuing clot beyond this early phase. Further resolution beyond this point is likely to rely on a process of clot organization and neovascularization, during which the obstructed vessel becomes recanalized and vessel patency is restored (Fig. 13.1).[16]

13.3.1 Abnormalities of Coagulation and Fibrinolysis

Most studies have failed to show any significant association between hereditary thrombophilia and CTEPH,[6,10] despite the prevalence of these in deep vein thrombosis (DVT) and VTE. The results of one large study comparing the prevalence of hereditary thrombophilias in DVT, CTEPH, and IPAH are shown in Table 13.1 [10]

Elevated Factor VIII levels have also been demonstrated in CTEPH, in a study involving 122 CTEPH, 88 PAH, and 82 healthy controls.[12] In this study, levels were significantly greater in CTEPH patients than either of the two other groups. Furthermore, levels failed to normalize following successful PEA surgery. Factor VIII is a 2,332 amino acid polypeptide that circulates bound to von Willebrand factor (vWF). Following activation by either thrombin or factor Xa, factor VIIIa complexes with activated factor IXa, calcium, and phospholipids to precipitate activation of factor X. This leads to sustained thrombin generation and clot formation. Elevated factor VIII levels have been associated with both single and recurrent

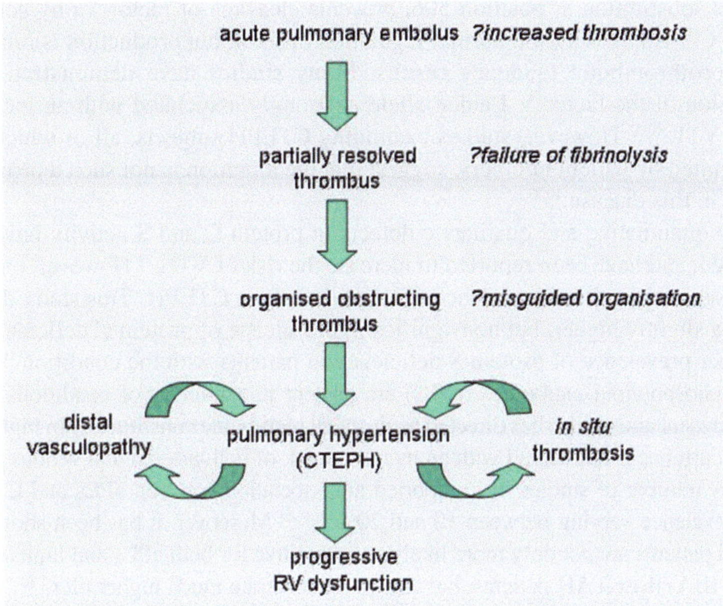

Fig. 13.1 Summary of potential pathogenic mechanisms in chronic thromboembolic pulmonary hypertension (CTEPH)

Table 13.1 Frequencies of inherited thrombophilias in IPAH subjects, CTEPH subjects, and controls. Results presented as number of subjects affected/number of subjects in cohort, and as percentage affected. Taken from reference 10

	Antithrombin III	Protein C deficiency	Protein S deficiency	Factor V mutation	Prothrombin mutation
IPAH	0/64 (0%)	0/26 (0%)	0/26 (0%)	1/64 (1.5%)	1/61 (1.6%)
CTEPH	0/46 (0%)	1/46 (2%)	0/46 (0%)	3/46 (6.5%)	1/40 (2.5%)
Controls	0/100 (0%)	1/100 (1%)	0/100 (0%)	3/100 (3%)	2/100 (2%)

venous thromboembolic events.[17,18] The persistent elevation of Factor VIII following PEA surgery suggests that elevated factor VIII levels may play a causal role in the development of CTEPH and are not simply a nonspecific secondary response to abnormal pulmonary artery pressures.

During platelet activaton, factor V is released from platelet granules and is subsequently cleaved by thrombin to form activated factor Va. This in turn acts as a co-factor with factor Xa in the production of thrombin from prothrombin. Although a number of genetic variants of factor V exist, the most clinically relevant is factor V Leiden. This mutation (G1691A), which results in a glutamine for arginine substitution at position 506, prevents cleavage of factor Va by activated protein C.[19] As a result, the normal regulation of thrombin production is impaired, and a prothrombotic tendency results. Many studies have demonstrated that possession of the factor V Leiden allele is strongly associated with an increased risk of VTE.[20-22] However, studies examining CTEPH subjects, all of which have been limited by patient numbers, suggest that the mutation is not such a prominent feature in this disease.[10,23]

Both quantitative and qualitative defects in protein C and S activity have been described, and have been reported to increase the risk of VTE.[24] However, only one study has examined the prevalence of these defects in CTEPH. This study demonstrated a slightly higher, but nonsignificant, prevalence of protein C deficiency and no higher prevalence of protein S deficiency in patients with the condition.[10]

Antiphospholipid antibodies (aPLs) are present in a number of conditions where they represent autoantibodies directed against cell membrane constituents. In high titers, their occurrence is associated with an increased risk of both arterial and venous thromboses. A number of studies have reported an association between aPLs and CTEPH, with prevalence varying between 10 and 20%.[10,12,25] Moreover, it has been shown that CTEPH patients are not only more likely to be positive for both aPLs and lupus anticoagulant (LA) than IPAH patients, but also to demonstrate much higher titers.[10]

Fibrinogen plays a pivotal role in balancing hemostasis, by acting as a substrate for both the coagulation and fibrinolytic systems. A number of abnormalities of fibrinogen have been described, affecting not only fibrinogen levels but also fibrinogen function. Clearly, prothrombotic abnormalities of fibrinogen are of interest in CTEPH, as this offers one potential mechanism by which thrombus may persist following an acute embolic event. A recent *in vitro* study highlighted this possibility by demonstrating that fibrin-derived from patients with CTEPH was more resistant to lysis when compared with controls.[26] We recently reported an association of a polymorphism of the fibrinogen Aα (Thr312Ala) chain and CTEPH.[27] The same polymorphism is more closely associated with PE, rather than DVT perhaps indicating that this polymorphism alters the structure of fibrin and predisposes clot to fragmentation and embolism.[28-30]

Plasminogen activator inhibitor (PAI) is the principal inhibitor of tissue- and urokinase-plasminogen activator (tPA and uPA), and is therefore a key factor in the regulation of fibrinolysis. Early studies in CTEPH failed to show any systemic imbalances in PAI-1 levels or activity,[31] but did report increased local expression of PAI-1 in the material removed at PEA surgery.[32] As such, it is possible that abnormal cellular PAI-1 expression may affect the handling of thrombus within the pulmonary vasculature and predispose to CTEPH.

Lipoprotein (a) (Lp(a)) is a cholesterol-rich lipoprotein, which shares structural and immunological properties with plasminogen. As a result, Lp(a) provides a competitive substrate for plasminogen activation. Raised Lp(a) levels have previously been reported in patients with acute pulmonary embolic disease and have also been demonstrated in patients with CTEPH.[11,33] Although unclear whether this represents a primary or secondary phenomenon, this could provide a mechanism by which endogenous fibrinolysis is impaired in this disease.

13.3.2 Abnormalities of Thrombus Organization

Although CTEPH is characterized by persistent thrombus, the material obstructing the pulmonary arteries is highly organized and cellular.[16] Organization is a natural part of wound healing and is pivotal to the "late" phase resolution of emboli that initially escape fragmentation and fibrinolysis. It is conceivable that this late phase may be disturbed and contribute to the vascular obstruction rather than assisting clearance. For example, it is known that conditions such as inflammatory bowel disease are over-represented in patients with CTEPH.[34] These conditions associated with systemic inflammation may predispose to further thrombus organization rather than resolution. One other possibility is that inappropriate organization is driven by "misguided" neovascularization. Neovascularization is a key component of the organizational process and is vital for restoring vessel patency.[35] However, endothelialized channels have also been noted within obstructed vessels in CTEPH.[16,32,36] Although these channels may simply represent remnants of failed recanalization, they may also contribute to the maintenance of the organized material and prevent its resolution.

13.3.3 Secondary Arteriopathy

CTEPH patients often display severe pulmonary hypertension that cannot be fully explained purely by the degree of pulmonary vascular obstruction visible on conventional imaging.[37] In these cases, the increased pulmonary vascular resistance may be due to distal obstructive thrombotic lesions situated beyond the subsegmental level, but also due to arteriopathic changes at a precapillary level.[38] These latter changes are histologically identical to those seen in IPAH, although their extent and distribution do differ.[39] It was originally thought that distal arteriopathy occurred in the unobstructed vascular bed as a consequence of the increased pressure and flow. Subsequent studies, however, have shown that these changes can also develop in obstructed beds, possibly as a result of ischemia or retrograde blood flow from broncho-pulmonary anastamoses.[38,39] These distal changes are responsible for the proportion of patients who have persisting pulmonary hypertension following PEA and may be amenable to targeted drug therapy.

References

1. Ribeiro A, Lindmarker P, Johnsson H, et al. Pulmonary embolism: one-year follow-up with echocardiography doppler and five-year survival analysis. *Circulation*. 1999;99:1325-1330.
2. Becattini C, Agnelli G, Pesavento R, et al. Incidence of chronic thromboembolic pulmonary hypertension after a first episode of pulmonary embolism. *Chest*. 2006;130:172-175.
3. Miniati M, Monti S, Bottai M, et al. Survival and restoration of pulmonary perfusion in a long-term follow-up of patients after acute pulmonary embolism. *Medicine (Baltimore)*. 2006;85:253-262.

4. Pengo V, Lensing AW, Prins MH, et al. Incidence of chronic thromboembolic pulmonary hypertension after pulmonary embolism. *N Engl J Med.* 2004;350:2257-2264.
5. Guidelines on diagnosis and management of acute pulmonary embolism. Task force on pulmonary embolism, European society of cardiology. *Eur Heart J.* 2000;21:1301-1336.
6. Moser KM, Auger WR, Fedullo PF. Chronic major-vessel thromboembolic pulmonary hypertension. *Circulation.* 1990;81:1735-1743.
7. Lang IM. Chronic thromboembolic pulmonary hypertension – not so rare after all. *N Engl J Med.* 2004;350:2236-2238.
8. Owen WR, Thomas WA, Castleman B, et al. Unrecognized emboli to the lungs with subsequent cor pulmonale. *N Engl J Med.* 1953;249:919-926.
9. Colorio CC, Martinuzzo ME, Forastiero RR, et al. Thrombophilic factors in chronic thromboembolic pulmonary hypertension. *Blood Coagul Fibrinolysis.* 2001;12:427-432.
10. Wolf M, Boyer-Neumann C, Parent F, et al. Thrombotic risk factors in pulmonary hypertension. *Eur Respir J.* 2000;15:395-399.
11. Ignatescu M, Kostner K, Zorn G, et al. Plasma Lp(a) levels are increased in patients with chronic thromboembolic pulmonary hypertension. *Thromb Haemost.* 1998;80:231-232.
12. Bonderman D, Turecek PL, Jakowitsch J, et al. High prevalence of elevated clotting factor VIII in chronic thromboembolic pulmonary hypertension. *Thromb Haemost.* 2003;90:372-376.
13. Peacock A, Simonneau G, Rubin L. Controversies, uncertainties and future research on the treatment of chronic thromboembolic pulmonary hypertension. *Proc Am Thorac Soc.* 2006;3:608-614.
14. Suntharalingam J, Machado R, Sharples L, et al. Demographics, BMPR2 status and outcomes in distal chronic thromboembolic pulmonary hypertension. *Thorax.* 2007;62:617-622.
15. Nijkeuter M, Hovens MM, Davidson BL, et al. Resolution of thromboemboli in patients with acute pulmonary embolism: a systematic review. *Chest.* 2006;129:192-197.
16. Wagenvoort CA. Pathology of pulmonary thromboembolism. *Chest.* 1995;107:10S-17S.
17. O'Donnell J, Tuddenham EG, Manning R, et al. High prevalence of elevated factor VIII levels in patients referred for thrombophilia screening: role of increased synthesis and relationship to the acute phase reaction. *Thromb Haemost.* 1997;77:825-828.
18. Kyrle PA, Minar E, Hirschl M, et al. High plasma levels of factor VIII and the risk of recurrent venous thromboembolism. *N Engl J Med.* 2000;343:457-462.
19. Bertina RM, Koeleman BP, Koster T, et al. Mutation in blood coagulation factor V associated with resistance to activated protein C. *Nature.* 1994;369:64-67.
20. Ridker PM, Hennekens CH, Lindpaintner K, et al. Mutation in the gene coding for coagulation factor V and the risk of myocardial infarction, stroke, and venous thrombosis in apparently healthy men. *N Engl J Med.* 1995;332:912-917.
21. Middeldorp S, Meinardi JR, Koopman MM, et al. A prospective study of asymptomatic carriers of the factor V Leiden mutation to determine the incidence of venous thromboembolism. *Ann Intern Med.* 2001;135:322-327.
22. Price DT, Ridker PM. Factor V Leiden mutation and the risks for thromboembolic disease: a clinical perspective. *Ann Intern Med.* 1997;127:895-903.
23. Lang IM, Klepetko W, Pabinger I. No increased prevalence of the factor V Leiden mutation in chronic major vessel thromboembolic pulmonary hypertension (CTEPH). *Thromb Haemost.* 1996;76:476-477.
24. Martinelli I, Mannucci PM, De Stefano V, et al. Different risks of thrombosis in four coagulation defects associated with inherited thrombophilia: a study of 150 families. *Blood.* 1998;92:2353-2358.
25. Auger WR, Permpikul P, Moser KM. Lupus anticoagulant, heparin use, and thrombocytopenia in patients with chronic thromboembolic pulmonary hypertension: a preliminary report. *Am J Med.* 1995;99:392-396.
26. Morris TA, Marsh JJ, Chiles PG, et al. Fibrin derived from patients with chronic thromboembolic pulmonary hypertension is resistant to lysis. *Am J Respir Crit Care Med.* 2006;173:1270-1275.
27. Suntharalingam J, Goldsmith K, van Marion V, et al. Fibrinogen Aalpha Thr312Ala polymorphism is associated with chronic thromboembolic pulmonary hypertension. *Eur Respir J.* 2008;31:736-741.

28. Behague I, Poirier O, Nicaud V, et al. ß Fibrinogen gene polymorphisms are associated with plasma fibrinogen and coronary artery disease in patients with myocardial infarction: the ectim study. *Circulation*. 1996;93:440-449.
29. Carter AM, Catto AJ, Grant PJ. Association of the alpha-fibrinogen Thr312Ala polymorphism with poststroke mortality in subjects with atrial fibrillation. *Circulation*. 1999;99:2423-2426.
30. Carter AM, Catto AJ, Kohler HP, et al. alpha-fibrinogen Thr312Ala polymorphism and venous thromboembolism. *Blood*. 2000;96:1177-1179.
31. Olman MA, Marsh JJ, Lang IM, et al. Endogenous fibrinolytic system in chronic large-vessel thromboembolic pulmonary hypertension. *Circulation*. 1992;86:1241-1248.
32. Lang IM, Marsh JJ, Olman MA, et al. Expression of type 1 plasminogen activator inhibitor in chronic pulmonary thromboemboli. *Circulation*. 1994;89:2715-2721.
33. Csaszar A, Karadi I, Juhasz E, et al. High lipoprotein(a) levels with predominance of high molecular weight apo(a) isoforms in patients with pulmonary embolism. *Eur J Clin Invest*. 1995;25:368-370.
34. Bonderman D, et al. Medical conditions increasing the risk of chronic thromboembolic pulmonary hypertension. *Thromb Haemost*. 2005;93:512-516.
35. Modarai B, Burnand KG, Humphries J, et al. The role of neovascularisation in the resolution of venous thrombus. *Thromb Haemost*. 2005;93:801-809.
36. Arbustini E, Morbini P, D'Armini AM, et al. Plaque composition in plexogenic and thromboembolic pulmonary hypertension: the critical role of thrombotic material in pultaceous core formation. *Heart*. 2002;88:177-182.
37. Azarian R, Wartski M, Collignon MA, et al. Lung perfusion scans and hemodynamics in acute and chronic pulmonary embolism. *J Nucl Med*. 1997;38:980-983.
38. Moser KM, Bloor CM. Pulmonary vascular lesions occurring in patients with chronic major vessel thromboembolic pulmonary hypertension. *Chest*. 1993;103:685-692.
39. Yi ES, Kim H, Ahn H, et al. Distribution of obstructive intimal lesions and their cellular phenotypes in chronic pulmonary hypertension. A morphometric and immunohistochemical study. *Am J Respir Crit Care Med*. 2000;162:1577-1586.

Chapter 14
Surgical Management of Chronic Thromboembolic Pulmonary Hypertension (CTEPH)

David P. Jenkins

14.1 History of Surgery

We should distinguish carefully operations for acute pulmonary embolism – embolectomy from the true endarterectomy procedures for chronic thromboembolic pulmonary hypertension (CTEPH). Trendelenburg famously described a salvage operation for acute massive pulmonary embolus at the turn of the twentieth century, at this time without cardiopulmonary bypass or indeed anesthesia. Despite reporting no survivors, the operation still bears his name. It was only in the 1960s with the development of cardiopulmonary bypass that embolectomy became a more reliable and safe operation that can be accomplished by any cardiothoracic surgeon. Interestingly, with the advent of thrombolysis, it is performed less frequently today.

Although acute embolic occlusion of the pulmonary arteries was relatively easily understood, the concept of chronic occlusion after repeated embolism took some time to become established. By the 1930s, case reports had begun to describe the chronic form of the disease discovered at postmortem. One of the first reports of CTEPH in the UK was by Dr Petch senior, a single patient, the case record described in detail in the Lancet.[1] By the end of the 1950s, a few cases of operative intervention, usually by thoracotomy, for conditions that were subsequently found to be chronic pulmonary artery occlusion, had been reported. The first successful planned pulmonary endarterectomy was by Dr Hufnagel in 1962. The physician caring for the patient at that time was Dr Moser, who subsequently moved to University of California in San Diego (UCSD) and with Dr Braunwald reported the first case of successful pulmonary endarterectomy at UCSD.[2] Cabrol and Dor reported successful operations in Europe, but by the mid-1980s, less than 100 procedures had been performed worldwide. When Dr Jamieson moved to UCSD in 1990, 189 endarterectomies had already been performed.[3] Since then, the operation has been refined with an improvement in outcome as experience has increased, and when I visited UCSD in November 2000, they had recently operated on their 2000th patient.

D.P. Jenkins (✉)
Cardiac Surgery, Papworth Hospital, Cambridge, Cambridgeshire, UK
e-mail: david.jenkins@papworth.nhs.uk

D. Abraham et al. (eds.), *Advances in Vascular Medicine*,
DOI 10.1007/978-1-84882-637-3_14, © Springer-Verlag London Limited 2010

In the UK, Papworth hospital commenced a formal pulmonary endarterectomy program in 1997, initially treating patients originally referred for consideration of heart lung transplantation. In 2000, a joint application was made by Freeman hospital in Newcastle and Papworth hospital to the National Specialist Commissioning Advisory Group (NSCAG) to fund a national pulmonary endarterectomy service. It was decided that expertise should be concentrated in one hospital, and Papworth was commissioned to provide the service for the UK. In 2001, NSCAG also designated five national pulmonary hypertension centers at Glasgow, London, Newcastle, Papworth, and Sheffield. Many of the current referrals for pulmonary endarterectomy arise through this network of specialist centers.

14.2 Pathophysiology

CTEPH is the result of intraluminal thrombus organization and fibrous stenosis and/or complete obliteration of the pulmonary artery lumen. It is very likely that acute pulmonary embolism, whether symptomatic or asymptomatic, serves as the initiating event. Following a pulmonary embolism, many patients recover as natural lysis of the occluding thrombus occurs. However, in a small percentage of patients, the thrombus does not clear and CTEPH develops due to progressive pulmonary vascular occlusion and the development of a generalized hypertensive pulmonary arteriopathy in the pulmonary arterial bed spared from thromboembolic occlusion. Moser and Braunwald first postulated this concept of a two-compartment pulmonary vascular bed after the first patient underwent pulmonary endarterectomy at UCSD in 1970. This model explains why some CTEPH patients have severe pulmonary hypertension out of proportion to the degree of vascular obstruction visible in imaging investigations. It is not understood why some patients do not recanalize their vessels completely after surviving the acute embolic event nor why the remodeling occurs in the smaller nonobstructed arteries. The pathophysiology of CTEPH is thus far more complex than simple vascular obstruction. It is interesting to note that patients losing 50% of their pulmonary vascular bed acutely after a pneumonectomy do not develop PH.

14.3 Investigation and General Management

Patients with suspected CTEPH are discussed at a weekly multidisciplinary meeting with surgeons, pulmonary hypertension physicians, and radiologists. It is vital to distinguish CTEPH from other causes of pulmonary hypertension because patients are potentially curable by endarterectomy. Once the diagnosis is confirmed, the next decision is whether the patient would benefit from endarterectomy, taking into account symptomatic state and expected prognosis. Judging whether there is sufficient surgically accessible ("clearable") disease to recommend the procedure can be extremely difficult in some patients.

The initial phase of investigation includes transthoracic echocardiography, 6-min walk test, full lung function tests, right heart catheterization, and at least two imaging modalities (CTPA, pulmonary angiography or magnetic resonance pulmonary angiography). Radiology for CTEPH is discussed in more detail in a separate chapter. Once accepted for surgery, patients over 50 years of age also require coronary angiography and carotid Doppler examination.

Patients with CTEPH should receive life-long anticoagulation with warfarin or in some cases low molecular weight heparin. It is accepted practice to insert an IVC filter in patients with CTEPH undergoing endarterectomy surgery although there has never been a randomized controlled trial. In the UK, an IVC filter is recommended for most patients and there have been no adverse consequences arising from this policy to date. Many patients require diuretics to control right heart failure. Increasingly, patients with NYHA class III symptoms receive targeted therapy with the endothelin receptor antagonist Bosentan or the phosphodiesterase five inhibitor Sildenafil. Patients with class IV symptoms benefit from prostacycline analogs in addition. Most patients requiring pulmonary endarterectomy are already on targeted therapy for "bridging" prior to surgery. It is our current policy for patients with residual PH postoperation to remain on targeted therapies. Further details of the modern medical therapy for CTEPH are discussed in detail in a previous chapter.

14.4 Determinants of Operability in Patients with CTEPH

The prognosis without surgery is poor for many patients with CTEPH, especially when mean PA pressure is >50 mmHg,[4] although we have recently found that survival may be improved a little with modern targeted therapy.[5] Therefore, in most patients who are thought to have operable disease after careful evaluation, there are major symptomatic and prognostic benefits from endarterectomy. Indeed, the operation may be curative in a large number of patients who regain pulmonary artery pressures in the normal range at rest after surgery. However, decision-making may be difficult in some patients with high pulmonary vascular resistance (PVR)\ and relatively distal disease on imaging. Ultimately, the ease of dissection and true extent of clearable disease can only determined at operation. The disease has been classified into four subgroups based on the operative findings (see Table 14.1) with patients with types 1 and 2 disease deriving more benefit from endarterectomy at less perioperative risk.[6] However, this is intraoperative classification scheme, and does not assist in the prospective selection of patients who would benefit most from surgery. Therefore, there is a need for a preoperative classification system and risk stratification score to aid in patient selection.

Table 14.1 Operative classification and outcome data from ref.[6]

Type of disease	Type 1	Type 2	Type 3	Type 4
	Thrombus in main/lobar PA branches	Organized thrombus or thickened intima/webs proximal to segmental vessels	Intimal thickening confined to distal segmental vessels only	Distal arteriolar vasculopathy only, no intraluminal disease
No patients	76	81	38	7
Age	51	48	48	47
Fall in PA systolic pressure mmHg post-op	36	36	19	0
Mortality	1%	3%	13%	14%

14.4.1 Pulmonary Vascular Resistance (PVR)

It is well established in the literature and reflected in our own experience that absolute PVR is a risk factor for poor outcome after endarterectomy, especially when >1,000 dyne/s/cm^{-5}. A number of reports from Europe and the USA have confirmed this finding. Mortality following endarterectomy was 20% in patients with a preoperative PVR>1,200 dyne/s/cm^{-5} [7], and in the UCSD detailed review of a consecutive series of 500 patients operated on in the modern era, mortality was almost tenfold higher in those with a PVR>1,000 dyne/s/cm^{-5}.[8] The actual PA systolic pressure also influences outcome after endarterectomy, but supra systemic PH should not be a contraindication to operation. At UCSD, patients with a preoperative systolic PA pressure >100 mmHg gained the greatest reduction in PA pressure and PVR.[9] As expected, this dramatic benefit was at the expense of a greater incidence of reperfusion injury and mortality that at 10% was double that observed in other patients.

Overall, the key determinant of operability is the correlation between the degree of visible disease in imaging studies and the hemodynamic dysfunction (absolute PA pressure, cardiac output, and the function of both – PVR). In particular, patients with a PVR disproportionately higher than the segmental obstruction visible by imaging have less benefit from endarterectomy and a much higher risk of postoperative mortality because they have a higher proportion of microvascular disease, contributing to PVR but not seen radiologically and not improved by endarterectomy. Attempts have been made to make assessment of microvascular disease more objective by partitioning the PVR into upstream and downstream components with analysis of the decay of the occlusion pressure waveform at right heart catheterization.[10] The upstream component is thought to represent the disease in larger vessels (that may be clearable at endarterectomy) and the downstream component the resistance due to smaller vessels (including microvascular disease). Preliminary analysis suggests that the higher the percentage of upstream PVR, preoperation correlates with better hemodynamic outcome and survival after surgery. We are also using this method in high-risk patients, but we have not yet confirmed a cut-off level of upstream PVR. Decisions regarding operability in some patients can therefore be subjective and be very difficult and often depend on the combined clinical experience of the multidisciplinary team, but will ultimately be made by the patient and surgeon. A minority of these patients with severe symptoms may be considered for lung transplantation instead of endarterectomy.

14.4.2 Other Risk Factors

Operative risk is almost totally dependent on CTEPH as discussed above, and concomitant procedures (CABG, valve replacement, etc.) can be performed as necessary during the rewarming phase of the operation without additional risk to the patient.[11] We screen for ischemic heart disease and valve disease during patient evaluation and positive findings would make us more rather than less likely to offer surgery. The only comorbidity that may influence the decision to operate is severe parenchymal lung disease. There are no absolute criteria for lung function abnormalities that preclude endarterectomy, but it makes physiological sense that improving perfusion to areas damaged by emphysema or interstitial lung disease will have little benefit as ventilation to those areas is reduced, so V/Q mismatch will remain. Other conditions that are associated with increased risk at surgery include a previous ventriculo-atrial shunt to treat hydrocephalus and a splenectomy probably as a result of a higher component of microvascular disease and inflammatory conditions such as osteomyelitis and inflammatory bowel disease.[12]

14.4.3 Current Policy

In our experience, most symptomatic patients with at least five segmental vessels diseased or occluded (25% of total) and a PVR of $<1{,}000$ dynes/s/cm^{-5} get a good result from surgery and are at relatively low risk of complications. Patients with higher PVR require more careful assessment, but may also benefit albeit at higher risk. The majority of patients in our experience and reported in most surgical series have NYHA class III or IV symptoms and a PVR of >300 dynes/s/cm^{-5}, but we also offer the procedure to patients with class II symptoms and some with normal PA pressure at rest and evidence of exertional PH.

14.5 Surgical Technique

The fundamental aim of the surgery is to perform a full endarterectomy (not embolectomy or thrombectomy) of both pulmonary vascular trees. The modern operation is performed via a median sternotomy with hypothermic cardiopulmonary bypass and right and left pulmonary arteriotomies within the pericardium. Adequate visualization for distal dissection necessitates reduction in bronchial arterial collateral return to the pulmonary arteries. Traditionally, this has been overcome by complete deep hypothermic circulatory arrest (DHCA) for periods of 20 min at 20°C and this technique remains the standard teaching.[13] More recently, a number of alternative techniques designed to avoid deep hypothermia and/or circulatory arrest have been advocated for performance of the endarterectomy.[14] This has generated much debate in the literature with some advocating that DHCA is essential for effective endarterectomy. At Papworth, we have devised a method of selective antegrade cerebral perfusion (SACP) to avoid total circulatory arrest.[15] My personal view is that dissection is technically easier with complete DHCA, but that it is not necessary in all patients and SACP appears to be technically feasible in the majority of patients. The SACP technique may provide some distinct advantages, with reference to both potentially optimizing cerebral protection and allowing completeness of endarterectomy. First, there is at least some theoretical advantage to maintaining antegrade cerebral perfusion. Second, the maintenance of this perfusion may allow a greater period of continuous time to complete the endarterectomy, thus optimizing disease clearance. Third, when necessary, easy and rapid recourse to DHCA is possible. We are currently conducting a randomized controlled trial to compare the techniques with respect to neurological and pulmonary outcomes.

14.5.1 Papworth Technique

Our current protocol and technique is outlined below. General anesthesia is induced with placement of right radial and femoral arterial, central venous, and pulmonary artery catheters for hemodynamic monitoring. A transesophageal echocardiograph probe is inserted. Nasopharyngeal and bladder thermometer probes are used for peripheral and central temperature monitoring. Near-infrared spectroscopy is used to monitor cerebral oxygenation (INVOS Cerebral Oximeter).

The chest is opened with a standard median sternotomy. The ascending aorta, main pulmonary artery, superior vena cava (SVC), and aortic arch between the left common carotid and left subclavian arteries are then isolated with tapes. These tapes are used to facilitate the application of the cross-clamps and exposure of the pulmonary arteries for endarterectomy.

Systemic heparin is administered to achieve an activated clotting time (ACT) of greater than 400 s. Previously aprotonin was used as an anti-inflammatory and hemostatic agent, but since publication of the BART trial, tranexamic acid has been substituted. Cardiopulmonary bypass is then established with venous drainage from two venous cannulae (one two-stage cannula inserted in the right atrial appendage and a straight superior vena caval cannula inserted through the body of the right atrium) and arterial return to the ascending aorta. Cooling is commenced to a core temperature of 20°C using pH-stat management. Vents are placed in the right superior pulmonary vein and main pulmonary artery. A cardioplegia cannula is placed in the aortic root.

A cooling water jacket is positioned over the right ventricle for additional myocardial protection. The patient's head is wrapped in a water cooled jacket and 1 g IV methylprednisolone administered.

Once a core temperature of 20°C is reached, the dissection can begin. The right pulmonary artery is exposed between the ascending aorta and SVC and opened with a longitudinal extended beneath the SVC laterally. Care is taken to avoid denuding the artery of adventitial tissue to facilitate secure hemostasis during closure.

The lumen of the right pulmonary arterial system is exposed and the endarterectomy plane raised using a scalpel and spatula. The dissection proceeds within the superficial media of the vessel wall that is only 1–2 mm in thickness. By the use of a nerve hook, forceps and a sucker–dissector, the plane can be extended circumferentially and then distally by careful traction as far as possible with the intention of tracing the endarterectomy into all the affected segmental or subsegmental vessels (see Fig. 14.1). A cast of the inner layer of the pulmonary arterial tree is then dissected free moving toward the periphery.

While the initial dissection can be performed with full CPB, once visualization becomes compromised, by return of bronchial collateral blood flow, dissection cannot proceed safely. At this stage, the circulation needs to be fully arrested using DHCA or partially arrested with continued SACP. When using the latter method, an angled clamp is placed between the left common carotid and left subclavian arteries and a cross-clamp is placed on the ascending aorta below the arterial inflow

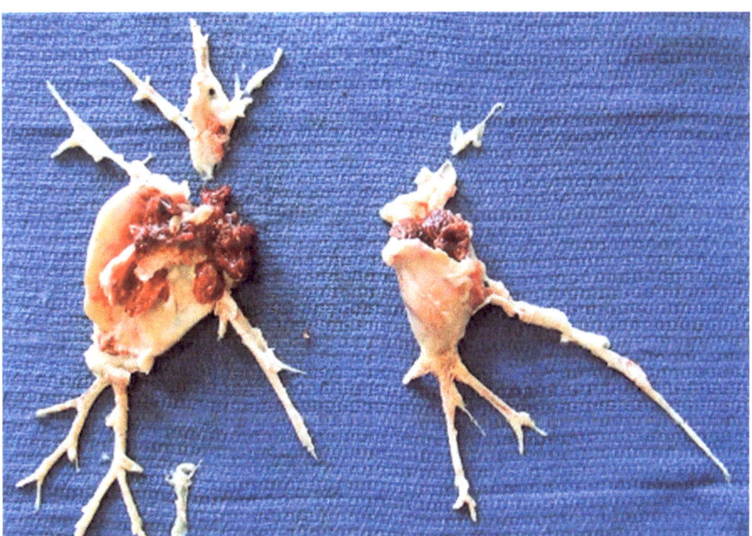

Fig. 14.1 Material removed from the right and left pulmonary arteries during pulmonary endarterectomy showing some proximal laminated thrombus indicating type 1 disease, and long fibrous segmental and subsegmental "tails"

cannula (CPB flow is then confined to the brachiocephalic and left carotid arteries). Cold blood cardioplegia into the aortic root is used for myocardial protection. SACP can be maintained at approximately 1.0–1.5 L/min with a perfusion pressure of 40–50 mmHg in the right radial artery for up to 60 min at 20°C. During this time, femoral arterial pressure is usually between 10 and 15 mmHg. If at any stage during SACP, collateral flow remains excessive and interferes with safe surgical dissection, the strategy can be switched to DHCA. DHCA is instituted for periods of up to 20 min at a time. A recovery period of 10 min is used between arrest periods when more dissection time is required.

The arteriotomy is closed with a 5/0 vascular suture (Prolene). The left pulmonary arteriotomy is extended from the vent insertion site in the main pulmonary artery into the main left pulmonary artery trending inferiorly. The left-sided endarterectomy can then be completed with the same surgical technique and perfusion strategy as for the right side.

After completion of the endarterectomies, the patient is rewarmed slowly on full CPB flow to a core temperature of at least 36.5°C. Any concomitant cardiothoracic surgical procedures can be completed during this phase of the operation. The patient is then weaned from CPB keeping right-sided filling pressures low, guided by invasive hemodynamic monitoring.

14.6 Postoperative Care

Following surgery, patients are transferred to the intensive care unit[16] and remain sedated and ventilated over the first night. Many patients can be extubated by the first postoperative day. Most of the general principles of postoperative cardiac surgical care apply, and in addition, we aim to avoid any factors that may increase PVR. Increasing PVR can initiate a series of adverse events because pulmonary artery pressure rises causing worsening of lung injury, right ventricular function is impaired as afterload increases and with reduced filling of the left heart cardiac output falls.

14.6.1 Cardiac

The majority of patients are maintained on a low-dose dopamine infusion over the first 3 days (3–5 mcg/kg/min). PA pressure monitoring continues over the first 24 h and it is usual to see a gradual further fall in PA pressure and PVR during this time. If further hemodynamic support is required, our first line is to increase the dopamine, followed by the addition of adrenaline and an intra-aortic balloon pump to optimize coronary blood flow in patients with right ventricular hypertrophy. Patients with severe reperfusion injury and significant residual PH may benefit from extracorporeal membrane oxygenation (ECMO) support.

14.6.2 Respiratory

A ventilation weaning strategy is employed to avoid episodes of hypoxia and resultant pulmonary vasoconstriction. Additionally, minimizing the mechanical impact of ventilation on PVR and on alveolar shear forces is important.

The general aims of ventilation are to maintain PaO2 value >12 kPa, with normocapnoea and normal acid–base balance and avoidance of peak airway pressures >30 cmH2O. Uncomplicated patients are usually extubated by the first postoperative day, but patients with reperfusion lung injury may require prolonged ventilation. These patients may be severely ill with a degree of residual pulmonary hypertension, reduced cardiac output, and very poor gas exchange resulting in hypoxia and CO_2 retention. Chest X-ray appearances are similar to those seen in acute lung injury. We have not found inhaled nitric oxide of any benefit in this patient group.

14.6.3 Renal/Fluid Balance

An aggressive diuresis is maintained in the first 24–48 h post-op. Many patients have edema pre–operation, and this is compounded by fluid retention after a long period of CPB. We use a combination of 40 mg IV furosemide and 12.5 g IV mannitol every 6 h for the first 36 h, followed by a reducing dose of IV or PO furosemide. If volume replacement is required, then human albumin solution or blood is used aiming for a hemoglobin concentration of 10 g/dl. Crystalloid solutions are usually avoided. In practical terms, the average patient is generally rendered 2.5–3.5 L negative on fluid balance in the first 48 h.

14.6.4 Anti-Coagulation

In the absence of postoperative bleeding, anticoagulation is usually commenced with low molecular weight heparin on the second evening post-op. Warfarin is started at the same time if the patient is extubated and absorbing with the therapeutic target range being 2.5–3.5. Selected patients with a specific procoagulant diathesis may benefit from a higher target INR range and the addition of antiplatelet agents.

Most patients are well enough to leave the ICU by the second postoperative day. Drains are normally removed by the third day and early mobilization encouraged. Oxygen therapy is gradually reduced, but as most patients required long-term oxygen therapy preoperation, this can be continued at home in the first few weeks. Many patients are well enough to be discharged directly to home within 2 weeks of surgery.

14.7 Surgical Outcome

14.7.1 Mortality

Many units have reported excellent early results after pulmonary endarterectomy in the last few years. As experience grows, operative mortality has fallen and there is increasing objective evidence that the early hemodynamic benefit is sustained and translates into both an improvement in reported symptoms and prognosis. The most comprehensive experience to date comes from UCSD with a review of lessons learnt after 1,500 operations.[8] It is accepted that a learning curve exists with this procedure, but with experience the in-hospital mortality is gradually reduced. At UCSD, the mortality was 17% for the first 200 patients prior to 1990, falling to <10% in the 1990s, and to 5% today with well over 2,000 operations in the series to date. Numerous small series are now in press (over 20 up until 2006) from around the world with mortality ranging from 4 to 24%, but few centers have experience of >200 patients. The largest series reported from Europe included 275 patients from Paris up to 2004, with an 11% hospital mortality.[7] Our own experience is similar with an in-hospital mortality of 15% for the 150 patients in the earlier part of our reported series,[15] with current mortality of only 4% for the latest 100 patients, and a total experience of >500 operations to date. As with all series of surgical patients, outcome is often determined by case mix once a learning curve for the procedure has been overcome.

14.7.2 Morbidity and Complications

As well as the general complications occurring after any cardiothoracic surgery, there are problems specific to pulmonary endarterectomy patients. The most serious problem is residual PH (usually due to a higher proportion of microvascular disease than anticipated) and/or reperfusion lung injury. The latter is a form of acute lung injury that appears as interstitial edema on X-ray imaging, and is present to some extent in up to 20% of patients and also seen in a similar proportion of lung transplant recipients. Often, a combination of both problems is present as residual PH exacerbates reperfusion damage, and the resulting hypoxia and hypercarbic acidosis further increases PVR so a vicious cycle develops. When severe, this accounts for the majority of the fatalities in the postoperative period. Technical operative problems should be rare in established programs, and in our own experience, there has only been one pulmonary artery perforation in the last 350 procedures.

There may be some benefit in supporting severely compromised patients in the early postoperative period with ECMO provided an adequate endarterectomy has been obtained. At UCSD, they have reserved ECMO support for a small subset of patients (1% of their series) with stable hemodynamics, but severe reperfusion lung injury and used peripheral veno-venous circuits.[17] This technique has delivered an in-hospital survival of 30%. The only other group to report ECMO use after pulmonary endarterectomy have, like us, relied on veno-arterial support and included patients with hemodynamic compromise as well.[18] With this method, relatively more patients were supported (9% of their series), with a 3 of 8 being weaned to survival. We have also used veno-arterial ECMO in approximately 5% of our recent patients in the last 2 years with 4 of 7 patients surviving to discharge from hospital.[19]

14.7.3 Hemodynamic Changes

Most reports detail early postoperative hemodynamic improvement with an immediate fall in PA pressure, and reduction in PVR to approximately one-third of the preoperative level. Our own results for a recent cohort of 100 consecutive patients are included in Table 14.2. There is increasing evidence that a postoperative residual PVR of >500 dynes/s/cm^{-5} is associated with a significant increased risk of in-hospital mortality, with a stark difference in survival for the 500 patients examined in detail in the largest UCSD experience; 31% mortality above this value and 1% below.[8]

14.7.4 Long-Term Follow-Up and Survival

Less data is available on longer-term follow-up. We have tried to correct this by routinely reviewing all survivors at 3 months and 1 year after endarterectomy and ensuring ongoing follow-up in their local PH unit. The unique nature of the dedicated PH service and single center commissioned for pulmonary endarterectomy surgery in the UK has facilitated comprehensive patient follow-up. The 6-min walk distance has been used to compare exercise tolerance objectively and the NYHA status has been determined independently by the PH physicians. A right heart catheter is also repeated at 3 months.

In a recent review of 230 patients surviving to 3-month follow–up,[20] we demonstrated a significant increase in 6-min walk distance when compared with preoperation (276.3 ± 17–375.8 ± 14 m, $p < 0.001$). At 12 months, there was a further increase in

Table 14.2 Actual data for the last 100 patients undergoing surgery at Papworth Hospital. Preoperation data is in theater immediately prior to cardiopulmonary bypass, PVR calculated with an assumed wedge pressure of 10 mmHg. Day 1 postoperation data is from the PA catheter, again with an assumed wedge pressure of 10 mmHg to allow comparison with preoperation data. Both sets of measurements were taken during general anesthesia

	Preoperation	Day 1 postoperation
Mean PA mmHg mean	52	23
PVR dynes s cm^{-5} mean	744	217

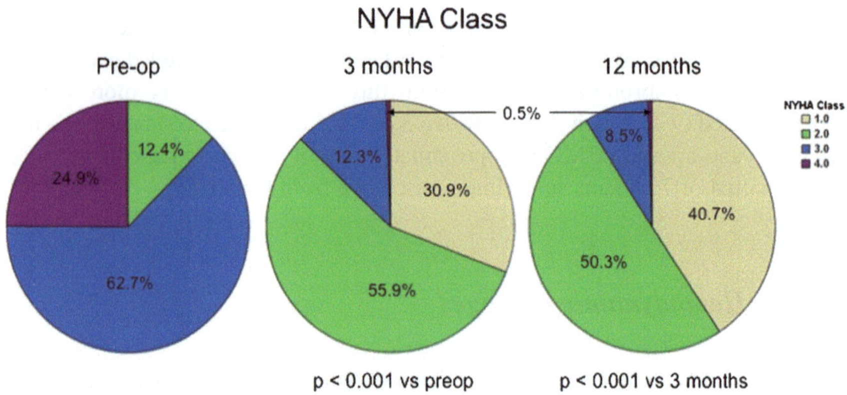

Fig. 14.2 NYHA class pre operation and 3months and 1 year post endarterectomy

6-min walk distance (375.8 ± 14–394.7 ± 15 m, $p > 0.004$). At 3 months, 30.9% were in NYHA class I, 55.9% were in NYHA class II, 12.3% were in NYHA class III, and 0.5% were in NYHA class IV ($p < 0.001$ vs. pre-op). At 12 months, there was further improvement in NYHA class: 40.7% were in NYHA class I, 50.3% were in NYHA class II, 8.5% were in NYHA class III, and 0.5% were in NYHA class IV ($p < 0.001$ vs. 3 months) (see Fig. 14.2).

We have also been able to investigate longer-term survival in this same group of 230 patients who survived to 3-month follow-up; there were only 14 late deaths. Causes of death included sepsis, stroke, intracerebral hemorrhage, ovarian cancer, metastatic carcinoma, and one patient died in the postpartum period. The cause of death could not be identified in 7 patients. Conditional survival was 94% at 3 years, 92.5% at 5 years, and 88.3% at 10 years follow-up (see Fig. 14.3). This is the largest series to report outcome after discharge from hospital and the first to report extended follow-up of a complete national patient cohort. A previous report from Japan comprehensively followed up 102 patients and demonstrated an actual survival rate (including in-hospital deaths) of 84% at 5 years with a similar significant increase in 6-min walk distance of >100 m, from 358 m preoperation to 490 m at 1 year.[21]

One of the most extensive and complete investigations of the outcome for patients with CTEPH, treated medically and by endarterectomy comes from the UK experience over the last 5 years.[22] This comprehensive follow-up of a national case series demonstrated that with optimum medical treatment, the prognosis in nonsurgical (distal CTEPH) disease has improved when compared with the previous publications, with 3-year survival of 70%. Patients treated by endarterectomy had an additional further survival benefit, and a sustained improvement in functional and hemodynamic parameters. Some patients (35%) had residual PH postendarterectomy (mean PA>30 mmHg at 3 months) and experienced less functional improvement than those with normalized hemodynamics, but interestingly the long-term prognosis for these patients who had persistent pulmonary hypertension at 3 months following surgery was very good with a 5-year conditional survival of 95%.

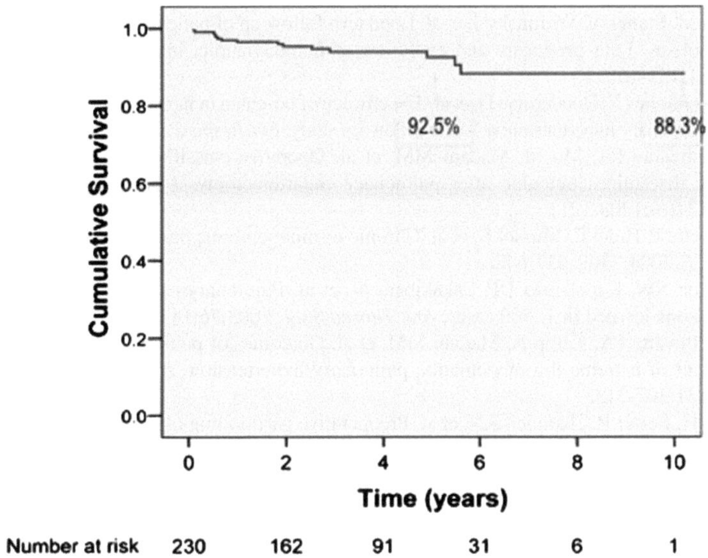

Fig. 14.3 Survival following pulmonary endarterectomy conditional from 3 month post surgery reassessment

14.8 Summary

In summary, there is now widespread acceptance that the mortality from surgery can be reduced to 5% in most experienced units reflecting improved perioperative care and better understanding of patient selection. Patients derive significant reduction in PA pressure and PVR immediately following complete endarterectomy. This early hemodynamic improvement is sustained at 3 months and translates into improved exercise capacity and reduced symptoms at 1 year following surgery. Longer-term survival (to 5 years at least) appears substantially better than that expected in patients with untreated CTEPH, and disease recurrence is very rare if anticoagulation is continued for life. The recently observed improvement in outcome during the modern treatment era reinforces the importance of early identification of patients with this treatable condition. Further information on CTEPH and pulmonary endarterectomy surgery can be found in two comprehensive review papers.[23, 24]

References

1. Petch CP. Cor pulmonale from recurrent pulmonary embolism. *Lancet*. 1951;23(June):1346.
2. Moser KM, Braunwald NS. Successful surgical intervention in severe chronic thromboembolic pulmonary hypertension. *Chest*. 1973;64:29-35.
3. Jamieson SW. Historical perspective: surgery for chronic thromboembolic disease. *Semin Thorac Cardiovasc Surg*. 2006;18:218-222.

4. Riedel M, Stanek V, Widimsky J, et al. Lond tern follow up of patients with pulmonary thromboembolism. Late prognosis and evolution of hemodynamic and respiratory data. *Chest.* 1982;81:151-158.
5. Hughes RJ, Jais X, Bonderman D, et al. The efficacy of bosentan in inoperable chronic thromboembolic pulmonary hypertension: a 1-year follow-up study. *Eur Respir J Jul.* 2006;28(1):138-143.
6. Thistlethwaite PA, Mo M, Madani MM, et al. Operative classification of thromboembolic disease determines outcome after pulmonary endarterectomy. *J Thorac Cardiovasc Surg.* 2002;124(6):1203-1211.
7. Dartevelle P, Fadel E, Mussot S, et al. Chronic thromboembolic pulmonary hypertension. *Eur Respir J.* 2004;23(4):637-648.
8. Jamieson SW, Kapelanski DP, Sakakibara N, et al. Pulmonary endarterectomy: experience and lessons learned in 1, 500 cases. *Ann Thorac Surg.* 2003;76:1457-1462.
9. Thistlethwaite PA, Kemp A, Madani MM, et al. Outcomes of pulmonary endarterectomy for treatment of extreme thromboembolic pulmonary hypertension. *J Thorac Cardiovasc Surg.* 2006;131:307-313.
10. Kim NH, Fesler P, Channick RN, et al. Preoperative partitioning of pulmonary vascular resistance correlates with early outcome after thromboendarterectomy for chronic thromboembolic pulmonary hypertension. *Circulation.* 2004;109(1):18-22.
11. Thistlethwaite PA, Auger WR, Madani MM, et al. Pulmonary endarterectomy combined with other cardiac operations: indications, surgical approach, and outcome. *Ann Thorac surg.* 2001;72:13-17.
12. Bonderman D, Skoro-Sajer N, Jakowitsch J, et al. Predictors of outcome in chronic thromboembolic pulmonary hypertension. *Circulation.* 2007;115(16):2153-2158.
13. Madani MM, Jamieson SW. Technical advances of pulmonary endarterectomy for chronic thromboembolic pulmonary hypertension. *Semin Thorac Cardiovasc Surg.* 2006;18:243-249.
14. Macchiarini P, Kamiya H, Hagl C, et al. Pulmonary endarterectomy for chronic thromboembolic pulmonary hypertension: is deep hypothermia required? *Eu J Cardiothoracic Surg.* 2006;30(2):237-341.
15. Thomson B, Tsui S, Duning J, et al. Pulmonary endarterectomy is possible and effective without the use of complete circulatory arrest – the UK experience in over 150 patients. *Eu J Cardiothoracic Surg.* 2008;33(2):157-163.
16. Thomson B, Jenkins DP. Chronic thromboembolic pulmonary hypertension and pulmonary endarterectomy. In: Klein A, Vuylsteke A, Nashef S, eds. *Core topics in cardiothoracic critical care.* Cambridge, MA: Cambridge University press; 2008.
17. Thistlethwaite PA, Madani MM, Kemp AD, et al. Venovenous extracorporeal life support after pulmonary endarterectomy: indications, techniques, and outcomes. *Ann Thorac Surg.* 2006;82:2139-2146.
18. Ogino H, Ando M, Matsuda H, et al. Japanese single-centre experience of surgery for chronic thromboembolic pulmonary hypertension. *Ann Thorac Surg.* 2006;82:630-636.
19. Berman M, Tsui S, Vuylsteke A, et al. Successful extra corporeal membrane oxygenation support after pulmonary thromboendarterectomy. Society for cardiothoracic surgery in GB & I annual meeting, abstract 2008;86:119.
20. Freed D, Thomson B, Tsui S, et al. Functional and haemodynamic outcome 1 year after pulmonary thromboendarterectomy. *Eu J Cardiothoracic Surg.* 2008;34:525-530.
21. Matsuda H, Ogino H, Minatoya K, et al. Long-term recovery of exercise ability after pulmonary endarterectomy for chronic thromboembolic pulmonary hypertension. *Ann Thorac Surg.* 2006;82:1338-1343.
22. Condliffe R, Kiely D, Gibbs J, et al. Improved outcomes in medically and surgically treated chronic thromboembolic pulmonary hypertension. *Am J Crit Care Med.* 2008;177:1122-1127.
23. Hoeper M, Mayer E, Simonneau G, et al. Chronic thromboembolic pulmonary hypertension. *Circulation.* 2006;113(16):2011-2020.
24. Jamieson SW, Kapelanski DP. Pulmonary endarterectomy. *Curr Probl Surg.* 2000;37(3):165-252.

Chapter 15
Imaging in Acute and Chronic Pulmonary Thromboembolic Disease

Charlotte Cash and Joanne Cleverley

15.1 Imaging a Patient with Suspected Acute Pulmonary Embolism

The spectrum of clinical manifestations with which a pulmonary embolus (PE) may present is wide. As a result, imaging is frequently used to exclude rather than to diagnose a PE. However, the consequence of missing a diagnosis of pulmonary embolism carries a significant mortality and as a result radiology departments find that an increasing part of their workload is dedicated to making or excluding this diagnosis. At the Royal Free Hospital, in 2006, 282 ventilation perfusion (V/Q) scintograms and 433 computed tomographic pulmonary angiograms (CTPA) were performed of which approximately 17% were positive for acute PE.

15.1.1 Chest Radiograph (CXR)

A Chest Radiograph (CXR) is a mandatory baseline investigation of any patient with suspected PE as it helps exclude other pathologies that may mimic the clinical presentation of PE. In the case of PE itself, the CXR is frequently unhelpful. The CXRs of patient with proven PE generally show nonspecific signs.[1] The often quoted Westermark's sign of increased transradiancy of the affected lung due to pulmonary oligemia is rarely, if ever, seen. It is much more likely to be due to technical factors in obtaining a CXR in the acute setting. The Hampton's Hump describes an area of infarct seen as a wedge-shaped opacity adjacent to the visceral pleural surface with its convex apex directed toward the hilum. Although a rare finding, the Hampton's Hump on a CXR indicating an infarct is a useful clue to the diagnosis. Importantly, a normal CXR does not exclude the diagnosis of PE, but

C. Cash (✉) and J. Cleverley
Department of Radiology, Royal Free Hospital, Pond Street, London, NW3 2QG, UK
e-mail: charlotte.cash@royalfree.nhs.uk

D. Abraham et al. (eds.), *Advances in Vascular Medicine*,
DOI 10.1007/978-1-84882-637-3_15, © Springer-Verlag London Limited 2010

may influence the choice of next investigation. If the CXR is normal, the chances of obtaining a diagnostic V/Q scintogram increase[2] and in many institutions will be the next investigation of choice.

15.1.2 Ventilation Perfusion Scintigraphy

Ventilation perfusion (V/Q) scintigraphy is a relatively low-radiation dose examination, with an average effective dose of 1.2–2.0 mSv.[3] A diagnosis of PE is made by identifying a mismatch between the inhaled radioisotope (e.g., krypton-81m), representing the ventilation component of the examination and the injected radioisotope (technetium-labeled, albumin macro-aggregates) representing the perfusion component of the examination (Fig. 15.1). Defects in perfusion are caused by a variety of lung pathologies other than PE, such as infection, bullous emphysema, previous infarction and scarring, etc. In these cases, however, there will be a matched defect on the ventilation scintogram. Unfortunately, a PE may result in a matched defect, particularly if the PE results in an area of infarction, and therefore the diagnosis of PE by ventilation-perfusion imaging is not clear cut. Many institutions report V/Q studies according to probability criteria set out by a prospective study that looked into the sensitivity and specificity of V/Q scintigraphy – the PIOPED I study (Prospective Investigation of Pulmonary Embolism Diagnosis). This compared V/Q scintigraphy

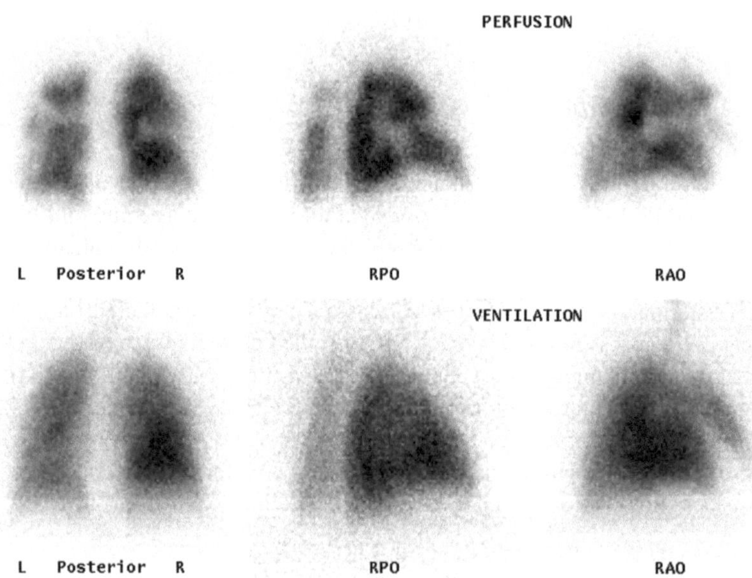

Fig. 15.1 Ventilation/perfusion scintogram showing multiple defects on the perfusion images with near normal ventilation images. This represents a high probability examination for acute pulmonary embolus (PE)

with pulmonary angiography. One thousand four hundred and ninety three patients were recruited into the multicenter study. Results indicate that a high-probability V/Q scintogram has a positive predictive value of 88%. The normal or near-normal scintogram carries a negative predictive value of 91%. The positive predictive value of a high probability scintogram rose when accompanied by high clinical probability, but was reduced in patients with a prior history of PE. Although a high probability and a normal/near normal scintogram carry high predicative values for the diagnosis or exclusion or PE, PIOPED showed that these results occurred in only a minority, with over 70% of studies returning an intermediate or low-probability result. The incidence of PE in patients with an intermediate or low-probability result was found to be 33 and 12%, respectively.[4] To reduce the likelihood of an intermediate or low probability result, patients with abnormalities on their CXR should not be considered for a V/Q scintogram. These patients and any patient with an intermediate or low probability result should be further investigated. Until the early 1990s, further investigation meant a conventional pulmonary angiogram; however, this has now been superseded by CTPA as discussed below.

15.1.3 Conventional Pulmonary Angiography

Pulmonary angiography is an invasive technique performed by selective catheterization of the pulmonary arteries. As with any fluoroscopic technique, radiation dose varies from examination to examination. Effective doses quoted vary from 3.2 to 30.1 mSv.[3] The femoral venous approach is used via a Seldinger technique. Angiographic diagnosis of an acute PE is made on the identification of an intraluminal filling defect or the trailing edge of a thrombus partially filling a vessel.[1] In an acute embolus, the filling defect makes a concave margin with respect to the contrast material.[5] The technique is not without its risks. In the PIOPED I study, there was a 0.5% mortality and a 6% morbidity rate.[1]

15.1.4 CT Pulmonary Angiography

During the 1990s, pulmonary angiography was steadily replaced by the less invasive CT pulmonary angiography. The CTPA technique involves the patient lying supine. A bolus of between 120 and 150 ml of nonionic contrast medium is injected at a rate of 4–5 mls/min, preferably via an antecubital fossa vein. The CT images are acquired on suspended inspiration either after 12–15 s following injection or using a bolus tracking device. Precise image protocols vary from institution to institution; however, in general, images are acquired from the level of the aortic arch to 2 cm below the level of the inferior pulmonary veins. A CTPA examination does *not* routinely examine the whole thorax. It is therefore *not* a complete screening tool for chest diseases.

In the early 1990s, with the slower single slice CT machines, the diagnosis was limited to central pulmonary arterial thrombus detection. The relatively slow image acquisition times resulted in poor-quality images in the dyspneic patient, and image reconstruction was limited to 5 mm. CT technology has advanced beyond recognition within the last decade, and the advent of multidetector or multislice CT (MDCT) has vastly improved image resolution as a result of faster acquisition times and very thin collimation down to 0.5–1 mm depending on the scanner used.

A meta-analysis using data from 1990 to 2000 found that the sensitivity and specificity of single slice CTPA in diagnosing PE was 74 and 90%, respectively.[6] With just a four-detector CT, the sensitivity and specificity improved to 100 and 89%, respectively.[7] It is now well recognized that with MDCT, emboli can be detected down to the subsegmental level.[8-11]

Results from the PIOPED II study have recently been published. The aim of PIOPED II was to formulate up-to-date recommendations for the investigation of suspected PE. As part of this study, the accuracy of CTPA alone and CTPA combined with venous phase imaging of the abdomen, pelvis, thighs, and calves to detect deep vein thrombosis was assessed. This combined study is referred to in the literature as either CTA-CTV (CT angiography-CT venography) or perhaps more precisely as CTVPA (CT venography and CT pulmonary angiography). The same bolus of contrast medium is used for the two phases, although clearly there is an increase in the radiation dose. The effective dose from a CTPA alone is 1.6–8.3 mSv; a CTV would add 5.7 mSv.[3] The PIOPED II study found that by combining the two examinations, there was an increase in sensitivity for diagnosis of thromboembolic disease from 83% for CTPA alone, to 90% when combined with a venous survey.[11] Numerous studies have shown that CTV of the lower limbs is highly accurate in the diagnosis of thrombus[12,13] and is diagnostically equivalent to compression sono graphy.[14,15] However, there is a significant additional radiation burden to the patient with relatively little additional diagnostic yield and it is therefore not routinely recommended in women of childbearing age.

A large meta-analysis of 15 studies assessed the validity of a negative CTPA study. Altogether 3,500 patients investigated with suspected PE, but with negative CTPA studies, were followed up for a minimum of 3 months. The results found a negative predictive value of 99.1% for a negative CTPA. This is comparable with a negative conventional pulmonary angiogram.[16] As result of these findings, the Fleischner Society in a recent statement conclude that in the current era, MDCT can replace conventional pulmonary angiography as the reference standard for the diagnosis of acute PE.[17] Some centers have abandoned the use of V/Q scintigraphy as a first-line investigation altogether, at the expense however of a greater radiation burden to those patients under investigation. At the Royal Free Hospital, over the past 6 years referral practice for acute pulmonary embolus (PE) diagnosis has changed with a decline in V/Q and an increase in CTPA referral (Fig. 15.2). It is well recognized now that MDCT provides an accurate diagnosis, is available out of hours (*vs.* limited availability of radiopharmaceuticals out of hours), and provides addition information over and above the pulmonary arterial circulation.

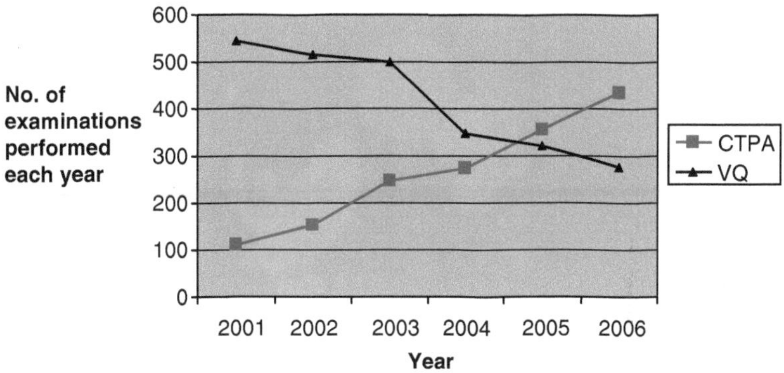

Fig. 15.2 Changes in referral pattern for investigation of acute PE at The Royal Free Hospital from 2001 to 2006

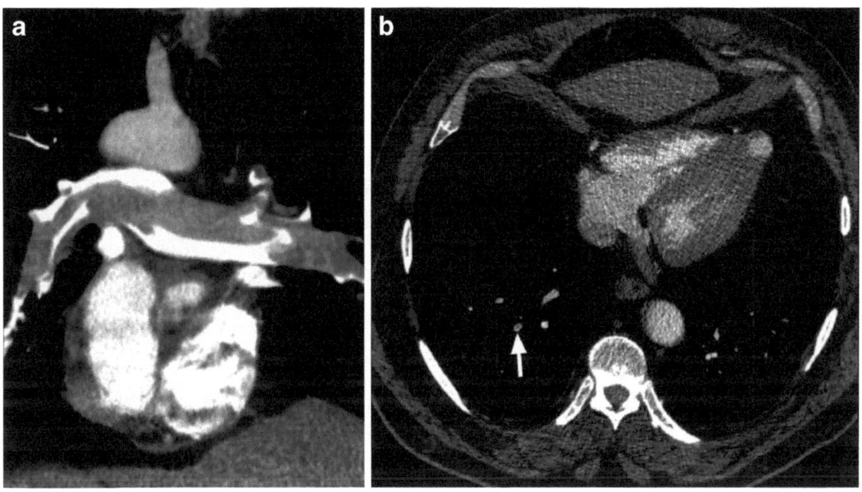

Fig. 15.3 A saddle embolus is demonstrated in (**a**), a reconstructed image from a computed tomographic pulmonary angiogram (CTPA) examination. Less conspicuous is a small central filling defect (*white arrow*) within a subsegmental artery (**b**) on an axial CTPA image

The signs of an acute PE on a CTPA include failure of contrast to fill the entire lumen of an artery due to a central filling defect (Fig. 15.3). This artery may be enlarged when compared with the accompanying arteries at the same level. The filling defect may be nonocclusive and only partial, allowing contrast medium to surround the filling defect akin to the conventional angiographic railway track or trailing edge sign (Fig. 15.4).[5,18] In acute PE, the thrombus forms an acute angle with the vessel wall (Fig. 15.5).[5]

In addition to mapping the pulmonary arterial tree, the majority of the lung parenchyma, pleural spaces, and mediastinum are also imaged on a CTPA. Additional signs of a PE may include an infarct, usually seen as a peripheral wedge shaped area

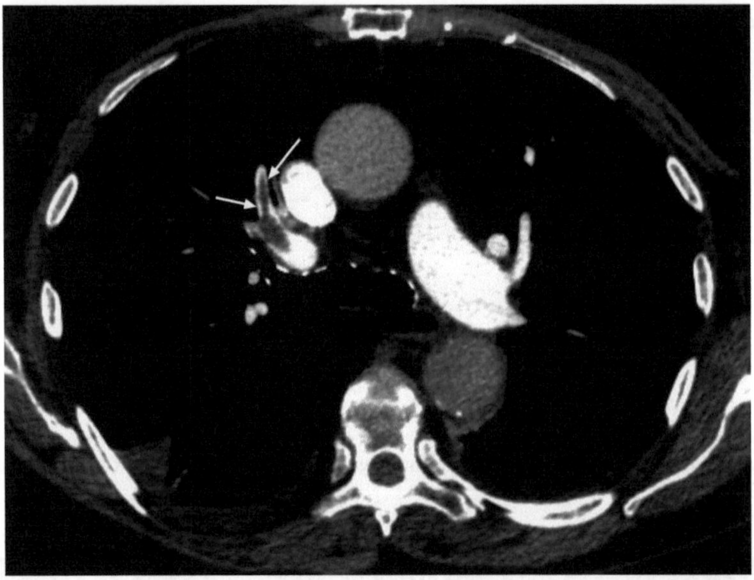

Fig. 15.4 Axial CTPA image. A nonocclusive thrombus with contrast media seen to pass either side of the thrombus (*white arrows*); the CT equivalent of the angiographic railway track sign

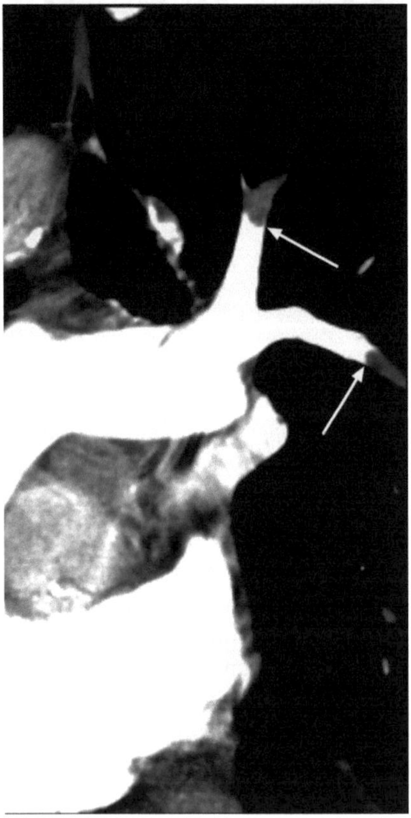

Fig. 15.5 The white arrows in this reconstructed image from a CTPA demonstrate the acute angle that the thrombus makes with the vessel wall

of consolidation with or without an accompanying pleural effusion. The examination may show signs of right ventricular strain. Obstruction of more than 30% of the pulmonary circulation is sufficient to raise the pulmonary vascular resistance resulting in pulmonary hypertension seen on the CTPA as dilation of the right ventricle (RV), reflux of contrast medium down the IVC, and deviation of the interventricular septum to the left.[18] If the CTPA is negative for PE, the examination may reveal other causes of the patient's clinical presentation and improve patient care by minimizing diagnostic delay, that might have occurred had other investigations been used. One study of over a thousand patients undergoing a CTPA found an alternate significant diagnosis requiring immediate or specific treatment (e.g., aortic aneurysm or mass suggestive of malignancy) in approximately 10%.[19]

There are a number of pitfalls in the diagnosis of PE from a CTPA. Technically, the examination may be suboptimal with poor opacification of the arterial tree. This may be due to contrast delivery or to patient factors such as obesity. Breathing and cardiac motion artifact may result in volume averaging with surrounding air-filled lung resulting in an apparent intra-luminal defect. Vessels coursing in and out of the plane of the axially acquired images may result in partial volume averaging artifact.[18] These defects occur less with thin-collimation CT, given the ability to reconstruct the data in the sagittal and coronal plane. High concentration of contrast medium in the SVC can result in streak artifacts in the adjacent right main pulmonary artery, and increased pulmonary vascular resistance due to consolidation may lead to false-positives.[20]

Complications of a CTPA include contrast reactions, extravasation of contrast material into the antecubital fossa, and contrast-related renal impairment.[11] Screening of those patients most at risk of a contrast reaction, accurate venous access for contrast delivery and management of patients likely to suffer renal impairment reduces the incidence and severity of these complications. In patients with contrast medium allergy or with renal failure, contrast-enhanced MR angiography (MRA) can be performed.

15.1.5 MR Pulmonary Angiography

Advances in MR technology have resulted in shortened data acquisition times making the images less susceptible to motion artifact and with improved spatial resolution. Contrast-enhanced MRA can be used for the detection of pulmonary emboli. A number of studies quote sensitivities and specificities in the range of 77–100% and 62–98%, but the study sizes were all small.[21-25] Sensitivities are higher for central and lobar vessels, with reduced sensitivities quoted for segmental and subsegmental emboli as a result of limitations in spatial resolution.

Contrast-enhanced MRA may have a role in the diagnosis of PE in patients who cannot undergo CTPA due to renal impairment or contrast allergy. There is no radiation involved in MRA; however, its use in pregnancy is still controversial. This technique and its role in the investigation of PE is currently under evaluation in the PIOPED III study.

15.1.6 Sonography of the Lower Limbs

Conventional venography has been replaced by compression and Doppler ultrasound (US). The sensitivity and specificity of US is reported as greater than 95% in symptomatic patients, using venography as a reference standard. The sensitivity and specificity of sonography is reduced in asymptomatic patients. US in asymptomatic patients as a screening tool remains controversial.[15] Doppler US only examines the deep venous system of the legs and is therefore only of diagnostic value, in the investigation of acute PE, if it is positive. A negative result is of no value at all.

15.1.7 Investigation of Acute PE in the Pregnant Patient

To avoid the relatively high radiation dose to metabolically active breast tissue and to the fetus during pregnancy, it is suggested that pregnant women suspected of having an acute PE undergo CXR to exclude other etiologies, D-dimer testing, bilateral leg sonography followed by perfusion only scintigraphy if no deep venous thrombosis found.[3] In patients with indeterminate scintigraphy results, modified dose CTPA should be performed.[26]

15.1.8 An Algorithm for the Investigation of PE Based on Recommendations from PIOPED II Investigators

An algorithm for the investigation of acute pulmonary embolism in the current era is shown (Fig. 15.6). It must be emphasized, however, that the diagnosis is frequently not clear cut. Discordances between clinical probability and imaging results will exist, necessitating re-assessment of the clinical probability and further or repeat imaging based on clinical judgment. The value of the pretest clinical probability and D-dimer assay has not been reviewed in this chapter; however, there is strong evidence for its routine use, particularly in excluding acute PE in patients with a low pretest clinical probability and negative D-dimer test.[27]

15.2 Imaging the Patient with Suspected Chronic Thromboembolic Disease

It is estimated that 1–3.8%[28,29] of patients with an acute pulmonary embolic event will go on to develop chronic thromboembolic disease (CTED). It is postulated that acute PE may be one of several potential insults to the endothelium of the pulmonary

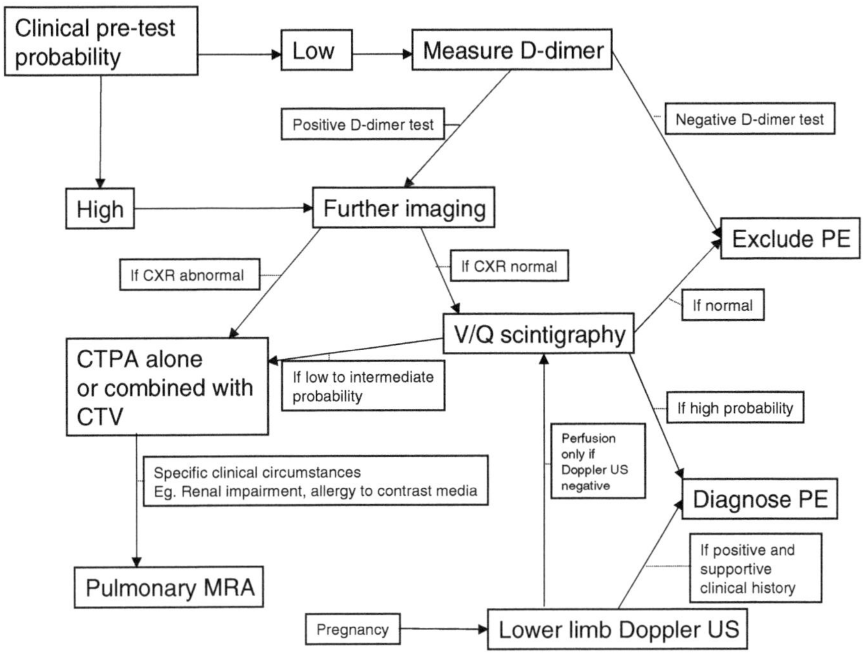

Fig. 15.6 An algorithm for the investigation of acute PE

arteries that may result in remodeling, atherosclerosis, in situ thrombus, and consequential development of CTED.[30] Only a third of patients with confirmed CTED have a previous history of an acute event.[31] The clinical history is therefore an unreliable pointer to the diagnosis of CTED in patients with pulmonary hypertension (PH). Imaging is pivotal in the evaluation of these patients in both the diagnosis and assessment for potentially curable surgery. Unlike for acute pulmonary embolism, there have been no prospective studies to evaluate the most effective diagnostic approach to CTED.

15.2.1 The Chest Radiograph (CXR)

Patients with CTED may have normal CXRs, but frequently show signs of pulmonary hypertension with enlarged main and hilar pulmonary vessels with "pruning" or tapering of the vessels peripherally (Fig. 15.7). There may be cardiomegaly, atelectasis, pleural effusion, and/or pleural thickening. These features are all nonspecific, and are found in many patients with pulmonary hypertension. The CXR is therefore not a helpful discriminator, but a necessary part of a patients' work-up and provides a baseline for future management.[32]

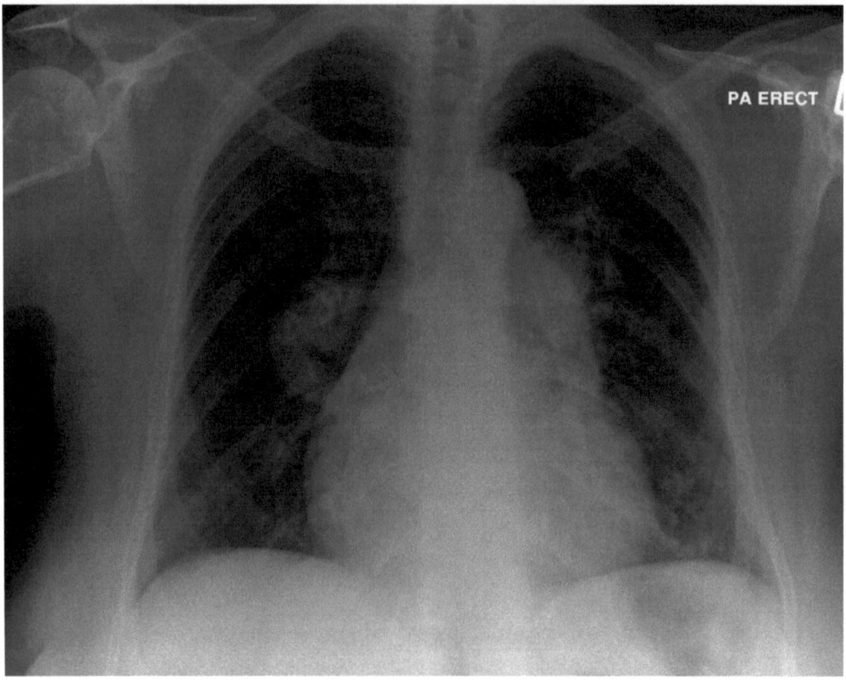

Fig. 15.7 Enlarged dilated main, right, and left pulmonary arteries on a CXR in a patient with pulmonary hypertension

15.2.2 Echocardiography

Transthoracic echocardiography is the most widely used screening tool for patients with suspected pulmonary hypertension[32] and is necessary in delineating cardiac anatomy and assessing right ventricular function.

15.2.3 Ventilation Perfusion Scintigraphy

Having confirmed pulmonary hypertension on echocardiography, the first diagnostic test in the work-up of a patient with suspected CTED is V/Q scintigraphy. There is general agreement among experts that a normal V/Q scintogram excludes the diagnosis of CTED.[33-35] In patients with CTED, the V/Q scintogram shows one or more mismatched, segmental, or larger defects.[36,37] Matched defects may coexist, possibly reflecting previous infarcts (Fig. 15.8). V/Q scintigraphy has been shown to have a high sensitivity (96–97.4%) and specificity (90–95%) for CTED.[38] It is, however, recognized that there are rare causes of multiple perfusional defects in this subgroup of patients presenting with pulmonary hypertension, such as pulmonary

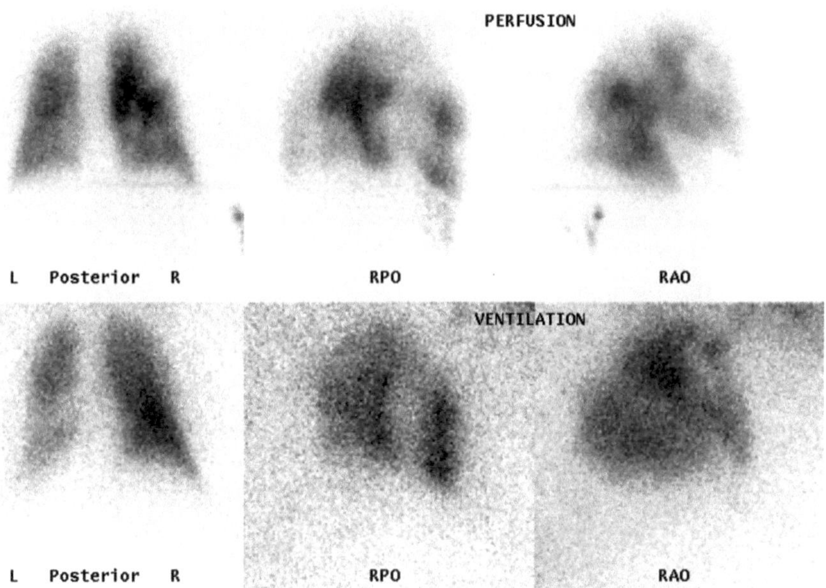

Fig. 15.8 Ventilation/perfusion scintogram in a patient with chronic thromboembolic disease (CTED), demonstrating multiple mismatched defects, and also some matched defects possibly representing areas of infarct from previous embolic insult

vasculitis,[39] fibrosing mediastinitis,[40] pulmonary artery sarcoma,[41] pulmonary veno-occlusive disease,[42] and pulmonary capillary hemangiomatosis.[43] While a helpful clue to the diagnosis of CTED, V/Q scintigraphy frequently underestimates the severity and overall clot burden.[32]

15.2.4 CT Pulmonary Angiography and High-Resolution CT Chest

A number of abnormalities have been described indicating CTED on a CTPA. Because treatment of CTED differs considerably from other causes of pulmonary hypertension, an accurate diagnosis is essential.

15.2.4.1 Cardiac Abnormalities

There are nonspecific signs that relate to the pulmonary hypertension itself such as enlarged RV with deviation of the interventricular septum (Fig. 15.9). With increasing right ventricular pressures, dilation of the RV may result in tricuspid valve incompetence with reflux of contrast material into the hepatic veins (Fig. 15.10).

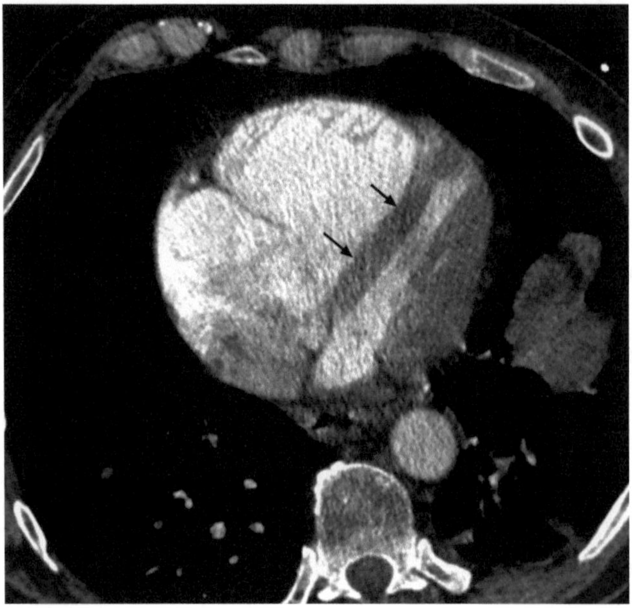

Fig. 15.9 Axial CTPA image. The right ventricle (RV) is dilated, and the interventricular septum is deviated toward the left (black arrows)

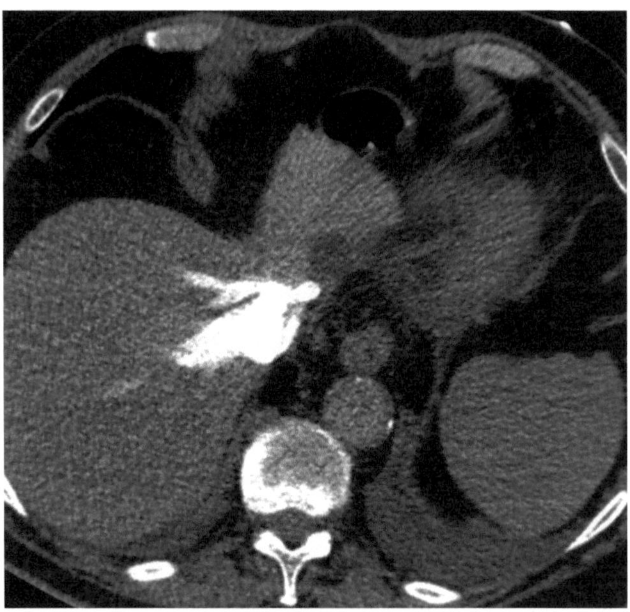

Fig. 15.10 Axial CTPA image. Contrast media refluxes into the hepatic IVC and veins in a patient with tricuspid regurgitation and pulmonary hypertension

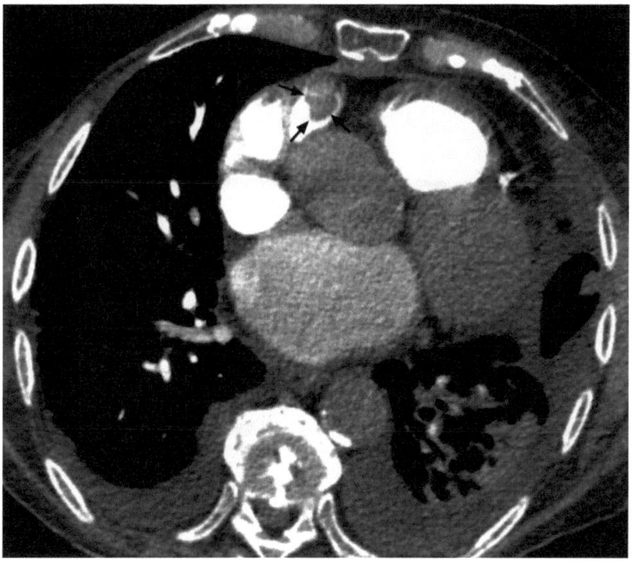

Fig. 15.11 Axial CTPA image. Black arrows indicate thrombus within the right atrial appendage

This may also be accompanied by right atrial enlargement and thrombus (Fig. 15.11) may develop within the right atrium, which may calcify.[44]

15.2.4.2 Vascular Abnormalities

Pulmonary Arteries

The main pulmonary artery and its branches dilate. The main pulmonary artery is frequently larger in diameter than the aorta (Fig. 15.12). An aortic/pulmonary diameter ratio of greater than 1[45] or a pulmonary artery diameter of greater than 2.86 cm[46] is considered abnormal and indicative or pulmonary hypertension. An increased diameter of the right and left pulmonary arteries greater than 1.6 cm measured in the scanning plane at the bifurcation of the main pulmonary artery is also considered enlarged[47] and more peripherally, pulmonary arteries are considered enlarged if they are larger than their accompanying bronchus with a ratio greater than 1.0 (Fig. 15.13). Signs specific to the diagnosis of CTED include obstructive filling defects characterized by complete occlusion of a pulmonary vessel with distal narrowing and stenosis, and nonobstructive filling defects as a result of organized thrombotic material. Such nonobstructive defects consist of mural thrombus (Fig. 15.14), pouch defects, irregularities of the intimal surface, calcified thrombus, webs, and bands. A band is a ribbon-like structure attached at both ends

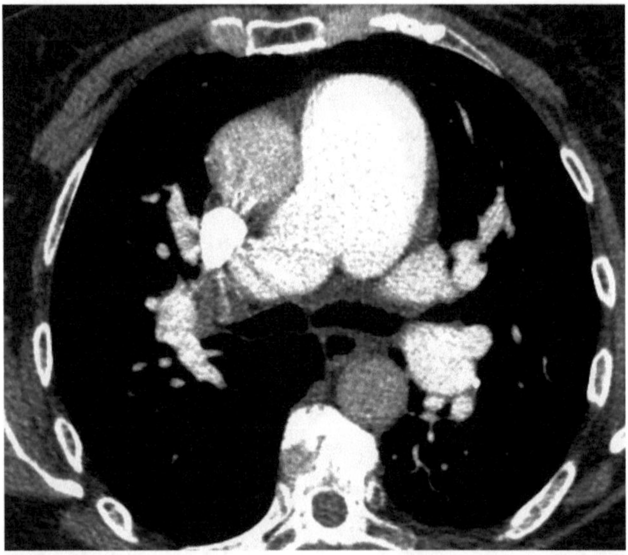

Fig. 15.12 Axial CTPA image. The pulmonary trunk is dilated. The pulmonary artery to aortic diameter ratio exceeds 1

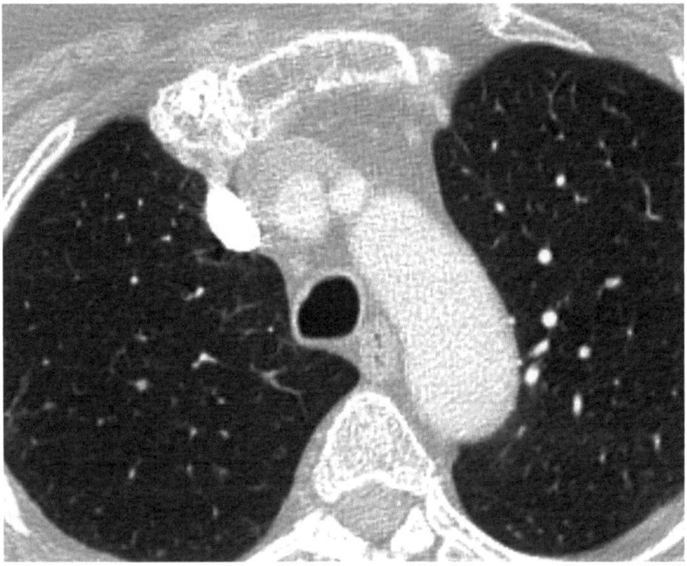

Fig. 15.13 Axial CTPA image. The pulmonary vessels to the apico-posterior segment of the left upper lobe are larger than the equivalent vessels to apical segment of the right upper lobe. Their accompanying bronchus is too small to appreciate

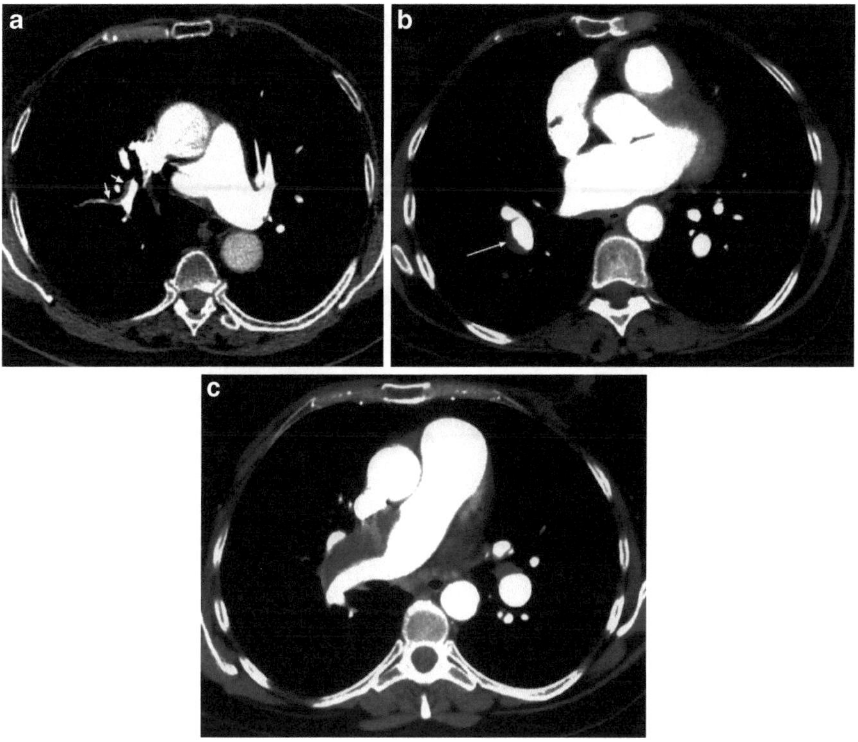

Fig. 15.14 Axial CTPA images depict mural thrombus within a segmental vessel (*white arrows*) in (**a**), within a dilated lobar vessel (*white arrow*) in (**b**) and within the right main pulmonary artery in (**c**)

to the vessel wall (Fig. 15.15). They range in size from 0.3 to 2 cm and from 0.1 to 0.3 cm in diameter. They are usually orientated in the direction of blood flow along the long axis of the vessel.[5] A web is defined as a network of bands. Variability in the size and distribution of pulmonary arteries is described in CTED[33,48](Fig. 15.16) and is the result of chronic vascular obstruction with thromboembolic material, which becomes organized and fibrotic resulting in the decrease in size of some vessels and the compensatory increase in size of others.[44] Variation in vessel size has been found to be more specific in distinguishing CTED from other causes of pulmonary hypertension and is seen more frequently than mosaic attenuation (see below), although there is a clear association between enlarged vessels and areas of increased attenuation.[49]

Bronchial and Nonbronchial systemic Arteries

Bronchial arteries do not normally participate in gaseous exchange, their role being one of nutritional support to the lungs. In the normal state, these arteries

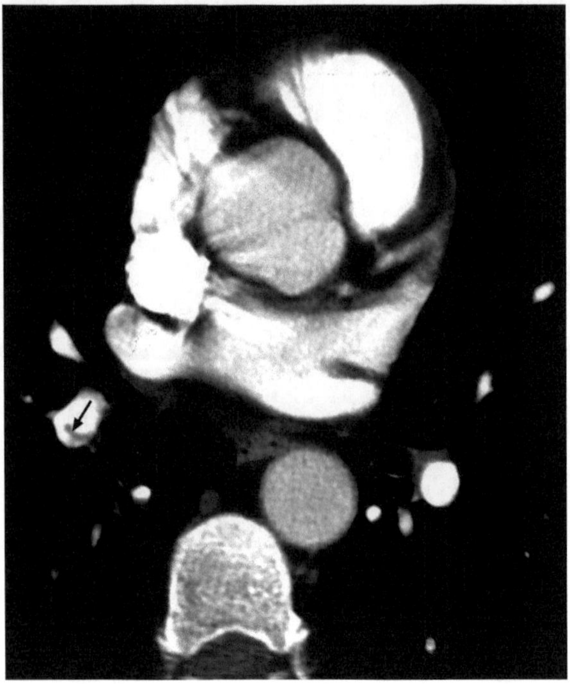

Fig. 15.15 Axial CTPA image. A web is demonstrated (black arrow) within a right lower lobar pulmonary artery

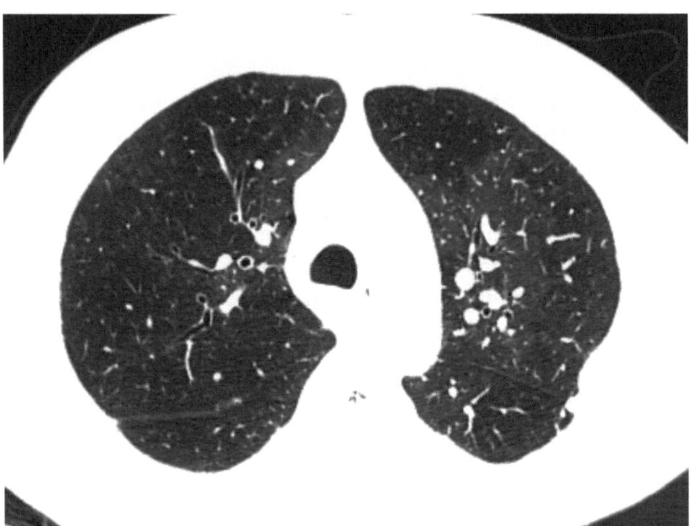

Fig. 15.16 Axial CTPA image. This reconstructed high-resolution image shows variation in vessel size, vessels greater in diameter than their accompanying bronchus, and mosaic attenuation of the lung parenchyma. The dilated arteries are associated with areas of increased parenchymal attenuation

have a maximum diameter of <1.5 mm and are not usually visible on CT. Bronchial arteries ≥1.5 cm are considered dilated.[50] In patients with CTED, there is increased flow through these arteries as a result of the development of systemic-to-pulmonary anastomoses. These vessels become visible on CT and may measure greater than 4 mm in diameter[47](Fig. 15.17). They are identified within the mediastinum and travel along the course of the bronchi. One study demonstrated that in 77% of patients with proven CTED, bronchial dilatation was present.[50] Nonbronchial systemic arteries may also contribute to systemic-pulmonary anastomoses. They do not parallel the course of the bronchi, but rather enter the parenchyma through the pulmonary ligament or through the pleura itself. These nonbronchial systemic arteries include the intercostal arteries, internal thoracic artery, and the inferior phrenic arteries. These are considered enlarged if their diameter exceeds 4 mm.[47]

15.2.4.3 Parenchymal Abnormalities

The lung parenchymal changes of CTED are best appreciated from either reconstructed data from the CTPA or from a separate dedicated high-resolution CT (HRCT).

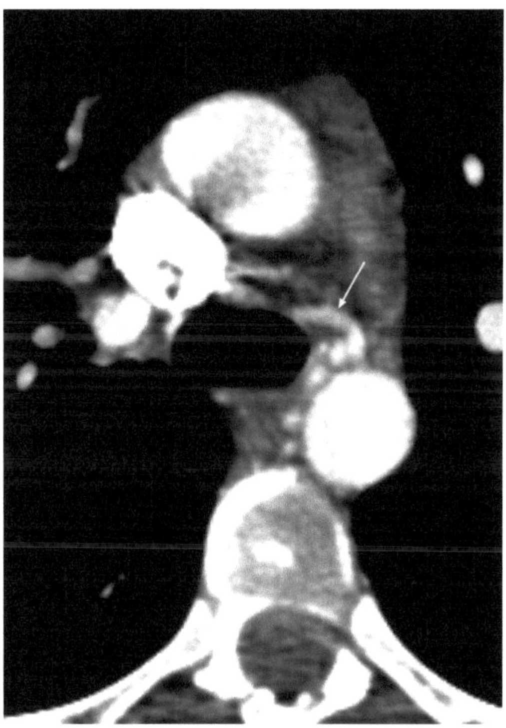

Fig. 15.17 Axial CTPA image. Within the mediastinum, dilated tortuous systemic bronchial arteries are shown (white arrow)

Mosaic Attenuation

A mosaic pattern of attenuation, with sharply demarcated areas of ground glass, is described in CTED and has been shown to be a useful discriminator, differentiating CTED from other nonembolic causes of pulmonary hypertension[49] (Fig. 15.18). Schwickert et al. found mosaic attenuation in 77% of patients with CTED.[48] Investigators have shown that this is due to mosaic perfusion and correlates well with perfusion scintigraphy.[33,48] This differential pattern of attenuation, as a result of regional variation in blood flow, must be distinguished from air-trapping, which also produces mosaic attenuation. However, air-trapping caused by small airways diseases such as extrinsic allergic alveolitis can be distinguished from mosaic perfusion by expiratory phase imaging, where the air-trapping is emphasized.

Peripheral Densities

Previous pulmonary infarcts from an acute event may appear on CT as parenchymal bands, or as subpleural wedge-shaped densities, which may or may not cavitate[44,48] (Fig. 15.19). Parenchymal scarring has been shown to be associated with a restrictive lung defect in patients with CTED.[51]

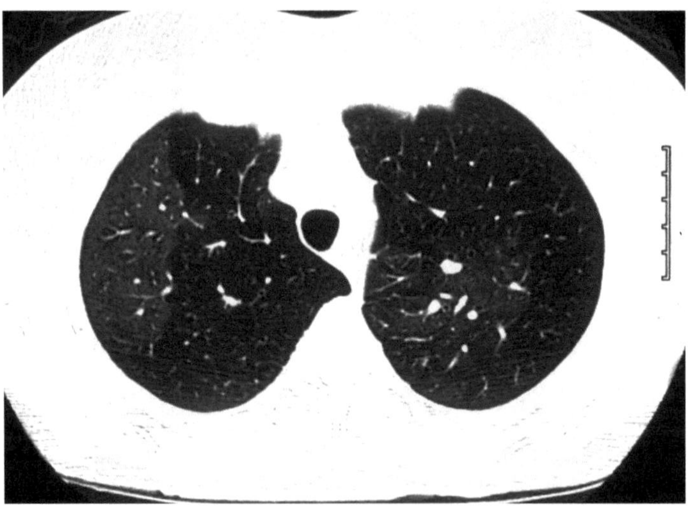

Fig. 15.18 A high-resolution CT (HRCT) image. Mosaic attenuation is demonstrated with areas of increased attenuation, reflecting increased perfusion, sharply demarcated against areas of decreased attenuation

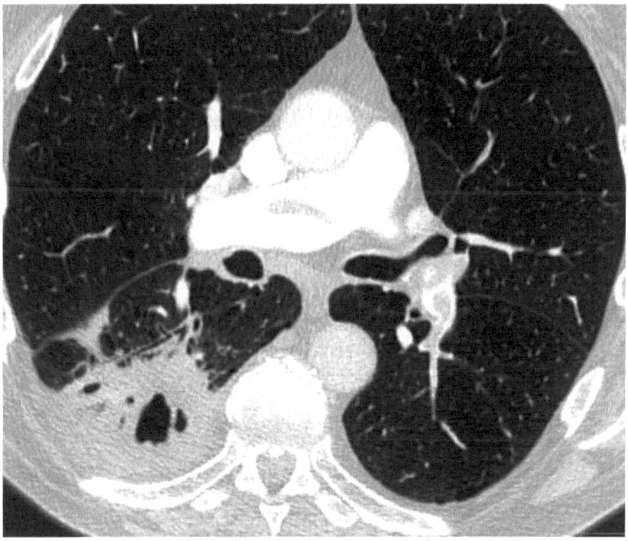

Fig. 15.19 Axial CTPA image. The reconstructed image displayed on lung windows demonstrates a large cavitating pulmonary infarct within the right lower lobe

15.2.5 MR Angiography

As MR techniques evolve to allow single breath-hold data acquisition and therefore improved spatial resolution, pulmonary MRA is emerging as a useful tool in the investigation of CTED. This particularly applies to patients who cannot receive iodinated contrast medium due to allergy or renal impairment, or in patients where radiation burden is a significant consideration. MRA demonstrates thrombus, webs, stenoses, and vessel cut-off as for CTPA; however, owing to limitations in spatial resolution despite recent advances, it is only capable of demonstrating these features to the segmental level.[52]

A significant advantage of MR imaging is that in addition to the information provided from MRA regarding thrombus, webs, stenoses, etc., further imaging sequences can provide functional assessment of disease, such as the effect on the right and left ventricles.[52] A pulmonary MRA, however, does not provide any useful diagnostic information regarding the lung parenchyma.

15.2.6 Conventional Pulmonary Angiography

There are a number of abnormalities seen on a conventional pulmonary angiogram suggesting the diagnosis of CTED. Experience has shown that in CTED, these abnormalities are usually multiple and bilateral.[53] On conventional pulmonary angiography, CTED may manifest as complete obstruction with complete vessel cutoff having a convex margin with respect to the contrast material, described as a pouch defect.[5] This differs from the concave margin seen with an acute PE. Arterial webs or bands are seen as lines of decreased opacification that cross the width of the vessel and may be associated with areas of narrowing and poststenotic dilatation. Irregularly organized mural thrombus manifests itself as intimal irregularity giving a scalloped appearance of the vessel wall on the angiogram[53] (Fig. 15.20). Pulmonary angiography is still considered the standard diagnostic tool in CTED and remains the most sensitive method of detecting segmental and subsegmental thrombus. It may, however, like V/Q underestimate central disease. Pulmonary angiography is frequently performed in conjunction with right heart catheterization. Right heart catheterization will confirm the degree of pulmonary hypertension, exclude pulmonary venous hypertension, and assess the degree of hemodynamic dysfunction.

15.2.7 Differentiating CTED from Other Causes of Pulmonary Hypertension

Other causes of pulmonary hypertension, listed above (see Sect. 15.2.3), may give similar V/Q scintographic or conventional angiographic appearances to those seen in CTED; however, cross-sectional imaging with CT or MRI may help in their

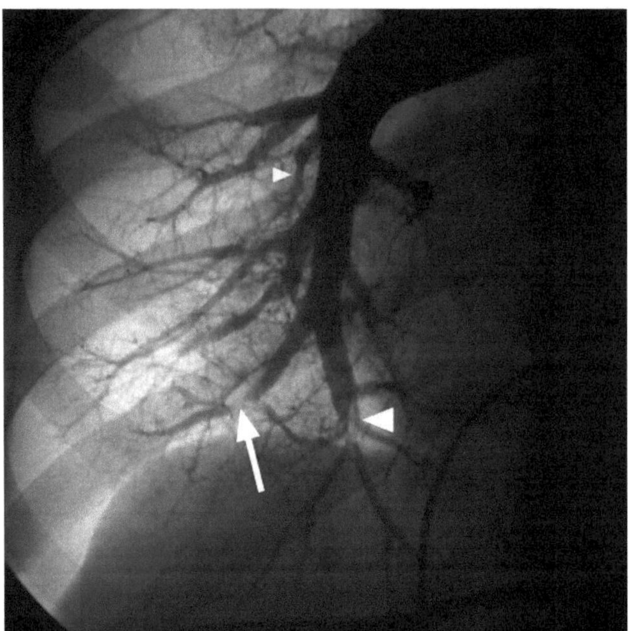

Fig. 15.20 A conventional pulmonary angiogram with catheterization of the right main pulmonary artery. Several abnormalities are present. The white arrow points to a complete occlusion, while the arrowheads outline areas of intimal irregularity in keeping with mural thrombus

differentiation. Primary pulmonary hypertension, pulmonary hypertension associated with scleroderma/systemic sclerosis, and capillary hemangiomatosis may show a centrilobular nodular ground-glass pattern on HRCT.[20,33,54] Primary pulmonary artery sarcomas are rare but may mimic CTED. The tumor may be seen as a mass resembling clot within the pulmonary arterial trunk or its proximal branches. This tumor is often associated with thrombus formation and therefore other signs typical of CTED may coexist. These tumors enhance with gadopentetate dimeglumine on T1-weighted MRI images and can be differentiated from thrombus.[20] In pulmonary veno-occlusive disease, the most common finding is smooth interlobular septal thickening, multifocal nodular ground-glass opacification, and pleural effusions.[55] Fibrosing mediastinitis characteristically causes narrowing of mediastinal structures and an associated mediastinal mass illustrated well by CT.[56] Pulmonary vasculitis or Takayasu's arteritis causes luminal narrowing, which may appear similar to CTED on conventional angiography; however, CT shows mural thickening and enhancement of the aorta, pulmonary artery, and their branches.[57] The presence of mediastinal lymphadenopathy is unusual in CTED and its presence suggests a different etiology such as primary pulmonary hypertension, capillary hemangiomatosis, or veno-occlusive disease.[33]

15.2.8 Assessment for Surgery

In addition to making the diagnosis of CTED and differentiating it from other causes of pulmonary hypertension, an important role of cross-sectional imaging is to determine the suitability for surgery and thrombo-endarterectomy. Because central thrombus becomes endothelialized, conventional angiography may underestimate central clot burden. CTPA and MRA are essential in this respect. Surgery is generally considered technically feasible if thrombus is detected centrally, at the level of the main or lobar arteries or at the origin of the segmental vessels to create a safe dissection plane. Central distribution of thrombus with little segmental involvement is a predictor of successful outcome.[58] High pulmonary vascular resistance and the absence of substantial central disease suggests concomitant small-vessel disease, which increases the risk of persistent pulmonary hypertension postoperatively and is also associated with a higher short- and long-term mortality.[59] The presence of dilated systemic bronchial arteries has been shown to be a good predictor of surgical outcome.[50] Age, hemodynamic function, and other coexisting morbidities are also taken into consideration during surgical work-up. Initial mortality rates were high at 17% between 1970 and 1990[60] however, as techniques have improved and selection criteria refined, a more recent mortality figure of 4.4% has been quoted.[59] A successful thrombo-endarterectomy has been shown to restore hemodynamic function to near-normal and the procedure is considered curative. Functional MR and MRA studies have shown significant improvements in right ventricular ejection fractions, reduction in the size of the main pulmonary artery, and re-opened segmental vessels.[52]

15.2.9 Diagnostic Pathway

As can be seen from Table 15.1, the accuracy of CT and MRI do not allow CTED to be made from single modality imaging. The investigation of a patient with suspected CTED requires multimodality input, particularly if surgery is to be considered. Previously invasive angiography was the reference imaging modality. Central thrombus may be underdiagnosed; however, angiography has the advantage of providing accurate segmental and subsegmental disease assessment and a surgical road-map for therapeutic planning. MDCT and pulmonary MRA are now playing a more dominant role in the diagnosis of this condition, not least because these are less invasive, less expensive, and generally more widely available than conventional angiography.

A suggested algorithm for the investigation of patients suspected of having CTED is provided in Fig. 15.21, which has been adapted from the Papworth algorithm[33] to include V/Q scintigraphy as an initial investigation and is in line with current recommendations from the American College of Chest Physicians.[32] A normal scintogram effectively rules out the diagnosis of CTED, however, and has been found to be more sensitive than CTPA in the diagnosis of CTED.[38] CTPA and MRA alone should not be used to exclude CTED. In the investigation of patients suspected of having CTED, the clinical status of the patient and their suitability for intervention should influence how far they proceed down the diagnostic pathway.

15.3 Conclusion

The diagnosis of acute pulmonary embolism has been intensely studied and modified over the past two decades, resulting in a readily accessible and minimally invasive investigation pathway. As a result, referral practices have changed and radiation

Table 15.1 The reported sensitivities and specificities of CTPA, MR angiography (MRA), and V/Q scintigraphy in the diagnosis of CTED

–	Sensitivity (%)	Specificity (%)
CTPA central vessels	77–86	50–92
Bergin et al.[61]	77	100
Schwickert et al.[48]	70	
Pitton et al.[62]		
CTPA segmental vessels	81–85	53–64
Bergin et al.	64	
Pitton et al.		
MRA central vesselsWolff et al.[63]	82–8327–44	93–9654–77
Bergin et al.		
MRA segmental vessels	68–76	93–95
Wolff et al.	59–60	54–64
Bergin et al.		
V/Q scintigraphy	96–97.4	90–95
Tunariu et al.[38]		

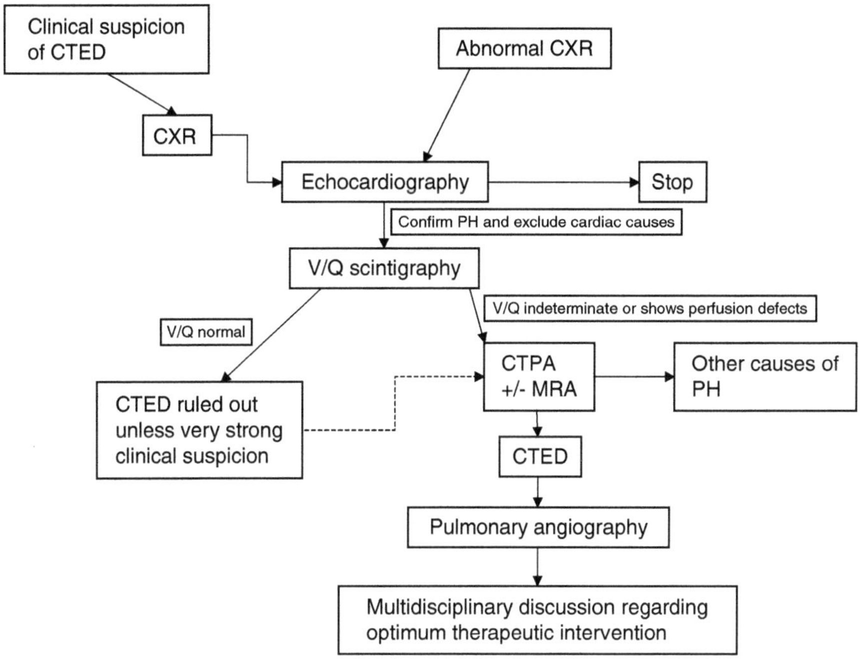

Fig. 15.21 An algorithm for the investigation of CTED

burden to the patient must be kept in mind. However, there is no doubt that recent advances in imaging, particularly in MDCT, have improved the efficiency with which this diagnosis is made and excluded, with associated cost benefits. In the very small minority of patients who develop CTED and in those patients in whom the disease appears to develop de novo, accurate imaging plays an increasingly important part in the management of a potentially curable cause of pulmonary hypertension, with MDCT and MRA developing as significant contributors to this diagnosis.

References

1. Worslcy DF, Alavi A. Comprehensive analysis of the results of the PIOPED Study. Prospective investigation of pulmonary embolism diagnosis study. *J Nucl Med*. 1995;36(12):2380-2387.
2. Daftary A, Gregory M, Daftary A, Seibyl JP, Saluja S. Chest radiograph as a triage tool in the imaging-based diagnosis of pulmonary embolism. *AJR Am J Roentgenol*. 2005;185(1):132-134.
3. Stein PD, Woodard PK, Weg JG, et al. Diagnostic pathways in acute pulmonary embolism: recommendations of the PIOPED II Investigators. *Radiology*. 2007;242(1):15-21.
4. Value of the ventilation/perfusion scan in acute pulmonary embolism. Results of the prospective investigation of pulmonary embolism diagnosis (PIOPED). The PIOPED Investigators. *JAMA* 1990;263(20):2753-2759.

5. Wittram C, Kalra MK, Maher MM, Greenfield A, McLoud TC, Shepard JA. Acute and chronic pulmonary emboli: angiography-CT correlation. *AJR Am J Roentgenol.* 2006;186 (6 Suppl 2):S421-S429.
6. Safriel Y, Zinn H. CT pulmonary angiography in the detection of pulmonary emboli: a meta-analysis of sensitivities and specificities. *Clin Imaging.* 2002;26(2):101-105.
7. Winer-Muram HT, Rydberg J, Johnson MS, et al. Suspected acute pulmonary embolism: evaluation with multi-detector row CT versus digital subtraction pulmonary arteriography. *Radiology.* 2004;233(3):806-815.
8. Ghaye B, Szapiro D, Mastora I, et al. Peripheral pulmonary arteries: how far in the lung does multi-detector row spiral CT allow analysis? *Radiology.* 2001;219(3):629-636.
9. Patel S, Kazerooni EA, Cascade PN. Pulmonary embolism: optimization of small pulmonary artery visualization at multi-detector row CT. *Radiology.* 2003;227(2):455-460.
10. Schoepf UJ, Holzknecht N, Helmberger TK, et al. Subsegmental pulmonary emboli: improved detection with thin-collimation multi-detector row spiral CT. *Radiology.* 2002;222(2):483-490.
11. Stein PD, Fowler SE, Goodman LR, et al. Multidetector computed tomography for acute pulmonary embolism. *N Engl J Med.* 2006;354(22):2317-2327.
12. Garg K, Kemp JL, Wojcik D, et al. Thromboembolic disease: comparison of combined CT pulmonary angiography and venography with bilateral leg sonography in 70 patients. *AJR Am J Roentgenol.* 2000;175(4):997-1001.
13. Katz DS, Loud PA, Bruce D et al. Combined CT venography and pulmonary angiography: a comprehensive review. *Radiographics* 2002; 22 Spec No:S3-19.
14. Goodman LR, Stein PD, Matta F, et al. CT venography and compression sonography are diagnostically equivalent: data from PIOPED II. *AJR Am J Roentgenol.* 2007;189(5):1071-1076.
15. Schwarcz TH, Matthews MR, Hartford JM, et al. Surveillance venous duplex is not clinically useful after total joint arthroplasty when effective deep venous thrombosis prophylaxis is used. *Ann Vasc Surg.* 2004;18(2):193-198.
16. Quiroz R, Kucher N, Zou KH, et al. Clinical validity of a negative computed tomography scan in patients with suspected pulmonary embolism: a systematic review. *JAMA.* 2005;293(16):2012-2017.
17. Remy-Jardin M, Pistolesi M, Goodman LR, et al. Management of suspected acute pulmonary embolism in the era of CT angiography: a statement from the Fleischner society. *Radiology.* 2007;245(2):315-329.
18. Remy-Jardin M, Mastora I, Remy J. Pulmonary embolus imaging with multislice CT. *Radiol Clin North Am.* 2003;41(3):507-519.
19. Richman PB, Courtney DM, Friese J, et al. Prevalence and significance of nonthromboembolic findings on chest computed tomography angiography performed to rule out pulmonary embolism: a multicenter study of 1, 025 emergency department patients. *Acad Emerg Med.* 2004;11(6): 642-647.
20. Remy-Jardin M, Remy J. Spiral CT angiography of the pulmonary circulation. *Radiology.* 1999;212(3):615-636.
21. Grist TM, Sostman HD, MacFall JR, et al. Pulmonary angiography with MR imaging: preliminary clinical experience. *Radiology.* 1993;189(2):523-530.
22. Gupta A, Frazer CK, Ferguson JM, et al. Acute pulmonary embolism: diagnosis with MR angiography. *Radiology.* 1999;210(2):353-359.
23. Meaney JF, Weg JG, Chenevert TL, Stafford-Johnson D, Hamilton BH, Prince MR. Diagnosis of pulmonary embolism with magnetic resonance angiography. *N Engl J Med.* 1997;336(20):1422-1427.
24. Oudkerk M, van Beek EJ, Wielopolski P, et al. Comparison of contrast-enhanced magnetic resonance angiography and conventional pulmonary angiography for the diagnosis of pulmonary embolism: a prospective study. *Lancet.* 2002;359(9318):1643-1647.
25. Stein PD, Woodard PK, Hull RD, et al. Gadolinium-enhanced magnetic resonance angiography for detection of acute pulmonary embolism: an in-depth review. *Chest.* 2003;124(6):2324-2328.
26. Scarsbrook AF, Evans AL, Owen AR, Gleeson FV. Diagnosis of suspected venous thromboembolic disease in pregnancy. *Clin Radiol.* 2006;61(1):1-12.

27. Schutgens RE, Ackermark P, Haas FJ, et al. Combination of a normal D-dimer concentration and a non-high pretest clinical probability score is a safe strategy to exclude deep venous thrombosis. *Circulation*. 2003;107(4):593-597.
28. Pengo V, Lensing AWA, Prins MH, et al. Incidence of chronic thromboembolic pulmonary hypertension after pulmonary embolism. *New Engl J Med*. 2004;350(22):2257-2264.
29. Tapson VF, Humbert M. Incidence and prevalence of chronic thromboembolic pulmonary hypertension: from acute to chronic pulmonary embolism. *Proc Am Thorac Soc*. 2006;3(7):564-567.
30. Egermayer P, Peacock AJ. Is pulmonary embolism a common cause of chronic pulmonary hypertension? Limitations of the embolic hypothesis. *Eur Res J*. 2000;15(3):440-448.
31. Lang IM. Chronic thromboembolic pulmonary hypertension – not so rare after all. *N Engl J Med*. 2004;350(22):2236-2238.
32. McGoon M, Gutterman D, Steen V, et al. Screening, early detection, and diagnosis of pulmonary arterial hypertension: ACCP evidence-based clinical practice guidelines. *Chest*. 2004;126(suppl 1):14S-34S.
33. Coulden R. State-of-the-art imaging techniques in chronic thromboembolic pulmonary hypertension. *Proc Am Thorac Soc*. 2006;3(7):577-583.
34. Hoeper MM. Diagnosis and treatment of pulmonary hypertension. *Pneumologie*. 2006;60(7):428-440.
35. Powe JE, Palevsky HI, McCarthy KE, Alavi A. Pulmonary arterial hypertension: value of perfusion scintigraphy. *Radiology*. 1987;164(3):727-730.
36. Fedullo PF, Auger WR, Kerr KM, Rubin LJ. Chronic thromboembolic pulmonary hypertension. *New Engl J Med*. 2001;345(20):1465-1472.
37. Lisbona R, Kreisman H, Novales-Diaz J, Derbekyan V. Perfusion lung scanning: differentiation of primary from thromboembolic pulmonary hypertension. *Am J Roentgenol*. 1985;144(1):27-30.
38. Tunariu N, Gibbs SJ, Win Z, et al. Ventilation-perfusion scintigraphy is more sensitive than multidetector CTPA in detecting chronic thromboembolic pulmonary disease as a treatable cause of pulmonary hypertension. *J Nucl Med*. 2007;48(5):680-684.
39. Kerr KM, Auger WR, Fedullo PF, Channick RH, Yi ES, Moser KM. Large vessel pulmonary arteritis mimicking chronic thromboembolic disease. *Am J Respir Crit Care Med*. 1995;152(1):367-373.
40. Berry DF, Buccigrossi D, Peabody J, Peterson KL, Moser KM. Pulmonary vascular occlusion and fibrosing mediastinitis. *Chest*. 1986;89(2):296-301.
41. Kauczor HU, Schwickert HC, Mayer E, Kersjes W, Moll R, Schweden F. Pulmonary artery sarcoma mimicking chronic thromboembolic disease: computed tomography and magnetic resonance imaging findings. *Cardiovasc Intervent Radiol*. 1994;17(4):185-189.
42. Bailey CL, Channick RN, Auger WR, et al. "High Probability" perfusion lung scans in pulmonary venoocclusive disease. *Am J Respir Crit Care Med*. 2000;162(5):1974-1978.
43. Rush C, Langleben D, Schlesinger RD, Stern J, Wang NS, Lamoureux E. Lung scintigraphy in pulmonary capillary hemangiomatosis. A rare disorder causing primary pulmonary hypertension. *Clin Nucl Med*. 1991;16(12):913-917.
44. King MA, Ysrael M, Bergin CJ. Chronic thromboembolic pulmonary hypertension: CT findings. *AJR Am J Roentgenol*. 1998;170(4):955-960.
45. Ng CS, Wells AU, Padley SP. A CT sign of chronic pulmonary arterial hypertension: the ratio of main pulmonary artery to aortic diameter. *J Thorac Imaging*. 1999;14(4):270-278.
46. Kuriyama K, Gamsu G, Stern RG, Cann CE, Herfkens RJ, Brundage BH. CT-determined pulmonary artery diameters in predicting pulmonary hypertension. *Invest Radiol*. 1984;19(1):16-22.
47. Remy-Jardin M, Duhamel A, Deken V, Bouaziz N, Dumont P, Remy J. Systemic collateral supply in patients with chronic thromboembolic and primary pulmonary hypertension: assessment with multi-detector row helical CT angiography. *Radiology*. 2005;235(1):274-281.
48. Schwickert HC, Schweden F, Schild HH, et al. Pulmonary arteries and lung parenchyma in chronic pulmonary embolism: preoperative and postoperative CT findings. *Radiology*. 1994;191(2):351-357.

49. Bergin CJ, Rios G, King MA, Belezzuoli E, Luna J, Auger WR. Accuracy of high-resolution CT in identifying chronic pulmonary thromboembolic disease. *AJR Am J Roentgenol.* 1996;166(6):1371-1377.
50. Kauczor HU, Schwickert HC, Mayer E, Schweden F, Schild HH, Thelen M. Spiral CT of bronchial arteries in chronic thromboembolism. *J Comput Assist Tomogr.* 1994;18(6):855-861.
51. Morris TA, Auger WR, Ysrael MZ, et al. Parenchymal scarring is associated with restrictive spirometric defects in patients with chronic thromboembolic pulmonary hypertension. *Chest.* 1996;110(2):399-403.
52. Kreitner KF, Ley S, Kauczor HU, et al. Chronic thromboembolic pulmonary hypertension: pre- and postoperative assessment with breath-hold MR imaging techniques. *Radiology.* 2004;232(2):535-543.
53. Auger WR, Fedullo PF, Moser KM, Buchbinder M, Peterson KL. Chronic major-vessel thromboembolic pulmonary artery obstruction: appearance at angiography. *Radiology.* 1992;182(2):393-398.
54. Ito K, Ichiki T, Ohi K, et al. Pulmonary capillary hemangiomatosis with severe pulmonary hypertension. *Circ J.* 2003;67(9):793-795.
55. Swensen SJ, Tashjian JH, Myers JL, et al. Pulmonary venoocclusive disease: CT findings in eight patients. *Am J Roentgenol.* 1996;167(4):937-940.
56. Devaraj A, Griffin N, Nicholson AG, Padley SP. Computed tomography findings in fibrosing mediastinitis. *Clin Radiol.* 2007;62(8):781-786.
57. Kim SY, Park JH, Chung JW, et al. Follow-up CT evaluation of the mural changes in active Takayasu arteritis. *Korean J Radiol.* 2007;8(4):286-294.
58. Bergin CJ, Sirlin C, Deutsch R, et al. Predictors of patient response to pulmonary thromboendarterectomy. *AJR Am J Roentgenol.* 2000;174(2):509-515.
59. Jamieson SW, Kapelanski DP, Sakakibara N, et al. Pulmonary endarterectomy: experience and lessons learned in 1, 500 cases. *Ann Thorac Surg.* 2003;76(5):1457-1464.
60. Mayer E, Klepetko W. Techniques and outcomes of pulmonary endarterectomy for chronic thromboembolic pulmonary hypertension. *Proc Am Thorac Soc.* 2006;3(7):589-593.
61. Bergin CJ, Sirlin CB, Hauschildt JP, et al. Chronic thromboembolism: diagnosis with helical CT and MR imaging with angiographic and surgical correlation. *Radiology.* 1997;204(3):695-702.
62. Pitton MB, Kemmerich G, Herber S, Schweden F, Mayer E, Thelen M. Chronic thromboembolic pulmonary hypertension: diagnostic impact of Multislice-CT and selective Pulmonary-DSA. *Rofo.* 2002;174(4):474-479.
63. Wolff K, Bergin CJ, King MA, et al. Accuracy of contrast-enhanced magnetic resonance angiography in chronic thromboembolic disease. *Acad Radiol.* 1996;3(1):10-17.

Part 2
Vascular Disease in Connective Tissue Diseases

Chapter 16
PAH in CTD – Clinical Trials Criteria and Performance

James R. Seibold

16.1 Introduction

Pulmonary arterial hypertension (PAH) is characterized by vascular proliferation and remodeling ultimately leading to right ventricular failure and death. PAH has emerged in recent years as the leading cause of death and late disease morbidity in systemic sclerosis and is an important complication of other connective tissue disorders. Pulmonary endothelial injury is associated with reduced production of vasodilating substances such as prostacyclin and nitric oxide and increased production of vasoconstrictive and proproliferative endothelin. Short term controlled trials have suggested benefits of endothelin receptor antagonism including improved exercise capacity and hemodynamics and increased time to clinical worsening. Improved survival is suggested from long term open label observational cohorts. As therapeutic options increase, it is important to consider the validity of outcome measures and the relative merits of endothelin receptor selectivity.

The prognosis for patients with PAH associated with CTD (PAH-CTD) is worse than that for patients with idiopathic PAH (iPAH) or PAH associated with congenital heart disease (PAH-CHD).[3,14] For example, the 3-year survival rate is 37% for patients with PAH-CTD when compared to 59% for patients with iPAH and 77% for patients with PAH-CHD (Fig. 16.1).[15] Similarly, the prognosis for patients with PAH-CTD is significantly worse than that for CTD patients without this complication.[7] The 1-year survival rates are >90% and <70% in patients with systemic sclerosis without PAH and with PAH, respectively, and the 5-year survival rates are ~85% in the absence of PAH and only 10% in the presence of PAH.[16]

The high prevalence of PAH in systemic sclerosis and the lack of symptoms in the initial stage of PAH have led the American College of Chest Physicians to recommend in their guidelines that CTD patients at risk should be screened at regular intervals. Annual echocardiogram assessments (or sooner, if symptomatic) and the easy to administer 6-min walk distance (6MWD) test can be scheduled with a concomitant consultation

J.R. Seibold (✉)
University of Michigan Scleroderma Program, University of Michigan Health System, Ann Arbor, MI, 48106, USA
jseibold@umich.edu

D. Abraham et al. (eds.), *Advances in Vascular Medicine*,
DOI 10.1007/978-1-84882-637-3_16, © Springer-Verlag London Limited 2010

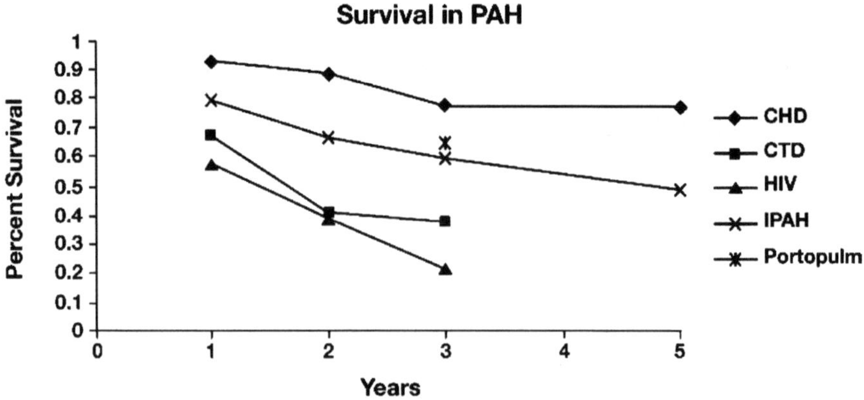

Fig. 16.1 Mean survival of patients with pulmonary arterial hypertension (PAH).[19] *CHD* congenital heart disease; *CTD* connective tissue disease; *HIV* human immunodeficiency virus; *IPAH* idiopathic pulmonary arterial hypertension; portopulm; portopulmonary hypertension

by a cardiologist or pulmonologist. Awareness in rheumatologists of the need to proactively screen patients with CTD should result in earlier diagnosis of patients with PAH. Early detection should permit implementation of a treatment strategy that is based on the condition's cause and severity, ultimately improving clinical outcome.[14,17]

The treatment of PAH-CTD is more complex than that of iPAH.[3,14] Therapeutic algorithms for PAH[18,19] and, specifically, for the management of PAH in systemic sclerosis[7,20] include a variety of clinically relevant endpoints, such as survival, exercise capacity, functional class, hemodynamic parameters, echocardiographic parameters, and quality-of-life measures.[21] The approach to treating PAH-CTD includes both general measures and targeted treatment of the underlying disease. General supportive care includes (1) oral anticoagulants, which are thought to reduce cellular proliferation by reducing thrombin levels in the pulmonary vasculature; (2) diuretics to treat edema and ascites; (3) oxygen for hypoxemia; and (4) digoxin for right ventricular failure.[14,18,22]

Three major pathways have been identified as playing critical roles in the abnormal proliferation and vasoconstriction of pulmonary arteriolar smooth muscle cells in PAH: (1) the prostacyclin pathway, (2) the nitric oxide pathway, and (3) the endothelin pathway. An understanding of these pathways has led to the identification of three targets that form the basis of current and emerging targeted therapies for PAH,[1] namely prostacyclin, phosphodiesterase type 5, and endothelin 1 (ET-1). This chapter examines the use of endothelin receptor antagonists (ETRAs) in the treatment of PAH in patients with CTDs.

16.2 The Role of Endothelin-1 in the Pathogenesis of PAH Associated with CTD

ET-1 is involved in the physiologic processes of vascular tone and endothelial cell mitogenesis and has an obligatory role in normal cellular proliferation, repair, and tissue development.[23,24] Under pathologic conditions, ET-1 leads to fibrosis, irreversible vascular remodeling of the lungs, smooth muscle cell hypertrophy, and inflammation.[25] Expression of ET-1 is dependent on tissue and cell type, as well as on the underlying disease entity and its severity, and appears to play a major role in the vascular dysfunction seen in both PAH and CTDs.[4,26,27] Elevated plasma and tissue levels have been demonstrated in iPAH, systemic sclerosis, and other CTDs. Immunohistochemical, autoradiographic, and molecular analyzes have shown increased ET-1 in the vasculature, pulmonary interstitium, and bronchial and alveolar epithelium regions.[28] The largest increase in ET-1 has been seen in areas of abnormal pulmonary vascular architecture.[29] Expression of total ET-1 receptors in scleroderma lung tissue appears to be localized to the alveolar epithelium and the pulmonary interstitium, which is primarily composed of fibroblastic cells with macrophages and some microvessels.[28] Circulating and pulmonary ET-1 levels have been correlated with hemodynamic parameters, disease severity, and disease outcome.[4]

The biologic effects of ET-1 are mediated by two membrane receptors, ET-1 receptor A (ET_A) and ET-1 receptor B (ET_B), which belong to the G-protein-coupled serpentine family. Both receptors are expressed differentially by various cell types as well as in multiple disease entities.[23,30] As shown in Fig. 16.2, activation of ET_A receptors, found on smooth muscle cells and cardiac myocytes, results in vasoconstriction, cell proliferation, and hypertrophy. In contrast, binding of ET-1 to ET_B receptors, which are located on vascular endothelial cells, results in the release of nitric oxide and prostacyclin thereby causing vasodilation. Endothelial ET_B receptors in the lung and kidney also have a role in clearing ET-1. Furthermore, activation of ET_B receptors results in inhibition of endothelin-converting enzyme-1 (ECE-1), the enzyme responsible for the biosynthesis of ET-1 from the inactive

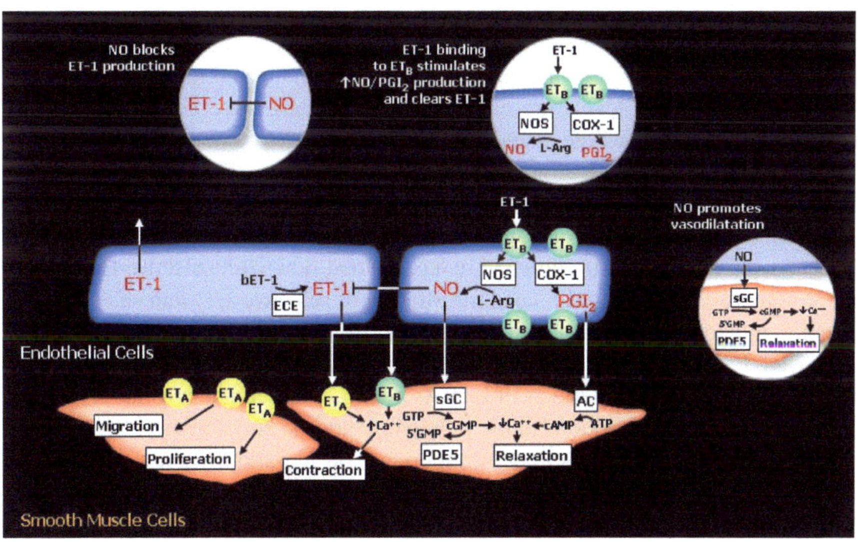

Fig. 16.2 ET-1 receptor A (ET_A) and ET-1 receptor B (ET_B) mediate activities of endothelin 1 (ET-1)

intermediate big ET-1.[31] ET_B receptors are also found on smooth muscle cells, where they are believed to play a role in vasoconstriction, proliferation, and hypertrophy similar to the effects of ET_A receptors on smooth muscle cells.

16.3 Endothelin-Receptor Antagonists in the Treatment of PAH Associated with CTD

The central role of ET-1 in the pathogenesis of PAH has led to the development of antagonists that block functional ET-1 receptors. Currently, two types of ETRAs have been studied in patients with PAH-CTD: nonselective ET_A/ET_B receptor antagonists and ET_A-selective receptor antagonists. Clinical trials with both ET_A-selective and nonselective ETRAs indicate that not only can these agents enhance the physical function of the patient, as measured by a change in exercise capacity and an improvement in functional class, but they can also improve hemodynamic parameters. In addition, initial data suggest that ETRAs confer a survival benefit in patients with PAH-CTD.[32-36]

The advantages and disadvantages of ET receptor type selectivity with regard to clinical outcomes have not been exhaustively investigated; however, experimental data suggest that ET receptor selectivity may be relevant to the efficacy and safety of ETRA therapy.[37] For example, blocking functional endothelial ET_B receptors with a nonselective ET_A/ET_B antagonist would be anticipated to result in increased levels of circulating ET-1, through reduced clearance via ET_B receptors and activation of ECE-1.[38,39] Increased levels of circulating ET-1 have been observed in patients with vascular disease after treatment with both nonselective and partially selective ETRAs.[39-45] In contrast, treatment with highly selective ET_A antagonists has been documented to decrease circulating ET-1 levels.[46] The pathophysiological or prognostic significance of increased ET-1 levels remains unknown; however, the observation that ET-1 levels are elevated in the disease state and correlate with aspects of the disease[4] would suggest that such increases are not beneficial.

Examination of the effects of selective and nonselective ETRAs on blood vessels in the systemic circulation of healthy subjects and those with vascular disease shows the anticipated responses of vasodilation and increased blood flow with ET_A receptor blockade and of vasoconstriction and reduced blood flow with selective ET_B blockade.[47-49] Interestingly, however, coadministration of selective ET_A and selective ET_B receptor antagonists was observed to attenuate the vasodilator response relative to selective ET_A receptor blockade alone.[47-49] These findings, however, should be interpreted with caution, as the effects of ETRAs on blood vessel tone in general may not necessarily reflect their actions in the pulmonary arterial circulation.

ETRA selectivity may also be an important consideration when evaluating the effects of these agents on vascular proliferation and remodeling. There is widespread evidence from animal models that both selective and nonselective ETRAs result in reduced expression of growth factors, reduced extracellular matrix deposition, and reduced matrix metalloproteinase activity.[50-56] Interestingly, while both ET_A-selective antagonists and nonselective antagonists have been shown to be effective in attenuating fibrosis and collagen production in short-term studies, ET_A-selective antagonism may provide greater benefits in terms of attenuating collagen secretion, cell proliferation, and interstitial area in the longer term.[53]

ET_A selectivity may also influence the incidence of certain adverse events following treatment with ETRAs, in particular edema. ET-1 is believed to play an important role in regulating sodium excretion through activation of ET_B receptors,[57] and it has recently been shown that blockade of the ET_B receptor in the rat renal medulla causes antidiuresis and antinatriuresis.[58]

16.4 Clinical Experience with ETRAs

To date, two ETRAs – bosentan and ambrisentan – have been approved by the US Food and Drug Administration for the treatment of PAH. A third ETRA – sitaxentan – has been approved in the European Union, Australia, and Canada. These agents differ in their selectivity for the ET_A and ET_B receptors and also with regard to their pharmacokinetic properties.

Bosentan is a nonselective ET_A/ET_B antagonist taken orally twice daily. It is currently indicated for the treatment of PAH (World Health Organization [WHO] Pulmonary Hypertension Classification Group I) in patients with WHO functional class III or IV symptoms to improve exercise capacity and decrease the rate of clinical worsening.[59] Ambrisentan is a moderately selective oral, once-daily ET_A receptor antagonist. It is indicated for PAH (WHO Pulmonary Hypertension Group I) in patients with WHO functional class II or III symptoms to improve exercise capacity and delay clinical worsening.[60] In vitro selectivity values (ET_A:ET_B) for ambrisentan range from 29:1 for ET-1-mediated contraction in the rat aorta to 4000:1 in myocardial membranes.[61,62] Increased ET-1 plasma levels have been observed following ingestion of ambrisentan, which suggests that the ET_B receptors – important in ET-1 clearance – are at least partially blocked in vivo.[63] Sitaxentan is an oral, once-daily, ET_A-selective ETRA. It is approved in the European Union and Australia for the treatment of primary PAH and PAH associated with collagen vascular diseases in patients with WHO functional class III PAH, and is awaiting approval in the USA for WHO functional class II, III, and IV iPAH and PAH-CTD. Sitaxentan has an ET_A:ET_B selectivity of 6,500:1[64] and should thus block the vasoconstrictive effects of ET-1 while maintaining the vasodilator and ET-1 clearance functions of ET_B receptors.[65] Functional selectivity is confirmed by the observation that sitaxentan decreases circulating plasma ET-1 in patients with congestive heart failure, indicating that the ET_B receptors remain operative.[46]

Sitaxentan inhibits cytochrome P450 2C9 enzyme, which is an important and manageable consideration for patients receiving concomitant warfarin. If sitaxentan and warfarin are coadministered, an 80% reduction in the warfarin dose is recommended followed by careful monitoring.

16.4.1 Clinical Findings with ETRAs in PAH

The safety and efficacy of bosentan, ambrisentan, and sitaxentan have been demonstrated in numerous controlled and open-label clinical trials involving patients with a range of PAH severities and including patients both with and without associated conditions such as CTD.

Bosentan was initially evaluated in two randomized placebo-controlled trials of 12 and 16 weeks' duration in a total of 235 patients with severe (WHO functional class III–IV) iPAH or PAH-CTD.[66,67] In these trials, bosentan was added to patients' current therapy, which could have included a combination of digoxin, anticoagulants, diuretics, and vasodilators. In both trials, twice-daily bosentan 250 mg and 125 mg resulted in significant improvements in exercise capacity (measured by the 6MWD), WHO functional class, and Borg Dyspnea Index rating, and significant reduction in time to clinical worsening. Long-term follow-up of patients included in the two studies and their open-label extensions also indicated improved survival with bosentan treatment.[68,69] Bosentan was generally well tolerated, although dose-related increases in aminotransferases to more than three times the upper limit of normal (ULN) were observed.[67] Elevated liver enzyme levels have also been documented through postmarketing surveillance of bosentan.[70] These were reported in ~10% of patients who received bosentan, leading to discontinuation in ~3% of patients.[70]

Ambrisentan was initially evaluated in two randomized placebo-controlled 12-week trials in a total of 393 patients with WHO functional class I–IV PAH, including iPAH, PAH-CTD, PAH with HIV or PAH associated with anorexigen use.[60] Similar to the bosentan trials, ambrisentan was added to patients' current therapy, which could have included a combination of anticoagulants, diuretics, calcium channel blockers, or digoxin. In these trials, ambrisentan 2.5, 5, and 10 mg once daily significantly improved exercise capacity (6MWD) and delayed clinical worsening.[60] The most significant adverse event reported in trials with ambrisentan was peripheral edema, occurring in 17% of ambrisentan-treated patients when compared with 11% of those receiving placebo.[60] In contrast to bosentan, aminotransferase elevations to >3 × ULN were a rare occurrence with ambrisentan, being observed in only 0.8% of patients over the 12 weeks.[60]

Sitaxentan was initially evaluated in two randomized placebo-controlled trials of 12 and 18 weeks' duration involving a total of 425 patients with PAH (iPAH, PAH-CTD, or PAH-CHD) of varying severities.[38,71] Treatment with once-daily sitaxentan 100 or 300 mg resulted in significant benefits for exercise capacity (6MWD), WHO functional class, and hemodynamic assessments. Increased liver enzyme levels were observed in 3% of patients who received sitaxentan 100 mg in one trial[38] and 0% in another.[71] Subanalysis of WHO functional class III and IV patients with PAH included in the 12-week trial indicated that improvements in efficacy parameters were even greater in more severely affected patients than observed for the overall patient group.[72]

16.4.2 ETRAs in PAH-CTD

The markedly poor prognosis for patients with PAH-CTD when compared with patients with iPAH or CTD without PAH makes it of paramount importance to identify the most appropriate interventions for this population. Although to date there are no published data from randomized controlled trials examining the efficacy of ETRAs in this population alone, post hoc subgroup analyzes are available from trials that recruited mixed populations.[32,73,74] In addition, limited open-label data are available for patients with PAH-CTD treated with some agents.[35,75,76]

16.4.2.1 Exercise Capacity

In general, available ETRAs provide varying degrees of efficacy with regard to exercise capacity in patients with PAH-CTD. In a subset of 66 patients with PAH-CTD (WHO functional class III–IV) included in clinical trials of the nonselective ET_A/ET_B antagonist, bosentan, exercise capacity was stable at the end of the studies (Week 12 or 16) in 44 patients who received bosentan (+19.5 m, 95% CI –3.2 to 42.2 m) compared with a reduction of 2.6 m (95% CI –54.0 to 48.7 m) among patients who received placebo. The absolute difference between the groups was 22.1 m (95% CI –32 to 76 m; $p>$NS).[32] Similarly, treatment with ambrisentan (2.5, 5, or 10 mg) in patients with PAH-CTD from the two pivotal clinical trials (ARIES-1 and ARIES-2) resulted in a nonsignificant increase in 6MWD (+19 ± 14.8 m; $p>0.056$)[77]. Further subanalysis, however, indicated that among patients receiving ambrisentan 5 or 10 mg, an improvement of 25 ± 14.2 m was observed ($p>0.020$).[77] In a subset of 42 patients with PAH-CTD from the sitaxentan STRIDE-1 clinical trial, exercise capacity had improved significantly after 12 weeks treatment with sitaxentan (+20 m; $p>0.037$) vs. a decrease of 38 m for patients who received placebo.[74] The absolute difference between the groups was 58 m ($p>0.027$). These findings suggest that selectivity for the ET_A receptor could provide advantages over nonselective agents in this patient population, but in the absence of randomized head-to-head trials, such a conclusion would be inappropriate. Indeed, similar beneficial therapeutic responses were also seen in STRIDE-2, a prospective, multicentr, randomized, double-blind, placebo-controlled, open-label trial examining the effects of sitaxentan 50 mg or 100 qd and bosentan 125 mg bid in PAH over 18 weeks, where up to 30% of patients had CTD-PAH.[38] In a 1-year open-label extension trial, STRIDE-2X, comparing a standard dose of twice-daily bosentan 125 mg ($n>25$) with the once-daily dose of sitaxentan 100 mg ($n>27$), although patients receiving sitaxentan showed a 2 m improvement in 6MWD and bosentan-treated patients showed a deterioration of 51 m at 1 year, the difference between the two groups was not statistically significant.[35] Furthermore, in a post hoc analysis of data from 39 patients with CTD-PAH functional class II–IV, from a combined analysis of STRIDE-1, -2, and -4, a placebo-corrected increase in 6MWD of 37.7 m was seen in patients treated with sitaxentan 100 mg ($p>0.042$).[78]

16.4.2.2 NYHA/WHO Functional Class

Data relating to decline in functional class in patients with PAH-CTD are difficult to compare between clinical trials due to differences in the way data are presented. Among patients with PAH-CTD treated with bosentan, functional class improved for 24%, remained stable for 69%, and worsened for 8% of patients after 16 weeks' treatment.[76] At Week 48, functional class improved in 27% of patients, remained stable in 57%, and worsened in 16%.[76] Deterioration in functional class was similar at Week 12 for the iPAH and PAH-CTD subgroups receiving ambrisentan (3.6 and 3.7%, respectively) and was less than that observed in the placebo group (16.5 and 20.9%, respectively).[77] At the end of an extension study (STRIDE-1X) with sitaxentan (median treatment duration 26 weeks), 39% of patients had an improvement in functional class, and more patients were in functional class I–II than class III–IV when compared with the start of active therapy ($p<0.001$).[74] In the open-label sitaxentan extension study STRIDE-2X, functional improvement was observed in 24 and 13% of sitaxentan- and bosentan-treated patients, respectively ($p>$NS). Based on the available data, it is not possible to comment on whether any particular treatment provides superior benefits for this endpoint.

16.4.2.3 Time to Clinical Worsening

In the subanalysis of patients with PAH-CTD included in the bosentan clinical trials, time to clinical worsening was delayed in those who received bosentan vs. placebo: 95.4% for bosentan and 90.9% for placebo at 12 weeks and 90.3% for bosentan and 86.4% for placebo at 16 weeks.[32] No data on time to clinical worsening were published in the subanalysis of the sitaxentan clinical trial; however, in the 1-year open-label STRIDE-2X extension trial comparing bosentan 125 mg twice daily with sitaxentan 100 mg once daily, the time to prospectively defined clinical worsening was improved for sitaxentan (88% at 1 year) vs. bosentan (52% at 1 year; $p<0.01$; Fig. 16.3).[35] No data on time to clinical worsening in patients with PAH-CTD have been published for ambrisentan to date.

16.4.2.4 Survival

In the long-term open extension phase of the bosentan trials, survival was 85.9% at 1 year and 73.4% at 2 years.[32] During the observational period, however, 16% of patients received epoprostenol as an add-on treatment, which may have influenced the survival findings.[32] For ambrisentan, 52-week survival in patients with PAH-CTD was 92% when compared with 73% as predicted by the National Institutes of Health registry formula.[73] No data on survival with sitaxentan were included for the subanalysis of patients with PAH-CTD from the 12-week clinical trial; however, in the 1-year open-label study, STRIDE-2X, survival was 96% for the sitaxentan group and 79% for the bosentan group ($p>0.0125$; Fig. 16.4).[35] No add-on therapies

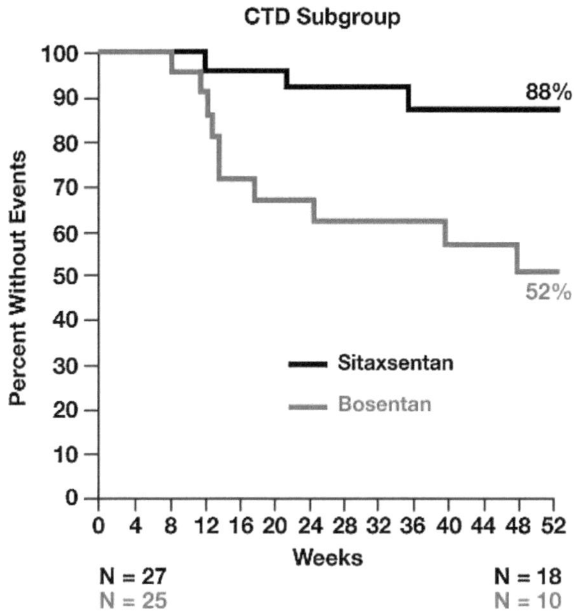

Fig. 16.3 Time to clinical worsening in an open-label extension trial comparing sitaxentan and bosentan in patients with PAH associated with CTD (PAH-CTD)[35]

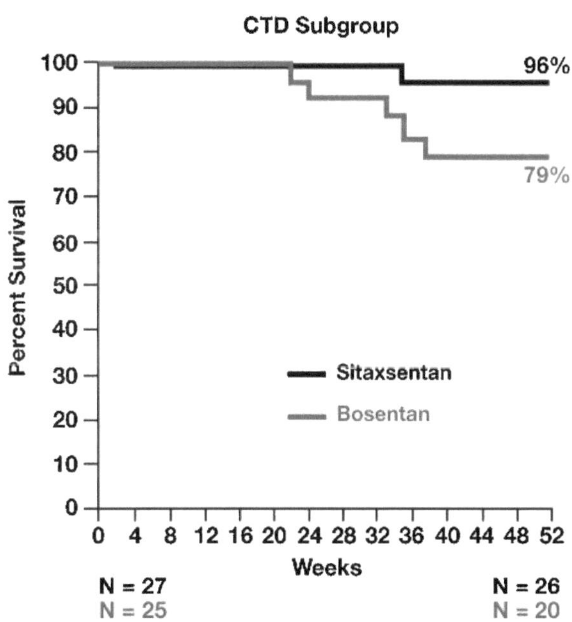

Fig. 16.4 One-year survival in an open-label extension trial comparing sitaxentan and bosentan in patients with PAH-CTD[35]

were required to achieve these effects with sitaxentan. These 1-year survival values are consistent with other reports.[28,36]

16.4.2.5 Safety

In the subanalysis of patients with PAH-CTD from the bosentan trials, the most frequent adverse events in the bosentan- vs. placebo-treated groups were dizziness (18.2% vs. 4.5%), lower limb edema (18.2% vs. 4.5%), headache (15.9% vs. 22.7%), and fatigue (13.6% vs. 0%). Abnormal hepatic function occurred in 11.4% of patients treated with bosentan vs. 9.1% of patients treated with placebo. In an analysis of long-term bosentan therapy in patients with PAH-CTD, the most frequently observed adverse events were peripheral edema (17%), liver enzyme elevations (17%), diarrhea (13%), exacerbated dyspnea (13%), and nausea (13%).[76] Interestingly, hepatic dysfunction with bosentan therapy appears to be greater (24%) among patients with PAH-CTD than among those with iPAH (11%) after 3 months of therapy.[79] Only limited data have been published for safety in patients with PAH-CTD treated with ambrisentan.[77] In the subanalysis of primary trials with ambrisentan, no patients receiving ambrisentan had aminotransferase elevations >3 × ULN; however, the incidence of peripheral edema was greater among patients receiving ambrisentan than in those who received placebo.[77] In the sitaxentan trials, the most frequent adverse events were peripheral edema, nasopharyngitis, headache, and nasal congestion. No increases in aminotransferases were observed during the 12-week trials, although two cases (5%) occurred during the extension phase.[74] Although data from blinded head-to-head trials are not available, in the open-label comparison of sitaxentan and bosentan, abnormal liver function tests occurred in 18% of patients who received bosentan, compared with 3% of patients treated with sitaxentan.[35]

16.5 Summary and Implications for the Future Treatment of PAH-CTD

Patients with PAH associated with CTD have a relatively poor prognosis. Rheumatologists, working together with pulmonologists and cardiologists, can institute early screening and detection of PAH in at-risk patients before significant damage to the microvasculature has occurred. This can then be followed by early treatment interventions, which may benefit patients with PAH-CTD. ET-1 appears to play an important role in the development of both CTDs and PAH, and the availability of ETRAs has increased treatment options and improved the management of PAH-CTD. Based on this review of the published literature to date, it is clear that treatment with both nonselective and selective ETRAs can lead to significant improvements in exercise capacity, WHO functional class, hemodynamic parameters, time to clinical worsening, and survival in patients with PAH-CTD. Despite a wide body of experimental evidence that indicates advantages for ET_A selectivity, no clear differences between selective and nonselective ETRAs have been identified for the majority of clinical endpoints examined in patients with either iPAH or PAH-CTD. The only apparent differences observed in patients with PAH-CTD were for time to clinical worsening and overall survival, with the selective ET_A receptor antagonist sitaxentan resulting in greater improvements in both parameters than bosentan; however, these observations were made in an open-label comparison involving only small numbers of patients. Further randomized controlled trials will be required to verify the findings. Another possible advantage of ET_A-selective sitaxentan and moderately selective ambrisentan compared with ET_A/ET_B-nonselective bosentan is the reduced occurrence of acute hepatotoxicity; however, again, further study and clinical experience are needed to verify this observation.

References

1. Humbert M, Sitbon O, Simonneau S. Treatment of pulmonary arterial hypertension. *N Engl J Med.* 2004;351:1425–1436.
2. Fagan KA, Badesch DB. Pulmonary hypertension associated with connective tissue disease. *Prog Cardiovasc Dis.* 2002;45:225–234.
3. Galie N, Manes A, Farahani KV, et al. Pulmonary arterial hypertension associated to connective tissue diseases. *Lupus.* 2005;14:713–717.
4. Hachulla E, Coghlan JG. A new era in the management of pulmonary arterial hypertension related to scleroderma: endothelin receptor antagonism. *Ann Rheum Dis.* 2004;63:1009–1014.
5. Denton CP, Black CM. Pulmonary hypertension in systemic sclerosis. *Rheum Dis Clin North Am.* 2003;29:335–349, vii.
6. Mukerjee D, St George D, Coleiro B, et al. Prevalence and outcome in systemic sclerosis associated pulmonary arterial hypertension: application of a registry approach. *Ann Rheum Dis.* 2003;62:1088–1093.
7. Ramirez A, Varga J. Pulmonary arterial hypertension in systemic sclerosis: clinical manifestations, pathophysiology, evaluation, and management. *Treat Respir Med.* 2004;3:339–352.
8. Ungerer RG, Tashkin DP, Furst D, et al. Prevalence and clinical correlates of pulmonary arterial hypertension in progressive systemic sclerosis. *Am J Med.* 1983;75:65–74.
9. Johnson SR, Gladman DD, Urowitz MB, Ibanez D, Granton JT. Pulmonary hypertension in systemic lupus. *Lupus.* 2004;13:506–509.
10. Shen JY, Chen SL, Wu YX, et al. Pulmonary hypertension in systemic lupus erythematosus. *Rheumatol Int.* 1999;18:147–151.
11. Tanaka E, Harigai M, Tanaka M, Kawaguchi Y, Hara M, Kamatani N. Pulmonary hypertension in systemic lupus erythematosus: evaluation of clinical characteristics and response to immunosuppressive treatment. *J Rheumatol.* 2002;29:282–287.
12. Winslow TM, Ossipov MA, Fazio GP, Simonson JS, Redberg RF, Schiller NB. Five-year follow-up study of the prevalence and progression of pulmonary hypertension in systemic lupus erythematosus. *Am Heart J.* 1995;129:510–515.
13. Keser G, Capar I, Aksu K, et al. Pulmonary hypertension in rheumatoid arthritis. *Scand J Rheumatol.* 2004;33:244–245.
14. Coghlan JG, Handler C. Connective tissue associated pulmonary arterial hypertension. *Lupus.* 2006;15:138–142.
15. McLaughlin VV, Presberg KW, Doyle RL, et al. Prognosis of pulmonary arterial hypertension: ACCP evidence-based clinical practice guidelines. *Chest.* 2004;126(suppl 1):78S–92S.

16. Steen V, Medsger TA Jr. Predictors of isolated pulmonary hypertension in patients with systemic sclerosis and limited cutaneous involvement. *Arthritis Rheum.* 2003;48:516–522.
17. Rubin LJ, Badesch DB. Evaluation and management of the patient with pulmonary arterial hypertension. *Ann Intern Med.* 2005;143:282–292.
18. Badesch DB, Abman SH, Ahearn GS, et al. Medical therapy for pulmonary arterial hypertension: ACCP evidence-based clinical practice guidelines. *Chest.* 2004;126(suppl 1):35S–62S.
19. Galie N, Branzi A. Pulmonary arterial hypertension: therapeutic algorithm. *Ital Heart J.* 2005;6:856–860.
20. Sanchez O, Humbert M, Sitbon O, Simonneau G. Treatment of pulmonary hypertension secondary to connective tissue diseases. *Thorax.* 1999;54:273–277.
21. Lee YK, Na SW, Kwak YL, Nam SB. Effect of pre-operative angiotensin-converting enzyme inhibitors on haemodynamic parameters and vasoconstrictor requirements in patients undergoing off-pump coronary artery bypass surgery. *J Int Med Res.* 2005;33:693–702.
22. Hoeper MM. Pulmonary hypertension in collagen vascular disease. *Eur Respir J.* 2002;19: 571–576.
23. Galie N, Manes A, Branzi A. The endothelin system in pulmonary arterial hypertension. *Cardiovasc Res.* 2004;61:227–237.
24. Goldie RG. Endothelins in health and disease: an overview. *Clin Exp Pharmacol Physiol.* 1999;26:145–148.
25. Clozel M. Effects of bosentan on cellular processes involved in pulmonary arterial hypertension: do they explain the long-term benefit? *Ann Med.* 2003;35:605–613.
26. Fagan KA, McMurtry IF, Rodman DM. Role of endothelin-1 in lung disease. *Respir Res.* 2001;2:90–101.
27. Highland KB, Silver RM. New developments in scleroderma interstitial lung disease. *Curr Opin Rheumatol.* 2005;17:737–745.
28. Abraham DJ, Vancheeswaran R, Dashwood MR, et al. Increased levels of endothelin-1 and differential endothelin type A and B receptor expression in scleroderma-associated fibrotic lung disease. *Am J Pathol.* 1997;151:831–841.
29. Giaid A, Yanagisawa M, Langleben D, et al. Expression of endothelin-1 in the lungs of patients with pulmonary hypertension. *N Engl J Med.* 1993;328:1732–1739.
30. Sticherling M. The role of endothelin in connective tissue diseases. *Rheumatology (Oxford).* 2006;45(suppl 3):iii8–iii10.
31. Naomi S, Iwaoka T, Disashi T, et al. Endothelin-1 inhibits endothelin-converting enzyme-1 expression in cultured rat pulmonary endothelial cells. *Circulation.* 1998;97:234–236.
32. Denton CP, Humbert M, Rubin L, Black CM. Bosentan therapy for pulmonary arterial hypertension related to connective tissue disease: a subgroup analysis of the pivotal clinical trials and their open-label extensions. *Ann Rheum Dis.* 2006;65:1336–1340.
33. Girgis RE, Frost A, Hill N, et al. Long-term treatment with sitaxsentan in patients with pulmonary arterial hypertension associated with connective tissue disease (PAH–CTD) [abstract]. Paper presented at the American Academy of Chest Physicians (ACCP) meeting; November 2, 2005; Seattle, Washington, DC.
34. Girgis RE, Mathai SC, Champion HC, et al. Determinants of survival in pulmonary hypertension associated with scleroderma. Paper presented at the 26th Annual Meeting of the International Society for Heart and Lung Transplantation; April 5–8, 2006; Madrid, Spain [Abstract No. 98].
35. Highland KB, Strange C, Girgis R, Black C. Comparison of sitaxsentan and bosentan in pulmonary arterial hypertension associated with connective tissue diseases. Paper presented at EULAR 2006: Annual European Congress of Rheumatology; June 21–24, 2006; Amsterdam, The Netherlands [Abstract No. 354].
36. Williams MH, Das C, Handler CE, et al. Systemic sclerosis associated pulmonary hypertension: improved survival in the current era. *Heart.* 2006;92:926–932.
37. Langleben D, Dupuis J, Langleben I, et al. Etiology-specific endothelin-1 clearance in human precapillary pulmonary hypertension. *Chest.* 2006;129:689–695.

38. Barst RJ, Langleben D, Badesch D, et al. Treatment of pulmonary arterial hypertension with the selective endothelin-A receptor antagonist sitaxsentan. *J Am Coll Cardiol.* 2006;47:2049–2056.
39. Leslie SJ, Spratt JC, McKee SP, et al. Direct comparison of selective endothelin A and non-selective endothelin A/B receptor blockade in chronic heart failure. *Heart.* 2005;91:914–919.
40. DiCarlo VS, Chen SJ, Meng QC, et al. ETA-receptor antagonist prevents and reverses chronic hypoxia-induced pulmonary hypertension in rat. *Am J Physiol.* 1995;269(5 pt 1):L690–L697.
41. Hiramoto Y, Shioyama W, Kuroda T, et al. Effect of bosentan on plasma endothelin-1 concentration in patients with pulmonary arterial hypertension. *Circ J.* 2007;71:367–369.
42. Kiowski W, Sütsch G, Hunziker P, et al. Evidence for endothelin-1-mediated vasoconstriction in severe chronic heart failure. *Lancet.* 1995;346:732–736.
43. Sütsch G, Kiowski W, Yan XW, et al. Short-term oral endothelin-receptor antagonist therapy in conventionally treated patients with symptomatic severe chronic heart failure. *Circulation.* 1998;98:2262–2268.
44. Weber C, Banken L, Birnboeck H, Schulz R. Effect of the endothelin-receptor antagonist bosentan on the pharmacokinetics and pharmacodynamics of warfarin. *J Clin Pharmacol.* 1999;39:847–854.
45. Williamson DJ, Wallman LL, Jones R, et al. Hemodynamic effects of bosentan, an endothelin receptor antagonist, in patients with pulmonary hypertension. *Circulation.* 2000;102:411–418.
46. Givertz MM, Colucci WS, LeJemtel TH, et al. Acute endothelin A receptor blockade causes selective pulmonary vasodilation in patients with chronic heart failure. *Circulation.* 2000;101:2922–2927.
47. Goddard J, Johnston NR, Hand MF, et al. Endothelin-A receptor antagonism reduces blood pressure and increases renal blood flow in hypertensive patients with chronic renal failure: a comparison of selective and combined endothelin receptor blockade. *Circulation.* 2004;109:1186–1193.
48. Love MP, Ferro CJ, Haynes WG, et al. Endothelin receptor antagonism in patients with chronic heart failure. *Cardiovasc Res.* 2000;47:166–172.
49. Verhaar MC, Strachan FE, Newby DE, et al. Endothelin-A receptor antagonist-mediated vasodilatation is attenuated by inhibition of nitric oxide synthesis and by endothelin-B receptor blockade. *Circulation.* 1998;97:752–756.
50. Ammarguellat F, Larouche I, Schiffrin EL. Myocardial fibrosis in DOCA-salt hypertensive rats: effect of endothelin ET_A receptor antagonism. *Circulation.* 2001;103:319–324.
51. Boffa JJ, Tharaux PL, Dussaule JC, Chatziantoniou C. Regression of renal vascular fibrosis by endothelin receptor antagonism. *Hypertension.* 2001;37:490–496.
52. Ergul A, Portik-Dobos V, Giulumian AD, Molero MM, Fuchs LC. Stress upregulates arterial matrix metalloproteinase expression and activity via endothelin A receptor activation. *Am J Physiol Heart Circ Physiol.* 2003;285:H2225–H2232.
53. Forbes JM, Hewitson TD, Becker GJ, Jones CL. Simultaneous blockade of endothelin A and B receptors in ischemic acute renal failure is detrimental to long-term kidney function. *Kidney Int.* 2001;59:1333–1341.
54. Park SH, Saleh D, Giaid A, Michel RP. Increased endothelin-1 in bleomycin-induced pulmonary fibrosis and the effect of an endothelin receptor antagonist. *Am J Respir Crit Care Med.* 1997;156:600–608.
55. Seccia TM, Belloni AS, Kreutz R, et al. Cardiac fibrosis occurs early and involves endothelin and AT-1 receptors in hypertension due to endogenous angiotensin II. *J Am Coll Cardiol.* 2003;41:666–673.
56. Yoshida J, Yamamoto K, Mano T, et al. Angiotensin II type 1 and endothelin type A receptor antagonists modulate the extracellular matrix regulatory system differently in diastolic heart failure. *J Hypertens.* 2003;21:437–444.
57. Vassileva I, Mountain C, Pollock DM. Functional role of ETB receptors in the renal medulla. *Hypertension.* 2003;41:1359–1363.
58. Guo X, Yang T. Endothelin B receptor antagonism in the rat renal medulla reduces urine flow rate and sodium excretion. *Exp Biol Med (Maywood).* 2006;231:1001–1005.

59. Tracleer prescribing information. South San Francisco, CA: Actelion Pharmaceuticals US, Inc.; 2007. <http://www.tracleer.com/pdf/PI_4pg_TR2454_032707_FINAL.pdf>;2007 Accessed 16.11.07.
60. Letairis prescribing information. Foster City, CA: Gilead Sciences, Inc.; 2007. <http://www.letairis.com/downloads/LETAIRIS_prescribing_information.pdf>;2007 Accessed 16.11.07.
61. Bolli MH, Marfurt J, Grisostomi C, et al. Novel benzo[1, 4]diazepin-2-one derivatives as endothelin receptor antagonists. *J Med Chem.* 2004;47:2776–2795.
62. Greene S, Nunley K, Weber S, et al. ETA vs. ETB receptor selectivity of endothelin-1 receptor antagonists. *J Am Coll Cardiol.* 2006;47:307A.
63. Food and drug administration center for drug evaluation and research. Letairis (Ambrisentan) tablets, clinical pharmacology biopharmaceutics reviews, Part 3; 2007:98–99. <http://www.fda.gov/cder/foi/nda/2007/022081s000_ClinPharmR_P3.pdf>; 2006 Accessed 14.11.06.
64. Thelin™ (sitaxsentan) Summary of Product Characteristics (SmPC) [online]. London, UK: Encysive (UK) Ltd; August 2004. <http://www.emea.europa.eu/humandocs/PDFs/EPAR/thelin/H-679-PI-en.pdf>; Accessed 7.02.08.
65. Widlitz AC, Barst RJ, Horn EM. Sitaxsentan: a novel endothelin-A receptor antagonist for pulmonary arterial hypertension. *Expert Rev Cardiovasc Ther.* 2005;3:985–991.
66. Channick RN, Simonneau G, Sitbon O, et al. Effects of the dual endothelin-receptor antagonist bosentan in patients with pulmonary hypertension: a randomised placebo-controlled study. *Lancet.* 2001;358:1119–1123.
67. Rubin LJ, Badesch DB, Barst RJ, et al. Bosentan therapy for pulmonary arterial hypertension. *N Engl J Med.* 2002;346:896–903.
68. McLaughlin VV, Sitbon O, Badesch DB, et al. Survival with first-line bosentan in patients with primary pulmonary hypertension. *Eur Respir J.* 2005;25:244–249.
69. Sitbon O, Badesch DB, Channick RN, et al. Effects of the dual endothelin receptor antagonist bosentan in patients with pulmonary arterial hypertension: a 1-year follow-up study. *Chest.* 2003;124:247–254.
70. Humbert M, Segal ES, Kiely DG, Carlsen J, Schwierin B, Hoeper MM. Results of European post-marketing surveillance of bosentan in pulmonary hypertension. *Eur Respir J.* 2007;30:338-344.
71. Barst RJ, Langleben D, Frost A, et al. Sitaxsentan therapy for pulmonary arterial hypertension. *Am J Respir Crit Care Med.* 2004;169:441–447.
72. Langleben D, Brock T, Dixon R, Barst R, for the STRIDE-1 study group. STRIDE 1: effects of the selective ETA receptor antagonist, sitaxsentan sodium, in a patient population with pulmonary arterial hypertension that meets traditional inclusion criteria of previous pulmonary arterial hypertension trials. *J Cardiovasc Pharmacol.* 2004;44(suppl 1):S80–S84.
73. Cada DJ, Levie T, Baker DE. Ambrisentan. *Hosp Pharm.* 2007;42:1145–1154.
74. Girgis RE, Frost AE, Hill NS, et al. Selective endothelin A receptor antagonism with sitaxsentan for pulmonary arterial hypertension associated with connective tissue disease. *Ann Rheum Dis.* 2007;66:1467–1472.
75. Cozzi F, Montisci R, Marotta H, et al. Bosentan therapy of pulmonary arterial hypertension in connective tissue diseases. *Eur J Clin Invest.* 2006;36(suppl 3):49–53.
76. Denton CP, Pope JE, Peter HH, et al. Long-term effects of bosentan on quality of life, survival, safety and tolerability in pulmonary arterial hypertension related to connective tissue diseases. *Ann Rheum Dis.* 2007;67:1222–1228.
77. Gilead Sciences, Inc. Gilead press release 2007: Gilead announces new Letairis(TM) (ambrisentan) data for the treatment of patients with pulmonary arterial hypertension (WHO Group 1). <http://www.gilead.com/pr_1066465>; 2008. Accessed 07.02.08.
78. Seibold JR, Badesch D, Galie N, et al. Sitaxsentan, a selective endothelin-A receptor antagonist, improves exercise capacity in pulmonary arterial hypertension (PAH) associated with connective tissue disease (CTD). American Academy of Chest Physicians (ACCP); November 2, 2005; Seattle, Wash; Abstract.
79. Suntharalingam J, Hodgkins D, Cafferty FH, Hughes RJ, Pepke-Zaba J. Does rapid dose titration affect the hepatic safety profile of bosentan? *Vascul Pharmacol.* 2006;44:508-512.

Chapter 17
Disease Progression in Systemic Sclerosis Associated Pulmonary Arterial Hypertension

Robin Condliffe

17.1 Introduction

Systemic sclerosis associated pulmonary arterial hypertension (SSc-PAH) is an important cause of morbidity and mortality in patients with systemic sclerosis (SSc). Although studies based on echocardiography-derived pulmonary artery pressures have resulted in estimates of the prevalence of pulmonary arterial hypertension (PAH) in patients with SSc as high as 35%, more recent studies, using catheter-based diagnoses, have produced estimates of between 7.8 and 12%.[1-3] Historically, SSc-PAH has had a poor outlook with rapid clinical deterioration and a 3-year survival of 30%.[4] This was worse than in idiopathic pulmonary arterial hypertension (iPAH) wherein median survival prior to disease-modifying therapy was 2.8 years.[5]

The treatment of PAH has, however, changed significantly over recent years with three main groups of disease-modifying therapies becoming widely available in the "modern treatment era." Although initial therapeutic studies tended to include only patients with iPAH, patients with associated causes including connective tissue disease (CTD), congenital heart disease, and HIV have also been recruited later. The first reported use of prostacyclin (epoprostenol) in iPAH was in a single patient in 1984, but it was not until 1996 that a randomized controlled trial (RCT) showed it to be effective in improving pulmonary hemodynamics, exercise tolerance, and survival.[6,7] Subsequently, more stable prostacyclin analogs such as iloprost and treprostinil have also entered clinical practice. Endothelin-1 is a potent vasoconstrictor and inducer of smooth muscle proliferation that is overexpressed in patients with pulmonary hypertension.[8] RCTs have shown that the nonselective endothelin-1 receptor antagonist, bosentan, and the selective endothelin-1 receptor antagonist, sitaxsentan, are also effective in improving pulmonary hemodynamics

R. Condliffe (✉)
Pulmonary Vascular Disease Unit, Royal Hallamshire Hospital, Sheffield, UK
e-mail: robin.condliffe@sth.nhs.uk

D. Abraham et al. (eds.), *Advances in Vascular Medicine*,
DOI 10.1007/978-1-84882-637-3_17,

and exercise tolerance in patients with PAH of various etiologies.[9,10] The third class of disease-modifying therapies to have gained widespread use is phosphodiesterase-5 inhibitors, which prevent the breakdown of cyclic GMP, the secondary messenger of the nitric oxide pathway. The efficacy of an oral phosphodiesterase-5 inhibitor, sildenafil, was investigated in an RCT involving 278 patients with both idiopathic and associated PAH.[11] Again, significant improvements in pulmonary hemodynamics and exercise tolerance were observed.

From both clinical experience and from the early observational studies looking at survival, it is clear that in the past SSc-PAH had an exceedingly poor outcome, worse even than in iPAH. Is this still the case in the modern treatment era and if so, why? This chapter explores this question by first reviewing data from the pivotal studies supporting the use of the disease-modifying therapies in PAH, and then examining whether there is evidence of a different short-term response to therapy in patients with SSc-PAH when compared to those with iPAH. Data on survival in the modern treatment era will then be reviewed to investigate whether a difference in outcome over the longer term is still observed. Finally, if SSc-PAH still has a more aggressive disease progression despite the available modern therapies, what are the possible reasons for this?

17.2 Short-Term Response to Treatment

With one exception, all RCTs in the field of pulmonary hypertension have, to date, not been adequately powered to demonstrate a survival benefit.[7] Instead, studies have been designed to test surrogate end-points. These have included measures of exercise capacity (most commonly the 6 min walk test – 6MWT), clinical status (World Health Organization (WHO) functional class), and pulmonary hemodynamics. Does the available evidence using these surrogate markers support the notion that response to therapy is less marked in SSc-PAH when compared with iPAH? To help answer this, the pivotal trials involving a separate analysis of patients with these two conditions will be explored.

17.2.1 Endothelin-1 Receptor Antagonists

The unselective endothelin-1 receptor antagonist, bosentan, was studied in a double-blind, placebo-controlled RCT involving 213 patients with PAH.[9] After 16 weeks of treatment, the 6 min walk distance (6MWD) in 150 patients with iPAH increased by 46 m, while it decreased by 5 m in the placebo group. A different picture was seen in patients with SSc-PAH; among the 14 patients in the placebo group, there was a decline of 40 m, while in the 33 patients in the bosentan group, deterioration was prevented with an improvement of 3 m. This suggested that SSc-PAH was less responsive to treatment with bosentan and that in this group of patients, disease stability could be viewed as a positive outcome. The efficacy of the selective endothelin-1 receptor antagonist, sitaxsentan, in connective tissue disease associated pulmonary arterial hypertension (CTD-PAH) has also been reported in abstract form.[12] Forty-two patients with CTD-PAH who were treated with sitaxsentan during an 18-week RCT were followed for 1 year in an extension trial. At this point, the 6MWD had increased by 2 m. Unfortunately, no similar data for 6MWD in patients with iPAH treated with sitaxsentan is available for comparison. In an RCT of the effect of the addition of bosentan to intravenous epoprostenol, only nonsignificant improvements in pulmonary hemodynamics were seen in the 22 patients in the treatment arm.[13] It is noteworthy that in this study, 1 (9%) of the placebo/epoprostenol arm had SSc-PAH when compared with 4 (18%) in the bosentan/epoprostenol arm. One of the SSc-PAH patients in the treatment arm died and one was withdrawn due to worsening PAH. The authors postulated that this uneven mix of SSc-PAH may have reduced the treatment efficacy in the bosentan/epoprostenol arm.

17.2.2 Prostacyclin and Prostacyclin Analogs

An unblinded RCT of intravenous epoprostenol involving 111 patients with moderate to severe SSc-PAH was performed.[14] A significant improvement in the 6MWD of 46 m after 12 weeks of treatment was observed in the active arm, while a decrease of 48 m in the placebo arm was seen. Improvements were also noted in pulmonary hemodynamics (mean pulmonary arterial pressure decreased by 5.03 mmHg in the active group and increased by 0.94 mmHg in the placebo group, while cardiac index increased by 0.5 L/min/m^2 in the active group and decreased by 0.1 L/min/m^2 in the placebo group) and functional status (functional class improved in 38% of the active and 0% of the placebo group). It would seem that, in the short term at least, significant improvements in SSc-PAH treated with intravenous epoprostenol are possible. It must be noted that, unlike in the pivotal RCT of epoprostenol in iPAH (where a survival benefit was observed over 12 weeks), no survival benefit was seen in the equivalent SSc-PAH study.[7,14] A major drawback of epoprostenol therapy is the fact that because of its short half-life, it is administered through an indwelling central line. Patients with SSc are more prone to infections because of their underlying CTD and immunosuppressive therapy in selected cases. In the SSc-PAH study sepsis, cellulitis, hemorrhage, and pneumothorax occurred each in 4% of the epoprostenol group. A subsequent prospective study of 17 patients with CTD-PAH treated with intravenous epoprostenol investigated the longer term outcome.[15] Three (18%) patients died of sepsis due to skin infections, while the overall rate of catheter infection was 0.64/patient/year. The authors concluded that although the majority of patients had shown short-term improvement after 6 weeks of treatment, there were several cases of major complications. The long-term benefit of epoprostenol in patients with SSc-PAH was therefore not certain.

Beraprost is an oral prostacyclin analog, which showed promise on the basis of a 12-week RCT involving 130 patients with PAH.[16] In this study, patients with iPAH increased their 6MWD by 45 m, while no significant change was seen in those with associated PAH, 19% of which was due to CTD-PAH. This drug is not currently licensed in the US and Europe because a subsequent 12-month RCT failed to show a sustained benefit.[17]

Treprostinil has a longer half-life than epoprostenol and so can be administered via a continuous subcutaneous infusion, removing the risk of line infection. Ninety patients with CTD-PAH (including 45 patients with SSc-PAH) from the original RCT (which also included 270 patients with iPAH and 109 patients with congenital heart disease) were analyzed separately.[18,19] In the CTD-PAH group, the between treatment group difference in median 6MWD, unlike in the original study, did not reach statistical significance (+25 m, $p>0.06$). In the original study, significant improvements were seen in the secondary end-points of Borg dyspnoea score, mean right atrial pressure, mean pulmonary artery pressure, cardiac index, pulmonary vascular resistance, and mixed venous oxygen saturation. Of these indices, significant improvements in the CTD group were only seen for cardiac index and pulmonary vascular resistance.

Iloprost is a prostacyclin analog, which was shown to be effective when administered via a nebulizer in an RCT involving 101 patients, of whom 13% had CTD-PAH, in the active arm.[20] After 12 weeks of treatment, the 6MWT increased by 58.8 m in those with iPAH, and 12 m among those with associated PAH. A small prospective study of the use of nebulized iloprost in patients with CTD-PAH had previously been performed.[20] In this study of only five patients, the overall change in 6MWD after 6 months of treatment was +85 m ($p>0.06$). Because of the small numbers in this study, it is difficult to make any firm conclusions based on these data.

17.2.3 Phosphodiesterase-5 Inhibitors

A subgroup analysis of the 84 patients with CTD-PAH (including 50 patients with SSc, 19 with SLE, and 8 with mixed CTD) who were included in the SUPER-1 trial has been reported.[21] In the original RCT of 278 patients, in which patients received 20, 40, or 80 mg tds of sildenafil, the mean placebo-controlled treatment effects on the 6MWD were +45, +46, and +50 m after 12 weeks.[9] Direct comparison of the subgroup analysis with the original study is difficult, as patients with SLE may well have a greater response to treatment. However, the treatment effect in the subgroup analysis tended to be smaller with the 6MWD increasing by +42, +36, and +15 m at the three different doses of sildenafil.

Much of the data discussed so far has been based on changes in the 6MWD after treatment. Over recent years, problems with using the 6MWT as a primary endpoint in PAH have become apparent.[22] Therefore, before we examine whether there is also a poorer long-term outcome in SSc-PAH compared with iPAH, evidence for the use of the 6MWT will be discussed.

17.3 Six Minute Walk Test

We have seen that, apart from the RCT of epoprostenol,[14] improvements in the 6MWD have been modest in drug studies in CTD-PAH when compared to the changes seen in iPAH. End-points used in trials involving patients with different complications of SSc were previously systematically reviewed using the OMERACT filter.[23] Of the commonly used end-points in PAH, only pulmonary hemodynamics were found to be robust enough to be recommended for use in clinical trials. Subsequently, an expert panel on outcomes measure in PAH related to systemic sclerosis (EPOSS) has been convened to develop a consensus document on clinical end-points in SSc-PAH, using a Delphi exercise.[24] In this anonymous consultation, exercise experts from a wide range of geographical and speciality backgrounds first answer a questionnaire regarding possible outcome measures in SSc-PAH and then are able to respond to the summarized results from the whole group. The results of this process have recently been reported and have concluded that in SSc-PAH, the 6MWT is the only adequately validated measure.[25]

Miyamoto et al studied 43 patients with iPAH and found only a weak correlation between 6MWD and cardiac output ($r>0.48$, $p<0.05$) and total pulmonary resistance ($r>-0.49$, $p<0.05$), and no correlation with mean pulmonary arterial pressure ($r>-0.32$, $p>$not significant; Fig. 17.1).[26] The authors of this paper did, however, find that a baseline 6MWD of >332 m (the median distance walked by subjects) was predictive of better survival (Fig. 17.2).[26] However, doubts regarding the ability of the 6MWT to detect improvements following drug treatment have been raised. Although the absolute value of the 6MWD in 178 patients with iPAH after 3 months of epoprostenol (>380 m) was predictive of better outcome, the actual change in 6MWD was not (Fig. 17.3).[27] Furthermore, several recent RCTs involving patients with various causes of pulmonary hypertension have demonstrated hemodynamic improvements, which have not been reflected in increases in the 6MWD. For example, the EARLY study, which assessed the effect of treating patients with mild disease (WHO functional class II – 34 of whom had CTD-PAH) with bosentan demonstrated a reduction of almost a quarter in pulmonary vascular resistance ($p<0.0001$) but no significant change in 6MWD.[28]

Patients with CTD-PAH often have musculoskeletal problems, which may also affect exercise capacity. Hence, the 6MWD may be even less reliable in this group of patients. Outside of pharmaceutical studies, there are no published studies examining the role of the 6MWT in patients with CTD-PAH. If we are to correctly interpret the available literature regarding the efficacy of therapies in CTD-PAH, it

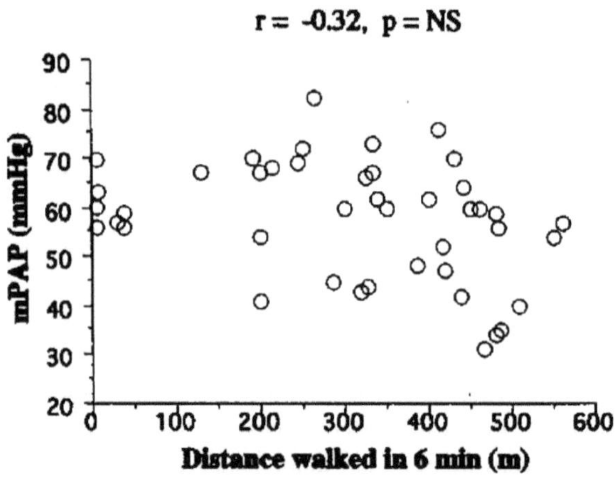

Fig. 17.1 Poor correlation between mPAP and 6 min walk test (6MWT) in idiopathic pulmonary hypertension[26]

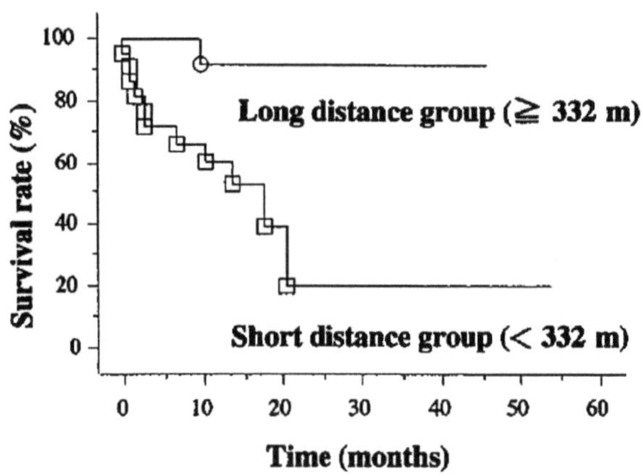

Fig. 17.2 Baseline 6MWT predicts outcome in idiopathic pulmonary arterial hypertension (iPAH)[26]

is important that this is rectified. Analysis of data from large cohorts of CTD-PAH patients, such as the United Kingdom national registry, may be helpful.[29] As of now, unpublished analysis from this cohort of 429 patients with CTD-PAH suggests that, as in iPAH, both baseline and absolute 6MWD at follow-up are indeed predictive of outcome. Further analysis of the predictive strength of the change in 6MWD is underway.

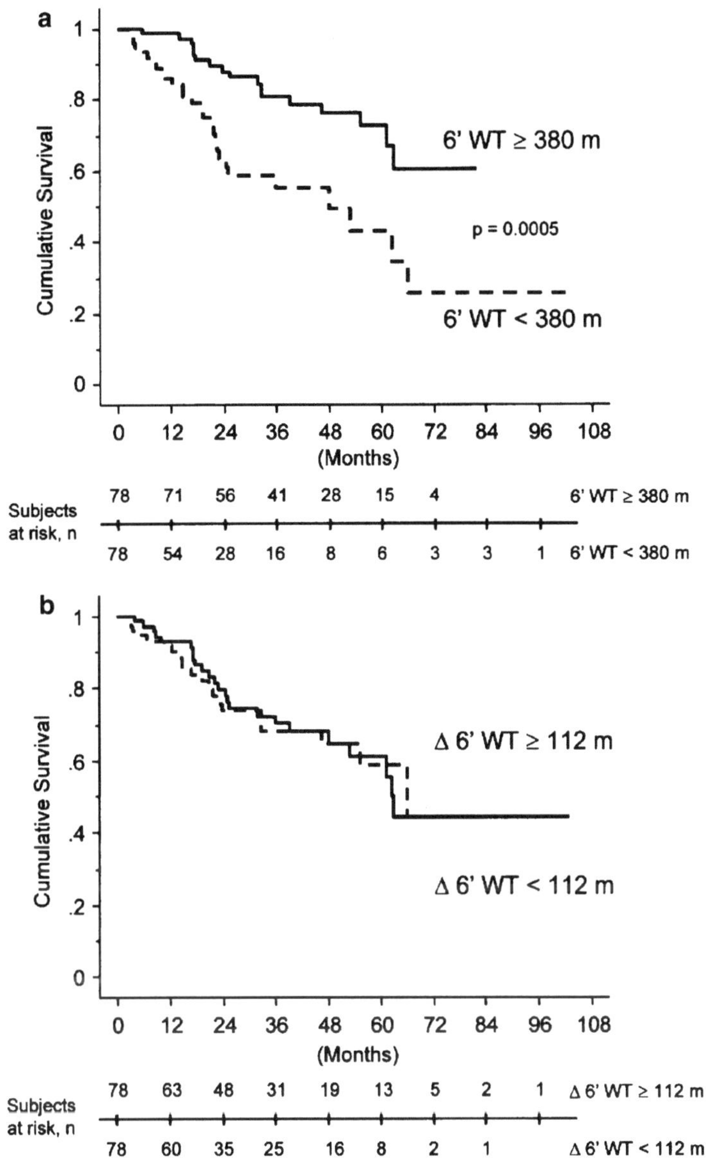

Fig. 17.3 Survival in iPAH patients treated with epoprostenol. (**a**) Absolute 6MWD after 3 months predicts outcome, whereas (**b**) Change from baseline does not[27]

17.4 Longer-Term Outcome

Is the poorer short-term response we have seen in SSc-PAH when compared with iPAH also reflected in a poorer longer-term outcome? As has already been discussed, SSc-PAH historically has had a poor long-term outcome with 1- and 3-year survival being as low as 45 and 30%.[4] Several studies have subsequently reported improved rates of survival in patients with SSc-PAH in the modern treatment era. In a group of 47 patients, all of whom were treated with bosentan, Williams et al found 1- and 2-year survival rates of 81 and 71% (Fig. 17.4).[30] Similar impressive results were seen in the open-label extension to the two pivotal bosentan studies in which comparable survival rates of 86 and 73% among the SSc-PAH group at 1 and 2 years were noted.[31] Both these studies, however, were relatively selective with patients excluded if they had significant underlying lung disease or severe pulmonary hemodynamics (cardiac index <2.1 L/min/m^2, right atrial pressure >11 mmHg, and mixed venous oxygen saturation <63% in the first study or a 6MWT of <150 m in the second study). Furthermore, prevalent cases were enrolled (i.e. patients were not newly diagnosed). This can bias survival statistics, as patients who have been stable enough to be included in a study are inherently more likely to have a better outcome than patients who are deteriorating rapidly at diagnosis. In an earlier study involving 89 incident (i.e. newly diagnosed) cases of SSc-PAH, approximately 50% of whom received treatment with a prostanoid, 81, 63, and 56% were alive at 1, 2, and 3 years following diagnosis.[3] Initial survival data from the United Kingdom national registry has recently been reported in abstract form. Overall 1- and 3-year

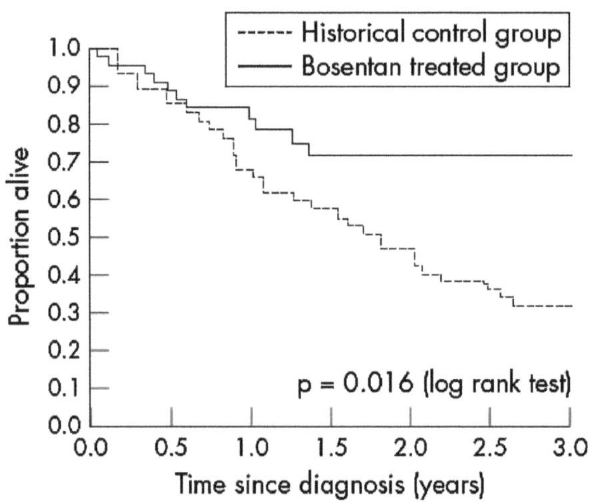

Fig. 17.4 Improved survival in systemic sclerosis (SSc) associated pulmonary arterial hypertension (SSc-PAH) in the endothelin receptor antagonist era[30]

survival from diagnosis for both isolated SSc-PAH and respiratory disease associated SSc-PAH was 75% and 45%.[29] Although these figures are slightly lower than in the more selected studies already discussed, we can see that, even allowing for different inclusion criteria, survival in SSc-PAH has improved in the modern treatment era. In the past, 45% of patients were alive at 1 year, whereas in the modern treatment era, this rate is now, at worst, the 3-year survival.

How do these survival rates compare with those observed in iPAH? One study has examined this directly. Kawut et al prospectively followed 33 patients with iPAH and 22 with SSc-PAH (Fig. 17.5).[32] There was no significant difference between the groups in terms of baseline hemodynamics (although a nonsignificant trend toward milder mean pulmonary arterial pressure, cardiac index, and pulmonary vascular resistance was seen in the SSc-PAH group) or proportion receiving intravenous epoprostenol (around 70% in both groups). Despite this, 2-year survival was 70% in the idiopathic group when compared with 45% in the SSc-PAH group ($p > 0.03$). In an open-label extension study of patients recruited in the two initial RCTs of bosentan, survival over the first 2 years in the 169 patients with iPAH was 96 and 89%, while in the 50 patients with SSc-PAH the authors found survival to be lower at 82 and 67%.[33] Two separate large studies have examined long-term outcome in iPAH patients treated with epoprostenol. In 178 patients, Sitbon et al found survival from commencement of therapy to be 85 and 63%, while in 162 patients, McLaughlin et al found 1- and 3-year survival to be 88 and 63%.[27,34] In the small long-term study of epoprostenol in SSc-PAH, only 10 (59%) patients were alive after an average of 1.5 years.[15] Registry data involving incident cases of iPAH is rather lacking. The French national registry did, however, include 56 incident cases of iPAH of whom 89% were still alive 1 year after diagnosis.[35]

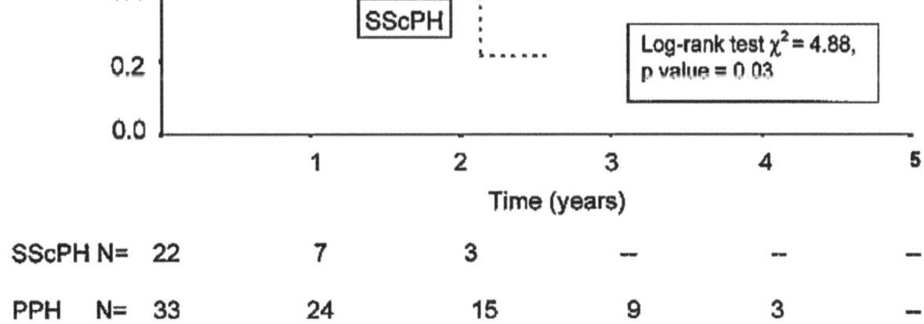

Fig. 17.5 Survival is worse in patients with SSc-PAH when compared with iPAH[32]

17.5 Underlying Reasons for the Poorer Prognosis in SSc-PAH

From the data reviewed here, we can conclude that although the outlook has improved during the modern treatment era, SSc-PAH patients are less responsive to treatment than iPAH patients, and also have a poorer longer-term outcome. Why might this be? SSc is a multisystem disease affecting the musculoskeletal system, heart, and lung parenchyma in addition to the pulmonary vasculature. It is therefore possible that coexisting pulmonary or cardiac disease may explain the worse prognosis.

Studies involving patients with SSc-PAH have generally excluded patients with significant pulmonary fibrosis. Exclusion criteria of forced vital capacity (FVC) or total lung capacity of less than 60–70% predicted have been used. It is apparent that minor degrees of pulmonary fibrosis are exceedingly common in SSc, occurring in up to 80% of patients,[36] and it is feasible that this may progress leading to a poorer outcome. However, several studies have demonstrated that patients at greatest risk of developing severe, end-stage, lung disease are those who develop significantly reduced FVC during the first few years after onset of SSc.[37-39] Morgan et al found that an FVC<80% predicted within the first 5 years after diagnosis of SSc was predictive of the subsequent development of significant pulmonary fibrosis.[38] Steen et al found that 62% of those who ever developed severe fibrosis (FVC<55%) did so in the first 5 years after diagnosis.[39] This makes the development of severe pulmonary fibrosis unlikely in patients who have SSc and who have adequate lung volumes when PAH is diagnosed. However, some SSc-PAH patients may still be at risk of severe fibrosis, especially if their FVC is <80% predicted or if they develop PAH within 5 years of diagnosis of SSc. Furthermore, even if severe pulmonary fibrosis does not develop, it is conceivable that a slower deterioration in parenchymal disease may negate some of the benefits of the PAH treatment. It does appear relatively unlikely that, in patients with adequate baseline lung function, parenchymal disease is alone responsible for the poorer prognosis and response to treatment seen in PAH.

What, then, is the role of myocardial involvement? Scleroderma heart disease has historically been grouped into primary or secondary disease.[40] Primary disease is due to myocardial, pericardial, or intramyocardial vessel involvement by SSc itself, while secondary disease occurs as a consequence of systemic hypertension or pulmonary vascular disease. Myocardial fibrosis unrelated to ischemia has been demonstrated in both ventricles in 70% of SSc patients at autopsy.[41] Using the technique of videodensitometry, where the echo reflectiveness of the myocardium is assessed using the posterior pericardium as a reference, Di Bello et al detected a pattern consistent with myocardial fibrosis in 90%.[42] It would therefore appear that myocardial fibrosis is common in SSc. What is the likely cause for this myocardial fibrosis and what functional impact does it have – could it be responsible for the poorer outcome we have observed?

Using tissue Doppler echocardiography (TDE) in 17 patients with SSc of less than 5 years duration, who had normal pulmonary arterial pressures and radionuclide left ventricular ejection fractions, Meune et al found that left ventricular strain-rates (a measure of regional myocardial velocity gradients) were significantly reduced in 11 patients during diastole and 10 during systole.[43] An earlier study by the same group using radionuclide ventriculography had demonstrated evidence of right ventricular systolic dysfunction in 42 patients with SSc and normal pulmonary artery pressures when assessed by standard echocardiography.[44] Of note, in both these studies, the authors did not find a relationship between strain-rate and pulmonary arterial pressures or pulmonary involvement. The authors therefore suggest that the observed abnormalities in cardiac function were due to primary myocardial involvement. Other studies, however, have found correlations between diastolic dysfunction and pulmonary pressures. Using standard Doppler echocardiography, Giunta et al found a reduced tricuspid early to late filling (E/A) ratio in 40% of 77 unselected SSc patients and 0% of 33 healthy controls.[45] The E/A ratio was found to be independently correlated to pulmonary arterial pressure and mitral valve E/A ratio. In a more recent study using TDE, Lindqvist et al compared 21 SSc patients with 21 controls.[46] Although left ventricular function, right ventricular systolic function, and pulmonary pressures were similar between the two groups, right ventricular diastolic function was significantly reduced in the SSc patients. Interestingly, pulmonary acceleration time and right ventricular thickness, which are indirect markers of pulmonary pressures, were both higher in the SSc patients. The authors suggest that the observed diastolic dysfunction could therefore be related to intermittent elevations in pulmonary pressures. A recent TDE study has explored the interaction between the pulmonary vasculature and diastolic dysfunction further.[47] There was no difference in mean systolic pulmonary arterial pressure when assessed by echocardiography between 25 SSc patients and 13 healthy controls. At rest, the tricuspid E/A ratio was reduced, pulmonary acceleration time increased, and right ventricular wall thickness increased in the SSc patients, when compared with the controls. There was no abnormality in the mitral E/A ratio or left ventricular systolic function, suggesting that left ventricular dysfunction was not present. On exercise, the majority of SSc patients developed an increase in dynamic pulmonary vascular resistance. The authors concluded that isolated right ventricular diastolic dysfunction is likely to be secondary to latent pulmonary hypertension.

From the above data, we can see that myocardial fibrosis and associated functional abnormalities are common in SSc patients. Different studies have found both diastolic and systolic dysfunction occurring in both ventricles in the absence of overt pulmonary hypertension. If these pathological and functional abnormalities seen in a general SSc population were of a great clinical significance, then one would expect survival in an unselected SSc population, in the absence of PAH, to be affected. A retrospective Greek study of 254 patients with SSc found an exceedingly high 4-year survival of 94%, suggesting that many of these abnormalities are not highly clinically relevant.[48] Several groups, however, have suggested that isolated right ventricular dysfunction is commonly seen and there is evidence that this may occur due to latent or exercise-induced pulmonary hypertension.[45-47] If this is true, then it would suggest that the predominant reason for the poorer outcome we have observed in SSc-PAH is less likely to be intrinsic myocardial involvement, but is more likely to be related to more aggressive progression of the underlying pulmonary vascular disease. It is interesting to note that a different pattern of intimal thickening has been noted by some authors in patients with CTD-PAH.[49] In these patients, circumferential, almost acellular intimal thickening, has been noted compared with the pattern of localized, eccentric thickening seen in other forms of PAH. Although there may be a difference in the natural history of the pulmonary vasculopathy in patients with CTD-PAH, it is important that reversible causes of potential diastolic dysfunction in these patients, such as systemic hypertension or ischemia, should still be appropriately diagnosed and treated.

An important factor, which has not been discussed so far, is age. SSc-PAH tends to present later in life than iPAH. In the therapeutic trials discussed above, the age of patients in both the main studies and the CTD-PAH subgroups were available for several studies.[7,9,14,18,19,31] Patients in the CTD-PAH subgroups were at least 10 years older than in the accompanying main study (range 53–58 years vs. 40–49 years), while in the previously discussed SSc-PAH registry, the mean age was 59.[30] Is it possible that the older age of SSc-PAH patients may explain at least some of the differences in outcomes that have been observed? Kawut et al found that in univariate analysis, age was a predictor of survival in patients with SSc-PAH, but not in patients with iPAH.[32,50] The older the patient, the more likely they are to have other significant comorbidities, so given the relatively young age of iPAH patients (mean of 42 years in Kawut's study[50]), it is perhaps not surprising that the effects of age are more pronounced in SSc-PAH patients. Having said that although SSc-PAH patients are on average older than iPAH patients, they are by no means extremely elderly. Shapiro et al found that mortality in patients with iPAH was increased in the elderly.[51] However, they defined elderly as ≥65 years in which case a significant number of SSc-PAH patients would still have been classed as being young. On balance, although it is possible that age may contribute to the poorer outcome that has been observed in SSc-PAH, it is very likely that other factors are involved. As has just been discussed, more aggressive pulmonary vascular disease progression in SSc-PAH is likely to be significant.

17.6 Conclusion

In patients with SSc-PAH, short-term response to disease-modifying therapy, as measured most commonly by the 6MWD, is less marked than in iPAH. As is the case in other forms of PAH, there are concerns regarding the robustness of the 6MWD as an end-point in patients with CTD-PAH and this requires further investigation. When the longer-term outcome in patients with SSc-PAH is examined, it is apparent that although survival has improved in the modern treatment, it is still probably worse than in patients with iPAH. Although coexisting direct pulmonary or myocardial involvement by SSc may be associated with some of this poorer outcome in SSc-PAH, there is no clear evidence that this is the case. Instead, more aggressive pulmonary vascular disease in SSc-PAH may be the culprit. If this is the case, then patients with SSc-PAH would be ideal candidates for early and aggressive introduction of combination disease-modifying therapies.

References

1. Battle R, Davitt M, Cooper SM, et al. Prevalence of pulmonary hypertension in limited and diffuse scleroderma. *Chest.* 1996;110:1515-1519.
2. Hachulla E, Gressin V, Guillevin L, et al. Early detection of pulmonary arterial hypertension in systemic sclerosis: a French nationwide prospective multicenter study. *Arthritis Rheum.* 2005;52:3792-3800.
3. Mukerjee D, St George D, Coleiro B, et al. Prevalence and outcome in systemic sclerosis associated pulmonary arterial hypertension: application of a registry approach. *Ann Rheum Dis.* 2003;62:1088-1093.
4. Koh ET, Lee P, Gladman DD, et al. Pulmonary hypertension in systemic sclerosis: an analysis of 17 patients. *Br J Rheumatol.* 1996;35:989-993.
5. D'Alonzo GE, Barst RJ, Ayres SM, et al. Survival in patients with primary pulmonary hypertension Results from a national prospective registry. *Ann Intern Med.* 1991;115:343-349.
6. Higenbottam T, Wheeldon D, Wells F, et al. Long-term treatment of primary pulmonary hypertension with continuous intravenous epoprostenol (prostacyclin). *Lancet.* 1984;1:1046-1047.
7. Barst RJ, Rubin LJ, Long WA, et al. A comparison of continuous intravenous epoprostenol (prostacyclin) with conventional therapy for primary pulmonary hypertension. The primary pulmonary hypertension study group. *N Engl J Med.* 1996;334:296-302.
8. Giaid A, Yanagisawa M, Langleben D, et al. Expression of endothelin-1 in the lungs of patients with primary pulmonary hypertension. *N Engl J Med.* 1993;328:1732-1739.
9. Rubin LJ, Badesch DB, Barst RJ, et al. Bosentan therapy for pulmonary arterial hypertension. *N Engl J Med.* 2002;346:896-903.
10. Barst RJ, Langleben D, Frost A, et al. Sitaxsentan therapy for pulmonary arterial hypertension. *Am J Respir Crit Care Med.* 2004;169:441-447.
11. Galie N, Ghofrani HA, Torbicki A, et al. Sildenafil citrate therapy for pulmonary arterial hypertension. *N Engl J Med.* 2005;353:2148-2157.
12. Highland KB, Strange C, Girgis R, et al. Comparison of sitaxentan and bosentan in pulmonary arterial hypertension associated with connective tissue diseases [abstract]. *Ann Rheum Dis.* 2006;65:A393.
13. Humbert M, Barst RJ, Robbins IM, et al. Combination of bosentan with epoprostenol in pulmonary arterial hypertension: BREATHE-2. *Eur Respir J.* 2004;24:353-359.
14. Badesch DB, Tapson VF, McGoon MD, et al. Continuous intravenous epoprostenol for pulmonary hypertension due to the scleroderma spectrum of disease. A randomized, controlled trial. *Ann Intern Med.* 2000;132:425-434.
15. Humbert M, Sanchez O, Fartoukh M, et al. Short-term and long-term epoprostenol (prostacyclin) therapy in pulmonary hypertension secondary to connective tissue diseases: results of a pilot study. *Eur Respir J.* 1999;13:1351-1356.
16. Galie N, Humbert M, Vachiery JL, et al. Effects of beraprost sodium, an oral prostacyclin analogue, in patients with pulmonary arterial hypertension: a randomized, double-blind, placebo-controlled trial. *J Am Coll Cardiol.* 2002;39:1496-1502.
17. Barst RJ, McGoon M, McLaughlin V, et al. Beraprost therapy for pulmonary arterial hypertension. *J Am Coll Cardiol.* 2003;41:2119-2125.
18. Simonneau G, Barst RJ, Galie N, et al. Continuous subcutaneous infusion of treprostinil, a prostacyclin analogue, in patients with pulmonary arterial hypertension: a double-blind, randomized, placebo-controlled trial. *Am J Respir Crit Care Med.* 2002;165:800-804.
19. Oudiz RJ, Schilz RJ, Barst RJ, et al. Treprostinil, a prostacyclin analogue, in pulmonary arterial hypertension associated with connective tissue disease. *Chest.* 2004;126:420-427.

20. Launay D, Hachulla E, Hatron PY, et al. Aerosolized iloprost in CREST syndrome related pulmonary hypertension. *J Rheumatol.* 2001;28:2252-2256.
21. Badesch D, Burgess G, Parpia T, et al. Sildenafil improves exercise ability and hemodynamics in patients with pulmonary arterial hypertension associated with connective tissue disease. *J Rheumatol.* 2007;34:2417-2422.
22. Rich S. The current treatment of pulmonary arterial hypertension: time to redefine success. *Chest.* 2006;130:1198-1202.
23. Merkel P, Clements PJ, Reveille JD, et al. Current status of outcome measure development for clinical trials in systemic sclerosis. Report from OMERACT 6. *J Rheumatol.* 2003;30:1630-1647.
24. Distler O, Behrens F, Huscher D, et al. Need for improved outcome measures in pulmonary arterial hypertension related to systemic sclerosis. *Rheumatology.* 2006;45:1455-1457.
25. Furst D. Measuring outcome in PAH: the gap between the measures that are used and their validity. *Ann N Y Acad Sci.* 2007;1107:410-416.
26. Miyamoto S, Nagaya N, Satoh T, et al. Clinical correlates and prognostic significance of six-minute walk test in patients with primary pulmonary hypertension Comparison with cardio-pulmonary exercise testing. *Am J Respir Crit Care Med.* 2000;161:487-492.
27. Sitbon O, Humbert M, Nunes H, et al. Long-term intravenous epoprostenol infusion in primary pulmonary hypertension: prognostic factors and survival. *J Am Coll Cardiol.* 2002;40:780-788.
28. Galie N, Rubin LJ, Hoeper MM, et al. Bosentan improves hemodynamics and delays time to clinical worsening in patients with mildly symptomatic Pulmonary Arterial Hypertension (PAH): results of the EARLY study [abstract]. *Eur Heart J.* 2007;28:A140.
29. Condliffe R, Kiely D, Peacock A, et al. Initial results of the United Kingdom audit of pulmonary arterial hypertension associated with connective tissue disease [abstract]. *Eur Heart J.* 2007;28:1010S.
30. Williams MH, Das C, Handler CE, et al. Systemic sclerosis associated pulmonary hypertension: improved survival in the current era. *Heart.* 2006;92:926-932.
31. Denton CP, Humbert M, Rubin L, et al. Bosentan treatment for pulmonary arterial hypertension related to connective tissue disease: a subgroup analysis of the pivotal clinical trials and their open-label extensions. *Ann Rheum Dis.* 2006;65:1336-1340.
32. Kawut SM, Taichman DB, Archer CCL, et al. Hemodynamics and survival in patients with pulmonary arterial hypertension related to systemic sclerosis. *Chest.* 2003;123:344-350.
33. McLaughlin VV. Survival in patients with pulmonary arterial hypertension treated with first-line bosentan. *Eur J Clin Invest.* 2006;36:10-15.
34. McLaughlin VV, Shillington A, Rich S. Survival in primary pulmonary hypertension: the impact of epoprostenol therapy. *Circulation.* 2002;106:1477-1482.
35. Humbert M, Sitbon O, Chaouat A, et al. Pulmonary arterial hypertension in France: results from a national registry. *Am J Respir Crit Care Med.* 2006;173:1023-1030.
36. Scully R, Mark E, McNeely W, et al. Case records of the Massachusetts general hospital. *N Engl J Med.* 1989;320:1333-1340.
37. Steen V, Conte C, Owens GR, et al. Severe restrictive lung disease in systemic sclerosis. *Arthritis Rheum.* 1994;37:1283-1289.
38. Morgan C, Knight C, Lunt M, et al. Predictors of end stage lung disease in a cohort of patients with scleroderma. *Ann Rheum Dis.* 2003;62:146-150.
39. Steen V, Medsger TA Jr. Severe organ involvement in systemic sclerosis with diffuse scleroderma. *Arthritis Rheum.* 2000;43:2437-2444.
40. Owens GR, Follansbee W. Cardiopulmonary manifestations of systemic sclerosis. *Chest.* 1987;91:118-127.
41. Follansbee W, Miller T, Curtiss E, et al. A controlled clinicopathologic study of myocardial fibrosis in systemic sclerosis (scleroderma). *J Rheumatol.* 1990;17:656-662.
42. Di Bello V, Ferri C, Giorgi D, et al. Ultrasonic videodensitometric analysis in scleroderma heart disease. *Coron Artery Dis.* 1999;10:103-110.
43. Meune C, Allanore Y, Pascal O, et al. Myocardial contractility is affected early in systemic sclerosis: A tissue Doppler echocardiography study. *Eur J Echocardiogr.* 2005;6:351-357.

44. Meune C, Allanore Y, Devaux J, et al. High prevalence of right ventricular systolic dysfunction in early systemic sclerosis. *J Rheumatol.* 2004;31:1941-1945.
45. Giunta A, Tirri E, Maione S, et al. Right ventricular diastolic abnormalities in systemic sclerosis. Relation to left ventricular involvement and pulmonary hypertension. *Ann Rheum Dis.* 2000;59:94-98.
46. Lindqvist P, Caidahl K, Neuman-Andersen G, et al. Disturbed right ventricular diastolic function in patients with systemic sclerosis. *Chest.* 2005;128:755-763.
47. Huez S, Roufosse F, Vachiery JL, et al. Isolated right ventricular dysfunction in systemic sclerosis: latent pulmonary hypertension? *Eur Respir J.* 2007;30:928-936.
48. Vlachoyiannopoulus P, Dafni U, Pakas I, et al. Systemic scleroderma in Greece: low mortality and strong linkage with HLA-DRB1*1104 allele. *Ann Rheum Dis.* 2000;59:359-367.
49. Tuder RM, Zaiman A. Pathology of pulmonary vascular disease. In: Peacock AJ, Rubin LJ, eds. *Pulmonary Circulation.* 2nd ed. London: Arnold; 2004:25-32.
50. Kawut SM, Horn E, Berekashvili K, et al. New predictors of outcome in idiopathic pulmonary arterial hypertension. *Am J Cardiol.* 2005;95:199-203.
51. Shapiro B, McGoon M, Redfield M. Unexplained pulmonary hypertension in elderly patients. *Chest.* 2007;131:94-100.

Chapter 18
Connective Tissue Disease Associated Pulmonary Arterial Hypertension: Special Cases?

Gerry Coghlan

18.1 Introduction

Is connective tissue disease (CTD) associated pulmonary arterial hypertension (CTPAH) a distinct entity or group of conditions, or merely a form of idiopathic pulmonary arterial hypertension (iPAH) perhaps triggered by the immunological perturbations seen in these conditions? In arguing for causality as is recognized for HIV-associated pulmonary arterial hypertension (PAH), fulfilling Hill's criteria for causation is quite complex,[1] especially because failure to meet any of the criteria does not preclude a cause–effect relationship.

What is clear is that the prevalence of PAH in patients with CTD is vastly higher than what could occur by chance: the CTD almost invariably precedes the development of PAH, the response to therapy is less favorable, the genetics are different, vasodilator reserve is less common and does not appear to predict prognosis and occasionally when there is a significant inflammatory component, and treatment of the CTD improves the PAH.

Hill's evidence for causation

1. Strength of association: Strong associations are more likely to be causal.
2. Consistency of association: Similar results emerge from several studies in different populations.
3. Specificity of the association: A single cause produces a single effect
4. Temporality: The cause must precede the effect.
5. Biological gradient: Increasing dose should lead to increasing disease frequency.
6. Plausibility: There should be some accepted reason for the cause to produce the effect.
7. Coherence: The association should not conflict with current knowledge about the disease.

G. Coghlan (✉)
The Royal Free and University College Medical School,
Gower Street, WC1E 6BT, London, UK
e-mail: Gerry.coghlan@royalfree.nhs.uk

D. Abraham et al. (eds.), *Advances in Vascular Medicine*,
DOI 10.1007/978-1-84882-637-3_18,

8. Experimental evidence: Introduction or removal of an agent should lead to a change in the effect.
9. Analogy: If a condition is already known to produce an effect, it is more likely that similar conditions produce the same effect.

18.2 Hill's First and Second Criteria

The incidence and prevalence of PAH in various forms of CTD are as yet incompletely resolved. Estimates of the prevalence of SScPAH vary from 2.6 to 60%,[2,3] and even in the last couple of years, estimates of clinical prevalence vary from 8 to 30%.[4,5] Rheumatoid arthritis (RA) is one of the commonest CTD populations affecting 5% of patients over 65 years of age. The prevalence of PAH in this population is less well-studied; the only sizable study ($N=146$)[6] suggests that the prevalence is very low (0.7% based on estimated PASP of 40 mmHg); however, significantly higher prevalences are suggested based on less certain criteria. Thus, in the same study,[6] Dawson et al suggested a prevalence of 20% using 30 mmHg as the diagnostic criterion, while Marasini et al also find a prevalence of 20% in 25 patients using an estimated PAP of 35, which relies on adding various possible right atrial pressures (5–15 mmHg) to the observed tricuspid gradient.[7] The only large-scale study to include dermatomyositis/polymyositis showed a prevalence of 0.56%.[2]

An overview of the epidemiological Studies of PAH in CTDs (Tables 18.1–18.3) shows that for CTDs associated with frequent organ-based complications, the prevalence is 0.5% or above in all studies in all populations. The prevalence in some conditions (systemic sclerosis (SSc) and mixed connective tissue disease (MCTD)) is much higher, but in any event, the prevalence is a minimum of 100-fold more common than in the normal population.

Autopsy evidence suggests an even higher presence of PAH than detected clinically.[8] Thus, neither all pulmonary vasculopathy leads to clinical disease nor is

Table 18.1 Systemic sclerosis (SSc)

Author	Year	Population size	Diagnostic criteria	Prevalence (%)
Mukerjee et al[9]	2003	722	Cardiac catheter	12
Yoshida[2]	2001	3,778	Echo or catheter findings	2.64
Wigley[6]	2004	586 (prospective cohort)	eRVSP>40 mmHg	13
			eRSVP>30 mmHg	42
Humbert[5]	2004	617	Cardiac catheter	10
Vlachoyiannopoulos[14]	2000	201	Catheter or echo	2.5
Battle[3]	1996	17	Catheter (PVR-based)	65
Marasini[7]	2005	51	ePASP>35 mmHg	20

Table 18.2 Systemic lupus erythematosus (SLE)

Author	Year	Population size	Diagnostic criteria	Prevalence (%)
Tanaka[61]	2002	194	ePASP	6
Pan[62]	2000	786	ePASP	4.3
Winslow[34]	1995	28	ePAP>30 mmHg	43
Li[24]	1999	419	ePASP>30 mmHg	4
Yoshida[2]	2001	9,015	Echo or catheter findings	0.9
Marasini[7]	2005	33	ePASP>35 mmHg	6.1

Table 18.3 Mixed connective tissue disease (MCTD)

Author	Year	Population size	Diagnostic criteria	Prevalence (%)
Yoshida[2]	2001	1,651	Echo or catheter	5
Michels[63]	1997	224	Various	1
Alpert[64]	1983	17	Catheter (PVR only)	64
Marasini[7]	2005	4	ePASP>35 mmHg	75
Wigley[6]	2004	89 (prospective cohort)	ePASP>40 mmHg	7

there a prolonged phase of progressive vascular damage, which would eventually lead to PAH if patients with SSc lived long enough.

It is therefore easy to show that these first two criteria are met for SSc and MCTD, and with reasonable certainty that this is also true for systemic lupus erythematosus (SLE) and Dermatomyositis in so far as a strong association has been shown in several populations by different researchers.

18.3 Hill's Third Criterion

In scleroderma, we see different forms of pulmonary hypertension, including fibrosis-associated pulmonary hypertension and post-capillary pulmonary hypertension as well as PAH.[9] The fact that different conditions can have the same cause does not negate the argument, as we see with smoking-associated lung cancer, emphysema, and obstructive airways disease.

Where tissue is available, medial hypertrophy, "onion bulb" lesions, and plexiform lesions are consistently demonstrated in the pulmonary vessels of patients with SScPAH[8,10]; thus, the pathological changes are specific and similar to those seen in iPAH. Further, perivascular inflammatory infiltrates are noted in involved vessels only[11] as with iPAH.[12] These findings confirm that SScPAH is a specific histological entity, similar, though not necessarily identical, to iPAH. Extension of this argument to other forms of CTD presently can only be managed by analogy (Hills' ninth criterion).

18.4 Hill's Fourth Criterion

PAH in limited scleroderma typically presents later in the course of SSc (Avg 6–14 years after diagnosis), when other symptoms such as Raynaud's phenomenon, are well established.[9,13]

While there are little data on the natural history of PAH associated with other CTDs, there are no reports of the development of CTD in patients previously classified as iPAH, though the underlying CTD is occasionally recognized after PAH has been diagnosed.

The criterion of temporality is therefore met, certainly by SSc and probably by other forms of CTD PAH.

18.5 Hill's Fifth Criterion

The dose–response criterion is not regarded as pivotal, and would indeed be difficult to demonstrate, until the precise aspect of disease activity that causes PAH is known. It can be argued that since PAH tends to present later in the course of the SSc, there is a dose–response in terms of duration of stimulus exposure and PAH development. One may also infer that the higher prevalence seen in the sickest patients (those cared for in tertiary centers[5,9]) when compared to population studies[2,14] supports this concept.

Another consistent observation has been the greater rate of progression of SScPAH when compared to iPAH. The higher mortality associated with this condition[15-17] has been underpinned by observations from the pivotal trials[18-23] and registry data[24] that patients with SScPAH in placebo/historical arms show a higher rate of rise of pulmonary pressures and a greater rate of clinical deterioration than what is observed in iPAH. This may suggest that drivers for pulmonary vasculopathy are present to a greater degree in patients with SScPAH.

Specific immunological perturbations associated with SScPAH (anticentromere antibody, and anti-U3-RNP antibodies) are believed not to have a pathogenic role.[25] However, antibodies seen in several autoimmune conditions and iPAH (anti-endothelial antibodies, which may include ACA) have been reported to be toxic to endothelial cells,[26] and it is conceivable that further work may show a dose–response relationship. However, it is equally possible that no single immunological perturbation is responsible for the development of PAH in a proportion of patients with CTD and that until the environmental triggers or genetic predispositions that are pivotal to the development of PAH in individuals with CTD are understood, the demonstration of a dose–response will remain elusive.

18.6 Hill's Sixth Criterion

In terms of PAH, several mechanisms have been proposed that could plausibly lead to the changes in the pulmonary vasculature observed in CTD PAH.

18.6.1 Immune Dysregulation and Pulmonary Arterial Hypertension

Dorfmuller et al and Nicholls et al have recently reviewed the evidence supporting this contention.[27,28] The increased prevalence of PAH in other conditions associated with immune dysfunction such as HIV,[29] and POEMS[30] supports this concept. In addition, patients with iPAH have been shown to have an increased prevalence of autoimmune thyroid disease,[31] antinuclear antibodies,[32] and increased levels of IL-1 and IL-6.[33] Finally, substantial reductions of pulmonary pressures in patients with SLE, MCTD, and POEMS syndrome have been observed in a proportion of patients treated with immune modulating therapy.[34-37]

The precise role of the immune system in the pathobiology of PAH remains unclear. It appears from the monocrotaline model of pulmonary hypertension that in extreme circumstances intense inflammatory stimuli can lead to proliferative vasculopathy.[38] However, no single antibody or inflammatory mediator appears to be sufficient to cause pulmonary hypertension, though clear associations have been identified.[39-45] It has been proposed that defective immune regulation due to CD4 deficiency, or a relative imbalance of CD4/CD8 or CD4+/CD25+ lead to autoimmune phenomena (as seen in HIV, HHV-8, and CTDs) and in the presence of other triggers leading to pulmonary endothelial cell damage triggers an autoimmune reaction against these cells and thus to the development of pulmonary hypertension.[28]

The active drivers appear to be somewhat different in SScPAH when compared to SLE and other CTD-associated PAH. It is particularly noteworthy that SSc, a primarily vascular disorder, associated with an overexuberant tissue injury response, does not exhibit a response to immunosuppressive therapy when administered for PAH[37] – suggesting that the response to damage may be of equal significance to the initiator, in determining who ultimately develops PAH and who does not.

18.6.2 Transforming Growth Factor Beta Superfamily in Pathogenesis

The absence of an association between BMPR II mutations and SScPAH is intriguing,[46] if CTDs simply reflected conditions in which a normal tendency to develop PAH was facilitated, one would anticipate a stronger correlation between the presence of the BMPR2 mutations and CTD PAH than is present in the normal population. However, dysfunction of the TGFβ superfamily pathways (which includes BMPR2) is probably pivotal in the pathogenesis of SSc[47] and possibly the associated PAH. To date, evidence of dysregulation of this pathway has been sought in fibroblasts and endothelial cells. There is substantial evidence that TGFbeta signaling is perturbed in SSc, but data about the precise nature of the abnormalities are conflicting. Studies have demonstrated upregulation of TGFβ1 and TGFβ2 mRNA and protein in SSc skin and lung, and skin and lung fibroblast genetic analysis supports a role for TGFbeta.[47] Upregulation of TGFbeta receptors and altered expression of downstream Smad signaling proteins is described. The inhibitory effects of Smad7 may be reduced in SSc fibroblasts[48] leading to increased collagen type 1 production,[49] and these effects can be inhibited by increasing available Smad7 or endoglin.[48,50] The precise nature of the abnormalities remains unclear as others have reported elevated levels of Smad7 in fibroblasts cultured from active SSc skin lesions.[51] The recent discovery that similar pathways are critical to BMPR2-mediated pulmonary hypertension, certainly increases the plausibility of the argument that TGFβ1 contributes to the causation of pulmonary hypertension in patient with SSc.

18.6.3 Endothelin Overproduction

RANTES have been shown to be increased in the lungs of patients with iPAH, and have been shown to increase endothelin production.[52] In PAH, endothelin may be a key pathogenic mediator, influencing vasoconstriction, fibrosis, vascular hypertrophy, and inflammation.[53] Abnormalities of endothelin homeostasis are widespread in CTD, and thus provide a potential further link between CTD-PAH and the pathogenesis of PAH.[54]

Endothelial cell activation in some collagen vascular diseases, causing dysfunctional endothelin activity is now recognized as an important contributor to this disease area.[54,55] Both clinical and preclinical investigations have reported elevated endothelin levels in primary and secondary Raynaud's phenomenon, SLE, and other collagen vascular diseases.[54]

Abnormalities of circulating endothelin levels have been demonstrated in SSc.[55] Endothelin has been implicated in vasoconstrictor and profibrotic activity and the increased extracellular matrix substances seen in the dermis and internal organs of scleroderma patients.[54,56,57] Endothelial cell damage leading to increased endothelin production may play an important role in the early-stage disease, underpinning such manifestations as Raynaud's phenomenon, and chronic elevation of endothelin levels may play an important role in later-stage organ fibrosis. Endothelin levels are elevated in lung tissue of scleroderma patients with pulmonary disease, including PAH, and the correlation between endothelin levels and the severity and prognosis of PAH support the evidence surrounding endothelin dysfunction.[54,57,58]

At least in the context of SSc , it is reasonable to propose that the biological plausibility criterion is met.

18.7 Hill's Seventh Criterion

In terms of coherence, the same arguments for causation that are put forward for iPAH, especially in terms of endothelin overproduction and the intracellular mechanisms influenced by BMPR2 mutations apply equally to SSc associated PAH.

18.8 Hill's Eight Criterion

Arguments for causation are considered considerably strengthened if backed by experimental evidence. As yet, no specific antibody, serum factor, or perturbation of T-cell regulation has been shown to induce PAH in experimental animals. Further, no animal model of CTD PAH has been produced. Some tantalizing evidence has been produced. Antiendothelial cell antibodies as stated above can induce endothelial cell apoptosis; anti-U1-RNP and antidsDNA induce upregulation of adhesion molecules on human pulmonary endothelial cells.[59] IgG antiendothelial cell antibodies from patients with iPAH and SScPAH bind endothelial cell extracts, while those from normal patients and SSc without PAH do so to a much lesser extent,[60] The fact that different endothelial cell extracts were bound in iPAH and SScPAH simply underlines the point, that as seen with differences in BMPRII, the initiating events are probably different in iPAH and CTD PAH, but that the vascular response to chronic injury is the same – a proliferative vasculopathy.

To date, mouse models of SSc have not been specifically studied to see if there are vascular changes suggestive of PAH present. We are currently evaluating a dominant-negative TGFβ1 mutation mouse model of SSc, and increased medial thickness has been observed in the pulmonary vasculature.

18.9 Hill's Ninth Criterion

Analogy per se is a relatively weak criterion; however, as discussed earlier, there are many analogies between the factors deranged in SScPAH and iPAH. For other CTDs, the evidence base is much weaker, and here one would have to rely heavily on analogy to SSc to regard these as causing PAH. One might argue that some of the reluctance to accept SSc as causative of PAH is due to the obvious necessity of having to accept other autoimmune conditions as causative once the first is accepted. As I have pointed out earlier, there are several plausible mechanisms by which SSc might cause PAH, not all of which are equally applicable to other autoimmune conditions.

18.10 Conclusion

As long as we have no definitive animal model of CTD PAH or extract from patients with CTD PAH that produces a similar clinical syndrome, only circumstantial inferences can be drawn. It is apparent that most of Hill's criteria are met for SSc-associated PAH. It is probable that proliferative PAH is not a homogeneous condition, but rather an injury response pattern common to a number of mechanisms of endothelial cell injury. It is reasonable to propose that CTDs are among these "causes" whether as initiators of or drivers in response to an initiating event is as yet unresolved. The obvious differences between SScPAH and iPAH in terms of natural history and genetics merely strengthen the argument for a causative role. Resolution of the precise drivers will ultimately permit us to identify patients with CTD at the highest risk of developing PAH and may help direct research toward preventative therapy.

References

1. Hill AB. The environment and disease: association or causation? *Proc R Soc Med.* 1965;58:295–300.
2. Yoshida A, Katayama M. Pulmonary hypertension in patients with connective tissue diseases. *Nippon Rinsho.* 2001;59(6):1164–1167.
3. Battle RW, Davitt MA, Cooper SM, et al. Prevalence of pulmonary hypertension in limited and diffuse scleroderma. *Chest.* 1996;110:1515–1519.
4. Humbert M, Carouat A, Bertocchi M, et al. ItinerAIR-HTAP a French national prospective registry of pulmonary arterial hypertension. *Am J Respir Crit Care.* 2004;169:A169.
5. Wigley F, Mayes M, Limia J, et al. The point prevalence of undiagnosed pulmonary arterial hypertension in patients with connective tissue disease (CTD) attending community based rheumatology clinics (Uncover). ACR Annual Scientific Meeting 2004 Presentation No 1057.
6. Dawson JK, Goodson NG, Graham DR, Lynch MP. Raised pulmonary artery pressures measured with Doppler echocardiography in rheumatoid arthritis patients. *Rheumatology.* 2000;39(12): 1320–1325.
7. Marasini B, Massarotti M, Cossutta R, et al. Pulmonary hypertension in autoimmune rheumatic diseases. *Rheumatismo.* 2005;57:114–118.
8. Young RH, Mark GJ. Pulmonary vascular changes in scleroderma. *Am J Med.* 1978;64: 998–1004.
9. Mukerjee D, St. George D, Coleiro B, et al. Prevalence & outcome in systemic sclerosis associated pulmonary arterial hypertension: application of a registry approach. *Ann Rheum Dis.* 2003;62:1088–1093.
10. Pietra G, Capron F, Stewart S, et al. Pathological assessment of vasculopathies in pulmonary hypertension. *J Am Coll Cardiol.* 2004;43 suppl 12:25S–32S.
11. Cool C, Kennedy D, Voelkel N, Tuder R. Pathogenesis and evolution of plexiform lesions in pulmonary hypertension associated with scleroderma and human immunodeficiency virus infection. *Hum Pathol.* 1997;28:434–442.
12. Tuder R, Groves B, Badesch D, Voelkel N. Exuberent endothelial cell growth and elements of inflammation are present in plexiform lesions of pulmonary hypertension. *Am J Pathol.* 1994;144:275–285.
13. Mayes MD, Lacey JV Jr, Beebe–Dimmer J, et al. Prevalence, incidence, survival, and disease characteristics of systemic sclerosis in a large US population. *Arthritis Rheum.* 2003;48: 2246–2255.
14. Vlachoyiannopoulos PG, Dafni UG, Pakas I, et al. Systemic scleroderma in Greece: low mortality and strong linkage with HLA-DRB1*1104 allele. *Ann Rheum Dis.* 2000;59(5):359–367.
15. Koh ET, Lee P, Gladman DD, Abu-Shakra M. Pulmonary hypertension in systemic sclerosis: an analysis of 17 patients. *Br J Rheum.* 1996;35:989–993.
16. Kawut SM, Taichman DB, Archer-Chicko CL, Palevsky HI, Kimmel SE. Hemodynamics and survival in patients with pulmonary arterial hypertension related to systemic sclerosis. *Chest.* 2003;23:344–350.
17. Kuhn KP, Byrne DW, Arbogast PG, Doyle TP, Loyd JE, Robbins IM. Outcome in 91 consecutive patients with pulmonary arterial hypertension receiving epoprostenol. *Am J Respir Crit Care Med.* 2003;167:580–586.

18. Rubin LJ, Badesch DB, Barst RJ, et al. Bosentan therapy for pulmonary arterial hypertension. *N Engl J Med.* 2002;346:896–903.
19. Simonneau G, Barst RJ, Galiè N, et al. Continuous subcutaneous infusion of treprostinil, a prostacyclin analogue, in patients with pulmonary arterial hypertension: a double-blind, randomised, placebo-controlled trial. *Am J Respir Crit Care Med.* 2002;165:800–804.
20. Olschewski H, Simonneau G, Galiè N, Higenbottam T, Naeije R, Rubin LJ. Inhaled iloprost for severe pulmonary hypertension. *N Engl J Med.* 2002;347:322–329.
21. Galiè N, Umbert M, Vachiery JL, et al. Effects of beraprost sodium, an oral prostacyclin analogue in patients with pulmonary arterial hypertension: a randomised, double-blind, placebo-controlled trial. *J Am Coll Cardiol.* 2002;39:1496–1502.
22. Barst RJ, Langleben D, Frost A, et al. Sitaxsentan therapy for pulmonary arterial hypertension. *Am J Respir Crit Care Med.* 2004;169:441–447.
23. Ghofani A for the Sildenafil 1140 study group. Efficacy and safety of sildenafil in pulmonary arterial hypertension: results of a multinational randomized, double blind placebo controlled trial. *Am Coll Chest Phys.* 2004.
24. Williams MH, Das C, Handler CE, et al. Systemic sclerosis associated pulmonary hypertension: improved survival in the current era. *Heart.* 2006;92(7):926–932.
25. Steen VD. Autoantibodies in systemic sclerosis. *Semin Arthritis Rheum.* 2005;35(1):35–42.
26. Bordron A, Dueymes M, Levy Y, et al. The binding of some human antiendothelial cell antibodies induces endothelial cell apoptosis. *J Clin Invest.* 1998;101:2029–2035.
27. Dorfmuller P, Perros F, Balabanian K, Humbert M. Inflammation in pulmonary arterial hypertension. *Eur Respir J.* 2003;22:358–363.
28. Nicholls M, Taraseviciene-Stewart L, Rai P, et al. Autoimmunity and pulmonary hypertension: a perspective. *Eur Respir J.* 2005;26:1110–1118.
29. Humbert M, Nunes H, Sitbon O, Parent P, Hervé P, Simonneau G. Risk factors for pulmonary arterial hypertension. *Clin Chest Med.* 2001;22:459–475.
30. Lesprit P, Godeau B, Authier FJ, et al. Pulmonary hypertension in POEMS syndrome: a new feature mediated by cytokines. *Am J Respir Crit Care Med.* 1998;157:907–911.
31. Chu JW, Kao PN, Faul JL, Doyle RL. High prevalence of autoimmune thyroid disease in pulmonary arterial hypertension. *Chest.* 2002;122:1668–1673.
32. Isern RA, Yaneva M, Weiner E, et al. Autoantibodies in patients with primary pulmonary hypertension: association with anti-Ku. *Am J Med.* 1992;93:307–312.
33. Humbert M, Monti G, Brenot F, et al. Increased interleukin-1 and interleukin-6 serum concentrations in severe primary pulmonary hypertension. *Am J Respir Crit Care Med.* 1995;151:1628–1631.
34. Winslow TM, Ossipov MA, Fazio GP, et al. Five year follow up study of the prevalence and progression of pulmonary hypertension in systemic lupus erythematosus. *Am Heart J.* 1995;129:510–515.
35. Mukerjee D, Kingdon E, VanDerPump M, Coghlan J. Pathophysiological insights from a case of reversible pulmonary hypertension. *J Royal Soc Med.* 2003;96:403–404.
36. Pacioccio G, Bossone E, Erba H, Rubenfire M. Reversible pulmonary hypertension in POEMS syndrome – another ethiology of triggered pulmonary vasculopathy. *Can J Cardiol.* 2000;16:1007–1012.
37. Sanchez O, Humbert M, Sitbon O, Simonneau G. Treatment of pulmonary arterial hypertension secondary to connective tissue diseases. *Thorax.* 1999;54:273–277.
38. Nishimura T, Faul JL, Berry GJ, Veve I, Pearl RG, Kao PN. 40–O–(2-Hydroxyethyl)-rapamycin attenuates pulmonary arterial hypertension and neointimal formation in rats. *Am.J Resp Crit Care Med.* 2001;163:498–502.
39. Tormey VJ, Bunn CC, Denton C, et al. Anti-fibrillarin antibodies in systemic sclerosis. *Rheumatology.* 2001;40:1157–1162.
40. Asherton RA, Oakley CM. Pulmonary hypertension and systemic lupus erythematosis. *J Rheumatol.* 1986;13:1–5.
41. Negi VS, Tripathy NK, Misra R, et al. Antiendothelial cell antibodies in scleroderma correlate with severe digital ischaemia and pulmonary arterial hypertension. *J Rheumatol.* 1998;25: 462–466.

42. Fritzler MJ, Hart DA, Wilson D, et al. Antibodies to fibrin bound tissue type plasminogen activator antibodies in systemic sclerosis. *J Rheumatol*. 1995;22:1688–1693.
43. Morse JH, Barst RJ, Fotino M, et al. Primary pulmonary hypertension, tissue plasminogen activator antibodies and HLA-DQ7. *Am J Respir Crit Care Med*. 1997;155:274–278.
44. Martin L, Paulus JD, Ryan P, et al. Identification of a subset of patients with scleroderma with severe pulmonary and vascular disease by the presence of autoantibodies to centromere and histone. *Ann Rheum Dis*. 1993;52:780–784.
45. Grigolo B, Mazzeti I, Meliconi R, et al. Anti-topoisomerase II alpha autoantibodies in systemic sclerosis-association with pulmonary hypertension and HLA-B35. *Clin Exp Immunol*. 2000;121:539–543.
46. Morse J, Barst R, Horn E, et al. Pulmonary hypertension in scleroderma spectrum of disease: lack of bone morphogenetic protein receptor 2 mutations. *J Rheumatol*. 2002;29:2379–2381.
47. Susol E, Rands AL, Herrick M, et al. Association of markers for TGFbeta3, TGFbeta2 and TIMPI1 with systemic sclerosis. *Rheumatology*. 2000;39:13332–13336.
48. Dong C, Zhu S, Wang T, et al. Deficient Smad7 expression: a putative molecular defect in scleroderma. *Proc Nat Acad Sci U S A*. 2002;99(6):3908–3913.
49. Yamane K, Ihn H, Kubo M, Tamaki K. Increased transcriptional activities of transforming growth factor beta receptors in scleroderma fibroblasts. *Arthritis Rheum*. 2002;46(9):2421–2428.
50. Leask A, Abraham DJ, Finlay DR, et al. Dysregulation of transforming growth factor beta signaling in scleroderma: overexpression of endoglin in cutaneous scleroderma fibroblasts. *Arthritis Rheum*. 2002;46(7):1857–1865.
51. Kubo M, Ihn H, Yamane K, Tamaki K. Upregulated expression of TGFβ receptors in dermal fibroblasts of skin sections from patients with systemic sclerosis. *J Rheumatol*. 2002;29: 2558–2564.
52. Dorfmüller P, Zarka V, Durand-Gasselin I, et al. Chemokine RANTES in severe pulmonary arterial hypertension. *Am J Resp Crit Care Med*. 2002;165:534–539.
53. Kim NHS, Rubin LJ. Endothelin in health and disease: endothelin receptor antagonists in the management of pulmonary artery hypertension. *J Cardiovasc Pharmacol Therapeut*. 2002;7:9–19.
54. Yamane K. Endothelin and collagen vascular disease: a review with special reference to Raynaud's phenomenon and systemic sclerosis. *Intern Med*. 1994;33:579–582.
55. Maeda M, Kachi H, Takagi H, Kitajima Y. Is there circadian variation of plasma endothelin (ET-1) in patients with systemic scleroderma (SSc)? *J Dermatol Sci*. 1997;16:38–44.
56. Vancheeswaran R, Azam A, Black C, Dashwood MR. Localisation of endothelin-1 and its binding sites in scleroderma skin. *J Rheum*. 1994;21:1268–1276.
57. Giaid A, Yanagisawa M, Langleben D, et al. Expression of endothelin-1 in the lungs of patients with pulmonary hypertension. *N Engl J Med*. 1993;328:1732–1739.
58. Yamane K, Miyauchi T, Suzuki N, et al. Significance of plasma endothelin-1 levels in patients with systemic sclerosis. *J Rehum*. 1992;19:1566–1571.
59. Okawa-Takatsuji M, Aotsuka S, Fujinami M, et al. Upregulation of intercellular adhesion molecule-1 (ICAM-1), endothelial leucocyte adhesion molecule-1 (ELAM-1) and class II MHC molecules on pulmonary artery endothelial cells by antibodies against U1-ribonucleoprotein. *Clin Exp Immunol*. 1999;116:174–180.
60. Tamby M, Chanseaud Y, Humbert M, et al. Anti-endothelial cell antibodies in idiopathic and systemic sclerosis associated pulmonary arterial hypertension. *Thorax*. 2005;60:765–772.
61. Tanaka E, Harigai M, Tanaka M, et al. Pulmonary hypertension in systemic lupus erythematosus: evaluation of clinical characteristics and response to immunosuppressive treatment. *J Rheumatol*. 2002;29(2):282–287.
62. Pan TL, Thumboo J, Boey ML. Primary and secondary pulmonary hypertension in systemic lupus erythematosus. *Lupus*. 2000;9(5):338–342.
63. Michels H. Course of mixed connective tissue disease in children. *Ann Med*. 1997;29(5):359–364.
64. Alpert MA, Goldberg SH, Singsen BH, et al. Cardiovascular manifestations of mixed connective tissue disease in adults. *Circulation*. 1983;68(6):1182–1193.

Chapter 19
Advances in Vascular Medicine

Vascular Disease in Scleroderma

Anna Nuttall, Emma Derrett-Smith, and Richard Stratton

19.1 Background

Scleroderma is a severe systemic disorder affecting 240 per million of population, characterized by fibrosis of the skin and internal organs, by autoimmune phenomena, and by vascular injury (for review, see[1]). Scleroderma patients form a heterogeneous group with a very wide spectrum of disease severity. In the most mildly affected individuals, Raynaud's phenomenon and barely noticeable skin thickening may be the only manifestations of the disease. At the opposite end of the disease spectrum are patients with total skin encasement and life-threatening pulmonary, cardiac, or renal complications.[2]

The American College of Rheumatology proposed a set of criteria for the diagnosis of scleroderma.[3] The diagnosis depends on the presence of one major criterion, the presence of skin thickening proximal to the metacarpophalangeal joints with at least two of the following minor criteria: sclerodactyly, pitting scars, loss of finger pad substance, and bibasal pulmonary fibrosis. A pragmatic system of classification was proposed by a committee of workers in the field of scleroderma.[4] The subsets of the condition were defined by the pattern of skin involvement. Those with skin sclerosis distal to the elbow flexures and knees and facial involvement were defined as limited scleroderma, and those with more extensive skin sclerosis, involving skin proximal to the elbow and knee flexures or with the involvement of the trunk, were defined as diffuse scleroderma. This classification system is simple to apply and has important implications for prognosis, risk of internal organ involvement, and enrollment in clinical studies.

Internal organ involvement due to fibrosis or vascular disease accounts for the increased mortality associated with scleroderma.[5] Pulmonary involvement in the form of fibrosis or pulmonary vascular disease is now the leading cause of mortality in the disease. Gut involvement is also common in scleroderma, manifesting as esophageal dysmotility or malabsorbtion syndrome due to small bowel involvement.

A. Nuttall, E. Derrett-Smith, and R. Stratton (✉)
Centre for Rheumatology, The Royal Free Hospital, London, UK
e-mail: r.stratton@medsch.ucl.ac.uk

D. Abraham et al. (eds.), *Advances in Vascular Medicine*,
DOI 10.1007/978-1-84882-637-3_19, © Springer-Verlag London Limited 2010

A wealth of data support autoimmunity in scleroderma. Most notably, hallmark autoantibodies against nuclear antigens are found to be specific to scleroderma and predictive of the clinical course of the disease.[6] Anticentromere antibody is associated with limited cutaneous involvement,[7-9] whereas the antiScl 70 (anti-topoisomerase I) antibody correlates with diffuse skin involvement and a high risk of lung fibrosis.[7,10] Antibodies to RNA polymerase I and III appear specific for scleroderma and are associated with high prevalence of scleroderma renal crisis (SRC).[11-14] Clinically, apparent vascular disease is present in the great majority of the patients with scleroderma.[15]

19.2 Microvascular Changes

Abundant evidence indicates abnormalities of microvascular endothelial cell function in scleroderma.[16] Microvascular abnormalities can be observed in the nailfolds of scleroderma patients showing dilation, disruption of normal architecture, and areas where the nailfold capillaries appear deficient (Fig. 19.1).

The nailfold capillaries appear dilated, broken, reduced in density, or may form enlarged loops. These changes are seen in early disease and may precede the development of skin fibrosis. Such changes are not seen in healthy individuals and are unusual outside the context of scleroderma, but can be seen in patients with dermatomyositis or systemic lupus erythematosus, both conditions associated with immune-mediated vascular injury. Patients with the primary form of Raynaud's phenomenon may have mild abnormalities of nailfold capillaries, but changes are usually limited to moderate dilation, and the extreme changes associated with scleroderma are not seen. These observations provide evidence of endothelial cell damage in scleroderma and suggest that the endothelium may be abnormal before the development of skin changes or internal organ involvement.

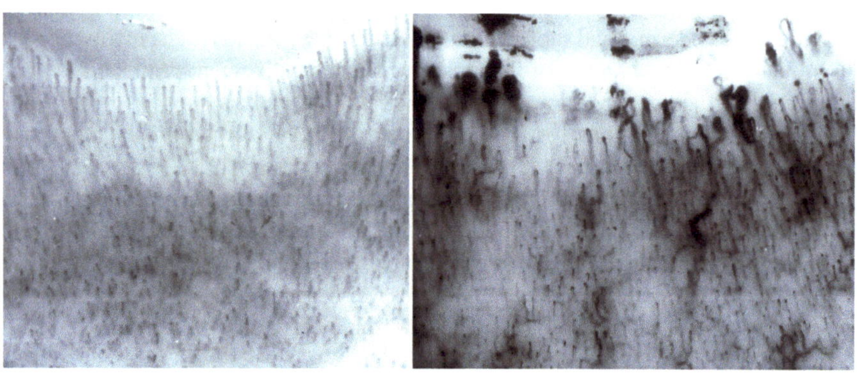

Fig. 19.1 Nailfold skin viewed under microscopy showing normal nailfold capillaries in healthy individual (*left*) and severe microvascular injury in scleroderma (*right*). Note the abnormal dilation, loss of uniformity, and areas of vascular depletion in the scleroderma nailfold

Electron microscopy studies of lung biopsy material from recent onset scleroderma patients have confirmed endothelial cell injury, revealing endothelial cell apoptosis in areas not affected by fibrotic changes.[17] In a study of scleroderma skin fibrosis, TUNEL stain was used to detect cells undergoing apoptosis in skin biopsy material from scleroderma patients, from keloid scars, and from the UCD-200/206 chicken, an animal model of systemic fibrosis.[18] Apoptotic endothelial cells were detected in the skin biopsy material from all the scleroderma patients with early disease (less than 20 months disease duration), but in none of the patients with late stage disease. Keloid scars did not stain positive for apoptotic endothelial cells, suggesting that endothelial cell apoptosis is not general to the fibrotic response. In the UCD 200/206 model of fibrosis, endothelial cell apoptosis was seen as an early event and was associated with positive staining for IgG. These observations demonstrate that endothelial cell damage occurs early in scleroderma and precedes fibrotic changes, and confirm that endothelial cells are undergoing apoptosis in early disease.

However, the nature of the insult to endothelial cells in scleroderma is not fully understood.

Measurement of circulating proteins released by damaged endothelial cells has been used to indirectly study the role of endothelial damage in scleroderma. Von Willebrand factor is a procoagulant glycoprotein made by endothelial cells. Plasma levels of Von Willebrand factor are seen to increase in conditions where there is damage to the vascular endothelium, such as glomerulonephritis,[19] toxemia of pregnancy,[20] and vasculitis.[21] A number of studies have found raised plasma Von Willebrand factor in the sera of patients with scleroderma.[22-25] Data regarding Von Willebrand factor levels in the subsets of the disease are conflicting, with one study showing higher levels in diffuse scleroderma,[25] while another study demonstrating equally raised levels in diffuse and limited subsets, with the highest levels in patients with the major vascular complications of scleroderma, SRC, and pulmonary hypertension (PHT).[26] These studies support the notion that endothelial cell injury is occurring in scleroderma, and also suggest that there is a pro-coagulation phenotype with increased production of pro-coagulant factors.

Further indirect evidence of the endothelial cell injury in scleroderma comes from the measurement of soluble adhesion molecules shed from the surface of activated endothelial cells. In vitro unstimulated endothelial cells express the adhesion molecule ICAM-1, and at low levels, VCAM-1. When endothelial cells are activated by IL-1 or TNFα, the expression of ICAM-1 and VCAM-1 is enhanced and E-selectin appears on the cell surface.[27] Endothelial cells activated in this way have been shown in vitro to shed soluble forms of these molecules from the cell surface,[28] and raised serum concentration of these soluble forms are seen in conditions where there is activation of endothelium.[29] Among the adhesins studied, E-selectin has the greatest specificity for cytokine-activated endothelial cells.[30] A number of studies on adhesion molecule expression in scleroderma have been performed.[31-39] Skin biopsies taken from the affected skin in patients during the first twelve months of diffuse scleroderma show evidence of activation of the vascular endothelium, with the expression of the adhesion molecule E-selectin in 5 of the 8 lesional skin biopsies, when compared with none out of the 4 control biopsies from healthy individuals.[38] Endothelial expression of ICAM-1, VCAM-1, and E-selectin is increased in skin biopsy specimens from patients with early progressive scleroderma, in contrast to biopsies from stable sclerotic skin, where endothelial adhesion molecule expression is limited to moderate ICAM-1 expression.[33] Another study that included lesional skin biopsies from diffuse scleroderma patients showed enhanced expression of E-selectin, ICAM-1, and VCAM-1 by endothelial cells, and enhanced ICAM-1 expression by fibroblasts.[35] Fibroblasts from scleroderma patients showed enhanced expression and shedding of ICAM-1 in vitro when compared with the control fibroblasts.[37] ICAM-1 was overexpressed by fibroblasts in the early progressive scleroderma lesions, but not in chronic stable scleroderma lesions, or by fibroblasts in healthy control tissue.[34] The ICAM-1 positive fibroblasts were located in the perivascular areas at sites of infiltration by T helper lymphocytes expressing mRNA for IFNγ and TNFα. Endothelial cells in the same areas showed increased expression of ICAM-1.

sE-selectin, sICAM-1, and sVCAM-1 levels were found to be raised in the sera from patients with early progressive scleroderma, when compared with patients with stable fibrotic skin involvement.[33] The serum levels appear to correlate well with the expression of these molecules by vascular endothelium in lesional skin.[33] In this study, the endothelial adhesion molecules were found in areas of perivascular lymphocyte accumulation. VCAM-1 and E-selectin were not found at other sites in the dermis, but confirming other studies, dermal fibroblasts from early lesional skin expressed ICAM-1. The implication here is that the activated endothelial cells contribute to the recruitment and migration of lymphocytes that migrate into the dermis and might be involved in the induction of the fibrotic response. A more recent study confirmed the elevation of sE-selectin in early severe scleroderma.[40] Interestingly, the levels were found to decrease to the normal range following 12 months of immunosuppression with cyclophosphamide and prednisolone therapy, possibly supporting a role for autoimmunity in endothelial cell activation in the disease.

Taken as a whole, these studies show that expression and shedding of E-selectin is increased in the vascular endothelial cells in early scleroderma. Therefore, the mechanism by which the endothelial cells are being activated in early disease needs to account for increased E-selectin and increased shedding of the molecule. Resting endothelial cells do not produce E-selectin, but after exposure to inflammatory mediators such as IL1, TNFα, or endotoxin, E-selectin is produced and expressed[30,41]. E-selectin production is transient, reaching a peak at 4 h and disappearing by 24 h, and mediates the adhesion of neutrophils,[30] monocytes,[42] and resting memory T-helper cells.[43,44] Following a period of stimulation, a soluble form of E-selectin (sE-selectin) appears in the supernatant of the cultured endothelial cells.[28] Analysis of sE-selectin shows a molecular weight of slightly less than the total extracellular portion of the molecule, and it appears to be produced by the cleavage of the intact molecule at a site adjacent to the plasma membrane. The mechanism of shedding is not known, but one polymorphism in E-selectin has been associated with reduced shedding and lower levels of circulating sE-selectin. The single nucleotide polymorphism (SNP) Ser128Arg in the

E-selectin gene is overrepresented in certain patient groups with atherosclerosis or re-stenosis.[45,46] Low levels of sE-selectin are found in serum taken from healthy individuals. Raised levels of sE-selectin have been observed in the serum of patients with vasculitides including those with Wegener's granulomatosis, polyarteritis nodosa, and systemic lupus.[29,47] In these conditions, the vascular walls are infiltrated with inflammatory cells, whereas in scleroderma, the vessels, although abnormal, show no inflammatory infiltrate. One possibility, therefore, is that in scleroderma, the endothelial cell activation resembles the phenotype seen following the stimulation of endothelial cells with inflammatory cytokines such as IL-1 and TNFα.

Thrombomodulin is an anticoagulant glycoprotein expressed by the endothelial cells, which binds thrombin and alters its substrate specificity, so that thrombin subsequently activates protein C.[48] A soluble form of the molecule appears to be shedding from the endothelial cells and can be measured in the plasma and urine.[49] Thrombomodulin expression is of interest in scleroderma-associated PHT for a number of reasons. Thrombosis of pulmonary arteries is an important factor in the development of severe PHT.[50,51] Outside the context of scleroderma, plasma soluble thrombomodulin levels are decreased in patients with PHT, suggesting that deficient expression of the molecule by endothelial cells contributes to thrombus formation within the pulmonary arteries.[52,53] Therefore, we measured the thrombomodulin concentration in the circulation of scleroderma PHT patients and found it to be increased above the normal range.[54] This effect was most marked in patients with early PHT and the levels decreased with time. The elevation of thrombomodulin in scleroderma has been confirmed by a more recent study showing increased thrombomodulin in the recent onset of the diffuse disease.[40] Circulating thrombomodulin increases in conditions favoring endothelial cell death, including exposure to bacterial cell-wall lypopolysaccharide and direct toxicity from exposure to hydrogen peroxide.[55] Thrombomodulin expression is decreased by shear stress and hypoxic environments.[56-58] The elevation of thrombomodulin in early scleroderma is found to be consistent with severe endothelial cell injury or cell death in early disease.

Collectively, these studies confirm endothelial cell injury in early scleroderma and suggest that endothelial cells are taking on an activated phenotype, and that endothelial cells undergo apoptosis in early disease.

A number of mechanisms have been proposed to account for the endothelial cell injury in scleroderma. Early studies showed that scleroderma sera have a direct cytotoxic effect on the endothelial cells in vitro.[16] This effect appeared to be due to functional deficiency of a plasma protease inhibitor. In another early study, the sera from around 40% of the patients with scleroderma were found to be cytotoxic in vitro, and this effect appeared to be mediated by a trypsin sensitive molecule of 67 kDa.[59] Cytotoxicity was not restricted to human umbilical vein endothelial cells (HUVEC), but was also seen with pulmonary arterial endothelial cells, fibroblasts, and neurofibroma cells. It was subsequently found that the cytotoxicity of scleroderma sera develops after storage and is associated with an increase in lipid peroxidation.[60] Increased oxidized lipoprotein, as measured by thiobarbituric acid reactive substances, is found in scleroderma sera and in sera from patients with primary Raynaud's.[61] A low molecular weight (<5 kD) fraction of scleroderma sera had in vitro cytotoxicity for endothelial cells and when given to rabbits, produced raised Von Willebrand factor and caused arterial endothelial cell proliferation.[22]

19.3 Anti-Endothelial Cell Antibodies in Scleroderma

It is widely believed that autoimmune mechanisms contribute to the development of scleroderma, and a number of groups have studied the antibody-mediated vascular injury in that condition. Since their first description,[62] anti-endothelial cell antibodies (AECA) have been described in a wide range of autoimmune and alloimmune conditions.[63-65] AECA are found in the sera of 30–50% of scleroderma patients.[66-69] In one study using a HUVE-based enzyme-linked immunosorbent assay,[69] it was shown that AECA binding was Fab mediated and that AECA appear to be distinct from the classical autoantibodies associated with scleroderma. In general, AECA have not been found to be directly cytotoxic to endothelial cells (except in Kawasaki disease) and not to mediate complement-mediated cell damage.

In keeping with this, the AECA found in the scleroderma sera do not appear to be directly cytotoxic to human endothelial cells, but peripheral blood mononuclear cells are found to kill the cultured endothelial cells in the presence of serum from a proportion of scleroderma patients, and this effect is dependent on the antibody fraction of serum.[67-69] Pretreatment of HUVEC with AECA-positive sera from scleroderma patients enhanced the binding of a human monocyte cell line.[70] These authors showed the in vitro increased adhesion of U937 monocytic cells to AECA-pretreated HUVEC. This effect was associated with the increased expression of E-selectin, VCAM-1, and ICAM-1, and with the enhanced production of IL-1 by endothelial cells. The sera alone were not cytotoxic to umbilical vein or microvascular endothelial cells, suggesting that the mechanism of cytotoxicity of AECA positive scleroderma sera was by antibody-dependent cellular cytotoxicity. AECA may thus promote adhesion of leukocytes to endothelium via stimulation of IL-1 production by endothelial cells.

In a more recent study, the human dermal microvascular endothelial cells (HDMEC) or HUVEC were cultured with the sera from scleroderma patients or controls with or without interleukin-2-activated natural killer (NK) cells or peripheral blood mononuclear cells.[71] Sera alone had no effect, but apoptosis induction was observed on HDMEC, but not on HUVEC, in the presence of AECA-positive scleroderma sera and activated NK cells. The apoptosis could be inhibited by anti-Fas ligand antibody. Sections of the scleroderma skin biopsy material revealed Fas expression by endothelial cells, supporting the tissue culture findings. Overall, these findings suggest that in scleroderma, AECA induce microvascular endothelial cell apoptosis by antibody-dependent cell-mediated cytotoxicity acting via Fas.

One limitation of the studies of AECA in scleroderma pathogenesis is that a large proportion of scleroderma patients test negative for AECA. Also, the target antigen(s) for AECA have proved difficult to define. AECA from patients with SLE have been shown to bind to a variety of molecules with molecular weights ranging from 15 to 200 kDa.[72] In a similar study, it was shown that AECA from patients with Wegener's granulomatosis displayed a consistent immunoprecipitation pattern of five endothelial surface proteins (180, 155, 125, 68, and 25 kDa), whereas AECA from SLE patients were found to bind to a wider range of proteins ranging from 25 to 200 kDa.[64,65] Another potential confounding factor with the assays for AECA is that the endothelial cells used as target antigen may express different HLA and blood group antigens from the patients whose sera are being tested raising the possibility of antibody recognition of alloantigens as the explanation for positive binding.

19.4 Macrovascular Changes

Further evidence of vascular involvement in scleroderma comes from the development of complications due to adverse remodeling of arteries and arterioles, resulting in clinically important macrovascular disease. Raynaud's phenomenon affects the majority of scleroderma sufferers, and can be extremely severe resulting in critical ischemia and loss of digits[73] (Fig. 19.2). In the limited form of the disease,

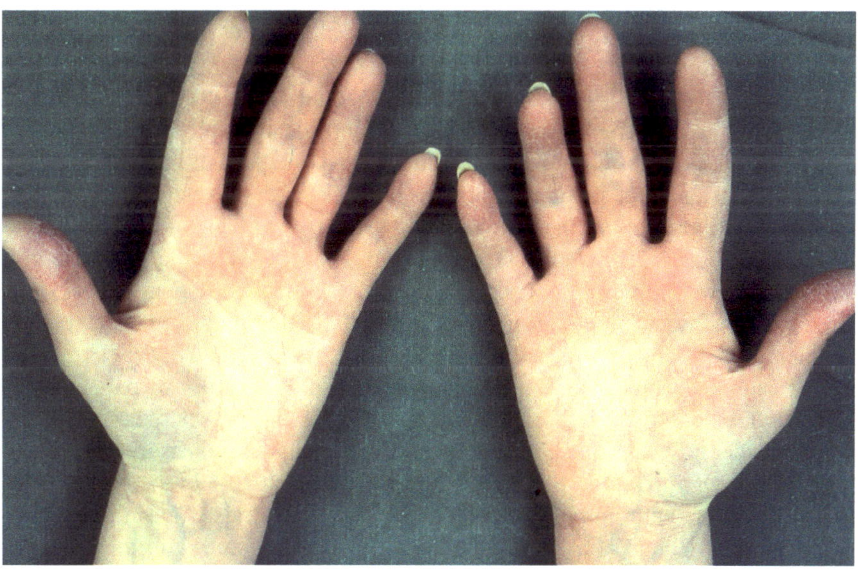

Fig. 19.2 Raynaud's phenomenon in a patient with limited scleroderma

Raynaud's phenomenon may precede the development of skin disease by years or decades, whereas in diffuse disease, the development of Raynaud's and skin thickening are often synchronous.[2] Attacks of Raynaud's can be unusually severe and frequent, particularly in the limited form of the disease. Critical digital ischemia is frequently seen in limited scleroderma patients, but only rarely seen in patients with the primary form of Raynaud's phenomenon. Critical digital ischemia may be the dominant feature of the disease in patients with limited scleroderma.[74]

Pulmonary artery remodeling resulting in PHT is a leading cause of mortality in scleroderma. In a post mortem study, abnormalities of the pulmonary arteries were found in 17/58 scleroderma patients when compared with 1/58 controls.[75] The majority of the scleroderma patients included in the study had no clinical evidence of PHT in life. Specifically, medial hypertrophy and concentric intimal proliferation with narrowing of the lumen appeared in scleroderma patients, but not in controls. These changes were not correlated with the presence of interstitial pulmonary fibrosis and therefore did not appear to be secondary to hypoxemia, destruction of lung architecture, or secondary activation of the vascular endothelium by soluble factors produced in the interstitium. The histological changes seen in the pulmonary arteries closely resembled those seen in the renal arteries of patients who died from SRC (see below). Scleroderma patients with severe renal artery involvement and those with severe pulmonary artery involvement formed distinct subgroups. The histological changes in the pulmonary arteries were confirmed by a later study,[76] in which the pulmonary artery abnormalities were found in 17/30 scleroderma autopsies. The histologic abnormalities were combined intimal and medial hyperplasia. None of the plexiform endothelial cell lesions were classically seen in the idiopathic form of primary PHT,[77] and the pulmonary vascular lesions appeared independent of the fibrosing alveolitis. At about the same time, it was recognized that a subset of patients with limited scleroderma developed an illness with signs of rapidly progressive PHT culminating in the right ventricular failure and death.[78,79] Patients included in these series had intimal and medial hyperplasia of the pulmonary arteries at postmortem. A more detailed morphometric study showed increased cross-sectional area of the intima, media, and increased percent luminal occlusion of the pulmonary arteries of the scleroderma patients when compared with the controls.[80] The changes that affected the pulmonary arteries were classified as small (<200 μm diameter), medium (200–350 μm), or large (>350 μm). The changes were more pronounced in limited scleroderma patients when compared with diffuse scleroderma patients. Particularly, the pulmonary artery measurements of 21 patients with diffuse scleroderma and SRC did not differ from the control samples. Clinically, overt PHT appears to affect 5–10% of patients with limited scleroderma and less commonly complicates diffuse scleroderma.[78,79] By using Doppler echocardiography, a noninvasive technique for measuring the mean pulmonary artery pressure, it has been shown that around 40% of the limited scleroderma patients with no clinical evidence of PHT have raised mean pulmonary artery pressure (Battle et al 1996). This figure correlates well with the observed frequency of pulmonary artery changes at postmortem in the scleroderma patients. In this study of subclinical PHT, the estimated mean pulmonary artery pressure was in the range of 30–50 mmHg for all but one patient. In clinically overt idiopathic primary PHT patients, the mean pulmonary artery pressure was 61 ± 2 mmHg in one study.[81] In another study of clinically overt disease, the mean pulmonary artery pressure was 58 mmHg and only 3/17 patients had mean pulmonary artery pressure of less than 45 mmHg, and 9/17 had a mean pulmonary artery pressure of greater than 60 mmHg (Rich et al 1992). It appears that PHT is usually clinically silent when the mean pulmonary artery pressure is less than 45 mmHg, and usually clinically overt as the mean pressure rises above 50 mmHg. Once clinical evidence of PHT develops, the condition appears to progress rapidly and carries a high mortality.

SRC is characterized by an abrupt rise in the arterial blood pressure and a rapid deterioration in the renal function. There may be hypertensive retinopathy, encephalopathy, microangiopathic hemolytic anemia, or heart failure[82,83]. The most striking histological changes are seen in the renal cortical arterioles with intimal proliferation, obliteration of the lumen, and fibrinoid necrosis[75-84] (Fig. 19.3).

Detailed morphometry of the renal arteries of SRC patients, diffuse scleroderma controls, limited scleroderma patients, and healthy controls, showed intimal proliferation and narrowing of the lumen in all the groups.[85] The most severe changes were seen in the small renal arteries (<200 μm diameter) of the SRC patients. Renal arteries of limited scleroderma patients were less severely affected than diffuse scleroderma patients. A previous study found SRC in 8% of the scleroderma patients seen

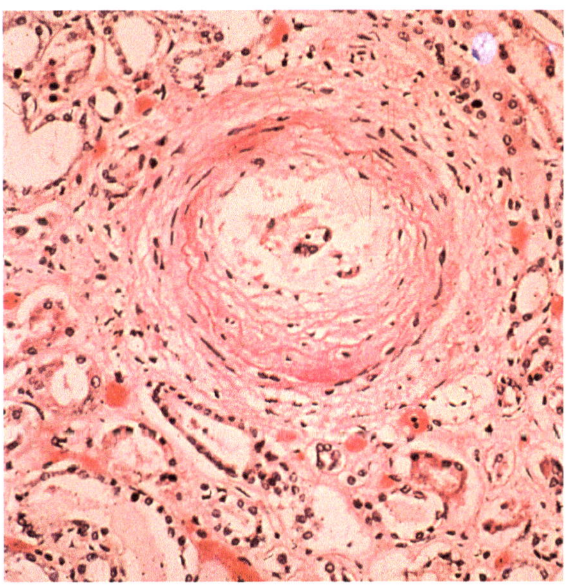

Fig. 19.3 Renal biopsy specimen from scleroderma patient showing scleroderma vascular disease affecting the renal arteriole. The lumen of the arteriole has been replaced by thickening of the intimal cell layer. There is no inflammatory cell infiltrate

over a 25-year period.[86] Before the introduction of angiotensin converting enzyme (ACE) inhibitors, the outcome was universally poor, with all patients progressing to end-stage renal failure.[87] A retrospective study showed the use of ACE inhibitors to have improved survival at 1 year from 15 to 76%, and ACE inhibitors are now the treatment of choice for blood pressure control during renal crisis.[88] Even then, 44% of those treated with ACE inhibitors went on to require dialysis. Symptoms may occur late in SRC, so that patients continue to present with hypertensive encephalopathy or with advanced renal failure. Clear identification of the risk factors for SRC would aid early diagnosis of the condition.

There is a strong association between the development of SRC and the presence of circulating antibodies against RNA polymerase I and III, so that surveillance for the condition can be focused on this group of patients.[13] ACE inhibitors may prevent further deterioration of renal function in patients with chronic renal impairment due to scleroderma,[89] but they have not been shown to protect against the development of SRC. A clearly defined subset of scleroderma patients at risk from SRC could form a cohort for a prospective placebo-controlled trial of ACE inhibitor as prophylaxis against SRC.

In healthy individuals, endothelial cells release endogenous vasodilators, nitric oxide (NO)[90] and prostacyclin (PGI_2),[91] and also release the endogenous vasoconstrictor, polypeptide endothelin-1 (ET-1).[92] Endothelial cell injury in scleroderma might therefore explain abnormal vascular tone seen in the disease, because injury to endothelial cells might alter the production of these endogenous factors controlling the vascular tone. However, it is still controversial whether the levels of serum NO, a strong vasodilator, are increased or decreased in SRC patients when compared with healthy donors.

In one study, NO metabolites were measured in the sera of the scleroderma patients and were found to be greatly elevated when compared with the controls, but further analysis revealed that these findings could be explained by increased NO production by scleroderma fibroblast due to expression of NO synthase (NOS) in scleroderma but not in normal fibroblasts.[93] Furthermore, in another study, fibroblasts, inflammatory cells, and endothelial cells were all found to overexpress inducible NO synthase (iNOS) in the disease.[94]

In order to eliminate the confounding effects of NO production by other tissues, one research group isolated the dermal endothelial cells from patients and controls, and measured the NO production and eNOS expression.[95] Expression of eNOS mRNA and protein was reduced in the dermal microvascular endothelial cells cultured from the scleroderma patients when compared with the controls, and the NO levels in the media of scleroderma cells were also reduced. These findings indicate that the injured microvascular cells in the scleroderma skin have reduced eNOS and NO, and this might explain the vasoconstriction/failure of vasodilation in the disease. Results from another group showed that as the disease progresses, microvascular cells switch from eNOS to iNOS, and speculated that iNOS induction leads to elevated NO that combines with reactive oxygen species released by inflammatory cells to form peroxynitrate.[96] Peroxynitrate combines with and oxidizes lipids, nucleic acids, and carbohydrates, resulting in degeneration of the tissues, and might contribute to chronic organ dysfunction in scleroderma.

Decreased endothelial PGI_2 synthesis might also account for vasoconstriction/increased vascular resistance in scleroderma. Early studies showed elevation of metabolites of PGI_2 in the scleroderma sera.[97] In our own studies, we found that scleroderma fibroblasts greatly overproduced PGI_2 when compared with the controls[98] and therefore the finding of elevated PGI_2 in scleroderma patients might reflect this overproduction by fibroblasts rather than the endothelial cells. In one study, which included patients with scleroderma, the pulmonary vessels were studied for PGI_2 synthase activity in PHT.[99] They hypothesized that a decrease in the expression of the critical enzyme, PGI_2 synthase, in the lung may represent an important manifestation of the pulmonary endothelial dysfunction in severe PHT. They found evidence of decreased PGI_2 production in the primary PHT and HIV-associated PHT, but the results were variable in the scleroderma patients. It remains unresolved whether endothelial-cell PGI_2 is deficient in scleroderma; however, PGI_2 is used clinically for the treatment of scleroderma vascular disease (see below).

19.5 Impaired Neuronal Control of Peripheral Vascular Tone

As patients with scleroderma report neuropathic symptoms such as parasthesia and dysasthesia, the peripheral nerve function has been formally studied using quantitative sensory testing to reveal increased vibration and cold detection thresholds, and reduced sensory action potentials in the disease. In addition, the autonomic function assessed by the tests of cardiovascular reflexes has also been found to be abnormal in scleroderma, showing evidence of both sympathetic and parasympathetic neuropathy.[100]

These findings are important, because nerves supplying blood vessels produce a number of vasodilatory substances. Calcitonin gene-related peptide (CGRP), an endogenous neuropeptide vasodilator, released from the sensory afferents, has been the most studied, and has been found to be depleted in the sensory nerve terminals in scleroderma skin biopsy material.[101] We previously used CGRP infusions to treat Raynaud's in scleroderma with some apparent benefit, but the agent has been withdrawn from clinical use because of the production costs and limited demand.

In addition, autonomic nervous system exerts regulation of peripheral vascular tone via catecholamine stimulated vasoconstriction. The α_2-adrenoceptor which mediates these effects gets altered upon persistent cold exposure, and the protein migrates to the plasma membrane of the vascular smooth muscle cells.[102] In scleroderma, the α_2-adrenoceptor levels are elevated in the dermal smooth muscle cells, and scleroderma arterioles exhibit increased constriction when exposed to α_2-adrenoceptor agonists when compared with the controls.[103]

Taken collectively, these data show that in scleroderma, arteries undergo a process of adverse remodeling, characterized by thickening of the intima and narrowing of the lumen, and culminate in stenosis and reduced perfusion of the affected organ. The vascular bed involved varies among the patient groups, so that the overall majority of the patients have digital artery involvement; patients with pulmonary artery involvement are mainly from the limited scleroderma subgroup, and those with renal artery involvement are mainly from diffuse scleroderma with positive antibodies against RNA polymerase I and III.

19.6 Endothelin 1 in Scleroderma

One interesting problem is how to link the immune-mediated endothelial cell injury with the remodeling pathology of the arteries. One possibility is that soluble factors released by the damaged endothelial cells within the lumen of the arteries and within the vasa vasorum, contribute to the adverse remodeling seen. One potential candidate factor for this process is endothelin 1 (ET-1), because ET-1 is released by injured endothelial cells and is a mitogen for vascular smooth muscle cells, as well as is a potent vasoconstrictor.[96] Therefore, ET-1 has been extensively studied in scleroderma. ET-1 levels are elevated in the scleroderma sera[25,26,104] and the levels increase following cooling of the extremities.[105] Immunostaining of the sections of skin biopsy material revealed the presence of ET-1 in the endothelial cells in the upper dermis as well as the expression in the dermal fibroblasts in scleroderma skin biopsy material, but not in the control biopsies.[106] In scleroderma fibroblasts, autocrine stimulation by ET-1was found to maintain the pro-fibrotic phenotype.[107] Use of the ET_A and ET_B receptor antagonist, bosentan, in scleroderma has been associated with clinical improvement in Raynaud's phenomenon,[108] healing of digital ulcers,[109] and improvement in hemodynamics and exercise tolerance in scleroderma PHT.[110,111] Therefore, ET-1 is considered as an important factor in scleroderma vascular pathology and in persistent fibroblast activation in the disease.

19.7 Treatment of Scleroderma Vascular Disease

The treatment of scleroderma vascular disease has become a priority for clinicians treating scleroderma patients, because of the severe morbidity and mortality associated with these complications. At the moment, lack of clear understanding regarding the pathogenesis of the vascular complications hinders specific therapies. Initially, calcium channel-blocking drugs that antagonize the peripheral vasoconstriction were studied in the scleroderma-associated Raynaud's and found to confer moderate benefit.[112] Following these studies, PGI_2 and the more stable synthetic PGI_2 derivative, Iloprost, became available and have been widely used in the treatment of scleroderma vascular disease. PGI_2 and its derivatives improve Raynaud's in scleroderma, improve exercise tolerance and hemodynamics in scleroderma-associated PHT, and improve renal plasma flow and reduce blood pressure in SRC.[13,113,114] Disadvantages of these therapies include high cost, need for intravenous use, and intrusive side effects such as nausea and headache during infusion. Despite the short plasma half life of Iloprost, improvement in digital perfusion lasts for up to 12 weeks following a 3-day course of intravenous therapy with the drug.[115] These findings imply that the drug has some slowly evoked beneficial effect on vascular remodeling. In our own studies, we showed that Iloprost at very low levels blocks the induction of CTGF and type-I collagen in human cells, and hence, one possibility is that the drug improves the vascular remodeling in scleroderma via suppression of these factors in the vessel wall.[100] In addition, PGI_2 agonists inhibit the vascular smooth muscle cell proliferation possibly acting via PPARδ signaling.[116-118]

Sildenafil is a specific inhibitor of phosphodiesterase-5 that elevates the cGMP in vascular smooth muscle cells, and has been studied as a therapy for scleroderma vascular disease.[119] Sildenafil, in case reports, was found to improve the hemodynamics in scleroderma PHT[120] and Raynaud's and digital ulcers in the disease.[121] In a placebo-controlled trial, sildenafil at a dose of 20 mg tds was found to improve hemodynamics, exercise tolerance, and functional class in patients with PHT secondary to connective tissue disorders including scleroderma.[122] Sildenafil also has antiproliferative effects, inhibiting vascular smooth muscle cell proliferation, in addition to its vasodilator properties.[123]

In the UK, the National Institute for Clinical Excellence (NICE) has recently reviewed treatments for PHT. By assessing the efficacy and cost, they compared five agents: intravenous epoprostinol, inhaled iloprost, oral bosentan, the ET_A antagonist sitaxentan, and the phosphodiesterase inhibitor sildenafil. All the agents have some evidence base for their use in PHT. NICE issued preliminary guidelines (March 1, 2008), suggesting that sildenafil should be the first line treatment and that bosentan or sitaxsentan should be used in patients intolerant or unresponsive to sildenafil. One strong factor influencing this choice is the relative cost with sildenafil costing around £4,000 per annum when compared with around £18,000 for the ET receptor antagonists.

Additional treatments in scleroderma-associated Raynaud's with critical ischemia include anti-platelet agents that are used empirically, because platelet activation has been shown to occur in scleroderma.[124] In addition, antioxidants have been studied in scleroderma vascular disease, because of the potential contribution of oxidation by free radicals to vascular injury in the disease. The synthetic antioxidant, probucol, was shown to improve Raynaud's phenomenon in scleroderma patients.[125]

19.8 Summary

Endothelial cell function is disturbed in scleroderma and in major vascular complications of the condition. The overall picture that emerges from the abovementioned studies is of the microvascular injury in early scleroderma, where the endothelial cells are in a highly activated state, expressing and shedding E-selectin, and showing evidence of Fas-dependent apoptosis. These effects may be partly due to the activation of the endothelial cells by autoantibodies and by antibody-dependent

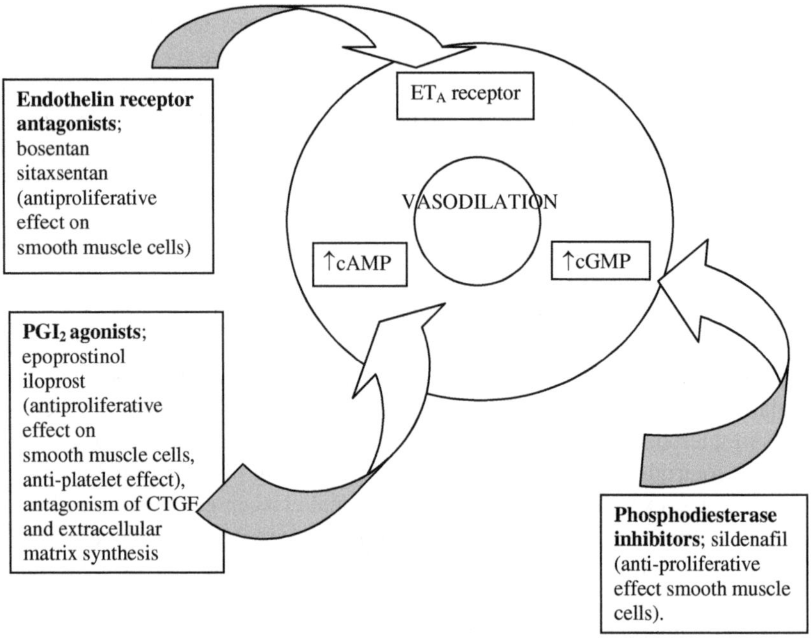

Fig. 19.4 Treatments for vascular disease in scleroderma. Effective treatments acting via antagonism of ET_A receptor expressed by vascular smooth muscle cells, and via elevation of cAMP or cGMP in vascular smooth muscle cells. In general, these agents have been shown to have antiproliferative effects on vascular smooth muscle cells and may therefore combine vasodilation with a beneficial effect on vascular remodeling

cellular cytotoxicity. These changes are accompanied by chronic adverse vascular remodeling, intimal hyperplasia, and stenosis of the arteries and arterioles leading to end-organ ischemia. Treatments for scleroderma vascular disease confer moderate benefits, and have common vasodilator properties combined with antiproliferative effects on vascular smooth muscle cells (Fig 19.4). Possible future therapeutic strategies include small molecule inhibitors of the signaling pathways activated in the endothelial cells in the disease, antibody-depleting strategies, detection of adverse vascular remodeling at the earliest stage before the onset of hemodynamic compromise, and more specific therapies against vascular cell hyperplasia (Fig 19.4).

References

1. Varga J, Abraham D (2007) Systemic sclerosis: a prototypic multisystem fibrotic disorder. J Clin Invest 117:557–567
2. Black CM, Stephens CO (1993) Scleroderma-systemic sclerosis. In: Maddison PJ, Isenberg D, Woo P, Glass DN (eds) Oxford Textbook of Rheumatology. Oxford University Press, Oxford, pp 771–789

3. Subcommittee for scleroderma criteria of the American Rheumatism Association diagnostic and therapeutic criteria committee (1980) Preliminary criteria for the classification of systemic sclerosis (scleroderma). Arthritis Rheum 23:581–590
4. LeRoy EC, Black CM, Fleischmajer R, Jablonska S, Krieg T, Medsger TA Jr, Rowell N, Wolheim F (1988) Scleroderma; classification, subsets, and pathogenesis. J Rheumatol 15:202–205
5. Steen VD, Medsger TA Jr (2007) Changes in causes of death in systemic sclerosis, 1972–2002. Ann Rheum Dis 66:940–944
6. Harris ML, Rosen A (2003) Autoimmunity in scleroderma: the origin, pathogenetic role, and clinical significance of autoantibodies. Curr Opin Rheumatol 15:778–784
7. Steen VD, Powell DL, Medsger TA Jr (1988) Clinical correlations and prognosis based on serum autoantibodies in patients with systemic sclerosis. Arthritis Rheum 31:196–203
8. Tan EM, Rodnan GP, Garcia I, Moroi Y, Fritzler MJ, Peebles C (1980) Diversity of antinuclear antibodies in progressive systemic sclerosis: anti-centromere antibody and its relationship to CREST syndrome. Arthritis Rheum 23:617–625
9. Weiner ES, Earnshaw WC, Senecal JL, Bordwell B, Johnson P, Rothfield NF (1988) Clinical associations of anti-centromere antibodies and antibodies to topoisomerase I: a study of 355 patients. Arthritis Rheum 31:378–385
10. Catagio LJ, Bernstein RM, Black CM, Hughes GRV, Maddison PJ (1983) Serological markers in progressive systemic sclerosis: clinical correlations. Ann Rheum Dis 42:23–27
11. Kuwana MJ, Kaburaki J, Mimori T, Tojo T, Homma M (1993) Autoantibody reactive with three classes of RNA polymerases in sera from patients with systemic sclerosis. J Clin Invest 91:1399–1404
12. Okano Y, Steen VD, Medsger TA Jr (1993) Autoantibody reactive with RNA polymerase III in systemic sclerosis. Ann Intern Med 119:1005–1013
13. Penn H, Howie AJ, Kingdon EJ, Bunn CC, Stratton RJ, Black CM, Burns A (2007) Denton CP Scleroderma renal crisis: patient characteristics and long-term outcomes. QJM 100:485–494
14. Reimer G, Rose KM, Scheer U, Tan EM (1987) Autoantibody to RNA polymerase I in scleroderma sera. J Clin Invest 79:65–72
15. Kahaleh MB (2004) Raynaud phenomenon and the vascular disease in scleroderma. Curr Opin Rheumatol 16:718–722
16. Kahaleh MB, LeRoy EC (1979) Endothelial injury in scleroderma. J Exp Med 149:1326–1335
17. Harrison NK, Myers AR, Corrin B, Soosay G, Dewar A, Black CM, Du Bois RM (1991) Turner-Warwick M.Structural features of interstitial lung disease in systemic sclerosis. Am Rev Respir Dis 144(3 Pt 1):706–713
18. Sgonc R, Gruschwitz MS, Dietrich H, Recheis H, Gershwin ME, Wick G (1996) Endothelial cell apoptosis is a primary pathogenetic event underlying skin lesions in avian and human scleroderma. J Clin Invest 98(3):785–792
19. Echberg MR, Nilsson IK, Linell F (1981) Significance of increased factor VIII in early glomerulonephritis. Ann Intern Med 94:482–484
20. Boneu B, Bierme R, Fournier A, Ponkonner G (1977) Factor VIII complex, fetal growth retardation and toxaemia. Lancet 1:263
21. Nusinow SR, Federici AB, Zimmerman TS, Curd JG (1984) Increased Von Willebrand factor antigen in the plasma of patients with vasculitis. Arthritis Rheum 27:1405–1409
22. Drenk F, Deicher HR (1988) Pathophysiological effects of endothelial cytotoxic activity derived from sera of patients with progressive systemic sclerosis. J Rheumatol 15:468–474
23. Gordon JL, Pottinger BE, Woo P, Rosenbaum J, Black CM (1987) Plasma von Willebrand factor in connective tissue disease. Ann Rheum Dis 46:491–492
24. Kahaleh MB, Osborn I, LeRoy EC (1981) Increased factor VIII von Willebrand factor antigen and von Willebrand factor activity in scleroderma and Raynaud's phenomenon. Ann Int Med 94:482–484
25. Yamane K, Miyauchi T, Suzuki N, Miyauchi T, Yanagisawa M, Goto K, Masaki T (1991) Elevated levels of plasma endothelin-1 in systemic sclerosis. Arthritis Rheum 34:243–244
26. Vancheeswaran R, Magoulas T, Efrat G, Wheeler-Jones C, Olsen I, Penny R, Black CM (1994) Circulating endothelin-1 levels in systemic sclerosis subsets-a marker of fibrosis or vascular dysfunction. J Rheumatol 21:1838–1844

27. Springer TA (1990) Adhesion molecules of the immune system. Nature 346:425–433
28. Pigott R, Dillon LP, Hemingway IH, Gearing AJH (1992) Soluble forms of E-selectin, ICAM-1, and VCAM-1 are present in the supernatants of cytokine activated cultured endothelial cells. Biochem Biophys Res Commun 187:584–589
29. Gearing JH, Newman W (1993) Circulating adhesion molecules in disease. Immunology Today 14:506–512
30. Bevilaqua MP, Stengelin S, Gimbrone MA Jr, Seed B (1989) Endothelial leukocyte adhesion molecule 1: an inducible receptor for neutrophils related to complement regulatory proteins and lectins. Science 243:1160–1165
31. Blann AD, Herrick A, Jayson MIV (1995) Altered levels of soluble adhesion molecules in rheumatoid arthritis, vasculitis, and systemic sclerosis. Br J Rheumatol 34:814–819
32. Claman HN, Giorno RC, Seibold JR (1991) Endothelial and fibroblast activation in scleroderma. Arthritis Rheum 34:1495
33. Gruschwitz MS, Hornstein OP, Von Den Driesch P (1995) Correlation of soluble adhesion molecules in the peripheral blood of scleroderma patients with their in-situ expression and with disease activity. Arthritis Rheum 38:184–189
34. Gruschwitz MS, Vieth G (1997) Up-regulation of class II major histocompatibility complex and intercellular adhesion molecule 1 expression on scleroderma fibroblasts and endothelial cells by interferon gamma and tumour necrosis factor alpha in the early disease stage. Arthritis Rheum 40:540–550
35. Jones SM, Mathew CM, Dixey J, Lovell CR, McHugh NJ (1996) VCAM-1 expression on endothelium in lesions from cutaneous lupus erythematosus is increased compared with systemic and localised scleroderma. Br J Dermatol 135:678–686
36. Sfikakis PP, Tesar J, Baraf H, Lipnick R, Klipple G, Tsokos GC (1993) Circulating ICAM-1 in patients with systemic sclerosis. Clin Immunol Immunopathol 68:88–92
37. Shi-Wen X, Panesar M, Vancheeswaran R, Mason J, Haskard DO, Black CM, Olsen IO, Abraham DJ (1994) Expression and shedding of ICAM-1 and LFA-3 by scleroderma and normal fibroblasts. Arthritis Rheum 37:1689–1697
38. Solberg S, Peltonen J, Uitto J, Jiminez S (1992) Elevated expression of beta 1 and beta 2 integrins, intercellular adhesion molecule 1, and endothelial leukocyte adhesion molecule 1 in the skin of patients with systemic sclerosis of recent onset. Arthritis Rheum 35:290–298
39. Stratton RJ, Coghlan JG, Pearson JD, Burns A, Sweny P, Abraham DJ, Black CM (1998) Different patterns of endothelial cell activation in renal and pulmonary vascular disease in scleroderma. QJM 91(8):561–566
40. Apras S, Ertenli I, Ozbalkan Z, Kiraz S, Ozturk MA, Haznedaroglu IC, Cobankara V, Pay S, Calguneri M (2003) Effects of oral cyclophosphamide and prednisolone therapy on the endothelial functions and clinical findings in patients with early diffuse systemic sclerosis. Arthritis Rheum 48:2256–2261
41. Pober JS, Bevilacqua MP, Mendrick DL, Lapieree LA, Fiers W, Gimbrone MA (1986) Two distinct monokines interleukin-1 and tumour necrosis factor, each independently induce biosynthesis and transient expression of the same antigen on the surface of the same antigen on the surface of cultured human vascular endothelial cells. J Immunol 136:1680–1687
42. Hakkert BC, Kuijpers TW, Leeuwenberg JF, Van Mourik JA, Roos D (1991) Neutrophil and monocyte adherence to and migration across monolayers of cytokine activated endothelial cells: the contribution of CD 18, ELAM-1, and VLA-4. Blood 78:2721–2726
43. Picker LJ, Warnock RA, Burns AR, Doerschuk CM, Berg EL, Butcher EC (1991) The neutrophil selectin LECAM-1 presents carbohydrate ligands to the vascular selectins ELAM-1 and GMP-140. Cell 66:921–933
44. Shimuzu Y, Newman W, Gopal TV, Horgan KJ, Graber N, Beall LD, van Seventer GA, Shaw S (1991) Four molecular pathways of T cell adhesion to endothelial cells: roles of LFA-1, VCAM-1, and ELAM-1 and changes in pathway hierarchy under different activation conditions. J Cell Biol 113:1203–1212
45. Ghilardi G, Biondi ML, Turri O, Guagnellini E, Scorza R (2004) Ser128Arg gene polymorphism for E-selectin and severity of atherosclerotic arterial disease. J Cardiovasc Surg (Torino) 45:143–147

46. Rauchhaus M, Gross M, Schulz S et al (2002) The E-selectin SER128ARG gene polymorphism and restenosis after successful coronary angioplasty. Int J Cardiol 83:249–257
47. Boehme MWJ, Schmitt WH, Youinou P, Stremmel WR, Gross WL (1996) Clinical relevance of elevated serum thrombomodulin and soluble E-selectin in patients with Wegener's granulomatosis and other systemic vasculitides. Am J Med 101:387–394
48. Dittman WA, Majerus PW (1990) Structure and function of thrombomodulin: a natural anticoagulant. Blood 75:329–336
49. Ishi H, Majerus PW (1985) Thrombomodulin is present in human serum and urine. J Clin Invest 76:2178–2184
50. Chaouat A, Weitzenblum E, Higenbottam T (1996) The role of thrombosis in severe pulmonary hypertension. Eur Resp J 9:356–363
51. Eisenberg PR, Lucore C, Kaufman L, Sobel BE, Jaffe AS, Rich S (1990) Fibrinopeptide A levels indicative of pulmonary vascular thrombosis in patients with primary pulmonary hypertension. Circulation 82:841–847
52. Cacoub P, Karmochkine M, Dorent R, Nataf P, Piette JC, Godeu P, Gandjbakhch I, Boffa MC (1996) Plasma levels of thrombomodulin in pulmonary hypertension. Am J Med 101:160–164
53. Welsh CH, Hassell KL, Badesh DB, Kressin DC, Marlar RA (1996) Coagulation and fibrinolytic profiles in patients with severe pulmonary hypertension. Chest 110:710–717
54. Stratton RJ, Pompon L, Coghlan JG, Pearson JD, Black CM (2000) Soluble thrombomodulin concentration is raised in scleroderma associated pulmonary hypertension. Ann Rheum Dis 59:132–134
55. Ishii H, Uchiyama H, Kazama M (1991) Soluble thrombomodulin antigen in conditioned medium is increased by damage of endothelial cells. Thromb Haemost 65:618–623
56. Malek AM, Jackman R, Rosenberg RD, Izumo S (1994) Endothelial expression of thrombomodulin is reversibly regulated by fluid shear stress. Clin Res 74:852–860
57. Shreeniwas R, Ogawa S, Cozzolino F, Torcia G, Braunstein N, Butura C, Brett J, Lieberman HB, Furie MB, Joseph-Silverstein J (1991) Macrovascular and microvascular endothelium during long term hypoxia: alterations in cell growth, monolayer permeability, and cell surface coagulant properties. J Cell Physiol 146:8–17
58. Shreeniwas R, Ogawa S, Cozzolino F, Torcia G, Braunstein N, Butura C et al (1991) Macrovascular and microvascular endothelium during long term hypoxia: alterations in cell growth, monolayer permeability, and cell surface coagulant properties. J Cell Physiol 146:8–17
59. Cohen S, Johnson R, Hurd E (1983) Cytotoxicity of sera from patients with scleroderma. Effects on human endothelial cells and fibroblasts in culture. Arthritis Rheum 26:170–178
60. Blake DR, Winyard P, Scott DGI, Brailsford S, Blann A, Lunec J (1991) Endothelial cell cytotoxicity in inflammatory vascular diseases-the possible role of oxidised lipoproteins. Ann Rheum Dis 44:176–182
61. Blann AD, Illingworth K, Jayson MI (1993) Mechanisms of endothelial cell damage in systemic sclerosis and Raynaud's phenomenon. J Rheumatol 20(8):1325–1330
62. Lindqvist RJ, Osterland CK (1971) Human antibodies to vascular endothelium. Clin Exp Immunol 9:753–760
63. Cervera R, Khamashta MA, Hughes GRV (1993) Antibodies to endothelial cells and vascular damage. CRC Press, Boca Raton, FL, pp 83–184
64. Del Papa N, Conforti G, Gambini D, La Rosa L, Tincani A, D'Cruz D, Khamashta M, Hughes GRV, Balastriere G (1994) and Meroni PL Characterisation of the endothelial surface proteins recognised by anti endothelial antibodies in primary and secondary autoimmune vasculitis. Clin Immunol Immunopathol 70:211–216
65. Del Papa N, Gambini D, Meroni PL (1994) Anti-endothelial cell antibodies in autoimmune disease. Clin Rev Allergy 12:275–286
66. Hashemi S, Smith CD, Izaguirre CA (1987) Anti-endothelial cell antibodies: detection and characterisation using a cellular enzyme-linked immunosorbent assay. J Lab Clin Med 109:434–440
67. Holt CM, Lindsey N, Moult J, Malia RG, Greaves M, Hume A, Rowell N, Hughes P (1989) Antibody-dependent cellular cytotoxicity of vascular endothelium: characterisation and pathogenic associations in systemic sclerosis. Clin Exp Immunol 78:359–365

68. Penning CA, Cunningham J, French MAH, Harrison G, Rowell NR, Hughes P (1984) Antibody dependent cytotoxicity of human vascular endothelium in systemic sclerosis. Clin Exp Immunol 57:548–556
69. Rosenbaum JR, Pottinger BE, Woo P, Black CM, Byron MA, Pearson JD (1988) Measurement and characterisation of circulating anti-endothelial cell IgG in connective tissue diseases. Clin Exp Immunol 72:450–456
70. Carvalho D, Savage CO, Black CM, Pearson JD (1996) IgG antiendothelial cells from scleroderma patients induce leukocyte adhesion to human vascular endothelial cells in vitro. Induction of adhesion molecule expression and involvement of endothelium derived cytokines. J Clin Invest 97:111–119
71. Sgonc R, Gruschwitz MS, Boeck G, Sepp N, Gruber J, Wick G (2000) Endothelial cell apoptosis in systemic sclerosis is induced by antibody-dependent cell-mediated cytotoxicity via CD95. Arthritis Rheum 43(11):2550–2562
72. Van Der Zee JM, Seigert CEH, De Vreede TA, Daha MR, Breedveld FC (1991) Characterisation of anti-endothelial cell antibodies in systemic lupus erythematosus. Clin exp Immunol 84:238–244
73. Tuffanelli DL, Winkelmann RF (1961) Systemic scleroderma : a clinical study of 727 cases. Arch Dermatol 84:359–371
74. Wigley FM (1993) Raynaud's phenomenon. Curr Opin Rheumatol 5:773–784
75. D'Angelo WA, Fries JF, Masi AT, Shulman LE (1969) Pathologic observations in systemic sclerosis. A study of fifty eight autopsy cases and fifty eight matched controls. Am J Med 46:428–440
76. Young RH, Mark GJ (1978) Pulmonary vascular changes in scleroderma. Am J Med 64:998–1004
77. Rubin LJ (1993) Primary pulmonary hypertension. Chest 104:236–250
78. Salerni R, Rodnan GP, Leon DF (1984) Shaver JA Pulmonary hypertension in the CREST syndrome variant of progressive systemic sclerosis (scleroderma). Ann Intern Med 86:394–399
79. Stupi AM, Steen VD, Owens GR, Barnes EL, Rodnan GP, Medsger TA Jr (1986) Pulmonary hypertension in the CREST syndrome variant of systemic sclerosis. Arthritis Rheum 29:515–524
80. Al-Sabbagh MR, Steen VD, Zee BC, Nalesnik M, Trostle DC, Bedetti CD, Medsger TA Jr (1989) Pulmonary artery histology and morphometry in systemic sclerosis: a case-control autopsy study. J Rheumatol 16:1038–1042
81. Barst RJ, Rubin LJ, Long WA, McGoon MD, Rich S, Badesh D, Groves B, Tapson VF, Barge RC, Brundage BH, Koerner SK, Langleben D, Keller CA, Murali S, Uretsky BF, Clayton LM, Jobsis MM, Blackburn SD, Shortino D, Crow JW (1996) A comparison of continuous intravenous epoprostinol (prostacyclin) with conventional therapy for primary pulmonary hypertension. N Engl J Med 334:296–301
82. Rodnan GP, Benedek TG (1962) An historical account of the study of progressive systemic sclerosis(diffuse scleroderma). Ann Intern Med 57:305–319
83. Shapiro AP, Medsger TA. Renal Involvment in Systemic Sclerosis. In Diseases of the Kidney.4th ed. Schreiner R and Gottschalk C, eds. USA: Little, Brown; 1988:2273-2283
84. Cannon PJ, Hassar M, Case DB, Casarella WJ, Sommers SC, LeRoy EC (1974) The relationship of hypertension and renal failure in scleroderma to structural and functional abnormalities of the renal cortical circulation. Medicine 53:1–46
85. Trostle DC, Bedetti CD, Steen VD, Al-Sabbagh MR, Zee B, Medsger TA Jr (1988) Renal vascular histology and morphometry in systemic sclerosis. A case-control autopsy study. Arthritis Rheum 31:393–400
86. Traub YM, Shapiro AP, Rodnan GP, Medsger TA, McDonald RH, Steen VD, Osial T Jr, Tolchin SR (1983) Hypertension and renal failure (scleroderma renal crisis) in progressive systemic sclerosis. Medicine 62:335–352
87. Medsger TA, Masi AT, Rodnan GP, Benedek TG, Robinson H (1971) Survival with systemic sclerosis. Ann Int Med 75:369–376
88. Steen VD, Costantino JP, Shapiro AP, Medsger TA (1990) Outcome of renal crisis in systemic sclerosis:relation to the availability of angiotensin converting enzyme inhibitors. Ann Int Med 113:352–357

89. Beckett VL, Donadio JV, Brennan LA, Comm DL, Osmundson PJ, Chao EY, Holley KE (1985) Use of captopril as early therapy for renal scleroderma: a prospective study. Mayo Clin Proc 60:763–771
90. Palmer RM, Ashton DS, Moncada S (1988) Vascular endothelial cells synthesize nitric oxide from L-arginine. Nature 333:664–666
91. DeWitt DL, Day JS, Sonnenburg WK, Smith WL (1983) Concentrations of prostaglandin endoperoxide synthase and prostaglandin I2 synthase in the endothelium and smooth muscle of bovine aorta. J Clin Invest 72:1882–1888
92. Yanagisawa M, Kurihara H, Kimura S, Tomobe Y, Kobayashi M, Mitsui Y, Yazaki Y, Goto K, Masaki T (1988) A novel potent vasoconstrictor peptide produced by endothelial cells. Nature 332:411–415
93. Takagi K, Kawaguchi Y, Hara M, Sugiura T, Harigai M, Kamatani N (2003) Serum nitric oxide (NO) levels in systemic sclerosis patients: correlation between NO levels and clinical features. Clin Exp Immunol 134:538–544
94. Yamamoto T, Katayama I, Nishioka K (1998) Nitric oxide production and inducible nitric oxide synthase expression in systemic sclerosis. J Rheumatol 25(2):314–317
95. Romero LI, Zhang DN, Cooke JP, Ho HK, Avalos E, Herrera R, Herron GS (2000) Differential expression of nitric oxide by dermal microvascular endothelial cells from patients with scleroderma. Vasc Med 5(3):147–158
96. Cotton SA, Herrick AL, Jayson MI, Freemont AJ (1999) Endothelial expression of nitric oxide synthases and nitrotyrosine in systemic sclerosis skin. J Pathol 189:273–278
97. Belch JJ, McLaren M, Anderson J, Lowe GD, Sturrock RD, Capell HA, Forbes CD (1985) Increased prostacyclin metabolites and decreased red cell deformability in patients with systemic sclerosis and Raynauds syndrome. Prostaglandins Leukot Med 17:1–9
98. Stratton R, Shiwen X, Martini G, Holmes A, Leask A, Haberberger T, Martin GR, Black CM, Abraham D (2001) Iloprost suppresses connective tissue growth factor production in fibroblasts and in the skin of scleroderma patients. J Clin Invest 108:241–250
99. Tuder RM, Cool CD, Geraci MW, Wang J, Abman SH, Wright L, Badesch D, Voelkel NF (1999) Prostacyclin synthase expression is decreased in lungs from patients with severe pulmonary hypertension. Am J Respir Crit Care Med 159(6):1925–1932
100. Klimiuk PS, Taylor L, Baker RD, Jayson MI (1988) Autonomic neuropathy in systemic sclerosis. Ann Rheum Dis 47(7):542–545
101. Bunker CB, Terenghi G, Springall DR, Polak JM, Dowd PM (1990) Deficiency of calcitonin gene-related peptide in Raynaud's phenomenon. Lancet 336(8730):1530–1533
102. Chotani MA, Flavahan S, Mitra S, Daunt D, Flavahan NA (2000) Silent alpha(2C)-adrenergic receptors enable cold-induced vasoconstriction in cutaneous arteries. Am J Physiol Heart Circ Physiol 278(4):H1075–H1083
103. Flavahan NA, Flavahan S, Liu Q, Wu S, Tidmore W, Wiener CM, Spence RJ, Wigley FM (2000) Increased α_2-adrenergic constriction of isolated arterioles in diffuse scleroderma. Arthritis Rheum 43:1886–1890
104. Kuryliszyn-Moskal A, Klimiuk PA, Sierakowski S (2005) Soluble adhesion molecules (sVCAM-1, sE-selectin), vascular endothelial growth factor (VEGF) and endothelin-1 in patients with systemic sclerosis: relationship to organ systemic involvement. Clin Rheumatol 24(2):111–116
105. Danese C, Parlapiano C, Zavattaro E, Di Prima M, Campana E, Rota C, Tonnarini G, Di Siena G, Borgia MC (1997) ET-1 plasma levels during cold stress test in sclerodermic patients. Angiology 48(11):965–968
106. Tabata H, Yamakage A, Yamazaki S (1997) Cutaneous localization of endothelin-1 in patients with systemic sclerosis: immunoelectron microscopic study. Int J Dermatol 36(4): 272–275
107. Shi-Wen X, Chen Y, Denton CP, Eastwood M, Renzoni EA, Bou-Gharios G, Pearson JD, Dashwood M, du Bois RM, Black CM, Leask A, Abraham DJ (2004) Endothelin-1 promotes myofibroblast induction through the ETA receptor via a rac/phosphoinositide 3-kinase/Akt-dependent pathway and is essential for the enhanced contractile phenotype of fibrotic fibroblasts. Mol Biol Cell 15(6):2707–2719

108. Hettema ME, Zhang D, Bootsma H, Kallenberg CG (2007) Bosentan therapy for patients with severe Raynaud's phenomenon in systemic sclerosis. Ann Rheum Dis 66(10):1398–1399
109. Riccardi MT, Chialà A, Lannone F, Grattagliano V, Covelli M, Lapadula G (2007) Treatment of digital ulcers in systemtic sclerosis with endothelin-1 receptor antagonist (bosentan). Reumatismo 59(2):135–139
110. Williamson DJ, Wallman LL, Jones R, Keogh AM, Scroope F, Penny R, Weber C, Macdonald PS (2000) Hemodynamic effects of Bosentan, an endothelin receptor antagonist, in patients with pulmonary hypertension. Circulation 102(4):411–418
111. Denton CP, Humbert M, Rubin L, Black CM (2006) Bosentan treatment for pulmonary arterial hypertension related to connective tissue disease: a subgroup analysis of the pivotal clinical trials and their open-label extensions. Ann Rheum Dis 65(10):1336–1340
112. Finch MB, Dawson J, Johnston GD (1986) The peripheral vascular effects of nifedipine in Raynaud's syndrome associated with scleroderma: a double blind crossover study. Clin Rheumatol 5:493–498
113. de la Mata J, Gomez-Sanchez MA, Aranzana M, Gomez-Reino JJ (1994) Long term iloprost infusion therapy for severe pulmonary hypertension in patients with connective tissue diseases. Arthritis Rheum 37:1528–1533
114. Kyle MV, Belcher G, Hazleman BL (1992) Placebo-controlled study showing therapeutic benefit of Iloprost in the treatment of Raynaud's phenomenon. J Rheumatol 19:1403–1406
115. Rademaker M, Cooke ED, Almond NE, Beacham JA, Smith RE, Mant TG, Kirby JD (1989) Comparison of intravenous infusions of iloprost and oral nifedipine in treatment of Raynaud's phenomenon in patients with systemic sclerosis: a double blind randomised study. BMJ 298(6673):561–564
116. Sinzinger H, Zidek T, Fitscha P, O'Grady J, Wagner O, Kaliman J (1987) Prostaglandin I2 reduces activation of human arterial smooth muscle cells in-vivo. Prostaglandins 33(6):915–918
117. Weber AA, Zucker TP, Hasse A, Bönisch D, Wittpoth M, Schrör K (1998) Antimitogenic effects of vasodilatory prostaglandins in coronary artery smooth muscle cells. Basic Res Cardiol 93(3):54–57
118. Lin H, Lee JL, Hou HH, Chung CP, Hsu SP, Juan SH (2008) Molecular mechanisms of the antiproliferative effect of beraprost, a prostacyclin agonist, in murine vascular smooth muscle cells. J Cell Physiol 214(2):434–441
119. Rybalkin S, Yan C, Bornfeldt K, Beavo J (2003) Cyclic GMP phosphodiesterases and regulation of smooth muscle function. Circ Res 93:280–291
120. Hayakawa I, Shirasaki F, Hirano T, Oishi N, Hasegawa M, Sato S, Takehara K (2006) Successful treatment with sildenafil in systemic sclerosis patients with isolated pulmonary arterial hypertension: two case reports. Rheumatol Int 26(3):270–273
121. Gore J, Silver R (2005) Oral sildenafil for the treatment of Raynaud's phenomenon and digital ulcers secondary to systemic sclerosis. Ann Rheum Dis 64(9):1387
122. Badesch DB, Hill NS, Burgess G, Rubin LJ, Barst RJ, Galiè N (2007) Simonneau G Sildenafil for pulmonary arterial hypertension associated with connective tissue disease. J Rheumatol 34(12):2417–2422
123. Tantini B, Manes A, Fiumana E et al (2005) Antiproliferative effect of sildenafil on human pulmonary artery smooth muscle cells. Basic Res Cardiol 100:131–138

Part 3
Cardiovascular Disease

Chapter 20
Coronary Heart Disease in Women

Nanette K. Wenger, Leslee J. Shaw, and Viola Vaccarino

Coronary heart disease (CHD) is the leading cause of morbidity and mortality among women in most developed nations. In the United States alone, almost 250,000 women die annually from CHD.[1] Contrary to popular view, a woman is >5 times more likely to die from CHD than from breast cancer. However, CHD has historically been considered a male disease. Only recently have the research community and the public begun to appreciate the large health burden that heart disease poses to women. In this chapter, we review the two most common clinical manifestations of CHD in women, angina pectoris and acute coronary syndromes, and discuss sex differences, treatment issues, and diagnostic modalities of CHD in women.

20.1 Angina Pectoris in Women

Women constitute more than half of the 8.9 million patients in the United States with chronic stable angina. However, gender differences are prominent, in that women are more likely to have atypical angina when compared with the male model, and are more likely to have a more complex symptom presentation. Women with angina report greater functional disability than do their male counterparts. It remains uncertain as to what extent the anatomic and pathophysiologic differences in the coronary vasculature contribute to these gender differences in presentation.

Angina is the predominant initial and subsequent presentation of CHD in women, in contrast to myocardial infarction (MI) and sudden cardiac death as the preeminent presentations among men. Among patients who present with an initial episode of MI, women are far more likely to have had antecedent stable angina. This challenges us to assess whether evaluation and risk stratification of women during this stable phase of their CHD and/or more intensive preventive and symptomatic therapies offer the potential to intervene and to avert the occurrence of MI.

N.K. Wenger (✉)
Emory University School of Medicine, Grady Memorial Hospital, Emory Heart and Vascular Center, Atlanta, GA, USA

D. Abraham et al. (eds.), *Advances in Vascular Medicine*,
DOI 10.1007/978-1-84882-637-3_20, © Springer-Verlag London Limited 2010

Women who present to their physician with angina,[2] compared with their male peers, tend to be older and are more likely to have hypertension, diabetes, and heart failure, with the latter characterized by intact ventricular systolic function. As suggested previously, women are less likely previously to have incurred MI or to have undergone a myocardial revascularization procedure. Nonetheless, in population-based studies in the United States, the prevalence of chronic stable angina is comparable in women and in men.

The prominence of angina as a symptom for women was documented in the early reports of the Framingham Heart Study describing the clinical manifestations of CHD by gender.[3] In the Framingham cohort, MI occurred in 43% of men vs. 29% of women, sudden death in 10% of men vs. 7% of women, uncomplicated angina in 26% of men vs. 47% of women, angina in association with MI in 13% of men vs. 18% of women, and unstable angina in 8 and 9% of the two populations, respectively.

20.1.1 Recent Data Characterizing Stable Angina Pectoris in Women

A recent systematic review and meta analysis revealed that women have a similar or slightly higher prevalence of angina than men across countries with widely differing mortality rates of MI.[4]

A 2006 study from Finland[5] involved the prospective evaluation of an ambulatory cohort for the years 1996–2001, encompassing patients 45 to 89 years of age. The fact that Finland and the U.S. have a comparable prevalence of chronic stable angina is of relevance. However, the Finnish medical record systems permit compilation and correlation of symptoms, medication use, test use, and disease event and mortality rates, as demonstrated in this report. The age-standardized annual incidence of angina was 1.89 for women and 2.03 for men. For the purposes of this study, angina was characterized as "nitrate prescription angina," i.e., indicating that the treating physician diagnosed the patient as having angina pectoris and prescribed a nitrate drug; alternatively, the angina was characterized as "test-positive angina," i.e., evidence that the treating physician ordered a diagnostic stress test, which had positive (abnormal) results. Nitrate prescription angina imparted a similar increase in coronary mortality for both genders, but the strong and graded relationship between the amount of nitrate used and event rates was striking. This defines the patients with frequent episodes of myocardial ischemia, clinically manifest by angina pectoris, as a high-risk population, worthy of consideration for further evaluation and intervention. Women younger than 75 years of age with test-positive angina had a coronary standardized mortality ratio greater than that for men, indicating the increased likelihood of an adverse outcome with test-positive angina in women. That diabetes particularly disadvantages women with CHD is evident by the age-standardized coronary event rates with both test-positive angina and diabetes of 9.9 per 100 per year for women, when compared with 9.3 per hundred per year for men.

The adverse impact of both nitrate prescription angina and test-positive angina for both genders was most prominent in the younger age groups, 45–54 and 55–64 years, with test-positive angina more predictive of coronary mortality than nitrate prescription angina.

Another European study of stable angina pectoris, the Euro Heart Survey,[6] encompassed 3,779 patients, 42% of them women, who were initially diagnosed by a cardiologist as having stable angina, with data collected for the subsequent year. Compared with men, women had less exercise electrocardiography (ECG) testing, 73% vs. 78%; less coronary angiography, 31% vs. 49%, despite a higher angina class; and had less application of statin and antiplatelet therapy, both initially and at 1 year. Not surprising, given the lesser coronary angiography, women also had prominently less coronary revascularization, 13% vs. 29% for men. The adjusted odds ratio for investigation by gender showed women less likely to have angiography (0.59) and an exercise ECG test (0.81), but more likely to have stress imaging (1.08). At 18-month follow-up, women had a doubled occurrence of death and nonfatal MI even when adjusted for age, the presence of diabetes, left ventricular function, and coronary disease severity. This gender difference persisted with adjustment for pharmacotherapy and revascularization, suggesting that the women were more likely initially encountered at a more severe and advanced stage of their disease, or that another nonmeasured variable was operative. The women whose CHD was confirmed at angiography were more likely than men to have angina at follow-up, 57% vs. 47%.

20.1.2 The Women's Ischemia Syndrome Evaluation (WISE) Study

The National Heart, Lung, and Blood Institute Women's Ischemia Syndrome Evaluation (WISE) Study[7,8] enrolled women presenting with chest pain and who had myocardial ischemia documented at noninvasive testing. At subsequent coronary angiography, about half of these women had no flow-limiting coronary obstructive disease. In clinical practice, this scenario would have likely been dismissed as a false-positive noninvasive test, but during clinical follow-up in the research protocol the WISE women were documented to have both persisting symptoms and a subsequent significant occurrence of coronary events. At 4 years of follow-up, the 330 women with significant (>50%) coronary disease had a 13.6% occurrence of death or MI; those with minimal obstructive disease (20–49%), 216 women, had a 6.9% occurrence of death or MI, whereas those without obstructive disease (lesions <20%), 317 women, had a 2.5% occurrence of death or MI. This computes to a dramatic 9.4% rate of death or MI for women with either no or minimal obstructive coronary disease at angiography.

Angina in the WISE cohort was also inversely related to quality of life.[9] The greater duration of symptoms, the increased intensity of symptoms, and the higher number of symptoms powerfully predicted a lesser quality of life.

These data support the concept of a specific vasculopathy in women with chest pain and documented myocardial ischemia, in the absence of obstructive disease of the epicardial coronary arteries.[7] In the subset studies, a substantial atherosclerotic burden was shown in these women by intravascular ultrasound (IVUS) examination that documented intramural atherosclerotic lesions with little or no protrusion into the coronary lumen. Another substudy identified a decrease in coronary flow reserve at cardiac catheterization, with the extent of impairment of coronary flow reserve independently predicting coronary events. A further substudy using magnetic resonance imaging (CMR) documented subendocardial ischemia. The likely culprit is coronary endothelial dysfunction of the microvasculature, but the roles of hormonal influences, of inflammatory markers, of oxidative stress, and other variables remain intriguing. Challenging is the clustering of risk factors, as seen in the metabolic syndrome, which was common in these women, with conventional coronary risk factor clustering also associated with oxidative stress. Other forms or mechanisms for microvascular dysfunction remain to be ascertained.

Nonetheless, in the interim, targets for intervention must be identified. It appears prudent to address symptoms of myocardial ischemia with conventional and novel antianginal therapies; and to implement precise control of coronary risk factors, given the potential of this approach to curtail endothelial dysfunction.

20.1.3 Clinical Coronary Outcomes by Gender

Women with clinical evidence of CHD, and in particular younger women,[10] have more adverse outcomes than men.[11] Women with angina have a doubled morbidity and mortality; once MI occurs, there is a 1.5 excess one-year mortality for women. Further, women have a doubled morbidity and mortality when they undergo coronary artery bypass graft surgery (CABG), and they incur a doubled incidence of heart failure. This occurs despite the documentation of less obstructive coronary disease in women when compared with men among patients undergoing elective diagnostic angiography for angina. In the National Cardiovascular Data Registry of the American College of Cardiology, at all ages, more women than men were likely to have <50% obstructive disease, with the gender difference lessening with advancing age.[12]

Based on the ACC/AHA 2002 Guidelines for the Management of Chronic Stable Angina,[13] the objectives of the treatment of angina pectoris are to reduce ischemia and relieve anginal symptoms, thereby improving the quality of life; and to prevent MI and death, thereby improving the quantity of life.

It is crucial to appreciate that the symptom of angina occurs at the end of the ischemic cascade.[14] In the initial seconds of ischemia, there is impairment of myocardial relaxation, i.e., diastolic dysfunction, followed by systolic dysfunction and a decrease in ventricular filling. Only then, among the abnormalities evolving during ischemia, is there evidence of ST segment depression on the electrocardiogram, followed by the occurrence of angina pectoris. Thus, many episodes of ischemia may never evoke pain, and about half of all patients with angina also experience episodes of asymptomatic myocardial ischemia.

The issues predominating in women involve accumulating evidence that implicate a microcirculatory dysfunction in the model of ischemia in women[15]; these issues have been inadequately investigated in men. Among these variables are a large disconnect between the severity of the luminal stenosis of the epicardial coronary arteries and the extent of impairment of coronary flow reserve, a wide variability in exercise tolerance over time, and a reduced flow response to stress in regions perfused by nonstenotic epicardial coronary arteries. Additionally, there is substantial variability in outcomes after successful revascularization interventions, with a sizeable population of women experiencing residual angina. Added to this is that about 25% of women with biomarker-positive acute coronary syndromes (ACS) have no flow-limiting stenoses demonstrated at subsequent coronary angiography. Added contributors include the prominent predictive value of an elevated brain naturetic peptide (BNP) and elevated high-sensitivity C-reactive protein (hsCRP) for adverse outcomes in women with ACSs. Challenge has been articulated as to whether women are more likely to have plaque erosion (vs. rupture of a vulnerable plaque), with microvascular embolization a consequence of this plaque erosion; the prominence of plaque erosion has been documented at autopsy among women who die suddenly.

A newer variable has been offered as explanatory for some of the ischemic symptoms in women.[16] In the presence of myocardial ischemia, there is an increase in the late inward sodium current, with resultant increased sodium entry into the cardiac cell. This causes sodium overload, and the sodium efflux in recovery is accomplished by the sodium/calcium exchange mechanism, which results in increased intracellular calcium and eventual calcium overload. Calcium overload is associated with an increase in diastolic wall tension (myocardial stiffness), with consequent increase in myocardial oxygen demand and a concomitant decrease in myocardial oxygen supply related to compression of the intramural small vessels; these worsen the myocardial ischemia and result in a vicious circle.[17]

20.1.4 Medical and Invasive Treatment Options for Chronic Stable Angina

The efficacy of traditional antiischemic therapies for chronic stable angina – beta blocking drugs, calcium channel blocking drugs, and nitrate drugs – relates to their favorable hemodynamic effects. Beta blockers do not alter myocardial oxygen supply, in that they lack effect on coronary blood flow; rather, they decrease myocardial oxygen demand via decreases in heart rate, arterial pressure, and myocardial contractility. The dihydropyridine calcium channel blockers increase oxygen supply by increasing coronary blood flow and concomitantly lessen myocardial oxygen demand by decreasing arterial pressure and myocardial contractility; save for amlodipine, these drugs increase heart rate and thus may increase myocardial oxygen demand. The nondihydropyridine calcium channel blockers also improve myocardial oxygen supply by increasing coronary blood flow and lessen myocardial oxygen demand by decreasing heart rate, arterial pressure, and myocardial contractility. Long-acting nitrates increase myocardial oxygen supply by increasing coronary blood flow and decrease myocardial oxygen demand by decreasing arterial pressure and venous return, but exert variable effects on heart rate and resultant myocardial oxygen demand, largely dependent on the adequacy of circulating blood volume.[13] The guidelines for pharmacotherapy for chronic stable angina include Class IA recommendations for beta blockers in patients with prior MI, and Class IB for patients without prior infarction.[13] No gender differences are present in these Guideline recommendations.

However, despite the use of traditional antianginal agents, many patients with chronic stable angina have a median of 2 anginal attacks per week, most likely owing to the significant percentage of patients who have relative intolerance to full doses of beta blockers, calcium channel blockers, and nitrate drugs, or who have clinical conditions that may limit their use.

Additional medical treatment options for patients with chronic angina can be characterized as vasculoprotective therapies and include aspirin, statin drugs, and ACE inhibitors. The ACC/AHA guidelines for pharmacotherapy in chronic stable angina cite a Class IA recommendation for aspirin, and a Class IA recommendation for lipid-lowering therapy in patients with suspected CHD and an LDL-C >130 mg/dl, with a target LDL-C <100 mg/dl; an optional LDL-C goal of <70 mg/dl is reasonable for patients at very high risk.[9] There is a Class IA recommendation for ACE inhibitors in all patients with coronary disease who have diabetes and/or left ventricular systolic dysfunction.

In patients for whom antianginal therapy is unsuccessful or where the coronary anatomy appears appropriate, myocardial revascularization with percutaneous coronary intervention (PCI) or CABG is an option. Data from the Arterial Revascularization Therapy Study document persistent angina despite optimal revascularization.[18] At 1 year, 78.9% of patients in the stenting group were free of angina, 21.1% were free of antianginal medication, and 19.1% were free of both angina and antianginal medication. In the surgery group, 89.5% were free of angina, 41.5% were free of antianginal medication, and 38.4% were free of both angina and antianginal medication. Nonetheless, in this optimally revascularized population, 60–80% were taking antianginal medication and 10–20% had angina at 1 year.

When patients experience recurrent ischemia and recurrent angina, management options include the titration of antianginal drug therapy or repeat revascularization. When these prove ineffective, additional therapeutic options[13] include transmyocardial revascularization, enhanced external counterpulsation, or spinal cord stimulation.

Cell therapy and gene therapy remain investigational approaches, and a new antianginal drug with a novel mechanism of action, ranolazine, will be discussed subsequently. Importantly, myocardial ischemia and attendant angina diminish the quality of life. In a follow-up study of almost 1,000 patients subsequent to percutaneous intervention and/or coronary bypass graft surgery,[19] excellent or very good health status at 1 year was reported by 40% of patients who were free of angina, but about 15% of patients with angina. In both groups, assessment of general health status declined over a span of 10 years, with almost 30% of those free of angina reporting excellent or good health status, when compared with about 5% of those with angina. Also relevant, a small study examining patient expectations about elective percutaneous intervention for stable angina[20] identified that 75% believed the procedure would prevent MI, and 71% thought it would prolong survival; gender-specific data are not available.

20.1.5 What Have We Learned from the COURAGE Trial?

COURAGE (Clinical Outcomes Utilizing Revascularization and AGgressive drug Evaluation) was designed to determine whether percutaneous intervention plus optimal medical therapy would improve the long-term prognosis of patients with stable coronary disease by reducing the risk of death or nonfatal MI compared with optimal medical therapy alone.[21] Two thousand two hundred and eighty seven patients were randomized and followed for a median of 4.6 years; the primary outcome was all-cause mortality and nonfatal MI. The angiographic criteria for entry were a stenosis of at least 70% in at least one proximal epicardial coronary artery and objective evidence of myocardial ischemia; or one coronary stenosis of at least 80% with classic angina without provocative testing. Nuclear imaging studies showed similar rates of single and multiple reversible perfusion defects at baseline in both arms of the study. The lifestyle interventions included smoking cessation, exercise, nutrition counseling, and weight control. The goal for coronary risk factors were an LDL-C of 60–85 mg/dl, an HDL-C $\geq$40 mg/dl, triglycerides <150 mg/dl, blood pressure <130/85 mmHg or <130/80 if diabetes or renal disease was present, and a hemoglobin A1C of <7.0. Pharmacologic therapy included antiplatelet therapy (aspirin and clopidogrel in accordance with established practice standards), simvastatin $\pm$ ezetimibe, extended release niacin or fibrates; angiotensin converting enzyme inhibitors, angiotensin receptor blockers, or diuretics; beta blockers, calcium channel blockers, and nitrates. At study end, there was no evidence that percutaneous intervention added to optimal medical therapy reduced the primary endpoint of death and nonfatal MI, compared with optimal medical therapy alone. Angina prevalence was substantially reduced during follow-up in both treatment groups. Reduction was greater in the percutaneous intervention group at years 1 and 3, although by 5 years there was no significant difference between groups in freedom from angina. The quality of life was slightly advantageous for percutaneous intervention to year 3, with similar results thereafter. The challenge to these data is the substantial (40%) crossover to percutaneous intervention during the course of the study, with greater crossover related to increased reversible perfusion defects at baseline. Reporting of gender-specific data is planned, but not yet accomplished.

20.1.6 Ranolazine: A Novel Antianginal (Antiischemic) Drug

Ranolazine is the first new antianginal (antiischemic) drug licensed for clinical use in decades. Previously discussed is the increase in late sodium entry into the cardiac myocyte in association with myocardial ischemia, which via the sodium/calcium exchange results in calcium overload and subsequent electrical and mechanical dysfunction, with potential worsening of myocardial ischemia. Ranolazine reduces the late inward sodium current.[16,17]

Major data for ranolazine derive from the MARISA, CARISA, and ERICA sentinel clinical trials.[22-24] Ranolazine improved exercise performance and symptoms in the absence of clinically significant hemodynamic effects. Gender comparisons[25] showed that exercise duration improved in both genders, with women showing less improvement than men. In the ERICA study, angina improved comparably in women and men.

The MERLIN-TIMI 36 study was designed to compare i.v./oral ranolazine with placebo in patients with non-ST segment elevation ACSs treated with standard medical and interventional therapies. The primary efficacy outcome was determined by a combined endpoint of cardiovascular death, MI, and recurrent ischemia. Ranolazine had no effect on this primary endpoint, with a hazard ratio of 0.92 ($p=0.11$); thus, the addition of ranolazine to standard treatments for ACSs was not effective in reducing major cardiovascular events. Nonetheless, there was difference in the components of the primary endpoint. Whereas there was no ranolazine effect on cardiovascular death or MI (HR 0.99, $p=0.87$), there was a significant benefit for recurrent ischemia, with an HR of 0.87 ($p=0.03$) favoring ranolazine. Importantly, ranolazine also reduced ventricular tachycardia, supraventricular events, and bradycardia events, based on 7-day Holter recordings, with no safety issues raised during the trial.

Specifically for women, although not a prespecified analysis, a presentation of MERLIN-TIMI 36 results by gender at the 2007 American Heart Association meeting[26] showed that, among the 2,291 women enrolled, there was a significant improvement in the primary endpoint for women with ranolazine when compared with placebo (relative risk reduction 17%, $p=0.03$) and in recurrent ischemia (relative risk reduction 29%, $p=0.002$), without significant effect on cardiovascular death or MI. As in prior studies, the women in MERLIN were older and more likely to have diabetes, hypertension, heart failure, prior angina, ECG ST-segment depression, and elevated BNP. However, they had lower rates of >50% obstructive epicardial coronary stenosis, and lower rates of elevated troponin. Whether these selectively beneficial effects for women with ranolazine therapy will translate into improved antianginal (antiischemic) outcomes for women with chronic stable angina remains conjectural, but promising. Nonetheless, ranolazine is an effective antianginal agent for women and men with chronic stable angina and should be added to the therapeutic armamentarium.

20.1.7 *Angina in Women: Summary*

Women and men differ in their initial manifestations of CHD, with women more likely to present with angina pectoris. Twice as many older women as men with CHD report chest pain.

Patients with angina, both women and men, commonly curtail their activities to avoid anginal episodes, hence recorded activity levels become important in correlating angina and antianginal therapy. Effective antianginal therapy may enable higher levels of activity, but accomplishing these increased levels of activity may then continue to evoke ischemia and angina, hence the compelling need for correlation between activity intensity and symptoms. Whether the increase in activity levels with ranolazine therapy translates into the lowered hemoglobin A1C levels with ranolazine use remains to be ascertained. Importantly, women with CHD, particularly symptomatic women, report worse quality of life outcomes than do men.

The comprehensive management of symptomatic myocardial ischemia (angina pectoris), therefore, includes symptom management, lifestyle modification, antiplatelet therapy, and aggressive coronary risk factor reduction, applied comparably for women and for men.

20.2 Acute Coronary Syndromes in Women

While uncomplicated angina is the most common initial manifestation of coronary artery disease in women, about one-third of women with CHD initially present with an ACS. As for stable angina, the clinical profile of women with ACS is different from men. Women hospitalized with ACS are more critically ill than men. They more often present with congestive heart failure and cardiogenic shock, have more cardiovascular risk factors, and are more likely to sustain major clinical events during the hospitalization, such as cardiac rupture, reinfarction, major bleeding, pulmonary edema, stroke, and death. Notably, more unfavorable outcomes occur among women with ACS than men despite less extensive coronary narrowing and smaller infarcts when compared with men.

The type of ACS, i.e., ST-segment elevation MI (STEMI), non-ST segment elevation MI, or unstable angina, plays a large role in sex-related outcome differences. Compared with men, women are less likely to present with an acute MI, particularly STEMI, as their initial ACS and more likely to present with unstable angina. Outcome differences track in the opposite direction: women fare worse than men after STEMI, have similar outcome after NSTEMI, and tend to have a better outcome after unstable angina.[27,28]

The age of the patient is important when evaluating outcome differences between women and men with ACS. Although acute MI is infrequent in younger women, younger women are the group with excess mortality when compared with men of similar age. Specifically, women less than 60–65 years old hospitalized with MI have higher mortality rates when compared with men of similar age, between 50% to two-times higher.[10,29,30] In contrast, no mortality differences, or even a better survival, are typically seen comparing older women with MI with older men. Older women admitted to the hospital with unstable angina, in particular, have a long-term survival advantage over older men, which mirrors the survival advantage of women in the general population.[28]

Among US women, African American women have higher mortality rates for cardiovascular diseases than white women[31] and account for a large share of the excess mortality of younger women with MI when compared with men. Because CHD develops earlier in African Americans than whites, there are proportionally more African Americans among younger than older MI patients. When race is considered as an additional stratification factor, the group that shows the highest mortality rate, compared with white males, is the younger African American women.[32]

20.2.1 Role of Comorbidity and Risk Factors

Because CHD onset is typically delayed in women when compared with men, women with ACS are on average older and have more comorbidity and risk factors such as diabetes, hypertension, renal insufficiency, and congestive heart failure. However, the higher prevalence of risk factors in women is not only due to their older age, but is seen virtually in every age interval. The difference with men is more marked among younger patients, the same group that shows the highest excess mortality in women when compared with men. One possible reason for this is that a higher "load" of risk factors may be needed to overcome the protective effect of female sex, since women, particularly if they are young, are normally "protected" against atherosclerotic cardiovascular disease, compared with their male counterparts. Another possible explanation is a different threshold for hospital admission, or for referral, in women than in men with myocardial ischemia. This may occur if more severely affected female patients are admitted or evaluated relative to men, while milder cases among women are delayed, deferred, or entirely missed. Indeed, missed diagnoses of acute cardiac ischemia in the emergency department are more frequent in women, particularly women younger than 55 years.[33]

It is also possible that men with high risk acute coronary events are more likely than women to die out of the hospital. This may create a cohort of survivors who eventually are admitted to the hospital among whom women have a higher risk status than men because the corresponding men have already died in the prehospital phase. Some large population-based MI registries support this hypothesis,[34] but data are not consistent, possibly due to variations and difficulties in ascertainment of prehospital causes of death.

20.2.2 Symptom Recognition, Diagnosis, and Referral

On average, women experience a longer time interval between onset of symptoms for acute coronary events and receipt of emergency care or hospitalization.[35] Many factors may contribute to this longer delay of women, such as lack of symptom recognition, lack of awareness about individual risk, provider misdiagnosis, or reluctance to call emergency care services. Among these, lack of recognizing symptoms has been suspected to play a major role, as lay people may not recognize heart disease as an important health issue for women. In community surveys, women often describe a heart attack as a male problem and more often attribute symptoms of ischemia to other chronic noncardiac conditions.[36] Although awareness of CHD risk in women has increased in recent years, a significant gap between perceived and actual risk remains.[37]

Differences in presentation of acute ischemia in women compared with men have long been implicated in the lower awareness of CHD symptoms among women and their delay in seeking emergency care for acute CHD. They are also thought to play a role in the inability of some providers to recognize an ACS in women, leading to an incorrect diagnosis and delays in the initiation of appropriate care. Indeed young age, female sex, and the absence of chest discomfort are among the most important predictors of a missed diagnosis of MI and inappropriate discharge from the emergency department.[33,38]

Historically, the description of symptoms associated with ACS has been based on the symptom presentation of men, which is put forward as the "classical" presentation. Women's presentation is instead often described as "atypical" and as being characterized by a constellation of associated and prodromal symptoms other than chest discomfort.[39,40] It is important to recognize, however, that the most common presentation of an ACS is chest pain in both women and men. A recent review of 69 studies of ACS showed that although more women (about 37%) presented without chest pain or discomfort than men (about 27%), chest pain or discomfort was the most common presenting symptom in both sexes.[41] Yet these data bring about another important consideration: the so-called "classical" chest pain presentation is not quite classical for either sex, since about one-third of patients overall present without it.

20.2.3 ACS Management

Sex differences in cardiac care have been described for at least 20 years, with women with symptomatic CHD being treated less aggressively than men. Recommended therapies for ACS, such as aspirin, beta blockers, fibrinolytic drugs, angiotensin-converting enzyme inhibitors, and primary angioplasty, have roughly a comparable benefit in women and men. Therefore, there is no empiric evidence supporting differences in treatment based on sex. Fortunately, sex-related variations in the use of evidence-based therapies tend to be small, while larger differences apply to more discretionary interventions with less proven benefit, such as revascularization procedures.[42,43]

An important consideration is that even though sex-related management differences are observed, they do not necessarily help explain outcome differences between women and men. A large study of MI patients found that cardiac care in the year following the MI became progressively less aggressive among older women relative to men, while survival tracked in the opposite direction, with older women clearly favoured.[29] Another study estimated that lower use of procedures after MI in women relative to men explained less than half a percentage point of the sex differences in mortality rate.[44] Based on these findings, differences in clinical care should not play a major role in explaining sex-related outcome differences after ACS. Differences in referral or intrinsic biological or psychosocial factors may be more important.

Fibrinolytic Therapy. Fibrinolytic therapy has significantly decreased the early mortality of STEMI in both women and men,[45] but the mortality risk reduction is somewhat lower in women, despite similar rates of successful coronary reperfusion after fibrinolysis. Hemorrhagic stroke and other major bleeding complications are more common in women, particularly elderly women. The risk of reinfarction after fibrinolysis is also greater in women. Unadjusted mortality rates were strikingly higher in women than in men in all fibrinolytic therapy trials, but the difference was partially accounted for by women's less favorable baseline characteristics. Because of these higher complication rates, women should be monitored closely after thrombolytic therapy. However, this lifesaving treatment should not be withheld or delayed in women when it is indicated.

Coronary Artery Bypass Graft Surgery (CABG). Whether female sex is a risk factor for adverse outcomes after CABG has been controversial for many years. Even recent studies have continued to report disparate results, showing either an increased risk of death in women, no mortality differences by sex, or even a better long-term survival in women. The bulk of the data, however, points to higher short-term mortality and complication rates after CABG in women when compared with men, and lower functional benefits. Again, when examining sex differences, younger women show the largest excess mortality when compared with men of similar age.[46] The issue is difficult to resolve because we can only rely on observational data, and women undergoing CABG are substantially different from men: they have more comorbidity, smaller coronary artery size, and more peri-procedural risk factors such as renal insufficiency, hemodilutional anemia, and hyperglycemia. In addition to comorbidity, it is likely that procedural factors play a role. For example, the use of internal thoracic artery graft and of off-pump CABG procedures may be especially beneficial in women. Unfortunately, although women account for about one-third of all CABG procedures, there are virtually no clinical trial data to help decide their risks and benefits for undergoing CABG because randomized trials have included almost exclusively men. Since CABG has become part of the standard of care, it is unlikely that new randomized trials in women will be feasible.

Since women live an average of 5 years longer than men, if they survive the operative period they should be expected to live at least as long, if not longer, than men after CABG. This has not been a consistent finding. However, an encouraging fact is that operative mortality has declined substantially in recent years in both sexes, but more so in women, resulting in narrowing of the sex-related mortality gap.[47]

Since CABG yields only a small absolute survival benefit relative to medical therapy, symptom relief and long-term improvement in function and quality of life are the most common indications for CABG. On average, women obtain less symptom relief and lower functional gains after CABG than men do.[48] Therefore, also for functional outcomes women fare less well than men. A way to help with women's recovery is to maximize their postoperative care and tailor it to women's needs, as well as encourage cardiac rehabilitation, for which women show much lower attendance than men.[49]

Percutaneous Coronary Interventions (PCI). More than one million PCI procedures are performed in the United States annually, of which about 33% are performed in women. As for CABG, the procedural outcomes in women tend to be less favorable than for men, with higher rates of short- and long-term mortality, cardiac events, and emergency CABG. The smaller body habitus of women has traditionally being implicated, because it is associated with smaller vessel diameter that, in turn, is a risk factor for periprocedural vascular complications.

While women's outcomes have improved in recent years and the sex gap in mortality has decreased, some recent series continue to report substantially higher short-term mortality in women than men.[50] Current hospital mortality after elective PCI, however, is quite low, less than 1%, in both women and men. Procedural complications are similarly rare, but they also tend to be more common in women, including contrast-induced nephropathy, bleeding, stroke, urgent CABG, and vascular complications. Again, younger women are at especially higher risk for PCI complications than men of similar age.

Glycoprotein IIb/IIIa inhibitors reduce major adverse outcomes equally in women and men undergoing PCI. These agents are underused in women, but they are also associated with an increased bleeding risk in women, which is partially related to excessive dosing. Women have almost a four-fold risk of receiving excess dosage of glycoprotein IIb/IIIa inhibitors than men.[51] Thus, some of the excess mortality and bleeding complications of women undergoing PCI could potentially be reduced by appropriate use and dosing of glycoprotein IIb/IIIa inhibitors.

Cardiac Rehabilitation. A key component of the management of patients recovering from ACSs is cardiac rehabilitation, which has documented benefits in promoting coronary risk reduction and improving functional status in both women and men. As for other management strategies, there is a differential use by sex, with fewer women than men being referred to cardiac rehabilitation programs. Referral of women to cardiac rehabilitation offers yet another opportunity for improving the clinical outcomes of women after ACSs.[49]

20.2.4 Psychosocial Factors

Women with ACS have a higher psychosocial burden than men, in particular depression, which is related to CHD prognosis either by influencing lifestyle and health behaviours, or through stress-related pathways implicated in myocardial ischemia, atherosclerosis, and arrhythmogenesis.[52] Depression is found in up to 40% of young women with MI.[53] The higher prevalence of depression in women compared with men contributes to women's higher rates of adverse outcomes, particularly rehospitalization and angina. Thus, it is important to screen for depressive symptoms at the time of hospitalization for ACS. However, intervention studies are needed to establish whether improved recognition and treatment of depression decrease sex differences in outcome of CHD.

20.2.5 ACS in Women: Summary

There are important differences in presentation, clinical profile, and outcome of women and men with ACS. An acute MI abolishes the survival advantage of women over men even though women have less extensive coronary atherosclerosis as assessed by angiography. There are also important sex-related differences in response to treatments, with women often showing greater propensity for complications. The substantial burden of comorbidity and psychosocial risk factors in women with acute coronary ischemia call for management strategies that are tailored to the unique needs of women.

20.3 Diagnostic Testing for Assessment of Cardiac Symptoms in Women

We have, in the prior sections, discussed the presentation and treatment of myocardial ischemia in women. Within the current section, we highlight the evidence for the role of noninvasive testing in the evaluation of suspected myocardial ischemia. Early evidence reported a consistently lower diagnostic accuracy for women undergoing many stress testing modalities including ECG, myocardial perfusion SPECT (MPS), and echocardiography.[54,55] The rationale for this diminished accuracy is varied but may be summarized as related to hormonal influences, a lower prevalence of CAD, diminished functional capacity, and differences in body habitus. More recent advances in cardiac imaging have improved many of these artifact challenges and will be discussed in more detail under related modality sections. However, at the core of comprehending test accuracy is a discussion of the principles of Bayesian theory. Bayesian theory dictates that the accuracy of a test is defined by a patient's pretest likelihood of CAD, such that the accuracy of stress testing in low risk male or female patients is diminished when compared with intermediate-high CAD likelihood patients. Thus, a first step in the test decision-making process is to define the likelihood of CHD for any female or male patient. As stated above, the prevalence of CAD is less in women when compared with men[56] and, as such, women are, in general, at lower risk. However, calculation of pretest risk is important to aid in test selection and there remain a sizeable proportion of at-risk women who are largely undertested.[55]

Women with chest pain symptoms are generally low risk prior to menopause (~age 50 years); with exception, diabetics or those with significant comorbidity, multiple risk factors, or those with noncardiac atherosclerosis. Current guidelines from the American Heart Association (AHA) and American College of Cardiology (ACC) do not recommend stress testing in patients due to the increased frequency of false-positive findings.[55,57,58] However, queries about a low-risk woman's provocative stressors for pain as well as her functional capabilities during activities of daily living can provide insight into her lifestyle accommodation and impairment due to symptoms.

Women who are at intermediate-high pretest CAD likelihood form the core of referrals to noninvasive stress testing. It is important to note that high-risk women with unstable symptom presentation should generally be evaluated with coronary angiography. The ACC guidelines for exercise testing have devised a table for identifying intermediate-high CAD likelihood women (Table 20.1).[57] Women capable of achieving at least 5 metabolic equivalents (METs) or higher of physical work should undergo exercise testing. The protocol that should be utilized in women should start at ~3 METs of work and increase linearly in stages or increments of 1–2 METs. This type of protocol is defined as a linear protocol and diminishes the problem of premature fatigue that is observed in most aggressive protocols such as the Bruce (stage 1 = 4.7 METs with per stage increases of 2–3 METs).

If a woman has a normal rest ECG and good exercise abilities, an exercise electrocardiogram is currently indicated based on the ACC/AHA guidelines.[57] If resting ST-T wave abnormalities exist on the 12 lead resting ECG, interpretation of peak exertional changes is difficult. In the latter case, referral to an imaging modality such as stress echocardiography or MPS is supported by current clinical practice guidelines.[55] Importantly, women evaluated with cardiac symptoms who are not capable of 5 METs of work, those with submaximal, premature fatigue on an exercise ECG, or those who have difficulties during the household chores should be referred to a pharmacologic stress test.

We recommend that a Duke Activity Status Index (DASI) questionnaire be used for women referred for a pharmacologic stress test or for those with suspected functional disability (due to limitations in everyday activities).[59] The DASI is a simple 12-item questionnaire that provides an estimate of METs.[59] Questions within the DASI include those involving her performance capabilities for household chores and recreation. Women who score <5 METs on the DASI should be referred to pharmacologic stress. When pharmacologic stress imaging is performed, the addition of a DASI estimate of METs may aid in discerning a patient's symptom burden and its impact on daily living. Data from the NIH-NHLBI-sponsored WISE trial reveals that 5-year cardiovascular death or MI-free survival ranges from 83 to 95% for MET levels ranging from ≤4.7 to >9.9 METs.[59] This publication evaluated

DASI subsets corresponding to stages of the Bruce protocol so that clinicians may interpret the questionnaire's result and project a patient's performance during the exercise treadmill test.

20.3.1 Exercise ECG

The exercise ECG is performed in approximately 12 million US patients each year with nearly half being women.[55] It is essential that physicians performing treadmill testing in their office acquire training in exercise physiology and ECG interpretation as it applies to women. This is often a difficult challenge for the busy clinician but the resulting knowledge base can aid dramatically in crafting accurate diagnostic pathways for women and men alike.

Given the increasing prevalence of obesity and related declines in functional capabilities, the potential candidate pool for exercise testing has gradually diminished over the last decade. An important consideration for the large proportion of obese patients is that orthopedic limitations and functional disability are common. Thus, in the obese patient, it is more difficult to achieve maximal levels of exercise and premature fatigue often precludes provocation of myocardial ischemia. Additionally, many treadmills have weight limits that constrain the evaluation of morbidly obese patients. In these cases, referral to pharmacologic stress imaging should be considered.

An exercise test includes performance of incremental stages of increasing speed and slope of the treadmill. Test stages usually last 1–3 min. As noted previously, a linear protocol with 1 MET increases in exercise is preferable for women. Testing is continued until maximal heart rate or volitional fatigue occurs. With little exception, the occurrence of marked ST segment depression, chronotropic incompetence, impaired systolic blood pressure response to increasing physical work, and ventricular tachycardia/fibrillation are high-risk findings and reason for immediate termination of testing. It should be noted that an impaired heart rate or dampened blood pressure response during testing is associated with an increased risk of CHD events.[60] For example, failure to increase heart rates (in patients not taking beta blockers) during exercise to more than 110–120 beats/min is associated with left ventricular dysfunction. As well, a minimal increase or drop in systolic blood pressure is associated with a greater frequency of multivessel CAD and a depressed ejection fraction.[60]

Should significant ST segment depression occur during exercise testing, this is defined as demand or work-related ischemia that generally occurs in the setting of a flow-limiting coronary stenosis. A primary aim of the treadmill test is to achieve adequate levels of exercise such that ischemia may be provoked. It is this premise of provoking ischemia that underlies the importance of choosing a more gentle protocol for women. Examples of female "friendly" protocols include the Balke or Asymptomatic Cardiac Ischemia Pilot (ACIP).[59]

Of the exercise test risk markers, the occurrence of exertional chest pain symptoms has a low diagnostic and prognostic accuracy in women.[61] Additionally,

ST segment changes (in general) are associated with diminished accuracy in women when compared with men; with approximately 10% lower sensitivity and specificity values. Meta-analysis reveal a modest diagnostic sensitivity (61%) and specificity (70%) for ST segment changes in women.[55] In particular, women have a higher rate of false-positive ECG results that may be due to hormonal influences, lower QRS voltage, or a lower CAD prevalence. With regard to the latter, using Bayesian principles, testing should exclude low-risk women where minimal shifts in post-test risk are possible and high false-positive test results likely.

Importantly, should testing be performed in premenopausal women, interpretation of ECG changes should be performed cautiously because of the low prevalence of CAD and the resulting high false-positive rate.[55] It has been postulated that endogenous estrogen may exert a digoxin-like effect on the ECG resulting in ST segment depression for the premenopausal women that is unrelated to underlying obstructive CAD. For premenopausal women, angina has been noted to vary with the menstrual cycle and to occur more often mid-cycle where estrogen levels are lowest. Physicians should take care to interview premenopausal women as to their anginal patterns with regard to their menstrual cycle as well as to schedule testing during mid-cycle. Moreover, for the postmenopausal women, poor performance on the treadmill with a failure to achieve maximal levels of physical work may also lead to false-negative findings.

There are scenarios where ST segment depression is an accurate marker of underlying obstructive CAD in women, particularly severe ST segment depression at lower workloads. Horizontal or downsloping ST segment depression ≥1.0 mm at an early workload (i.e., <5 METs) is a very sensitive and accurate marker of obstructive CAD in women and men alike. A similarly accurate marker of CAD for both women and men is the persistence of ST segment changes beyond 5 min of recovery. These high-risk markers occur less often in women but are harbingers for significant CAD. The slope of the ST segment that may provide important clues as to the accuracy of ECG changes in women. As heart rate increases, rapidly upsloping j point depression may occur and should only be considered abnormal if ≥1.5 mm of ST segment depression occurs (at 60–80 ms after the j point).

The Duke treadmill score (DTS) is a common index used to define risk during treadmill testing that is available on most ECG treadmill systems. It is calculated as exercise time – [5 * ST deviation] – [4 * chest pain index (0=none, 1=nonlimiting, 2=limiting)] where low-risk scores are 5 or higher and high-risk scores <–10.[62,63] The DTS has female-specific prognostic data in a large cohort of 976 women.[61] From this report, 5-year CHD death rates range from 5 to >10% for women with low to high DTS results. In 2,249 men, the CHD mortality rates ranged from <9 to >25% for low to high-risk DTS.

Although women are generally more functionally impaired and experience a greater age-related decline in physical capabilities than their male counterparts, exercise duration remains the strongest prognostic factor from the treadmill test.[64-66] Several recent publications have highlighted the prognostic accuracy of exercise METs. From the Lipid Research Clinics, a total of 2,994 women were enrolled and followed for more than 20 years.[64] Women achieving 9.3 METs or higher had the lowest mortality rates (<0.4%/year) while high-risk females included those unable to walk for >5.4 METs.[64] A second report from the St. James Women Take Heart Project ($n=5,721$) revealed similar findings.[65] This same investigative group devised normative age-specific standards for exercise METs in a cohort of nearly 10,192 asymptomatic and symptomatic women[66] (Fig. 20.1). These results can help define expected performance on a treadmill test for women of all ages. As seen in Fig. 20.1, physical work capacity declines with age; with this latter statement being true for women and men alike. But, importantly, reviewing the data for elderly women, many will be incapable of 5 METs of exercise and, therefore, the lion's share of this patient subset will be referred for pharmacologic stress imaging. The DASI, as noted previously, may aid in this decision point as to who are appropriate candidates for pharmacologic stress testing.

Heart rate recovery is another strong prognosticator that is measured by examining the time course of restoration of vagal tone during recovery. At the onset of exercise, there is a withdrawal of vagal tone and the sympathetic nervous system is activated to increase heart rate, allowing for performance of higher levels of physical work. During recovery, restoration of vagal tone and ensuing decline of heart rate should occur promptly within 1–2 min. However, a failure of heart rate to decline was associated with insulin resistance and an elevated mortality risk in large cohorts of women and men.[55]

Given the abundant evidence regarding exercise capacity, heart rate changes, as well as integrated scores (e.g., DTS), these measures should be integrated into a woman's exercise test interpretation. Moreover, these risk markers should be reported for the exercise portion of a stress imaging modality.

20.3.2 Stress Cardiovascular Imaging

The next portion of this chapter highlights the evidence base for the role of an array of stress imaging modalities and their accuracy in women. As a preamble to this discussion, it is important to understand the types of available risk markers and the

Age (Decade)	Typical Angina	Atypical Angina	Nonanginal Chest pain
50	**Intermediate**	**Intermediate**	Low
60	**High**	**Intermediate**	**Intermediate**
70 or Older	**High**	**Intermediate**	**Intermediate**

Fig. 20.1 Pretest coronary artery disease likelihood estimates based on age, gender, quality of chest pain symptoms

underlying pathophysiology associated with test abnormalities. As such, a discussion of the ischemic cascade is warranted as it relates to our understanding of accuracy and risk detection in women. Figure 20.2 provides details of the ischemic cascade including identifying the place and importance of imaging risk markers. Within this cascade, myocardial perfusion abnormalities occur early and can be associated with intermediate-severe coronary stenosis. As such, MPS will be able to detect women with less extensive CHD. Several modalities (i.e., SPECT, positron emission tomography [PET], and CMR) can also provide a measurement of myocardial perfusion reserve, a measure of endothelial function. Endothelial dysfunction occurs early in the atherosclerotic disease process and may be a helpful measure in women who are likely to have less extensive or mild CHD. However, caution should be applied to the use of nuclear imaging techniques with a substantive radiation burden in lower risk women. For this latter group, use of brachial artery reactivity testing or peripheral arterial tonometry may be more appropriate. More recent developments in the field of imaging now allow for discerning perfusion in the subendocardial region of the heart (i.e., PET, MR). Subendocardial ischemia is an early manifestation and may prove to be a useful method to assess risk in women.

A wall motion abnormality occurs later in the ischemic cascade and is generally associated with more severe coronary stenosis. Thus, for echocardiographic, magnetic resonance, or SPECT wall motional abnormalities, there is a greater likelihood of severe stenosis when stress-induced hypokinesis is noted. This likelihood is further accentuated in the setting of stress-induced akinetic or dyskinetic wall motion abnormalities. Thus, care should be taken when referring lower risk women

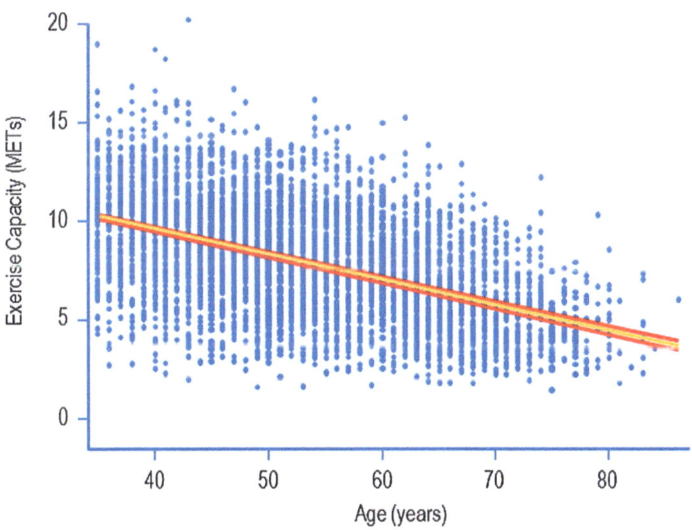

Fig. 20.2 Prognostic value of exercise capacity in womens: nomogram for female populations

to a test providing only a measurement of stress-induced wall motion abnormalities (e.g., echocardiography). However, tests such as echocardiography are generally low cost and provide an excellent test for women owing to its high negative predictive accuracy.

20.3.3 Stress Echocardiography

Stress echocardiography is a widely used tool for the assessment of ventricular function and wall motion abnormalities. The test has advantages over other imaging modalities, such as MPS, PET, or computed tomography (CT), in that it does not expose the patient to ionizing radiation and the procedural time is sizably shorter. This procedure "adds on" to an exercise ECG by providing an ultrasound visualization of rest and inducible wall motion abnormalities and left ventricular ejection fraction. Female patients who are obese or those with chronic lung disease may have suboptimal visualization of the left ventricle, in particular the endocardial border for determination of wall motion. More recent advances in the field have allowed for an intravenous contrast agent (e.g., Definity or Optison) that provides enhanced left ventricular opacification.[55] Contrast agents are encapsulated gas-filled microbubbles that allow for enhanced visualization of regional left ventricular function prior to exercise and in the poststress evaluation. However, following recent safety concerns, the FDA has added caution to the use of perflutren-containing ultrasound contrast agents in patients with ACSs, acute MI, and worsening or clinically unstable heart failure.[67] This label change, however, does not reflect the reported clinical utility of these agents for the assessment of stable chest pain.

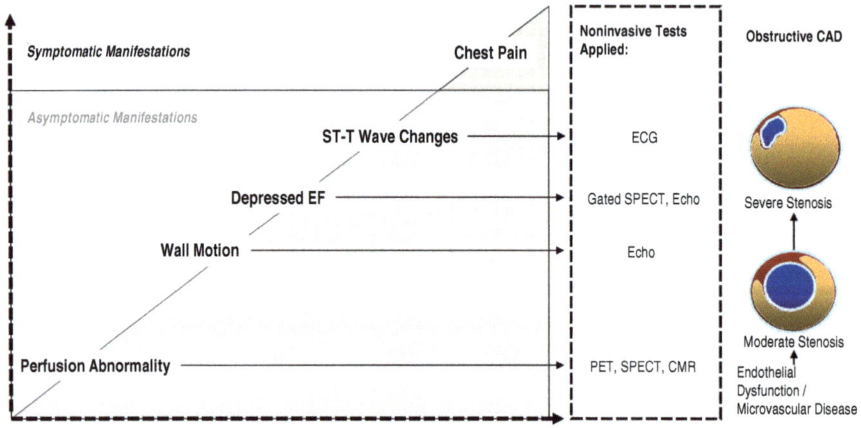

Fig. 20.3 Ischemic Cascade and associated noninvasive tests applied as well as associated obstructive coronary artery disease (CAD) states

Candidates for stress echocardiography include women presenting for evaluation of cardiac symptoms who have limited functional capabilities or have resting ST-T wave changes. Also included in this candidate pool are women with known CHD. For women with known CHD and resting wall motion abnormalities, discerning inducible changes provides a difficult challenge. For those with resting segmental wall motion abnormalities, ischemia is defined as deterioration from hypokinesis to akinesis or dyskinesis. Additionally, an abnormal or ischemic response is also defined as a biphasic response (i.e., initial improvement in wall motion followed by deterioration).

Data are available not only for exercise but also for pharmacologic stress echocardiography. In the US, the most common pharmacologic stress agent is intravenous dobutamine, with incremental doses given to achieve predicted maximal heart rate levels (similar to exercise testing). In most of Europe, the pharmacologic stress agent is intravenous dipyridamole, a vasodilating agent that in presence of impaired vascular function and significant obstructive CHD will elicit wall motion abnormalities. Vasodilating agents, such as dipyridamole and adenosine, are also commonly used for PET and SPECT techniques, as discussed later in this chapter.

There is a large body of evidence on the role of stress echocardiography in women including several meta-analyses on its diagnostic and prognostic accuracy.[68-77] Importantly, recent data support that stress echocardiography is a cost-effective choice for stable chest pain patients when compared with exercise ECG.[69,71] There is only a minimal added cost for the addition of echocardiography to an exercise ECG, with a decidedly improved diagnostic accuracy.[68,69,72-74] From one meta-analysis, the diagnostic sensitivity and specificity of exercise ECG was 61 and 70%. By comparison, there are three published meta-analysis on stress echocardiography reporting higher accuracy statistics (sensitivity range = 81–86%, specificity range = 73–79%).

There is also a robust body of evidence as to the prognostic accuracy of inducible wall motion abnormalities during stress echocardiography.[70,75,76] The results report a consistently low rate of CHD events following a negative stress echocardiogram.[70] From a recent systematic review,[70] the observed negative predictive value for stress echocardiography was 98.4% (97.9–98.9%). This included observations in 3,021 patients followed for an average of 33 months with a 1.56% (1.14–2.07%) summary rate of CHD death or nonfatal MI.[70] From this meta-analysis, the annualized event rate with a negative exercise echocardiogram was low: 0.75% for women and 1.24% for men. The CHD event rate increases with the number of territories with new or worsening wall motion abnormalities. As reported in a large series of 4,234 women undergoing stress echocardiography for evaluation of chest pain symptoms, CHD mortality rates were 3- (for dobutamine stress) to 10- (for exercise stress) fold higher for women with multiple vascular territories with worsening wall motion abnormalities.[76] In fact, annual CHD mortality was 0.1% for a negative exercise echocardiogram but increased to 1.0% per year for women with ischemia in multiple vascular territories ($p < 0.0001$). Women referred to dobutamine stress are at higher clinical risk with more comorbidity with resultantly elevated CHD event rates. Annual CHD mortality rates ranged from 1% for women with no wall motion abnormalities to 3% for those with multivessel ischemia during dobutamine echocardiography ($p < 0.0001$).

Thus, stress echocardiography remains a valuable test for the assessment of suspected myocardial ischemia with a robust body of evidence as to its utility in women and men. Although supported by clinical practice guidelines, the application of this modality is generally limited to lower risk patients with more atypical symptom presentation. In this manner, a negative exercise echocardiogram has a very high negative predictive accuracy and is evidence of a low CHD risk. Although stress echocardiography is certainly a specific diagnostic test, other modalities (e.g., MPS) if negative are associated with event rates that are decidedly lower than for negative echocardiography. For in women, a negative exercise echocardiogram is associated with an annual CHD death or MI rate of 0.75% when compared with a rate of 0.33% for women undergoing exercise MPS.[70] Of course, local expertise will ultimately guide test decision-making, but it appears from the literature that echocardiography may be optimally applied in lower risk women while stress MPS is preferred for higher risk women. There are several reasons for this, including limiting radiation exposure to those who clearly benefit from its application. Moreover, echocardiography, given its minimally added cost, may be viewed as an effective "add-on" to the exercise ECG. Data are available as to the incremental cost-effectiveness of echocardiography vs. MPS.[77] Marginal cost-effectiveness ratios were ~$20,000 per life year saved for echocardiography when compared with MPS when the annual death or MI rate was <2%. These results would favor the use of stress echocardiography for patients with suspected CHD capable of performing maximal exercise. By comparison, these results also favor stress MPS for higher risk patients including those with limited functional capacity, diabetics, or those with known CHD.

20.3.4 Stress Gated MPS and Positron Emission Tomography (PET)

MPS is the most commonly performed procedure, with more than 8 million scans done each year in the US. Of these, nearly 40% are for women. Stress MPS is a nuclear-based technique and its use should be limited to postmenopausal women or for selected high-risk premenopausal females. The first radioisotope commonly used over the past few decades was Thallium-201 (Tl-201). A higher rate of breast tissue artifact (i.e., false-positives) is reported with Tl-201 for women and obese men when compared with the newer technetium-99m (Tc-99m) agents.[55] Thus, Tc-99m results in an improved diagnostic specificity and is the recommended agent of choice for the evaluation of chest pain in women.[55] Breast tissue artifact is reduced when using the Tc-99m agents, but it remains the number one reason for a false-positive exam in women and obese men. For women, high diagnostic specificity measures can be achieved if image processing includes attenuation correction algorithms or prone imaging.[78,79] The use of a rest Tl-201 combined with stress Tc-99m MPS is associated with decidedly higher radiation exposure and should not be used in women.

Contemporary rest/stress MPS with Tc-99m now includes the possibility of gating images to acquire quantitative left ventricular volumes and ejection fraction and semi-quantitative measurements of wall motion during the rest and stress scans. Although diagnostic sensitivity has always been excellent with stress MPS (~85%), the addition of wall motion and global/regional ventricular function results in an improved diagnostic specificity (within the range of 85–90%).[55] The integration of multiple risk markers allows for a better detection of risk, especially for women. That is, in the setting of borderline myocardial perfusion abnormality, should wall motion, function, and thickness be normal, this scan should very likely be called normal.

Several meta-analyses are available as to the accuracy of stress MPS for risk stratification purposes.[70,76,80] Multiple large multicenter and single site registries reporting gender-specific data on prognostication have also been published.[81-83] Stress MPS has been shown to be a cost-efficient strategy for women when compared with invasive angiographic approaches.[84] The prognostic data also show a low rate of CHD events in patients with normal stress perfusion with annual death or MI rates of 0.33% for women.[70] In fact, in one large series including more than 3,402 women, the 3-year survival with normal stress perfusion findings was 98.5% for both male and female patients.[81] By comparison, moderate-severely abnormal stress MPS is associated with a 10-fold increase in CHD events.[76] In several series, the extent and severity of stress perfusion abnormalities was associated with increasing risk; however, MPS has been shown to be gender-neutral in terms of its interpretation.[76,81] That would mean that the interpretation of MPS abnormalities and their ensuing CHD risk is generally similar by gender. With one exception: several reports noted a substantially higher CHD mortality and event risk for female than male diabetics.[55,82-84] The AHA guidelines estimate that the overall CHD risk for ischemic abnormalities is approximately 50% higher for female diabetics when compared with women without diabetes and men.[55] In one recent report, women requiring insulin to manage their diabetes had a twofold higher CHD mortality risk when their MPS study was abnormal.[83]

Higher CHD death or MI rates can be observed in women with high-risk findings including perfusion abnormalities encompassing ≥10% of the myocardium, multiple vascular territories with stress defects, transient ischemic dilation, or poststress left ventricular ejection fraction measurements of 45% or lower.[55] Anterior defects encumbering a larger proportion of the myocardium can also accelerate risk in women and men.

Much of the evidence presented with regard to MPS also applies to perfusion imaging with PET. There is minimal prognostic literature with PET, but the data that are available reveal similar abilities to risk stratify based on the extent and severity of perfusion abnormalities.[85] However, several differences are notable with PET imaging; in particular, as they relate to female patients. First, image quality is generally superior with PET. This may be particularly helpful for women who are obese. It is unclear what percentage of obese patients referred to MPS have an indeterminate study but if followed by a second study using PET imaging, a large number of these studies elicit normal findings. Second, the overall radiation exposure is less with PET. Third, PET is able to provide estimation of absolute myocardial blood flow or an estimate of myocardial perfusion reserve. This may prove helpful for women whose cardiac symptoms are the result of endothelial dysfunction. An additional measure available with PET is that of contractile reserve.

Finally, most PET scanners sold today include hybrid CT. This is generally used for attenuation correction purposes and is helpful to improve diagnostic specificity. However, the addition of a CT scan now allows us to "add on" coronary artery calcium (CAC) or angiographic measurements. As we will discuss in more detail, CT angiography is associated with excessive radiation when combined with PET imaging (at this time), but the addition of a CAC scan can be very helpful to provide anatomic detail as to the underlying calcified plaque burden as it relates to a patient's ischemic burden or vascular function. The addition of a CAC scan is associated with an approximately 1 mSv of radiation and, for the right patient, the added benefit of this information may well be worth it. A CT scan can also be coupled with SPECT for hybrid imaging; although this application has not gained as much enthusiasm as the potential for PET-CT. Let us illustrate a case where combined imaging may be helpful for women. Given the high rate of false-positive findings, should a perfusion defect be observed, documentation of underlying CAC may aid in improving the precision of interpretation. Moreover, for higher risk diabetics, normal findings in the setting of significant CAC should foster more intensive management, despite the low-risk findings.

20.3.5 CT CAC and Angiography

As discussed above, a relatively new diagnostic modality is CT imaging. There are two applications that may be of value to the diagnostic and prognostic assessment of women, including defining the extent of CAC as well as utilizing a noninvasive assessment of angiographic disease burden. Several guidelines on this subject as well as a number of systematic reviews will be highlighted in this section.[86-88]

Most of the literature on CAC scanning is derived from asymptomatic, apparently healthy individuals revealing an exceptional ability to risk stratify women and men as to risk of CHD death or nonfatal MI.[89] A recent meta-analysis revealed that the extent of CAC conveys a similar prognosis in women and men.[88] In 6,481 women, annual rates of CHD death or MI ranged from 0.3 to 1.3% for low- to high-risk findings ($p<0.0001$). In this case, a low-risk CAC score is defined as <10 and a high-risk score is ≥400 using the Agatston score.[90,91]

The Agatston score is calculated by multiplying the calcified plaque area by a plaque density factor measured in Hounsfield units (scores up to 4). If a woman and man have the same CAC score, it will encumber a larger arterial surface in a female and potentially accelerate risk in this patient subset. As such, this score fails to consider arterial size and provides clues as to some intriguing recent gender-specific prognostic findings.[88,92] In an initial report by Raggi et al,[92] women with a given burden of CAC had a higher mortality risk when compared with their male counterparts. Interestingly, if we adjusted the CAC score by the average difference in epicardial coronary artery size, these mortality differences were attenuated.

In a related report, Bellasi and colleagues[88] noted that this higher CAC risk in women was apparent only for women with three or more risk factors. Thus, in those with a heavy risk factor burden, including women with the metabolic syndrome, a significant burden of CAC may be the key to defining at-risk asymptomatic women. Although few reports highlight the value of CAC testing in symptomatic women, a synthesis of evidence reveals that measureable CAC is a very sensitive marker for underlying atherosclerosis.[86]

CT is also capable of providing a noninvasive assessment of the extent and severity of obstructive CAD. Currently, the radiation exposure for CT angiography is roughly twice that of invasive angiography and limits its appeal for large segments of symptomatic patients. However, active research is ongoing and in the future, newer CT systems or protocols will provide an examination at around 3–4 mSv, equivalent to yearly background radiation. The 3 mSv is a goal for ongoing research providing an acceptable standard for this scan and is well below that of invasive coronary angiography. Given its current exposure, many hospitals limit the application of CT angiography to women of nonchildbearing age and, thus, it should be used cautiously in premenopausal women.

Several meta-analyses and systematic reviews on the diagnostic accuracy of CT angiography have been published,[93-97] the most recent of which reflected the updated information with 64-slice CT angiography.[96] Diagnostic sensitivity and specificity for the five published reports using 64-slice CT averaged 96 and 90%, establishing it as the most accurate of the diagnostic tests performed today. There is only one report in 52 women, which notes a diagnostic sensitivity of 95% and specificity of 93%.[98] There are also limited data on the prognostic accuracy of CT angiography including only five reports in 2,403 patients.[99-103] The pooled data reveal that a low-risk CT angiogram or no to mild CAD is associated with a 1 year CHD event rate of 0.6%. By comparison, obstructive CAD on CT angiography has a 1 year CHD event rate of 14.5%. No prognostic data are available for women, but it will be critical to evaluate the role of nonobstructive CAD, in particular for females with a greater prevalence of insignificant disease.[56]

20.3.6 CMR

The final stress imaging modality is CMR perfusion imaging, also advantageous due to a lack of ionizing radiation. CMR, as well as other modalities (e.g., echocardiography, PET), are capable of delineating epicardial from subendocardial perfusion. As initial manifestations of ischemia are first visualized within the subendocardium, it is likely that a failure to discriminate or identify ischemia in this region may underestimate risk, in particular for females. In a small yet intriguing report from the Royal Brompton Hospital,[104] women with normal coronary arteries and typical angina commonly had evidence of subendocardial ischemia. As such, it may be stated that these women may have "true" ischemia in the absence of obstructive CAD. For the nearly 60% of women undergoing coronary angiography with insignificant CAD, symptom etiology may be clarified and ensuing therapeutic options may be effectively targeted, should subendocardial ischemia be documented. Although this report has yet to be validated and no data are available on the prognostic relationship between subendocardial ischemia and outcomes, it provides hope for more gender-specific diagnostic testing patterns.

20.3.7 Optimal Patient Selection and Test Choice

In this section of the chapter, we have highlighted the current evidence base as to the role of a variety of diagnostic testing modalities for the evaluation of cardiac symptoms in women (Fig. 20.4). The guidelines provide information as to appropriate patient selection with testing limited to intermediate-high CAD likelihood women. Specifically, women with adequate exercise tolerance (e.g., DASI MET estimate of 5 or higher) and a normal rest ECG should be referred to a treadmill ECG test. There is a low likelihood of obstructive CAD and a low CHD event risk for women with a negative ECG and a good exercise workload (i.e., stage III of the Bruce protocol or >9 METs of exercise). Women with an abnormal resting ECG should be referred to a stress imaging modality. Should the woman have few risk factors, perhaps a more atypical symptom presentation, and be within the low-intermediate pretest risk range, then referral to stress echocardiography may be helpful. The addition of an echocardiographic evaluation provides minimal added cost to the

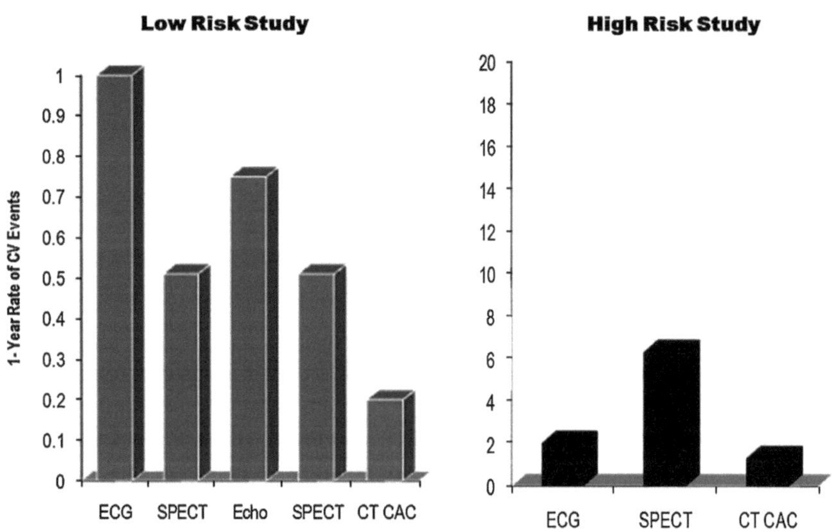

Abbreviations: ECG=Electrocardiography, Echo=Echocardiography, CT=Computed Tomography Coronary Artery Calcification.

Fig. 20.4 Comparative annual CV event rates for varying diagnostic testing modalities in women. The CT CAC data are in asymptomatic women

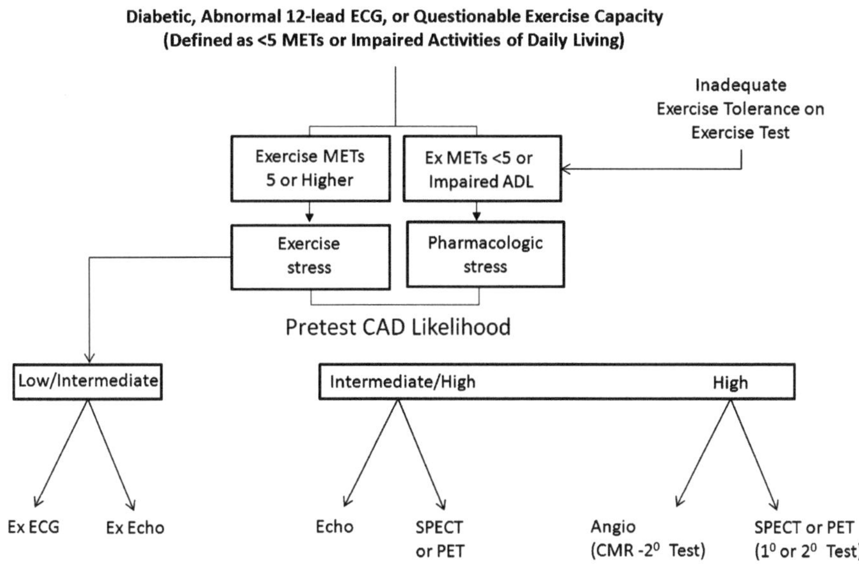

Fig. 20.5 Modified diagnostic testing algorithm for evaluation of women at intermediate-high CAD likelihood women with atypical or typical chest pain symptoms

work-up while providing information on ventricular function at rest and following stress testing. For intermediate-high risk women, the decision is to refer a woman to MPS or PET vs. angiography. A CT CAC scan may serve as an inexpensive method to discern the extent of underlying atherosclerosis and may soon provide an excellent complement to the physiologic data from PET or SPECT. In many cases, the extent and severity of perfusion ischemia will drive therapeutic decision-making for the women with angiographic CAD; as such, it may prove useful to have the patient undergo this procedure prior to an invasive procedure. However, should there be instability in her symptoms, including rest angina, or an accelerating pattern of angina, direct angiography is warranted. In the future, CT angiography may provide a viable noninvasive assessment of the extent and severity of CAD. But, as yet, the radiation exposure precludes its widespread utility. In the absence of obstructive CAD, an additional evaluation of vascular function or sub-endocardial ischemia may prove useful in discerning the etiology of symptoms.

We have provided a synthesis of the available evidence on the diagnostic evaluation of is ischemic symptoms in women. The utility of this information is constrained by the availability of advanced imaging techniques and local expertise. However, a key to optimal patient selection and test choice is to define the relationship between the atherosclerotic disease process and test markers, in particular as they relate to women. Given our unfolding knowledge on gender base differences, the marriage between physiologic and anatomic assessments provide the key to identifying at risk women.

References

1. American Heart Association. *Heart Disease and Stroke Statistics – 2008 Update*. Dallas, Texas: American Heart Association; 2008.
2. Pepine CJ, Abrams J, Marks RG, Morris JJ, Scheidt SS, Handberg E. Characteristics of a contemporary population with angina pectoris. TIDES Investigators. *Am J Cardiol*. 1994;74:226-231.
3. Lerner DJ, Kannel WB. Patterns of coronary heart disease morbidity and mortality in the sexes: a 26-year follow-up of the Framingham population. *Am Heart J*. 1986;111:383-390.
4. Hemingway H, Langenberg C, Damant J, Frost C, Pyorala K, Barrett-Connor E. Prevalence of angina in women versus men: a systematic review and meta-analysis of international variations across 31 countries. *Circulation*. 2008;117:1526-1536.
5. Hemingway H, McCallum A, Shipley M, Manderbacka K, Martikainen P, Keskimaki I. Incidence and prognostic implications of stable angina pectoris among women and men. *JAMA*. 2006;295:1404-1411.
6. Daly C, Clemens F, Lopez Sendon JL, et al. Gender differences in the management and clinical outcome of stable angina. *Circulation*. 2006;113:490-498.
7. Pepine CJ. Ischemic heart disease in women. *J Am Coll Cardiol*. 2006;47:S1-S3.
8. Johnson BD, Shaw LJ, Buchthal SD, et al. Prognosis in women with myocardial ischemia in the absence of obstructive coronary disease: results from the National Institutes of Health-National Heart, Lung, and Blood Institute-Sponsored Women's Ischemia Syndrome Evaluation (WISE). *Circulation*. 2004;109:2993-2999.
9. Olson MB, Kelsey SF, Matthews K, et al. Symptoms, myocardial ischaemia and quality of life in women: results from the NHLBI-sponsored WISE Study. *Eur Heart J*. 2003;24:1506-1514.
10. Vaccarino V, Parsons L, Every NR, Barron HV, Krumholz HM. Sex-based differences in early mortality after myocardial infarction. National Registry of Myocardial Infarction 2 Participants. *N Engl J Med*. 1999;341:217-225.
11. Pepine CJ. Ischemic heart disease in women: facts and wishful thinking. *J Am Coll Cardiol*. 2004;43:1727-1730.
12. Shaw LJ, Shaw RE, Merz CN, et al. Impact of ethnicity and gender differences on angiographic coronary artery disease prevalence and in-hospital mortality in the American College of Cardiology-National Cardiovascular Data Registry. *Circulation*. 2008;117:1787-1801.
13. Gibbons RJ, Abrams J, Chatterjee K, et al. ACC/AHA 2002 guideline update for the management of patients with chronic stable angina–summary article: a report of the American college of cardiology/American heart association task force on practice guidelines (Committee on the management of patients with chronic stable angina). *Circulation*. 2003;107:149-158.
14. Cohn PF, Fox KM, Daly C. Silent myocardial ischemia. *Circulation*. 2003;108:1263-1277.
15. Pepine CJ, Kerensky RA, Lambert CR, et al. Some thoughts on the vasculopathy of women with ischemic heart disease. *J Am Coll Cardiol*. 2006;47:S30-S35.
16. Belardinelli L, Shryock JC, Fraser F. The mechanism of ranolazine action to reduce ischemia-induced diastolic dysfunction. *Eur Heart J*. 2006;8(suppl A):A10-A13.
17. Belardinelli L, Shryock JC, Fraser H. Inhibition of the late sodium current as a potential cardioprotective principle: effects of the late sodium current inhibitor ranolazine. *Heart*. 2006;92(suppl 4):iv6-iv14.
18. Serruys PW, Unger F, Sousa JE, et al. Comparison of coronary-artery bypass surgery and stenting for the treatment of multivessel disease. *N Engl J Med*. 2001;344:1117-1124.
19. Hlatky MA, Boothroyd DB, Melsop KA, et al. Medical costs and quality of life 10 to 12 years after randomization to angioplasty or bypass surgery for multivessel coronary artery disease. *Circulation*. 2004;110:1960-1966.
20. Holmboe ES, Fiellin DA, Cusanelli E, Remetz M, Krumholz HM. Perceptions of benefit and risk of patients undergoing first-time elective percutaneous coronary revascularization. *J Gen Intern Med*. 2000;15:632-637.
21. Boden WE, O'Rourke RA, Teo KK, et al. Optimal medical therapy with or without PCI for stable coronary disease. *N Engl J Med*. 2007;356:1503-1516.

22. Chaitman BR, Skettino SL, Parker JO, et al. Anti-ischemic effects and long-term survival during ranolazine monotherapy in patients with chronic severe angina. *J Am Coll Cardiol.* 2004;43:1375-1382.
23. Chaitman BR, Pepine CJ, Parker JO, et al. Effects of ranolazine with atenolol, amlodipine, or diltiazem on exercise tolerance and angina frequency in patients with severe chronic angina: a randomized controlled trial. *JAMA.* 2004;291:309-316.
24. Stone PH, Gratsiansky NA, Blokhin A, Huang IZ, Meng L. Antianginal efficacy of ranolazine when added to treatment with amlodipine: the ERICA (Efficacy of ranolazine in chronic angina) trial. *J Am Coll Cardiol.* 2006;48:566-575.
25. Wenger NK, Chaitman B, Vetrovec GW. Gender comparison of efficacy and safety of ranolazine for chronic angina pectoris in four randomized clinical trials. *Am J Cardiol.* 2007;99:11-18.
26. Mega J, Hochman, JS, Scirica, BM, et al. *Anti-ischemic effects of ranolazine in women: Results from the randomized, placebo-controlled MERLIN-TIMI 36 Trial.* Presented at the 2007 Scientific sessions of the American heart association, Orlando, FL; 2007.
27. Hochman JS, Tamis JE, Thompson TD, et al. Sex, clinical presentation, and outcome in patients with acute coronary syndromes. *N Engl J Med.* 1999;341:226-232.
28. Chang WC, Kaul P, Westerhout CM, et al. Impact of sex on long-term mortality from acute myocardial infarction vs unstable angina. *Arch Intern Med.* 2003;163:2476-2484.
29. Alter DA, Naylor CD, Austin PC, Tu JV. Biology or bias: practice patterns and long-term outcomes for men and women with acute myocardial infarction. *J Am Coll Cardiol.* 2002;39:1909-1916.
30. Rosengren A, Spetz CL, Koster M, Hammar N, Alfredsson L, Rosen M. Sex differences in survival after myocardial infarction in Sweden; data from the Swedish national acute myocardial infarction register. *Eur Heart J.* 2001;22:314-322.
31. Rosamond W, Flegal K, Friday G, et al. Heart disease and stroke statistics–2007 update: a report from the American heart association statistics committee and stroke statistics subcommittee. *Circulation.* 2007;115:e69-e171.
32. Manhapra A, Canto JG, Vaccarino V, et al. Relation of age and race with hospital death after acute myocardial infarction. *Am Heart J.* 2004;148:92-98.
33. Pope JH, Aufderheide TP, Ruthazer R, et al. Missed diagnoses of acute cardiac ischemia in the emergency department. *N Engl J Med.* 2000;342:1163-1170.
34. MacIntyre K, Stewart S, Capewell S, et al. Gender and survival: a population-based study of 201, 114 men and women following a first acute myocardial infarction. *J Am Coll Cardiol.* 2001;38:729-735.
35. Goldberg RJ, Gurwitz JH, Gore JM. Duration of, and temporal trends (1994–1997) in prehospital delay in patients with acute myocardial infarction – The second national registry of myocardial infarction. *Arch Intern Med.* 1999;159:2141-2147.
36. Finnegan JR Jr, Meischke H, Zapka JG, et al. Patient delay in seeking care for heart attack symptoms: findings from focus groups conducted in five U.S. regions. *Prev Med.* 2000;31:205-213.
37. Mosca L, Ferris A, Fabunmi R, Robertson RM. Tracking women's awareness of heart disease – An American heart association national study. *Circulation.* 2004;109:573-579.
38. McCarthy BD, Beshansky JR, Dagostino RB, Selker HP. Missed diagnosis of acute myocardial infarction in the ED: results from a multicenter study. *Ann Emerg Med.* 1993;22:579-582.
39. Milner KA, Funk M, Richards S, Wilmes RM, Vaccarino V, Krumholz HM. Gender differences in symptom presentation associated with coronary heart disease. *Am J Cardiol.* 1999;84:396-399.
40. McSweeney JC, Cody M, O'Sullivan P, Elberson K, Moser DK, Garvin BJ. Women's early warning symptoms of acute myocardial infarction. *Circulation.* 2003;108:2619-2623.
41. Canto JG, Goldberg RJ, Hand MM, et al. Symptom presentation of women with acute coronary syndromes – Myth vs reality. *Arch Intern Med.* 2007;167:2405-2413.
42. Gan SC, Beaver SK, Houck PM, MacLehose RF, Lawson HW, Chan L. Treatment of acute myocardial infarction and 30-day mortality among women and men. *N Engl J Med.* 2000;343:8-15.
43. Vaccarino V, Rathore SS, Wenger NK, et al. Sex and racial differences in the management of acute myocardial infarction, 1994 through 2002. *N Engl J Med.* 2005;353:671-682.

44. Milcent C, Dormont B, Durand-Zaleski I, Steg PG. Gender differences in hospital mortality and use of percutaneous coronary intervention in acute myocardial infarction: microsimulation analysis of the 1999 nationwide French hospitals database. *Circulation*. 2007;115:833-839.
45. Fibrinolytic Therapy Trialists' (FTT) Collaborative Group. Indications for fibrinolytic therapy in suspected acute myocardial infarction: collaborative overview of early mortality and major morbidity results from all randomised trials of more than 1000 patients. *Lancet*. 1994;343:311-322.
46. Vaccarino V, Abramson JL, Veledar E, Weintraub WS. Sex differences in hospital mortality after coronary artery bypass surgery: evidence for a higher mortality in younger women. *Circulation*. 2002;105:1176-1181.
47. Humphries KH, Gao M, Pu A, Lichtenstein S, Thompson CR. Significant improvement in short-term mortality in women undergoing coronary artery bypass surgery (1991 to 2004). *J Am Coll Cardiol*. 2007;49:1552-1558.
48. Vaccarino V, Lin ZQ, Kasl SV, et al. Sex differences in health status after coronary artery bypass surgery. *Circulation*. 2003;108:2642-2647.
49. Narins CR, Ling FS, Fischi M, Peterson DR, Bausch J, Zareba W. In-hospital mortality among women undergoing contemporary elective percutaneous coronary intervention: a reexamination of the gender gap. *Clin Cardiol*. 2006;29:254-258.
50. Alexander KP, Chen AY, Newby LK, et al. Sex differences in major bleeding with glycoprotein IIb/IIIa inhibitors: results from the CRUSADE (Can Rapid risk stratification of Unstable angina patients Suppress ADverse outcomes with Early implementation of the ACC/AHA guidelines) initiative. *Circulation*. 2006;114:1380-1387.
51. Balady GJ, Jette D, Scheer J, Downing J. Changes in exercise capacity following cardiac rehabilitation in patients stratified according to age and gender: results of the Massachusetts association of cardiovascular and pulmonary rehabilitation multicenter database. *J Cardpulm Rehabil*. 1996;16:38-46.
52. Rozanski A, Blumenthal JA, Davidson KW, Saab PG, Kubzansky L. The epidemiology, pathophysiology, and management of psychosocial risk factors in cardiac practice: the emerging field of behavioral cardiology. *J Am Coll Cardiol*. 2005;45:637-651.
53. Mallik S, Spertus JA, Reid KJ, et al. Depressive symptoms after acute myocardial infarction: evidence for highest rates in younger women. *Arch Intern Med*. 2006;166:876-883.
54. Wenger NK, Shaw LJ, Vaccarino V. Coronary heart disease in women: update 2008. *Clin Pharmacol Ther*. 2008;83(1):37-51.
55. Mieres JH, Shaw LJ, Arai A, Budoff M, Hundley G, Flamm SD, Marwick TH, Mosca L, Patel AR, Redberg RF, Taubert K, Thomas G, Wenger NK, for the cardiovascular imaging committee. American heart association – cardiac imaging committee consensus statement: The role of cardiac imaging in the clinical evaluation of women with known or suspected coronary artery disease. *Circulation* 2005;111:682-696.
56. Shaw LJ, Shaw RE, Bairey Merz CN, et al. Impact of ethnicity and gender differences on angiographic coronary artery disease prevalence and in-hospital mortality in the American college of cardiology – National cardiovascular data registry (ACC-NCDR). *Circulation*. 2008;117:1787-1801.
57. <http://www.acc.org/qualityandscience/clinical/guidelines/exercise/exercise_clean.pdf.>;2008 Accessed 09.04.08.
58. <http://www.acc.org/qualityandscience/clinical/guidelines/stable/stable_clean.pdf.>;2008 Accessed 09.04.08.
59. Shaw LJ, Olson MB, Kip K, et al. The value of estimated functional capacity in estimating outcome: results from the NHLBI-sponsored women's ischemia syndrome evaluation. *J Am Coll Cardiol*. 2006;47:S36-S43.
60. Shaw LJ, Bax JJ, Marwick TH, Berman DS. Noninvasive testing for myocardial ischemia, In: O'Rourke, Fuster, Alexander, eds. *Hurst's The Heart Manual Of Cardiology*. 11th ed. New York: McGraw-Hill; 2005:31-52.
61. Alexander KP, Shaw LJ, Shaw LK, DeLong ER, Mark DB, Peterson ED. Diagnostic and prognostic value of the Duke treadmill score in women. *J Am Coll Cardiol*. 1998;32(6):1657-1664.

62. Mark DB, Shaw L, Harrell FE, et al. Prognostic value of treadmill exercise score in patients with suspected coronary artery disease. *N Engl J Med*. 1991;325:849-853.
63. Shaw LJ, Peterson ED, Kesler KL, et al. Use of a prognostic treadmill score in identifying diagnostic coronary disease subgroups and altering patient management. *Circulation*. 1998;98(16):1622-1630.
64. Mora S, Redberg RF, Cui Y, et al. Ability of exercise testing to predict cardiovascular and all-cause death in asymptomatic women: a 20-year follow-up of the lipid research clinics prevalence study. *JAMA*. 2003;290(12):1600-1607.
65. Gulati M, Pandey DK, Arnsdorf MF, et al. Exercise capacity and the risk of death in women: the St James women take heart project. *Circulation*. 2003;108(13):1554-1559.
66. Gulati M, Black HR, Shaw LJ, et al. The prognostic value of exercise capacity in women: nomogram for the female population. *New Eng J Med*. 2005;353(5):468-475.
67. <http://www.fda.gov/CDER/drug/InfoSheets/HCP/microbubbleHCP.htm.>;2008 Accessed 09.04.08.
68. Kim C, Kwok YS, Heagerty P, Redberg R. Pharmacologic stress testing for coronary disease diagnosis: A meta-analysis. *Am Heart J*. 2001;142(6):934-944.
69. Kim C, Kwok YS, Saha S, Redberg RF. Diagnosis of suspected coronary artery disease in women: a cost-effectiveness analysis. *Am Heart J*. 1999;137(6):1019-1027.
70. Metz LD, Beattie M, Hom R, Redberg RF, Grady D, Fleischmann KE. The prognostic value of normal exercise\imaging and exercise echocardiography: a meta-analysis. *J Am Coll Cardiol*. 2007;49(2):227-237.
71. Marwick TH, Shaw LJ, Case C, Vasey C, Thomas JD. Clinical and economic impact of exercise electrocardiography and exercise echocardiography in clinical practice. *Eur Heart J*. 2003;24(12):1153-1163.
72. Fleischmann KE, Hunink MG, Kuntz KM, Douglas PS. Exercise echocardiography or exercise SPECT imaging? A meta-analysis of diagnostic test performance. *JAMA*. 1998;280(10):913-920.
73. Kwok Y, Kim C, Grady D, Segal M, Redberg R. Meta-analysis of exercise testing to detect coronary artery disease in women. *Am J Cardiol*. 1999;83(5):660-666.
74. Redberg RF, Shaw LJ. Diagnosis of coronary artery disease in women. *Prog Cardiovasc Dis*. 2003;46(3):239-258.
75. Arruda-Olson AM, Juracan EM, Mahoney DW, McCully RB, Roger VL, Pellikka PA. Prognostic value of exercise echocardiography in 5,798 patients: is there a gender difference? *J Am Coll Cardiol*. 2002;39(4):625-631.
76. Shaw LJ, Vasey C, Sawada S, Rimmerman C, Marwick TH. Impact of gender on risk stratification by exercise and dobutamine stress echocardiography: long-term mortality in 4,234 women and 6,898 men. *Eur Heart J*. 2005;26(5):447-456.
77. Shaw LJ, Marwick TH, Berman DS, et al. Incremental cost effectiveness of exercise echocardiography versus SPECT imaging for the evaluation of stable chest pain. *Eur Heart J*. 2006;27(20):2448-2458.
78. Berman DS, Kang X, Nishina H, et al. Diagnostic accuracy of gated Tc-99m sestamibi stress myocardial perfusion SPECT with combined supine and prone acquisitions to detect coronary artery disease in obese and nonobese patients. *J Nucl Cardiol*. 2006;13(2):191-201.
79. Slomka PJ, Nishina H, Abidov A, et al. Combined quantitative supine-prone myocardial perfusion SPECT improves detection of coronary artery disease and normalcy rates in women. *J Nucl Cardiol*. 2007;14(1):44-52.
80. Shaw LJ, Iskandrian AE. Prognostic value of stress gated SPECT in patients with known or suspected coronary artery disease. *J Nucl Cardiol*. 2004;11(2):171-185.
81. Marwick TH, Shaw LJ, Lauer MS, et al. The noninvasive prediction of cardiac mortality in men and women with known or suspected coronary artery disease. *Am J Med*. 1999;106(2): 172-178.
82. Berman DS, Kang X, Hayes SW, et al. Adenosine myocardial perfusion SPECT in women compared with men: Impact of diabetes mellitus on incremental prognostic value and effect on patient management. *J Am Coll Cardiol*. 2003;41(7):1125-1133.

83. Giri S, Shaw LJ, Murthy DR, et al. Impact of diabetes on the risk stratification using stress single-photon emission computed tomography myocardial perfusion imaging in patients with symptoms suggestive of coronary artery disease. *Circulation*. 2002;105:32-40.
84. Shaw LJ, Heller GV, Travin MI, et al. Cost analysis of diagnostic testing for coronary artery disease in women with stable chest pain. *J Nucl Cardiol*. 1999;6(6):559-569.
85. Yoshinaga K, Chow BJ, Williams K, et al. What is the prognostic value of myocardial perfusion imaging using rubidium-82 positron emission tomography? *J Am Coll Cardiol*. 2006;48(5):1029-1039.
86. Budoff MJ, Achenbach S, Blumenthal RS, et al. Assessment of coronary artery disease by cardiac computed tomography: a scientific statement from the American heart association committee on cardiovascular radiology and intervention, and committee on cardiac imaging, council on clinical cardiology. *Circulation*. 2006;114(16):1761-1791.
87. Greenland P, Bonow RO, Brundage BH, et al. American college of cardiology foundation clinical expert consensus task force (ACCF/AHA writing committee to update the 2000 expert consensus document on electron beam computed tomography); Society of atherosclerosis imaging and prevention; society of cardiovascular computed tomography. ACCF/AHA 2007 clinical expert consensus document on coronary artery calcium scoring by computed tomography in global cardiovascular risk assessment and in evaluation of patients with chest pain: a report of the American college of cardiology foundation clinical expert consensus task force (ACCF/AHA writing committee to update the 2000 expert consensus document on electron beam computed tomography) developed in collaboration with the society of atherosclerosis imaging and prevention and the society of cardiovascular computed tomography. *Circulation*. 2007;115(3):402-426.
88. Bellasi A, Lacey C, Taylor AJ, et al. Comparison of prognostic usefulness of coronary artery calcium in men versus women (results from a meta- and pooled analysis estimating all-cause mortality and coronary heart disease death or myocardial infarction). *Am J Cardiol*. 2007;100(3):409-414.
89. Detrano R, Guerci AD, Carr JJ, et al. Coronary calcium as a predictor of coronary events in four racial or ethnic groups. *N Engl J Med*. 2008;358(13):1336-1345.
90. Nasir K, Shaw LJ, Liu ST, et al. Ethnic differences in the prognostic value of coronary artery calcification for all-cause mortality. *J Am Coll Cardiol*. 2007;50(10):953-960.
91. Budoff MJ, Shaw LJ, Liu ST, et al. Long-term prognosis associated with coronary calcification: observations from a registry of 25,253 patients. *J Am Coll Cardiol*. 2007;49(18):1860-1870.
92. Raggi P, Shaw LJ, Berman DS, Callister TQ. Gender-based differences in the prognostic value of coronary calcification. *J Women's Health*. 2004;13(3):273-282.
93. Hamon M, Biondi-Zoccai GG, Malagutti P, et al. Diagnostic performance of multislice spiral computed tomography of coronary arteries as compared with conventional invasive coronary angiography: a meta-analysis. *J Am Coll Cardiol*. 2006;48(9):1896-1910.
94. Sun Z, Jiang W. Diagnostic value of multislice computed tomography angiography in coronary artery disease: a meta-analysis. *Eur J Radiol*. 2006;60(2):279-286.
95. Stein PD, Beemath A, Kayali F, Skaf E, Sanchez J, Olson RE. Multidetector computed tomography for the diagnosis of coronary artery disease: a systematic review. *Am J Med*. 2006;119(3):203-216.
96. Janne d'Othee B, Siebert U, Cury R, Jadvar H, Dunn EJ, Hoffmann U. A systematic review on diagnostic accuracy of CT-based detection of significant coronary artery disease. *Eur J Radiol*. 2008;65(3):449-461.
97. Schuijf JD, Bax JJ, Shaw LJ, et al. Meta-analysis of comparative diagnostic performance of magnetic resonance imaging and multislice computed tomography for noninvasive coronary angiography. *Am Heart J*. 2006;151(2):404-411.
98. Pundziute G, Schuijf JD, Jukema JW, et al. Gender influence on the diagnostic accuracy of 64-slice multislice computed tomography coronary angiography for detection of obstructive coronary artery disease. *Heart*. 2008;94(1):48-52.
99. Pundziute G, Schuijf JD, Jukema JW, et al. Prognostic value of multislice computed tomography coronary angiography in patients with known or suspected coronary artery disease. *J Am Coll Cardiol*. 2007;49(1):62-70.

100. Min JK, Shaw LJ, Devereux RB, et al. Prognostic value of multidetector coronary computed tomographic angiography for prediction of all-cause mortality. *J Am Coll Cardiol.* 2007;50:1161-1170.
101. Matsumoto N, Sato Y, Yoda S, et al. Prognostic value of non-obstructive CT low-dense coronary artery plaques detected by multislice computed tomography. *Circ J.* 2007;71(12):1898-1903.
102. Gilard M, Le Gal G, Cornily JC, et al. Midterm prognosis of patients with suspected coronary artery disease and normal multislice computed tomographic findings: a prospective management outcome study. *Arch Intern Med.* 2007;167(15):1686-1689.
103. Gaemperli O, Valenta I, Schepis T, et al. Coronary 64-slice CT angiography predicts outcome in patients with known or suspected coronary artery disease. *Eur Radiol.* 2008;18(6):1162-1173.
104. Panting JR, Gatehouse PD, Yang GZ, et al. Abnormal subendocardial perfusion in cardiac syndrome X detected by cardiovascular magnetic resonance imaging. *N Engl J Med.* 2002;346(25):1948-1953.

Chapter 21
Recent Strategies to Improve Graft Performance in Patients Undergoing Coronary Artery Bypass Surgery. Are Best Results Achieved by Improved Surgical Techniques of Graft Preparation?

Rosa Baron, Michael R. Dashwood, Mikael Arbeus, Derek Filbey, and Domingos S.R. Souza

21.1 Introduction

The patency rate of the saphenous vein (SV) used as a graft in patients undergoing coronary artery bypass surgery (CABG) is poor, with 15–25% grafts occluding within 1 year and over 50% patients requiring further (redo) surgery within 10 years.[1] In order to investigate the strategies to reduce vein graft failure in patients undergoing CABG, the underlying pathology of the disease must first be established. The high prevalence of coronary heart disease in Western society has prompted surgeons to develop procedures to improve myocardial blood flow, and subsequently relieve the symptoms of angina pectoris along with other myocardial crises.[2] One of the most significant advances in vascular surgery was the finding that venous conduits could be used as replacements for atherosclerotic arteries. Following the work of Alexis Carrel at the turn of the century, a venous graft was first used in 1906 to replace a popliteal aneurysm.[2] Promising experimental results encouraged surgeons to apply this method to the coronary vessels. By the 1950s, at the Cleveland Clinic,

M.R. Dashwood (✉)
Clinical Biochemistry, Royal Free and University College Medical School, London, NW3 2QG, UK
e-mail: m.dashwood@medsch.ucl.ac.uk

R. Baron
Faculty of Biomedical and Life Sciences, West Medical Building, University of Glasgow, Glasgow G12 8QQ, Scotland, UK
e-mail: rosic.jb@hotmail.com

M. Arbeus and D.S.R. Souza
Department of Thoracic and Cardiovascular Surgery, Örebro University Hospital, SE-701 85, Örebro, Sweden
e-mail: mikael.arbeus@orebroll.se; domingos.ramosdesouza@orebroll.se

D. Filbey
Transfuión Medicine, Örebro University Hospital, SE-701 85, Örebro, Sweden
e-mail: derek.filbey@orebroll.se

D. Abraham et al. (eds.), *Advances in Vascular Medicine*,
DOI 10.1007/978-1-84882-637-3_21, © Springer-Verlag London Limited 2010

Favaloro et al. had treated numerous cases of peripheral and renal artery reconstruction with venous conduits and employed this procedure in coronary vessels.[3] Research efforts by Favaloro[4] led to the development of CABG, a technique which has been used for almost four decades. The great SV of the leg is the conduit of choice for three main reasons. First, it is expendable as deeper vessels maintain blood flow to superficial tissues after its removal. Second, the extensive length of this vein allows for multiple grafts, and finally, its superficial position renders it easily accessible. A 10-year follow-up recatheterization of Favarolo's first operation showed that both the graft and the bypassed right coronary artery remained patent.[3] Such promising results reshaped the history of cardiac surgery and led to the rise of surgical revascularization in the treatment of ischemic heart disease.

21.2 Pathological Conditions Promoting Early Graft Failure in Saphenous Vein Grafts

Following the popular use of venous conduits for CABG in the 1970s, high graft failure rates became apparent, initiating studies into the postoperative events occurring in vein grafts and the pathophysiology of graft failure (Fig. 21.1). Studies on the

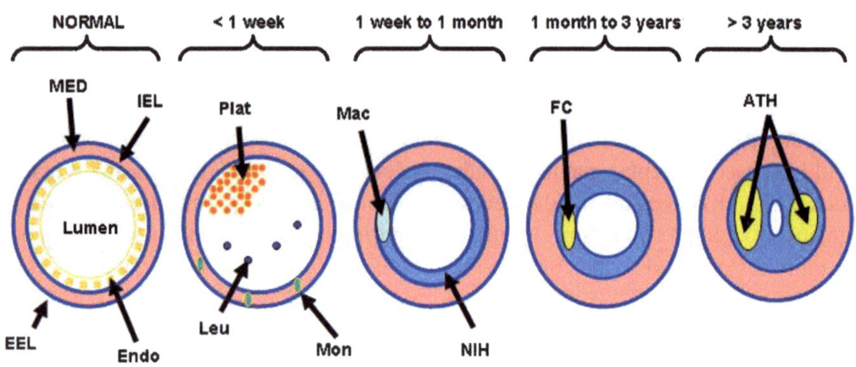

Fig. 21.1 Stages of vein graft failure. Diagrammatic representation of the stages of vascular remodeling after vein graft surgery. At time zero, the ungrafted/normal vein media (MED) lies between the internal elastic lamina (IEL) and external elastic lamina (EEL). The lining endothelium (Endo) is continuous and the lumen is open. Up to 1 week following graft implantation, the surgical preparation during harvesting results in endothelial injury and denudation. Subsequent thrombus formation, caused by platelet (Plat) aggregation occurs at regions where the intima is exposed. Adhesion of leukocytes (Leu) and monocytes (Mon) leads to the release of a range of factors stimulating vascular smooth muscle cell (VSMC) migration and proliferation. Between 1 week and 1 month, the medial thickening occurs as a result of VSMC proliferation. VSMCs also migrate through the IEL and this leads to neointimal hyperplasia (NIH) and possible appearance of macrophages (Mac) within this region. Between 1 month and 3 years, typical atheromatous lesions may appear which are rich in foam cells (FC). Beyond 3 years, progressive intimal thickening occurs with the formation of superimposed atheromatous plaques (ATH) resulting in narrowing of the graft lumen. Plaque rupture may occur leading to thrombotic occlusion

causes of graft failure led to the realization that many of the pathological factors contributing to vein graft failure stemmed from the denudation of endothelial cells in response to surgical trauma during harvesting.[3] Moreover, additional endothelial damage occurs in veins in response to the altered hemodynamics once placed into the arterial circulation. In response to vascular injury, vein grafts undergo a self-healing process in an attempt to recover from surgical trauma and exposure to arterial hemodynamics. This process, however, is poorly regulated and may be responsible for the high vein graft failure rate.[5] The two main causes of endothelial denudation are surgical trauma and the differences between arterial and venous hemodynamics.

21.2.1 Surgical Trauma

The conventional technique of harvesting the SV involves making a long, superficial leg incision from the calf to the groin, depending on the number of grafts required. Subsequently, the connective and adipose tissue surrounding the vein is removed, while all side branches of the vein are ligated to minimize bleeding and damage to the lower limb. The removal of such protective tissue results in sections of the vein going into spasm. This venospasm is typically overcome by high pressure saline distension (up to 560 mmHg) of the vein that results in further endothelial damage.[6] Although pharmacological relaxing agents such as the nitrovasodilators have been employed to overcome venospasm, high pressure distension with saline is still most often used. Furthermore, removal of, or damage to, the adventitia disrupts the vasa vasorum, the microvascular network supplying the vessel with oxygen and nutrients. The resultant ischemia coupled with endothelial denudation is suggested to contribute to high graft failure rates. In addition, the structural and functional differences between arteries and veins must also be considered as important factors in graft failure.

21.2.2 Difference Between Arteries and Veins

The basic structure of both arteries and veins shares a common arrangement, consisting of three distinct layers, the intima, the media, and the adventitia (Fig. 21.2). The innermost layer, the intima, consists of a single layer of endothelial cells with an underlying layer of connective tissue, the internal elastic lamina (IEL) that is com posed of type IV collagen, laminin, and heparin sulfate proteoglycans, all of which are fundamental to its function of support. Positioned adjacent to this is the tunica media, mainly comprising vascular smooth muscle cells (VSMC) and extracellular matrix (ECM). Additional support is provided by the external elastic lamina (EEL) adjoining the outermost layer of the vessel, the adventitia. This layer consists of collagen fibers, ECM, perivascular fat, and nerves which together provide a stabilizing cushion of the surrounding tissue.[7] Although these features are common to arteries and veins, subtle differences that are believed to play a role in graft failure are apparent.

First, the tunica media of arteries is generally thicker than that of veins of similar size.[8] Being densely packed with VSMC and elastic fibers, this layer resists the

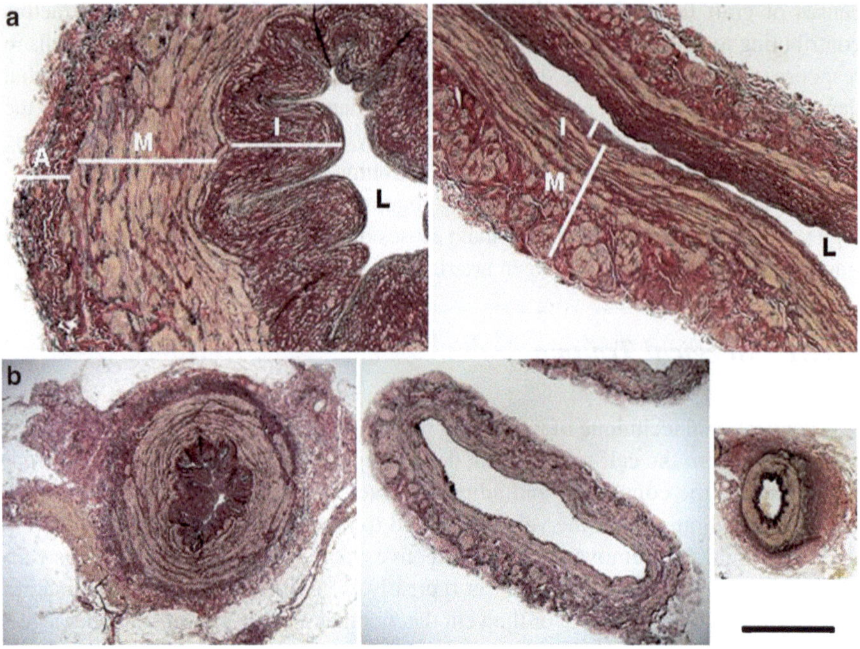

Fig. 21.2 Examples of conduits used as grafts in patients undergoing CABG. Representative transverse elastin van Gieson-stained sections. (**a**) *Left panel*. No-touch SV where the intima (*I*) is thrown into folds as such vessels have not been distended. *Right panel*. Conventionally-prepared vein with collapsed lumen (*L*) caused by high-pressure distension. Note how the intima and media (*M*) of the conventional vein are thinner than the no-touch vein due to distension/surgical trauma. Also, the adventitia (*A*) that is intact on the no-touch vein has been removed during conventional harvesting. (**b**) Transverse sections of SV (*left*), radial artery (RA) (*middle*) and internal mammary artery (*right*) comparing typical graft sizes. Scale bar = 200 μm for top panels and 500 μm for *lower panels*

pressure of the arterial system. The increased muscular and elastic components in the walls of arteries provide the capability of influencing vascular tone, enabling them to alter the vessel diameter in response to the locally-released factors or neuronal control mechanisms, offering the ability to control blood pressure. Moreover, the thicker nature of the arterial walls enables them to retain a stable structure even under increased pressure, unlike veins which, when subjected to high internal pressure are more likely to over distend and risk rupture.[9]

The different hemodynamics experienced by a vein once placed into the arterial system can account for many of the problems associated with vein grafts. Prior to grafting, a typical vein will be subjected to low pressures (~5–8 mmHg), nonpulsatile flow, and a shear stress of ~0.2 dyne/cm². Subsequent to grafting into the arterial system, however, the vein will be subjected to high pressures (~60–140 mmHg), pulsatile flow, and a shear stress of ~3–6 dyne/cm².[10] Furthermore, the intrinsic antithrombotic properties of veins are relatively weak when compared with those of the arteries. In arteries, the IEL is well developed and contains high levels of the proteoglycan heparin sulfate which mediates the actions of antithrombin III.[11] This feature is less well-established in the walls of the veins which may have a poorly developed IEL, lower proteoglycans levels, and reduced antithrombotic activity. It must be kept in mind, however, that this feature would only become apparent when the endothelium has been removed or damaged during surgery, as under basal conditions, the IEL would not be exposed to the blood. Morphological examination of the venous grafts from repeat operations or autopsy has contributed toward the understanding of the pathogenesis of occlusion and the regulation of cellular processes leading to the development of therapeutic interventions.[12] Three distinct phases of occlusion have been suggested: initial thrombosis, intimal hyperplasia, and accelerated atherosclerosis (Fig. 21.1).

21.2.3 Thrombosis

Within the first month following CABG, 3–12% of grafts exhibit early occlusion primarily caused by thrombus formation. In vein grafts, damage of the vessel wall during harvesting coupled with changes in the blood flow, upon grafting into the arterial circulation, initiates the blood coagulation cascade resulting in thrombus formation via trauma-induced stimulation of the extrinsic pathway of the blood coagulation cascade. This cascade involves the conversion of many inactive precursors present in the blood, to their active states.[13] Vascular trauma caused during surgery induces the expression of Tissue Factors involved in the conversion of prothrombin (II) to thrombin (IIa), the main effector protease of the cascade.[14]

21.2.4 Intimal Hyperplasia

It is now recognized that intimal hyperplasia is the leading cause of graft failure in the midterm postoperative period (2–24 months), and is responsible for between 10 and 30% of restenosis in CABG.[10] Intimal hyperplasia represents an increase in the number of VSMCs that subsequently migrate from the media to the intima resulting in luminal narrowing. In venous grafts, this is accompanied by an increase in ECM deposition, and the combination of these two factors results in intimal thickening.[15] Although intimal hyperplasia is rarely the main cause for repeat operations in vein graft patients, it lays the foundations for progressive disease states such as atheroma and atherosclerosis; therefore, its development must be carefully considered.[11] The wall of the vein consists of three layers: the tunica intima, the tunica media, and the tunica adventitia. It is believed that 80% of the intimal hyperplasia results from the deposition of ECM and the remaining 20% from migration of VSMCs from the media to the intimal layer of the vessel.[10] Following surgery, the SV is exposed to both increased blood flow and pressures of the coronary arterial system. The resulting changes in wall shear stress also play an important role in the development of intimal hyperplasia, along with the mechanical stress experienced during vein preparation.[16] The development of intimal hyperplasia is considered to have two distinct steps, the first being the VSMC proliferation in the media. It produces a diffused atherosclerotic-prone region, representing "soil for atheroma"[17] and subsequent formation of the atherosclerotic plaque, the final stage of the occlusion process.[11]

21.2.5 *Atherosclerosis*

Atherosclerotic symptoms of ischemia are not usually noted until around 3 years postoperatively. Atherosclerotic plaques in the venous grafts have been identified as early as 1 year after surgery. Even though some similarities exist between surgery-induced atherosclerotic plaques and those occurring naturally, there are fundamental differences between these two disease states. The principle difference is the greatly enhanced rate of atherosclerosis formation in vein grafts when compared with the native vessels, where atherosclerosis develops over a period of 50–60 years; in vein grafts, it develops over a much shorter period of 5–10 years. One of the fundamental causes of this accelerated disease state is the endothelial cell injury and dysfunction caused during surgery. Additionally, atherosclerotic plaques formed in the vein grafts contain higher levels of foam cells (FC) and inflammatory cells, as well as many different morphological features. Thus, it is evident that in vein grafts, the risk factors associated with atherosclerotic plaques are amplified by the loss of the anatomic and functional integrity of the endothelium.[11] Despite efforts to reduce intimal hyperplasia and its clinical sequelae in response to vein grafting, "the impact of intimal hyperplasia on all fields of vascular interventions remains high."[10]

21.3 Current Strategies to Reduce Vein Graft Hyperplasia

21.3.1 *Harvesting*

The SV is the most commonly used conduit for CABG, since its introduction by Favarolo in 1969.[3] It has been acknowledged for many years that the high pressure distension applied during harvesting of veins for grafting to surmount venospasm, is central to endothelial loss and medial damage. This resulting damage is a principle factor leading to a series of pathological events, namely initial thrombosis, intimal hyperplasia, and accelerated atherosclerosis. The combination of these pathological conditions results in the poor patency rates in SVs used in CABG. Attempts to improve the patency rate of vein grafts include the use of various harvesting techniques, immersion media, and intraluminal distension.

21.3.2 *Harvesting Technique*

An early technique of SV harvesting was described 30 years ago by Gottlob[18] (Fig. 21.3). This technique was developed using vein segments obtained from

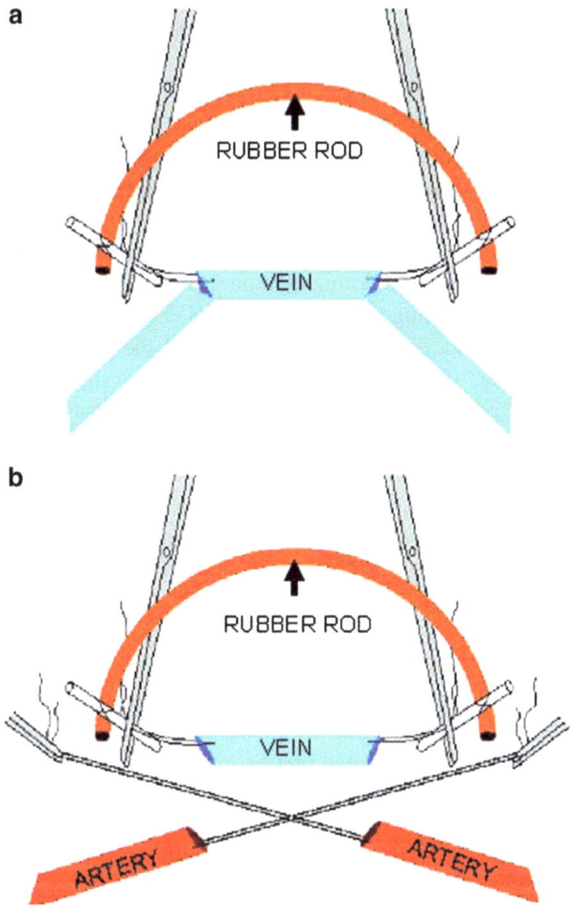

Fig. 21.3 Gottlob's method of "dissection without touching." (**a**) The SV is kept under tension by "holding sutures" that are attached to a piece of bent rubber before isolation of the segment to be grafted. (**b**) The severed ends of the vein remain under tension while suturing the anastomosis during insertion into the artery. Modified from Gottlob[18]

CABG patients or cadavers and canine femoral veins. The aim of this preparation of "dissection without touching" was to preserve the vein's endothelium using a method where "The venous graft is fixed by holding sutures to a bent rod of rubber."[18] Although the endothelium is protected using this method, other regions of the vein are damaged as the instructions continue, "For dissection only the connective tissue surrounding the vessel was grasped by forceps and sharply severed away from the vein." Gottlob's method of preparing the SV as a bypass graft is often cited, including its use in patients undergoing CABG.[19] The practicalities of doing so in an operation theater/sterile environment are intriguing. Although this technique may preserve the endothelium, it is clear from many illustrations of those purporting to use Gottlob's method that the outer vessel regions are often damaged.

When comparing the 10-year patency rate of SV grafts versus left internal mammary arteries (LIMA [also known as left internal thoracic artery, LITA]), the latter presents a markedly lower occlusion rate (50% and 10%, respectively).[20] One possible explanation for this discrepancy could be the differences in harvesting techniques used for the two vessels. Primarily, LIMAs are generally prepared in situ completely, with the surrounding tissue (pedicle) intact and are subjected to minimal handling throughout both preparation and grafting. In contrast, using conventional harvesting methods, SVs are stripped of their cushion of the surrounding tissue (Fig. 21.4). Furthermore, venous conduits are distended at high pressures to overcome venospasm experienced during conventional harvesting and pressures causing additional endothelial damage. Souza et al[21] described a randomized angiographic study comparing the early patency rates of three harvesting techniques of the SV: the conventional "intermediate," and "no touch" techniques. Here, preservation of the perivascular tissue, specifically the adventitia, using a "no touch" technique, significantly ($p<0.025$) improved vein graft patency. In a recent long-term follow-up study, this group described the patency rates of 90% for "no-touch"-harvested SVs at 8.5 years, and the results were comparable with the LIMA and significantly higher than those obtained using the conventional method of preparing the SV ($p=0.01$).[21] The "no-touch" technique avoids venospasm, thus obviating the need for distension.[21] However, the exact mechanisms underlying the superior results obtained by this technique remain unclear. One possibility presented by Tsui et al.[22] is that the pedicle of surrounding tissue provides the vein with an external "biological" stent providing support against pulsatile coronary arterial pressure. In addition, the surrounding tissue is rich in collagen and connective tissue that protects the vein from any damage caused by direct handling with surgical instruments during CABG.[20]

Moreover, it has been demonstrated that endothelial integrity of veins harvested by the "no-touch" technique is improved when compared with the conventional

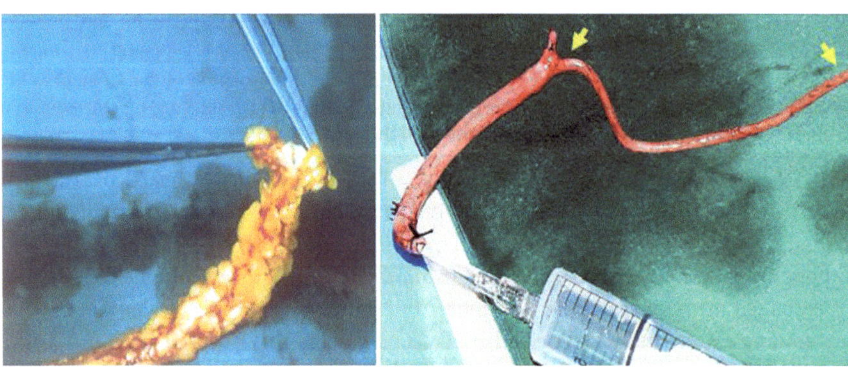

Fig. 21.4 "No-touch" and conventionally harvested saphenous vein (SV). Examples of SV segments at surgery harvested for coronary artery bypass surgery (CABG). *Left panel* shows a typical "no-touch" SV harvested with surrounding cushion of tissue (handled by surgical instruments). *Right panel* shows a conventionally-harvested segment of SV that has been stripped of surrounding tissue. This has caused the vein to go into spasm (between arrows) and this is overcome using high-pressure intraluminal saline distension

technique.[22] The preservation of endothelial nitric oxide (NO) synthesis, thus, represents an additional mechanism for the improved patency rates of these grafts when compared with the conventional harvesting, as venospasm and thrombosis is reduced. In addition to the role of the preserved endothelium, it has been suggested that the intact vasa vasorum plays an important role in the improved patency of "no-touch"-harvested SV grafts. The preservation of the vasa vasorum is thought to provide a restoration of blood flow to the vessel wall on implantation into the coronary arterial system, thus reducing ischemic injury. The vasa vasorum, a microvessel network surrounding the vein, penetrates deep into the media with evidence that terminations appear in the vein lumen.[23] Although vasa vasorum exist in all vessels, their principle role is in veins where luminal oxygen levels are low. As oxygenated blood passes through the lumen of the arteries, the endothelium and VSMCs receive sufficient oxygen and nutrients by diffusion; thus, the role of the vasa vasorum is less significant. Retrograde filling of the vasa vasorum upon implantation into the arterial circulation has been demonstrated, and this, coupled with the maintenance of endothelial integrity, has been suggested to contribute to the improved patency rate of "no-touch" venous grafts.[21]

21.3.3 Immersion Media

Although the factors outlined earlier are important in the preservation of the endothelial layer, there are other variables that could account for endothelial damage other than physical trauma. Principally, the immersion (or storage) media used between harvesting and implantation may also affect the vein's endothelial lining. Gundry et al[24] revealed that SVs immersed in warm saline (28°C) displayed extensive endothelial damage, including endothelial cell separation, cell loss, exposure of basement membrane, and luminal stenosis. Vein segments immersed in warm blood (28°C) and cold saline (4°C) exhibited only slight endothelial damage. In addition, SV segments immersed in cold blood (4°C) also displayed only light endothelial cell separation, and cell loss was comparable with the control veins. More recently, Souza et al. described storing the vein in heparinized blood at room temperature, extracted from the patient before extracorporeal circulation.[20] With an abundance of storage solutions available including heparinized blood, simple saline solution (0.9% NaCl), University of Wisconsin solution, Perfadex, Euro-Collins, and Papworth and Bretschneider's solution, much research is aimed at identifying the optimum conditions for the preservation of the vein wall prior to implantation.[25]

21.3.4 Distension Pressures

High-pressure intraluminal distension is routinely used to overcome the spasm that occurs in a high proportion of SVs during harvesting. Distension has a deleterious effect on the vein, as placement of the SV grafts into high pressure conditions induces expression of genes stimulating VSMC proliferation.[26] Furthermore, short-term distension, as used during routine SV harvesting in CABG, induces p38-MAPK, a factor involved in graft occlusion.[27] Apart from inducing cell proliferation, pressure distension stimulates the expression of endothelial adhesion molecules in the human SV, a process involved in early graft failure.[28] There is evidence that pharmacological relaxation of vein grafts is beneficial when compared with veins exposed to pressure distension, as pharmacological preparation of grafts using various vasodilators results in increased NO production after implantation.[29] In addition, in porcine jugular vein, pressure distension irreversibly overstretches the vessel and increases the matrix metalloproteinases (MMP-2 and -9). Neointimal formation, once grafted, is more pronounced in distended grafts than those pharmacologically treated.[30]

Finally, the application of vasodilators, both topically and intraluminally, during harvesting may be considered to prevent venospasm and the need for high pressure distension.[6] The alternative to the use of the SV is the use of arterial conduits for grafting. Arteries are commonly used in conjunction with SVG in multiple bypass operations. However, in certain cases, complete arterial revascularization has been employed in an attempt to improve the success rate of CABG.

21.3.5 Arterial Grafting

Arterial conduits have been considered an alternative to the SV graft for many years. Their mechanical characteristics are beneficial due to their ability to withstand high pressures and pulsatile stretch. It is also believed that the extensive elastic laminae in the vascular wall of the arteries contribute to improved arterial graft patency by preventing the invasion of smooth muscle cells and subsequent atherosclerotic alterations. Furthermore, the arterial endothelium can produce large amounts of NO aiding the regulation of vascular tone via its dilator effects. Vascular tone alters in response to blood flow demand, leading to the phenomenon known as arterial remodeling which is "considered one of the great advantages of a living arterial conduit."[31] Arterial conduits, in particular, the LIMA, have high patency rates of ~90% at 10 years when compared with the 50–60% patency rates of SV grafts over the same period.[32]

21.3.6 Left Internal Mammary (Thoracic) Artery (LIMA/LITA)

The long-term patency of the LIMA grafts is superior to the SV.[1,33] A number of different arterial conduits have been studied over the years, ranging from the LIMA and gastroepiploic artery to the more recent development of the radial artery (RA). Having produced patency rates of 95–98% at 10 years postoperatively, the LIMA is recognized as the "gold standard."

There are a number of drawbacks that may be associated with the LIMA. Primarily, there is an increased risk of sternal wound infection, a problem which is especially prominent in diabetic patients. Additionally, the length of the LIMA is often insufficient to graft to the "healthy" (nonatherosclerotic) part of the coronary artery, a problem not encountered with the free SV graft. However, skeletonization of the internal thoracic artery (ITA) has been suggested as a means of lengthening the graft. Here, the ITA is dissected away from the chest wall preserving the collateral sternal and internal venous blood flow. In this way, the length of the ITA is increased allowing a longer segment to be used, and maintaining sufficient sternal blood flow that reduces further complications especially those seen in diabetics.

Despite the complications discussed earlier, the LIMA is still the conduit of choice among most surgeons who generally believe that every CABG patient should receive LIMA grafts unless there are any of the following contraindications: (1) The patient has had radiotherapy to the chest wall due to previous cancer, (2) Subclavian stenosis is present resulting in inadequate blood flow to the ITA, and (3) Injury to the LIMA occurs during harvesting. If any of these exist, the use of an alternative arterial conduit may be considered, namely the RA.

21.3.7 Radial Artery (RA)

The RA was first introduced for use in CABG in the 1970s, but its success at this time was short-lived. Early studies showed spastic properties with a high propensity for accelerated intimal hyperplasia.[31] The popularity of the RA increased in the early 1990s, when 20-year follow-up studies of patients with RA grafts revealed low rates of progressive disease.[31] Subsequently, harvesting techniques were altered to prevent previous problems associated with endothelial damage. Also, pharmacological vasodilatation using papaverine and diltiazem, phenoxybenzamine, GTN, or calcium blockers was applied. Subsequently, many researchers reported the beneficial short- and medium-term results for the use of the RA.[31] In some cases, the RA is the second conduit of choice, relegating the SVG to the third choice as a bypass conduit. The RA has several potential disadvantages, the most significant of which is its hyper-responsiveness to vasoactive agents when compared with other arteries, such as the LIMA. Furthermore, the RA is highly susceptible to spasm, particularly, those frequently noted in postoperative angiography.[31] The RA has a thicker wall than the LIMA and the medial VSMCs play a major role in the increased severity of spasm in this vessel, particularly, when handled during CABG. Thus, many patients receiving an RA graft present with "hypoperfusion syndrome" where the artery goes into spasm resulting in myocardial ischemia and subsequent infarction. It must also be noted that it is not simply the choice of conduit that will determine the success rates; secondary prevention is of equal importance to the surgery itself. Postoperative use of statins or aspirin following CABG will play an important role in the potential outcome of the graft following CABG.

21.3.8 Complete Arterial Revascularization

As excellent results have been reported using the LIMA, it has been hypothesized that complete arterial revascularization without the use of SV grafts would result in improved long-term results following CABG. Direct comparison of SV grafting with complete arterial revascularization, however, has proved to be a difficult task, the main limitation being variations in patient characteristics during selection of the two different techniques for CABG. However, after carefully matching the patient variables, Legare et al[34] conducted a study to compare the safety and patency rates of composite arterial grafts with conventional SV grafting. An increased combined outcome of death including myocardial infarction, stroke, and prolonged ventilation was found in the complete arterial revascularization group when compared with the LIMA plus SV group ($p=0.007$), thus rendering complete arterial revascularization to be a high-risk alternative. Although the results of this study would appear to counter the use of complete arterial revascularization, it must be remembered that conclusions drawn from a retrospectively designed study must be taken as tentative until otherwise demonstrated. Thus, with careful consideration of the current data, it appears that a large-scale randomized control trial comparing complete arterial revascularization with conventional CABG (a combination of LIMA with SV) is required to fully validate the safety and efficacy of arterial grafting. One alternative to arterial grafting is the use of artificial stents.

21.3.9 Stenting

The application of luminal stents has become increasingly common of late, as cardiologists attempt not only to provide the best care for the patients, but also reduce the recovery time, hospital stay, and thus, cost.[35] The use of bare metal stents (BMS) revolutionized the treatment of cardiac disease; however, the emergence of in-stent restenosis has lead to the more recent development of drug-eluting stents (DES) aimed at preventing stenosis (Fig. 21.5).

21.3.10 Drug-Eluting Stents (DES)

Randomized control trials conducted in the first few years highlighted the therapeutic advantage associated with the use of the DES. Although short- and mid-term results have demonstrated results comparable with BMS, recent concerns have been raised regarding the long-term safety and efficacy of DES.

The main two pharmacological agents approved for clinical use with DES are Paclitaxel and Sirolimus. Such stents not only provide additional support to the graft against arterial pressures, but also slowly release the drugs, preventing cell proliferation. Paclitaxel inhibits cell replication by binding to the microtubules during cell division, whereas Sirolimus also prevents progression through the cell cycle by

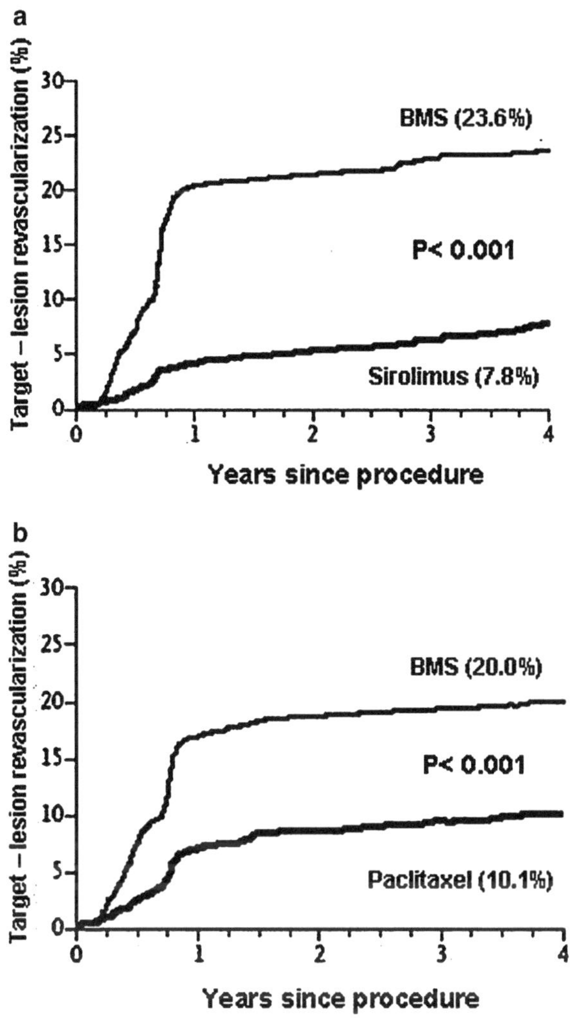

Fig. 21.5 Comparison of bare metal and drug-eluting stents (DES). Graphs comparing target-lesion revascularization in the 4 year postoperative period with Sirolimus (**a**) and Paclitaxel (**b**) with bare metal stents (BMS). A marked decrease in the percent revascularizations required can be seen with the use of DES in comparison with BMS (p<0.001) (adapted from Stone et al[77])

altering regulatory genes controlling the progression of VSMCs from the quiescent to the proliferative stage.[36]

21.3.11 DES Safety Issues

Concern regarding the occurrence of late thrombosis with DES was initially raised in 2003. This usually occurs before re-endothelialization has been completed; in this scenario, Paclitaxel or Sirolimus released from the stent inhibits proliferation of both VSMCs and endothelial cells. However, the prolonged exposure of the metal stent to the vessel lumen and circulating blood initiates the coagulation cascade resulting in potentially fatal thrombosis. A possible solution to these problems could be the development of a compound which, when released from the stent, would target only VSMCs without affecting the endothelial cells. In this way, promoting re-endothelialization would prevent thrombosis, and inhibition of VSMC proliferation would prevent intimal hyperplasia.

21.3.12 External Stents

Experimentally, external stents, both synthetic and biodegradable, have been shown to reduce neointima formation in a porcine SV-carotid artery interposition model, although, to date, there is no clinical evidence that this strategy is effective in CABG patients. For example, Mehta et al,[37] using an external Dacron ™ sleeve surrounding the vein, showed that external stenting reduces long-term medial and neointimal thickening and platelet derived growth factor expression in a pig model of arteriovenous bypass grafting.[37] More recently, this group have described the long-term reduction of medial and intimal thickening in porcine SV grafts using a polyglactin biodegradable external sheath.[38,39] Interestingly, it has been shown that perivenous polytetrafluroethylene (PTFE) support reduces endothelial damage and other early changes in the human blood perfused vein grafts.[40] In addition, this group demonstrated that perfusing human SVs under conditions mimicking arterial hemodynamics causes circumferential stretch of VSMCs and increased responsiveness to constrictors, effects that are prevented by external application of fibrin glue.[41] Using the porcine vein graft model, distension has been shown to reduce short-term patency by promoting platelet and leukocyte adhesion.[42] Furthermore, this model has been used to reveal the time-course of medial and intimal thickening and their relationship to endothelial injury and cholesterol accumulation.[43]

21.3.13 Pharmacological Agents

New strategies are under development based on recent advances in the understanding of the pathobiology of vein graft failure. Adjuvant pharmacological interventions have been introduced in an attempt to improve graft patency. Agents that are in clinical use include platelet inhibitors, anticoagulants, and prostanoids, although in some cases, as outlined in the following paragraphs, data from ongoing clinical trials are inconclusive.

Post-CABG, lipid-lowering agents, and antiplatelet agents are the established strategies for reducing vein graft occlusion. A recent trial demonstrated that aggressive reduction of low-density lipoprotein cholesterol with lovastatin significantly reduces the rate of vein graft occlusion assessed by angiographic follow-up, thus reducing the need for repeat revascularization.[44] Aspirin has been found to increase vein graft patency at 60 days and 1 year after CABG, when compared with the placebo, although the benefits of aspirin are only seen when treatment is started no later than the first postoperative day. However, after the first year post-CABG, aspirin no longer has any beneficial effect on graft patency, suggesting that its predominant action is on the reduction of early thrombosis.[45] Consistent beneficial effects of other antithrombotic agents are yet to be established.[46,47] The benefits of aspirin on graft patency in infrainguinal bypass surgery have been extensively studied. While there is conflicting evidence from various trials, it is generally accepted that aspirin confers a significant benefit on graft patency, especially when prosthetic grafts are used. In addition, aspirin reduces cardiovascular mortality in patients who have had infrainguinal bypasses. Of the other platelet inhibitors, ticlopidine has been observed to improve vein graft patency,[47] whilst no benefits from sulphinpyrazone, dipyridamole, and indobufen have been shown.[48] Evidence for the use of oral anticoagulants, most commonly warfarin, is also inconclusive. However, its use in high-risk vein graft patients is generally accepted. Other agents requiring further investigation include low molecular weight heparin and dextrans. Benefits of prostaglandin E_1 and iloprost appear to be short-lived, with no long-term effect on graft patency demonstrated so far.[49]

There is evidence that NO synthesis is impaired at areas of vascular injury and that the NO system is involved in graft failure. In particular, early vasospasm and thrombotic occlusion may be due to reduced endothelial NO activity in vein grafts. NO is an endothelium-derived factor synthesized by the enzyme, nitric oxide synthase (NOS), from L-arginine. It is a potent vasodilator with other vasoprotective properties. NO prevents platelet and leukocyte adhesion, inhibits VSMC proliferation and migration, and demonstrates antioxidant activity. NO donors, such as S-nitrosoglutathione, have been investigated in vein grafts and found to cause vasodilatation[50] and inhibit platelet deposition.[51]

Compounds that reduce neointimal hyperplasia (NIH) are also being investigated as potential pharmacological approaches for preventing vein graft occlusion. Thapsigargin is a compound that increases cytosolic Ca^{2+} by its action as an irreversible inhibitor of Ca^{2+}-ATPase. Intracellular calcium pools are important in regulating VSMC migration, a prerequisite for neointimal hyperplasia. Pre-treatment with thapsigargin ex vivo, has been found to reduce neointima formation in cultured human SV,[52] although the effects of exposing SV to thapsigargin before implantation have not been studied in vivo. Oral and intramuscular administration of rapamycin, a macrolide antibiotic with antimitotic properties, has also been found to reduce neointimal formation after balloon-induced vascular injury in porcine models.[53,54] While these approaches have been shown to reduce neointima formation in experimental models, their clinical potential is yet to be demonstrated.

21.3.14 Endothelin

The endogenous peptide, endothelin-1 (ET-1), is known to have both vasoconstrictor and mitogenic properties. Veins possess a greater density of ET-1 receptors when compared with the arteries, predisposing them to both ET-1-mediated spasm and vessel wall thickening.[55] A proliferative role for ET-1 has been demonstrated, and evidences demonstrate that ET-1 is involved in the graft occlusion process, thus presenting a potential target for pharmacological intervention.[53] As part of the adaptive response to grafting into the arterial system, both the expression and secretion of ET peptides are found to increase. Additionally, adherent platelets and leukocytes generate an increased amount of ET-1.[56] This is supported by the fact that ET-1-like immunoreactivity has been detected and localized in the endothelium and proliferating VSMCs of surgically removed failed SV grafts.[57] Additionally, an up-regulation of functional ET converting enzymes (ECEs), required for the synthesis of ET-1, has been demonstrated in atherosclerotic coronary arteries, which bear great similarities to the diseased SV grafts. Moreover, the use of receptor-subtype selective radioligands has shown that the expression of ET-1 receptors becomes altered in SV grafts and resembles the pattern of receptor expression in atherosclerotic SVs.[57] In vitro studies have shown that ET-1 stimulates DNA synthesis in cultured humans SV VSMCs, an action primarily stimulated by ET-1 binding to the ET_A receptor. As ET-1 appears to play a role in the development and progression of vein graft disease, including hyperplasia, ET receptor antagonists have been investigated as compounds with therapeutic potential to reduce vein graft hyperplasia. In 2004, Wan et al.[58] described the effect of the ET_A receptor antagonist, BSF 302146, on SV graft thickening in a porcine model, and concluded that oral administration of this compound inhibited neointima formation. However, there are limitations with this model and clinical trials are required before the therapeutic potential of ET_A antagonists in improving vein graft patency in CABG patients can be established.

21.3.15 The Role of Matrix Metalloproteinases (MMPs)

A major feature in the formation of intimal thickening involves the migration of VSMCs from the media to the intimal layer of the vessel. The prerequisite for this step is the degradation of the basement membrane, involving the action of MMPs and, in particular, the gelatinases MMP-2 and MMP-9, both of which are up-regulated in the venous grafts subjected to arterial conditions.[16] It has been suggested that the principle factors behind vascular remodeling such as altered hemodynamics, injury, and inflammation regulate MMP expression and activity.[59] MMPs degrade the ECM to form a pathway for the VSMCs to migrate to the intima,[60] and under basal conditions, the actions of MMPs are closely regulated by the tissue inhibitors of matrix metalloproteinases (TIMPs). A study on the differential expression of both TIMPs and MMPs in the normal veins and grafted veins revealed a significant increase in MMP-9 mRNA expression after implantation into the arterial circulation, although there was no statistical difference in the expression of MMP-2 or TIMPs. Based on these results, MMP-9 inhibitors seem to be the most likely candidates for improving vein graft patency. Furthermore, additional evidence showed that when comparing the intact cultured endothelium and denuded human SVs, neointimal formation was associated with increased MMP-9. Doxycyclin, a potent MMP inhibitor, reduced neointimal hyperplasia, suggesting a therapeutic potential of compounds affecting MMPs in reducing hyperplastic lesions.[61]

21.3.16 Gene Therapy

With poor clinical outcomes for re-intervention cases, it is clear that novel strategies are required which can be used at the time of grafting to reduce the incidence of graft failure. The rapid advances in molecular biology over the recent years have led to exciting novel techniques involving gene therapy in the quest to identify potential methods of reducing vein graft hyperplasia. Since the late 1990s, many prospective gene targets have been identified and considered, for example, E2F decoy, p53, TIMPs, and NOS.[62] Failure of pharmacological interventions targeting MMPs led the research into developing gene delivery vectors for MMP inhibition via TIMP and NOS manipulation (Fig. 21.6).

21.3.17 Gene Delivery Vectors

The initial phase in the development of gene delivery to cells in vein grafts requires the presence of a powerful vector system. In macaque monkeys, systemic application of gene therapy has demonstrated an uneven distribution in the body, resulting in 90% of recombinant genes accumulating in the liver.[63] Thus, a small window

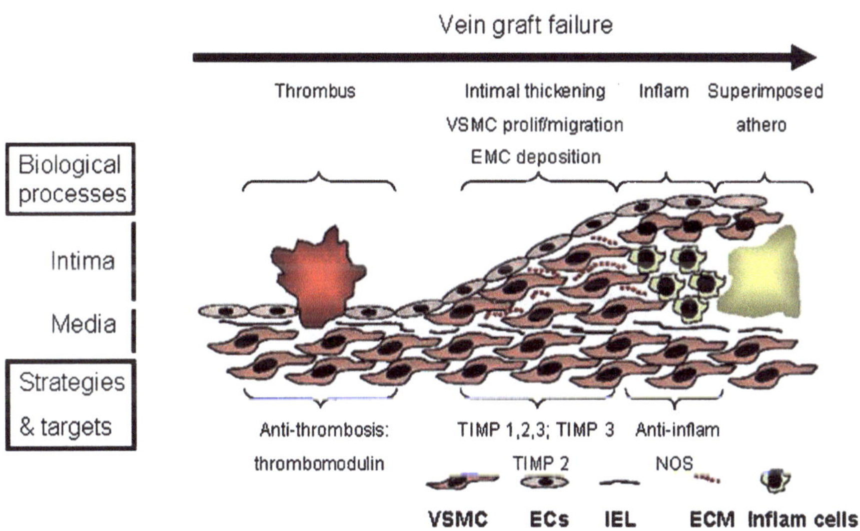

Fig. 21.6 Stages of vein graft failure and targets for gene therapy. Schematic diagram illustrating the stages involved in vein graft failure and potential targets for gene therapy aimed at reducing vein graft hyperplasia. The biological processes involved in vein graft failure are thrombosis (early phase), intimal thickening (VSMC proliferation/migration and extracellular matrix (ECM) deposition (mid stage) and inflammation/atherosclerosis (late stage). Potential strategies/targets range from antithrombotic factors (e.g. thrombomodulin) to tissue inhibitors of the matrix metalloproteases (TIMPs, 1, 2 and 3) and nitric oxide synthases (NOS) (adapted from George et al[64])

(30–60 min) exists, during which time, the vein can be genetically manipulated preceding implantation.[62] To date, many vectors have been explored for use in vein graft gene delivery systems, including adenoviruses (Ad), adeno-associated viruses (AAV), and lentiviruses, of which Ad has been demonstrated to be the most effective. The rationale behind this finding is that Ad provokes significant transduction of both endothelial and VSMCs within a short exposure time of the tissue to the virus (30 min). Moreover, Ad vectors have demonstrated a highly beneficial safety advantage over the alternative vector systems. Clinical trials on the use of gene therapy are warranted for "potential routine clinical practice to improve patency rates of bypass graft procedures involving SV."[62]

The three main processes involved in the development of graft failure, initial thrombosis, intimal hyperplasia, and accelerated atherosclerosis, all represent potential strategies, whereby vein graft hyperplasia may be reduced via agents possessing antithrombotic, antiproliferative, antimigratory, pro-apoptopic, and anti-inflammatory properties.[64]

21.3.18 Nitric Oxide Gene Therapy

NO possesses both vasodilator activity and an antiproliferative role, which prevents intimal hyperplasia via the inhibition of SMC proliferation, leukocyte adhesion, platelet adhesion, and aggregation. Previous studies have demonstrated a decrease in NO and NOS activity early after CABG.[65] The time of vessel harvesting provides an ideal opportunity for local application of vectors. The NOS isoenzymes convert L-arginine to NO and the potential of recombinant NOS gene therapy has been investigated. It has been suggested that increased recombinant NOS expression may provide a constant vascular supply of NO, inhibiting leukocyte and platelet adhesion, aggregation, and VSMC migration, resulting in increased blood flow and re-endothelialization of the graft material. Studies in which the local NO levels were raised by increasing the dietary arginine, the use of NO donors, or NOS gene transfer, demonstrated reduced intimal hyperplasia in vein grafts. All the three NOS isoforms, neuronal NOS (nNOS), endothelial NOS (eNOS), and inducible NOS (iNOS) have been investigated for improving vein graft performance.

21.3.19 NOS Transfer

A study using a rabbit jugular-carotid interposition bypass model has reported the effect of nNOS transfer on graft remodeling and the long-term effects of nNOS treatment on vein graft VSMC phenotype. Here, intimal hyperplasia was significantly reduced in vein segments treated with adenovirus neuronal NOS (Ad.nNOS). However, nNOS was no longer detected 28 days after surgery, indicating that the effects were on early graft remodeling rather than on the direct effects of long-term nNOS activity.[66] A beneficial effect of eNOS has also been suggested, as Ad-mediated gene transfer of eNOS causes increased NO production and reduces early graft failure.[67] Animal models of eNOS over-expression have also produced promising results in the regulation of pathological vascular remodeling. For example, in balloon-injured rat carotid arteries, eNOS transgene over-expression resulted in 70% reduction in neointima formation. Transgenic animals lacking eNOS have been described, where the luminal area is reduced and vessel wall thickness doubled due to VSMC proliferation, demonstrating the role of NO in the regulation of vascular remodeling.[68] In addition, it has been revealed that eNOS over-expression in cultured human SV segments increased NO production and NO-mediated relaxation, and inhibited intimal hyperplasia.[65] Furthermore, it has been demonstrated that adventitial eNOS gene delivery is more effective than intraluminal delivery at reducing neointimal thickening in experimental vein grafts.[69] iNOS gene expression increased in the human vein grafts, and the authors concluded that "manipulation of iNOS expression may lead to therapies to alleviate neointimal formation in graft failure."[70]

Although concerns were initially raised regarding the potential cytotoxicity of exogenous recombinant iNOS genes, presently-available data indicate that there are no differences in terms of safety and effectiveness between all the three isoforms.[68] Before the clinical application of NOS gene therapy can be seriously considered, additional issues need to be addressed, and the most effective isoform identified. Additionally, the most desirable duration of transgene expression needs to be considered; a transgene with a short duration of expression may be acceptable for the inhibition of thrombosis and intimal hyperplasia; however, for the prevention of atherosclerosis, a form of the transgene with more prolonged expression would be more desirable. Furthermore, the development of gene therapy in the prevention and treatment of thrombosis, intimal hyperplasia, and atherosclerosis in vein grafts following CABG represents a strategy with a potential for improving the treatment of vascular disease. Also, recombinant gene therapy allows local application of NOS to isolated graft material and reduces the potential side effects often associated with systemic administration related to NO donors.[68] Although promising results have been established in experimental models using gene therapy, further clinical trials are needed before this technique can be applied to patients. An additional area in which the role of NO has been implicated is in the improved patency rates of SV harvested atraumatically, preserving the cushion of the surrounding tissue, in particular the perivascular fat.

21.3.20 Perivascular Fat

The development of the "no-touch" technique, in which the vein is harvested completely with the surrounding pedicle of the tissue, has produced excellent long-term results comparable with those of the "gold standard," the LIMA (90% patency rates at 8.5 years).[21] Recently, a novel fat-derived compound, adipocyte-derived relaxing factor (ADRF), has been identified[71] and a study on the potential role of the perivascular fat and its components on improved patency rates observed with the "no-touch" method of harvesting has been reported. In particular, whether perivascular fat is a potential source of NO, contributing to the improved patency rates reported in SVs harvested with surrounding tissue intact has been investigated.

Human SV segments obtained from patients undergoing CABG were harvested by the "no-touch" technique, and perivascular fat was removed to study the eNOS distribution and activity (Fig. 21.7).

The data indicate that perivascular fat-derived eNOS may play an important role in the improved patency rates observed in veins harvested atraumatically.[72] Stripping of the perivascular fat, as performed in conventional harvesting methods, is found to decrease the eNOS levels substantially, contributing to the inferior patency rates generally associated with the SV. These results suggest that the surrounding cushion of tissue, including the perivascular fat, in veins harvested by the "no-touch" technique, provides not only structural support and minimizes

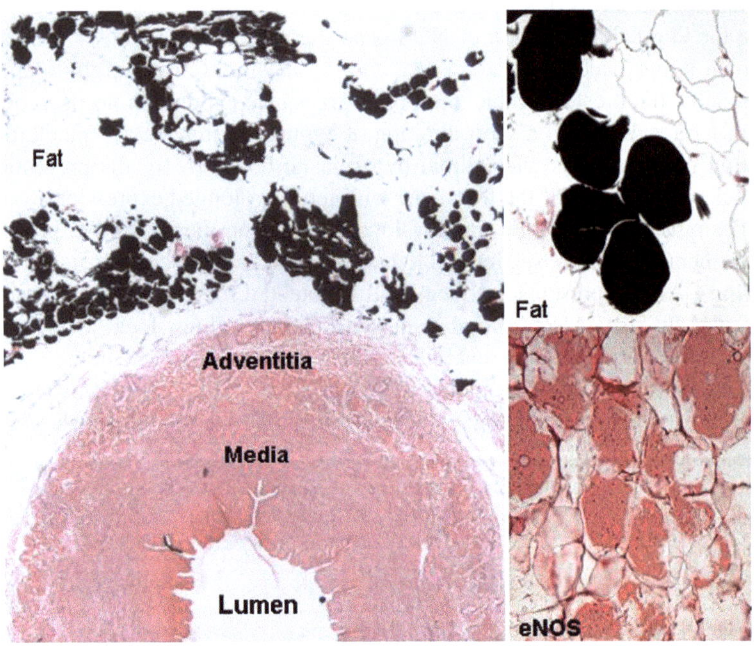

Fig. 21.7 Potential role of periadventitial fat in improved vein graft patency. *Left panel*. Part of a transverse section of "no-touch"-harvested human SV. The vein wall (adventitia and media) have been stained with eosin (red/pink) and surrounding cushion of tissue with osmium tetroxide using the Marchi technique (adipose tissue stains black). *Right panel*. Top perivascular fat shown at higher magnification (black staining) exhibits positive endothelial NOS (eNOS) immunostaining (*lower panel*, red staining). Extracts of SV perivascular fat show both eNOS mRNA and protein as well as NOS activity (adapted from Dashwood et al[72])

surgical damage, but also plays a role in modulating the vascular tone and reducing neointimal thickening.[72] Of particular relevance to this suggestion is an earlier study by Gao et al.[73] who reported that perivascular adipose tissue has the potential to play an important role in the improved patency rates of the LIMA when compared with SV grafts. These authors concluded that adipose tissue of arteries releases a relaxant factor which may explain the superior results obtained with the LIMA.

21.3.21 Prosthetic Grafts

To date, prosthetic grafts are rarely used in CABG and only considered as an alternative if no other conduit is available,[74] mainly, in patients requiring repeat operations (redo patients) in which both the LIMA and SVs have already been used. Prosthetic grafts may also be considered in patients whose remaining vessels are of poor quality, for example, elderly patients who are prone to varicose veins and atherosclerotic arteries.[75] Ideal prosthetic grafts for CABG must exhibit the same characteristics as the native vessel, such as durability, good handling characteristics, flexibility, ease of suturing, and similar viscoelastic and nonthrombogenic properties. Additionally, the biostability of the graft is vital as any degradation of the graft would result in irreversible altered graft characteristics, increasing the risk of aneurism formation.[74] The selection of the polymer for grafting is therefore essential. The most popular polymers used in cardiovascular surgery are Dacron™ and PTFE. Studies on the use of prosthetic grafts in CABG redo patients have proved to be disappointing primarily due to their thrombogenicity and subsequent intimal hyperplasia, particularly at regions of anastomosis. Patency rates using PTFE in CABG patients were only 14% at 45 months.[75] The formation of intimal hyperplasia is believed to be a result of compliance mismatch at the anastomoses between the viscoelastic blood vessel and the comparatively nonelastic grafts. These disappointing results have prompted the development of superior grafts made with polymers such as polyurethanes (PU), which are said to be more compliant and reseal after use. These materials are still under clinical trial so far, with no long-term in vivo data available in patients.

21.3.22 Seeding

Prosthetic grafts do not spontaneously endothelialize, except at the anastomotic regions, suggesting that the endothelial cell deficiency throughout the graft provides the basis for thrombus formation and graft failure.[75] A cellular engineering technique, "seeding," has been developed in an attempt to surmount thrombus formation. The superior method of seeding, the "two-stage" technique, extracts endothelial cells, usually from a patient's vein, that are then cultured in a laboratory before seeding. This technique ensures full endothelial coverage of the synthetic graft. Initial results from animal models are encouraging with patency rates of 100% observed in conjunction with antiplatelet therapy, and studies producing patency rates of 90.5% at 52 weeks.[75] The main limitation of this method of seeding is the long culture period required, leading to the investigation of an alternative "single-stage" seeding method, in which the endothelial cells are harvested and immediately transferred onto the graft material. The advantages of this technique are a decreased risk of infection and the ability of seeding to be performed in one surgical procedure. Disappointingly, results from studies using this method of seeding are poor with patency rates of only 60% at 4 years.[75]

21.3.23 Tissue Engineering

This approach involves the development of "fully engineered grafts made from a scaffold and mixtures of VSMCs/collagen and endothelial cells."[75] This technique takes several weeks to prepare a potential graft, and therefore, cannot be applied within a time frame suitable for emergency procedures. This method has not yet undergone clinical trial, and the long-term effectiveness and patency rates are not yet known. However, it has been suggested that this method of vascular tissue engineering will not be accepted until the results superior to autologous grafts have been demonstrated in clinical trials. At present, a combination of prosthetic grafts with two-stage seeding appears to produce optimal results. The development of a more rapid endothelial cell lining for the graft is needed before these grafts are a realistic option in emergency cases. The use of recombinant eNOS gene therapy has the potential to be combined with the "seeding" of the prosthetic grafts. Here, the resultant combination of prosthetic grafts for CABG coupled with the antiproliferative and relaxant role of NO promises greater patency rates. To date, only one such study has investigated this potential. The effect of vascular prosthesis seeded with lacZ gene-transduced mesenchymal stem cells (MSC) expressing eNOS was used, as these cells possess the ability to differentiate into vascular endothelial cells exerting their paracrine function when transduced into PTFE grafts. The results from this study demonstrated an increased eNOS activity with time. Furthermore, the addition of the NOS inhibitor, L-NAME, completely abolished this response, demonstrating that NOS activity had successfully been induced by eNOS gene transduction of MSCs.[76]

These results highlight the potential development of vascular prostheses for CABG. However, several issues still remain to be addressed. The graft must withstand in vivo blood flow for the endothelial cells to maintain their function at high pressures and changes in shear stress. Additionally, VMSCs have the ability to differentiate into bone, cartilage, fat, and muscle. It must therefore be established that over-expression of eNOS does not affect progenitor differentiation.[76] Also, these results only showed increased eNOS activity over short time periods and conclusions regarding the long-term effects of eNOS gene transduction of MSCs and graft patency require further investigation.

21.4 Conclusions

The introduction of the SV as a bypass conduit in CABG patients represents a major advance in cardiac surgery. However, Favarolo's suggestion that "Care must be taken to dissect only the vein, avoiding as much as possible the adventitia that surrounds it"[4] may have been a fundamental error, as recent evidence shows that the patency rate of the SV is dramatically improved if the vessel is removed with minimal trauma. Paradoxically, many attempts at improving vein graft performance, both in patients and in experimental models, may be seen as attempts to repair the effects of surgical damage or examining ways of replacing, restoring, or inhibiting factors released by vascular trauma during harvesting. Such strategies range from adventitial delivery of drugs and gene transfer to pharmacological interventions and methods of providing mechanical support. There is a need to reduce vascular damage during harvesting and the benefits of the "no-touch" technique have been recognized in a recent review.[32] By harvesting the SV with an intact cushion of the surrounding tissue, high patency rates, equivalent to those reported for the LIMA can be obtained. While the "no-touch" technique is gaining popularity in many cardiac centers, some surgeons seem to be resistant to the change. However, for this technique to become a routine method of harvesting, the organization of a multicenter trial is essential.

References

1. Mehta D, Izzat MB, Bryan AJ, Angelini GD. Towards the prevention of vein graft failure. *Int J Cardiol.* 1997;62:S55-S63.
2. Miller DW, ed. *The Practice of Coronary Artery Bypass Surg.* New York: Plenum Medical Book Company; 1977
3. Captur G. Memento for Rene Favaloro. *Tex Heart Inst J.* 2004;31:47-60.
4. Favarolo RG. Saphenous vein graft in the surgical treatment of coronary artery disease: operative technique. *J Thorac Cardiovasc Surg.* 1969;58:178-185.
5. Tsui JC, Dashwood MR. Recent strategies to reduce vein graft occlusion: a need to limit the effect of vascular damage. *Eur J Vasc Endovasc Surg.* 2002;23:202-208.
6. Roubos N, Rosenfeldt FL, Richards SM, Conyers RAJ, Davis BB. Improved preservation of saphenous vein grafts by the use of glyceryl trinitrate-verapamil solution during harvesting. *Circulation.* 1995;92:31-36.
7. Mitra AK, Gangahar DM, Agrawal DK. Cellular, molecular and immunological mechanisms in the pathophysiology of vein graft intimal hyperplasia. *J Aust Soc Immunol.* 2006;84:115-124.
8. Kierszenbaum AL, ed. *Histology and Cell Biology.* Missouri: Mosby; 2002
9. Martini FH, Ober WC, Garrison CW, Welch K, Hutchings RT, Ireland K, eds. *Fundamentals of Anatomy and Physiology.* New Jersey: Pearson Education; 2004
10. Lemson MS, Tordoir JHM, Daemen MJAP, Kitslaar PJEHM. Intimal hyperplasia in vascular grafts. *Eur J Var Endovasc Surg.* 2000;19:336-350.
11. Motwani JG, Topol EJ. Aortocoronary saphenous vein graft disease: pathogenesis, predisposition and prevention. *Circulation.* 1998;97:916-931.
12. Cox JL, Chiasson DA, Gotlieb AI. Stranger in a strange land: the pathogenesis of saphenous vein graft stenosis with emphasis on structural and functional differences between vein and arteries. *Prog Cardiovasc Dis.* 1991;34:45-68.

13. Rang HP, Dale MM, Ritter JM, Moore PK, eds. *Pharmacology*. 5th ed. Edinburgh: Churchill Livingstone; 2003
14. Coughlin SR. Protease-activated receptors in hemostasis, thrombosis and vascular biology. *J Thromb Haemost*. 2005;3:1800-1814.
15. Newby AC, Zaltsman AB. Molecular mechanisms in intimal hyperplasia. *J Pathol*. 2000;190:300-309.
16. Anstadt MP, Franga DL, Portik-Dobos V, et al. Native matrix metalloproteinase characteristics may influence early stenosis of venous versus arterial coronary artery bypass grafting conduits. *Chest*. 2004;125:1853-1858.
17. Schwarz SM, deBlois D, O'Brien ER. The intima. Soil for atherosclerosis and restenosis. *Circ Res*. 1995;77:445-465.
18. Gottlob R. The preservation of the venous endothelium by dissection without touching and by an atraumatic technique of vascular anastomosis. *Minerva Chir*. 1977;32:693-700.
19. Galea J, Armstrong J, Francis SE, Cooper G, Crossman DC, Holt CM. Alterations in c-*fos* expression, cell proliferation and apoptosis in pressure distended human saphenous vein. *Cardiovasc Res*. 1999;44:436-448.
20. Souza DS, Dashwood MR, Tsui JCS, et al. Improved patency in vein grafts harvested with surrounding tissue: results of a randomised study using three harvesting techniques. *Ann Thorac Surg*. 2002;73:1189-1195.
21. Souza DS, Johansson B, Bojo L, et al. Harvesting the saphenous vein with surrounding tissue for CABG provides long-term patency comparable to the left internal thoracic artery: results of a randomised longitudinal trial. *J Thorac Cardiovasc Surg*. 2006;132:373-378.
22. Tsui JCS, Souza DSR, Filbey D, Bomfim V, Dashwood MR. Preserved endothelial integrity and nitric oxide synthase in saphenous vein grafts harvested by a "no-touch" technique. *Br J Surg*. 2001;88:1209-1215.
23. Dashwood MR, Anand R, Loesch A, Souza DS. Hypothesis: a potential role for the vasa vasorum in the maintenance of vein graft patency. *Angiology*. 2004;55:385-395.
24. Gundry SR, Jones M, Ishihara T, Ferran VJ. Optimal preparation techniques for human saphenous vein grafts. *Surgery*. 1980;88(6):785-794.
25. Oto T, Griffiths AP, Rosenfeldt F, Levvey BJ, Williams TJ, Snell GI. Early outcomes comparing perfadex, euro-collins, and papworth solutions in lung transplantation. *Ann Thorac Surg*. 2006;82(5):1842-1848.
26. Moggio RA, Ding J-Z, Smith CJ, Tota RR, Stemerman MB, Reed GE. Immediate-early gene expression in human saphenous veins harvested during coronary artery bypass operation. *J Thorac Cardiovasc Surg*. 1995;110:209-213.
27. Cornelissen J, Armstrong J, Holt CM. Mechanical stretch induces phosphorylation of p38-MAPK and apoptosis in human saphenous vein. *Arterioscler Thromb Vasc Biol*. 2004;24(3):451-456.
28. Chello M, Mastroroberto P, Frati G, et al. Pressure distension at harvesting upregulates adhesion molecules causes increased neutrophil accumulation and early graft failure. *Ann Thorac Surg*. 2003;76:453-458.
29. Chung AW, Rauniyar P, Luo H, Hsiang YN, van Breeemen C, Okon EB. Pharmacological preparation of graft using various vasodilators results in increased eNOS expression and NO production after implantation; pharmacology better than distension. *J Thorac Cardiovasc Surg*. 2006;132:925-932.
30. Chung AW, Rauniya P, Luo H, Hsiang YN, van Breemen C, Okon EB. Pharmacologic relaxation of vein grafts is beneficial compared with pressure distention caused by upregulation of endothelial nitric oxide synthase and nitric oxide production. *J Vasc Surg*. 2005;42:747-756.
31. Manabe S, Sunamori M. Radial artery for coronary artery bypass surgery: biological characteristics and clinical outcome. *J Cardiothorac Surg*. 2006;21:102-114.
32. Shuhaiber JH, Evans AN, Massad MG, Geha AS. Mechanisms and future directions for prevention of vein graft failure in coronary bypass surgery. *Eur J Cardio-Thorac Surg*. 2002;22:387-396.
33. Goldman S, Zadina K, Moritz T, et al. Long-term patency of saphenous vein and left internal mammary artery grafts after coronary artery bypass surgery: results from a Department of Veterans Affairs Cooperative Study. *J Am Coll Cardiol*. 2004;44:2149-2156.

34. Legare JF, Buth KJ, Sullivan JA, Hirsch GM. Composite arterial grafts versus conventional grafting for coronary artery bypass grafting. *J Thorac Cardiovasc Surg*. 2004;127:160-166.
35. Eefting F, Nathoe H, van Dijk D, et al. Randomised comparison between stenting and off-pump bypass surgery in patients referred for angioplasty. *Circulation*. 2003;108:2870-2876.
36. Gershlick AH. Drug eluting stents in 2005. *Heart*. 2005;91:24-31.
37. Mehta D, George SJ, Jeremy JY, et al. External stenting reduces long-term medial and neointimal thickening and platelet derived growth factor expression in a pig model of arteriovenous bypass grafting. *Nat Med*. 1998;4:235-239.
38. Vijayan V, Shukla N, Johnson J, et al. Long-term reduction of medial and intimal thickening in porcine saphenous vein grafts with a polyglactin biodegradable external sheath. *J Vasc Surg*. 2004;40:1011-1019.
39. Jeremy JY, Bulbulia R, Johnson J, et al. A bioabsorbable (polyglactin), non-restrictive, external sheath inhibits porcine saphenous vein graft thickening. *J Thorac Cardiovasc Surg*. 2004;127:1766-1772.
40. Stooker W, Niessen HWM, Baidoshvili A, et al. Perivenous support reduces early changes in human vein grafts: studies in whole blood perfused human vein segments. *J Thorac Cardiovasc Surg*. 2001;121:290-297.
41. Stooker W, Gök M, Sipkema P, et al. Pressure-diameter relationship in the human greater saphenous vein. *Ann Thorac Surg*. 2003;76:1533-1538.
42. Angelini GD, Bryan AJ, Williams HMJ, Morgan R, Newby AC. Distension promotes platelet and leukocyte adhesion and reduces short-term patency in pig arterio-venous bypass grafts. *J Thorac Cardiovasc Surg*. 1990;99:433-439.
43. Angelini GD, Bryan AJ, Williams HMJ, et al. Time-course of medial and intimal thickening in pig venous arterial grafts: Relationship to endothelial injury and cholesterol accumulation. *J Thorac Cardiovasc Surg*. 1992;103:1093-1103.
44. The Post Coronary Artery Bypass Graft Trial Investigators. The effect of aggressive lowering of low-density lipoprotein cholesterol levels and low-dose anticoagulation on obstructive changes in saphenous-vein coronary-artery bypass grafts. *N Engl J Med*. 1997;336:153-162
45. Goldman S, et al. Saphenous vein graft patency 1 year after coronary artery bypass surgery and effects of antiplatelet therapy. Results of a Veterans Administration Cooperative Study. *Circulation*. 1989;80:1190-1197.
46. Stein PD, et al. Antithrombotic therapy in patients with saphenous vein and internal mammary artery bypass grafts. *Chest*. 1995;108:424S-430S.
47. Becquemin JP. Effect of ticlopidine on the long-term patency of saphenous-vein bypass grafts in the legs. Etude de la Ticlopidine apres Pontage Femoro-Poplite and the Association Universitaire de Recherche en Chirurgie. *N Engl J Med*. 1997;337:1726-1731.
48. Watson HR, Belcher G, Horrocks M. Adjuvant medical therapy in peripheral bypass surgery. *Br J Surg*. 1999;86:981-991.
49. Antiplatelet Trialists' Collaboration. Collaborative overview of randomised trials of antiplatelet therapy II: Maintenance of vascular graft or arterial patency by antiplatelet therapy. *BMJ*. 1994;308:159-168.
50. Sogo N, Campanella C, Webb DJ, Megson IL. S-nitrosothiols cause prolonged, nitric oxide-mediated relaxation in human saphenous vein and internal mammary artery: therapeutic potential in bypass surgery. *Br J Pharmacol*. 2000;131:1236-1244.
51. Salas E, et al. S nitrosoglutathione inhibits platelet activation and deposition in coronary artery saphenous vein grafts in vitro and in vivo. *Heart*. 1998;80:146-150.
52. George SJ, Johnson JL, Angelini GD, Jeremy JY. Short-term exposure to thapsigargin inhibits neointima formation in human saphenous vein. *Arterioscler Thromb Vasc Biol*. 1997;17:2500-2506.
53. Burke SE, et al. Neointimal formation after balloon-induced vascular injury in Yucatan minipigs is reduced by oral rapamycin. *J Cardiovasc Pharmacol*. 1999;33:829-835.
54. Jeremy JY, Dashwood MR. Microvascular repair. In: Shepro AM, ed. *Encyclopaedia of the Microvasculature*. New York: Elsevier; 2006:903-911.

55. Dashwood M, Anand R, Loesch A, Souza D. Surgical trauma and vein graft failure: further evidence for a role of et-1 in graft occlusion. *J Cardiovasc Pharmacol.* 2004;44:S16-S19.
56. Dashwood MR, Barker SG, Muddle JR, Yacoub MH, Martin JF. [125I]-Endothelin-1 binding to vasa vasorum and regions of neovascularization in human and porcine blood vessels: a possible role for endothelin in intimal hyperplasia and atherosclerosis. *J Cardiovasc Pharmacol.* 1993;22(suppl 8):S343-S347.
57. Dashwood MR, Sykes RM, Muddle JR, et al. Autoradiographic localization of [125I] endothelin binding sites in human blood vessels and coronary tissue: functional correlates. *J Cardiovasc Pharmacol.* 1991;17(suppl 7):S458-S462.
58. Wan S, Yim AP, Johnson JL, et al. The endothelin 1A receptor antagonist BSF 302146 is a potent inhibitor of neointimal and medial thickening in porcine saphenous vein-carotid artery interposition grafts. *J Thorac Cardiovasc Surg.* 2004;127(5):1317-1322.
59. Zorina SG, Kharti JJ. Matrix metalloproteinases in vascular remodeling and atherogenesis: the good, the bad, and the ugly. *Circulation.* 2002;90:251-262.
60. Shi Y, Patel S, Niculescu R, Chung W, Desrochers P, Zalewski A. Role of matrix metaloproteinases and their tissue inhibitors in the regulation of coronary cell migration. *Arterioscler Thromb Vasc Biol.* 1999;19:1150-1155.
61. Porter KE, Thompson MM, Loftus IM, et al. Production and inhibition of the gelatinolytic matrix metalloproteinases in a human model of graft stenosis. *Eur J Vasc Endovasc Surg.* 1999;17:404-412.
62. Baker AH, Yim APC, Wan S. Opportunities for gene therapy in preventing vein graft failure after coronary artery bypass surgery. *Diabetes Obes Metab.* 2005;8(2):119-124.
63. Lozier JN, Metzger ME, Donahue RE, Morgan RA. Adenovirus-mediated expression of human coagulation factor ix in the rhesus macaque is associated with dose-limiting toxicity. *Blood.* 1999;94(12):3968-3975.
64. George SJ, Channon KM, Baker AH. Gene therapy and coronary artery bypass grafting: current perspectives. *Curr Opinion Mol Ther.* 2006;8:288-294.
65. Chen AF, Ren J, Miao CY. Nitric oxide gene therapy for cardiovascular disease. *Jpn J Pharmacol.* 2002;89:327-336.
66. West NEJ, Qian HS, Guzik TJ, et al. Nitric oxide synthase (nNOS) gene transfer modifies venous bypass graft remodeling: effects on vascular smooth muscle cell differentiation and superoxide production. *Circulation.* 2001;104:1526-1532.
67. Cable DG, O'Brien T, Schaff HV, Pompili VJ. Recombinant endothelial nitric oxide synthase-transduced human saphenous veins: gene therapy to augment nitric oxide production in bypass conduits. *Circulation.* 1997;96(9 suppl)):II-173-II-178.
68. Von der Leyen H, Dzau VJ. Therapeutic potential of nitric oxide synthase gene manipulation. *Circulation.* 2001;103:2760-2765.
69. Kalra M, Jost CJ, Secerson SR, Miller VM. Adventitial versus intimal liposome-mediated ex vivo transfection of canine saphenous vein grafts with endothelial nitric oxide synthase gene. *J Vasc Surg.* 2000;32:1190-2000.
70. Dattilo JB, Dattilo MP, Spratt JA, Matsuura J, Yager DR, Makhoul RG. Inducible nitric oxide synthase expression in human vein grafts. *Am J Surg.* 1997;174:177-180.
71. Fernandez-Alfonso MS. Regulation of vascular tone: the fat connection. *Hypertension.* 2004;44:255-256.
72. Dashwood MR, Dooley A, Shi-Wen X, Abraham DJ, Souza DSR. Does periadventitial fat-derived nitric oxide play a role in improved saphenous vein graft patency in patients undergoing coronary artery bypass surgery? *J Vasc Res.* 2007;44:175-181.
73. Gao YJ, Zeng ZH, Teoh K, et al. Perivascular adipose tissue modulates vascular function in the internal thoracic artery. *J Thorac Cardiovasc Surg.* 2005;130:1130-1136.
74. Kannan RY, Salacinski HJ, Butler PE, Hamilton G, Seifalian AM. Current status of prosthetic bypass grafts: a review. *J Biomed Mater Res B Appl Biomater.* 2004;74B:570-581.
75. Vara DS, Salacinski HJ, Kannan RY, Bordenave L, Hamilton G, Seifalian AM. Cardiovascular tissue engineering: state of the art. *Pathol Biol.* 2005;53:599-612.

76. Kanki-Horimoto S, Horimoto H, Mieno S, Kishida K, Watanabe F, Furuya E. Synthetic vascular prosthesis impregnated with mesenchymal stem cells overexpressing endothelial nitric oxide synthase. *Circulation*. 2006;114:327-330.
77. Stone GW, Moses JW, Ellis SG, et al. Safety and efficacy of sirolimus and paclitaxel-eluting coronary stents. *N Engl J Med*. 2007;356:998-1008.

Chapter 22
Cardiovascular Risk Prediction by Measurement of Arterial Elastic Properties and Wall Thickness

Konstantinos Aznaouridis, Saurabh S. Dhawan, and Arshed A. Quyyumi

22.1 General Considerations

Arterial structure and function provide important and accurate representation of overall cardiovascular (CV) performance. Accumulating data suggest that arterial stiffness, central (aortic) hemodynamics, and arterial wall thickness are associated with the presence and extent of CV disease, and importantly, they are independent predictors of outcomes in several populations.[1-7] Recent studies suggest that these arterial characteristics may also be useful in stratifying risk and monitoring efficacy of treatment.[8] The 2007 European Society of Hypertension/European Society of Cardiology guidelines for the management of arterial hypertension suggest that carotid intima-media thickness (IMT) and aortic pulse wave velocity (PWV) can serve as tools to detect vascular damage in hypertensive patients.[9] The expectation is that characteristics of arterial function and structure will ultimately serve as valuable adjunct tools for clinical decision-making. Herein, we present an overview of the principal techniques used in the evaluation of arterial elastic properties and thickness, and their clinical correlates.

22.2 Arterial Elastic Properties

22.2.1 Pathophysiological Considerations

Several biophysical models of the human circulation have been developed to conceptualize the elastic properties of the arterial system. The current concept models the arterial system as a single distensible tube into which the heart ejects blood in systole. The more elastic proximal part of the tube consists of the aorta, and the

A.A. Quyyumi (✉)
Division of Cardiology, Department of Medicine, Emory University Hospital,
1364 Clifton Rd. NE, Suite D403C, Atlanta, GA 30322, USA
e-mail: aquyyum@emory.edu

D. Abraham et al. (eds.), *Advances in Vascular Medicine*,
DOI 10.1007/978-1-84882-637-3_22, © Springer-Verlag London Limited 2010

distal part is made up of the relatively inelastic resistance arteries (small arteries and arterioles). The arterial network combines two principal functions: a cushioning function, which is mainly mediated by the proximal large elastic arteries and a conductance function. Both are combined throughout the arterial system; however, elasticity (cushioning function) decreases and conduit function increases in a stepwise manner as one moves from the aorta to the periphery.[2-5,10,11]

The ejection of blood by the heart in systole generates rhythmic pressure waves, which propagate to the peripheral vasculature. These waves are reflected from several sites of the arterial tree, such as the major branches of the abdominal aorta and especially from the small resistance arteries that comprise the major reflecting sites in the periphery. Accordingly, the pressure waveform recorded at any site of the arterial system arises from the merging of an incident, forward-traveling wave generated by the heart with a backward-traveling wave that has been reflected from the periphery.[2-5,10,11] The contribution of the reflected wave to the final recorded wave depends on the amplitude and the timing of the reflected wave. The amplitude is primarily determined by the peripheral arterial tone, whereas the timing is mostly dependent on arterial stiffness. In subjects with normal arterial stiffness (compliant and elastic arteries), the PWV is low so the two waves (incident and reflected) are merged during the diastole in the aorta (Fig. 22.1). This increases the diastolic blood pressure (BP) of the aorta and thus facilitates coronary perfusion. On the other hand, increased arterial stiffness (noncompliant, inelastic arteries) leads to a higher PWV and faster propagation of both the incident and the reflected waves. At the aortic level, the reflected wave merges with the incident one earlier (in systole), increases the aortic systolic BP, and thus augments the afterload of the heart and impairs coronary perfusion and relaxation the left ventricle[2-5,10,11] (Fig. 22.1).

The amplitude of the pressure wave, the systolic BP, and the pulse pressure increase as one moves to the periphery. In contrast, diastolic and mean BP do not

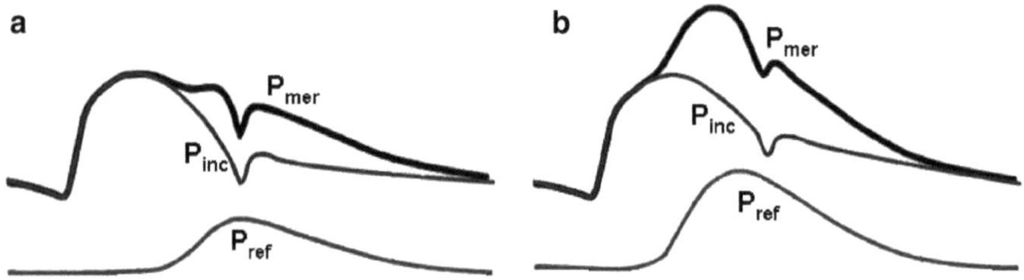

Fig. 22.1 The merged aortic pressure wave (P_{mer}) and its constituents. The incident, forward-traveling wave (P_{inc}) is produced by the ejection of the left ventricle during systole. The reflected, backward-traveling wave (P_{ref}) is produced by the reflection of the incident waveform from peripheral reflecting sites. With an elastic aorta and normal peripheral tone, the P_{inc} and P_{ref} merge during or after the closure of the aortic valve, in diastole (**a**). In the presence of a stiff aorta or peripheral vasoconstriction (**b**), the magnitude of the P_{ref} increases. Both P_{inc} and P_{ref} travel faster and merge earlier during the cardiac cycle (in late systole) resulting in augmentation of the central systolic pressure. Adapted from AtCor Medical Inc

change significantly along the arterial tree. This amplification of the pulse pressure is due to (1) the heterogeneity of arterial stiffness (progressive increase of stiffness – and PWV – in the periphery) and (2) the proximity of the peripheral arteries to the reflecting sites.[3-5] Consequently, the pressures obtained in the brachial artery are only crude estimates of the central hemodynamics, as they tend to be higher than the pressures in the aorta. Pressures in the central circulation are physiologically more important than the respective peripheral (brachial) pressures. Central systolic BP is a better index of the cardiac afterload, and central diastolic BP is a more accurate marker of coronary perfusion pressure than the respective brachial pressures. Similarly, carotid pressures are more relevant to brain perfusion.[3-5]

In large arteries, stiffness is primarily determined by the components and the turnover rate of the extracellular matrix. Elastin fibers are more abundant in the aorta and its large branches and account for elasticity of these vessels. The absolute quantity of elastin fibers and the elastin-to-collagen ratio decreases toward the periphery, so stiffness increases. On the other hand, hypertrophy and tone of the smooth muscle is a more important determinant of smaller arterial stiffness, such as the muscular type conduit and resistance arteries.[3,10] Furthermore, functional characteristics may influence arterial stiffness. For example, stiffness of large arteries increases in parallel with BP, as a higher distending pressure leads to recruitment of more inelastic collagen fibers. Endothelial function and the availability of nitric oxide also regulate large artery stiffness.[12] Finally, elastic properties are dependent in part on the expression of several genes that interfere with arterial homeostasis.[13]

22.2.2 Methods for Evaluation of Arterial Elastic Properties

Currently, there are many techniques that evaluate several aspects of arterial elasticity. Here, we describe the most widely applied techniques for clinical assessment of arterial stiffness.

22.2.2.1 Measurement of Local Arterial Stiffness

The *pulse pressure* is a simple and crude index of large artery stiffness, but it also depends on ventricular stroke volume and other physiological factors. The pulse pressure of the brachial artery is an easily obtained index, which has been used in large population studies. Nowadays, the pulse pressure of central arteries (aorta, carotid artery) can be estimated with noninvasive techniques (see Sect. 2.2.3). Although strictly speaking, central pulse pressure is an index of local stiffness, it is also dependent on the arterial elastic properties of the peripheral arterial system because there is a contribution of the reflected wave to this pressure.[2-5,8]

More sophisticated indices that express stiffness at a particular site of the arterial tree (*distensibility, compliance,* etc.) are listed in Table 22.1. All calculations

Table 22.1 Definition and calculation of indices of local arterial stiffness

Pulse pressure (PP)	Change of BP during systole PP = SBP–DBP
Strain or systodiastolic diameter change (S)	Relative increase in arterial diameter during systole. $S = [(D_s - D_d)/D_d] \times 100$
Distensibility (Dist)	Relative increase in cross-sectional area of the artery for a given increase in BP $Dist = (A_s - A_d)/(A_d \times PP)$
Compliance (C)	Absolute increase in arterial diameter for a given increase in BP $C = (A_s - A_d)/PP$
β-stiffness index (β)	Natural logarithm of the ratio of systolic to diastolic BP divided by the relative increase of arterial diameter during systole $\beta = \ln (SBP/DBP)/[(D_s - D_d)/D_d]$
Peterson's elastic modulus (PEM)	The pressure change required for a given change in relative cross-sectional area of the artery. Conceptually, this is the inverse of distensibility $PEM = PP \times A_d/(A_s - A_d)$
Incremental (Young's) elastic modulus (YEM)	The pressure per unit area required to produce a relative increase in arterial diameter. It expresses the ratio of stress to strain and represents the mechanical characteristics of the arterial wall proper $YEM = (D_d/IMT) \times \{PP/[(D_s - D_d)/D_d]\}$

BP blood pressure; *SBP* systolic BP; *DBP* diastolic BP; D_s arterial diameter in systole; D_d arterial diameter in diastole; A_s arterial cross-sectional area in systole; A_d arterial cross-sectional area in diastole; *IMT* arterial wall intima-media thickness
Cross sectional area A is calculated as: $A = \pi \times (D/2)^2$

require measurement of the change of diameter or cross-sectional area of the artery of interest during the cardiac cycle, and most of them also require estimation of the pressure difference that drives that change. For practical purposes, geometric changes from end diastole to peak systole and pulse pressure are used. The advantage of these methods is that stiffness is determined directly, and there are no assumptions made. However, they are technically demanding and require a higher degree of expertise than measurement of PWV.

A few studies have estimated local stiffness with *invasive methods,* which allow direct measurement of diameter and pressure changes at the same arterial site.[14,15] These methods are highly reproducible but cannot be applied widely because of their invasive nature. On the other hand, local stiffness can be evaluated noninvasively with *ultrasound-based techniques* (traditional ultrasonography or echo-tracking systems).[16,17] A drawback of the noninvasive methods for local stiffness is that most of them employ determination of changes in arterial dimensions at one site (i.e., the carotid artery or the aorta), and measurement of BP at a different site (usually the brachial artery).[17] This certainly introduces a systemic error, given that pressure is amplified across the arterial tree. To overcome this, several "*hybrid*" *methods* have been developed where vascular dimensions are measured in the carotid artery or the aorta and pressures at those sites are calculated noninvasively with applanation tonometry.

The aortic pressure waveform is reconstructed after tonometry of the radial artery by using a transfer function, whereas the carotid waveform can be directly recorded over the common carotid artery (CCA). Aortic and carotid pressures can be derived after calibrating the respective waveforms with the diastolic and mean pressures.[18-20]

Local stiffness of superficial and deep arteries can be also measured accurately by *tissue Doppler* imaging and *magnetic resonance imaging* (*MRI*) technology.[21-23] Furthermore, *arterial compliance of the thigh and calf* can be estimated by automated computer-controlled *air plethysmograph* (Vasogram, Vasocor).[24]

Pulse wave intensity analysis is a recently developed method that studies vascular hemodynamics in terms of traveling energy waves at several points in the circulation. Pressure waveforms are recorded with applanation tonometry (Millar Instrument) at several arterial sites and then flow velocity data at these arteries are acquired through Doppler probes. Then, the acquired pressure and flow-velocity data are ensemble-averaged with specialized software, giving rise to simultaneous pressure-flow velocity waveforms, which allow calculations of the *intensity of the wave* (power per unit cross-sectional area) and the absolute energy carried by a wave at an arterial site.[25]

22.2.2.2 Measurement of Regional Arterial Stiffness

PWV is a direct index of arterial stiffness that can be measured noninvasively using reproducible techniques. The principle for measuring PWV between two sites of the arterial tree is simple. PWV is calculated as the distance between the two sites divided by the travel time of the pulse (transit time) from the proximal to the distal site.[1-4,11] PWV can be calculated in several arterial segments. Carotid-femoral PWV represents the stiffness of the aorta and is the method most frequently employed in clinical studies that has also been shown to correlate more strongly with clinical variables and outcomes.

There are several ways of calculating PWV noninvasively. The pulse transit time between the carotid and the femoral artery can be measured with *mechanotransducers* and subsequent online recording of the pressure wave (Complior, Artech Medical). Two different pulse waves are obtained simultaneously at the base of the neck for the CCA and over the right femoral artery with two transducers. The transit time is measured from the point where the steep rise of each of the recorded waveform starts (foot-to-foot method, left panel in Fig. 22.2).[1-4,11] *Applanation tonometry* (SphygmoCor, AtCor Medical) is another popular method for measuring carotid-femoral PWV, which involves sequential (nonsimultaneous) detection and subsequent recording of pressure waveforms from the two arterial sites using sensitive tonometers (Millar instruments). An electrocardiogram is recorded simultaneously with the measurements, and the pulse transit time is calculated as the difference of the foot of the distal (femoral) waveform from the R wave minus the difference of the foot of the proximal (carotid) waveform from the R wave (right panel in Fig. 22.2). Owing to the nonsimultaneous recording of the waveforms, abrupt changes in heart rate or the contractility of the heart may theoretically decrease the accuracy of the method.

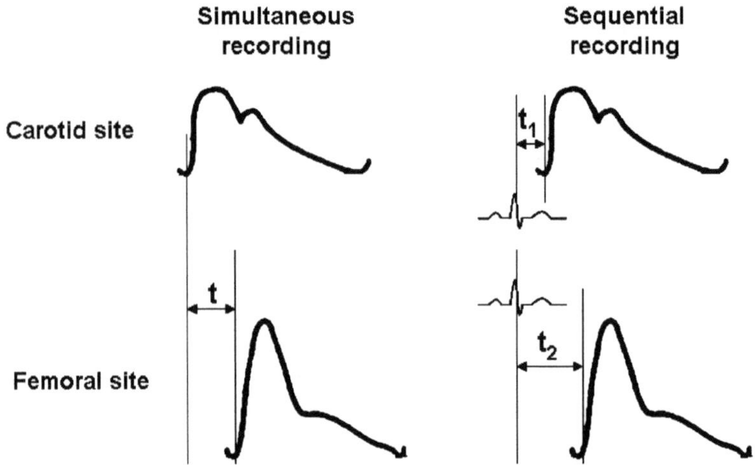

Fig. 22.2 Measurement of pulse transit time t for calculation of carotid-femoral pulse wave velocity. For simultaneous recordings of the pressure waveform, t is measured directly with the foot-to-foot method. For sequential recordings, the time from the R wave to the foot of the carotid (t_1) and the femoral site (t_2) is measured, and t is calculated as t_2–t_1

However, this issue is of rather minor importance, given that both arterial sites are examined during a single session only few minutes apart.[3,4,11]

The measurement of distance between the two arterial sites for calculating carotid-femoral PWV is an intriguing issue. Some investigators use the whole distance between the carotid and femoral sites, whereas others subtract the sternal-carotid distance from the sternal-femoral distance. Currently, there is no consensus on the best way to measure distance. However, given that the pulse propagates from the aorta to both the carotid and femoral arteries and does not travel from the carotid artery downward to the femoral artery, it seems that, from a physiological point of view, using the difference of the two distances would introduce a smaller error in the calculations.

Echotracking systems use a specialized software and high-resolution image analysis and detect changes of arterial diameter over the cardiac cycle. Distension waveforms in the arterial wall are recorded sequentially and the time from the peak of the R wave to the time point when a 10% increase of arterial diameter occurs (10% increase time) is measured at each arterial site. The pulse transit time is calculated as the difference of the 10% increase times between the two sites.[16]

Aortic PWV can also be measured with *Doppler ultrasonography*. With this technique, the pulse transit time is calculated with the "foot-to-foot" method by recording the flow waveforms in the proximal aorta and the distal aorta or the common femoral artery. These flow waves may be obtained simultaneously[26] or sequentially.[27] Recently, some studies have used echocardiography to evaluate the stiffness of the proximal aorta by calculating the PWV between the ascending and the proximal descending aorta.[28,29]

PWV in small or larger segments of the aorta can also be measured with *MRI*. This method involves acquisition of flow velocity signals along the ascending and the descending aorta, measurement of the pulse transit time and distance between sites, and subsequent calculation of PWV.[21,22] This modality allows calculation of PWV with high accuracy, but its use is limited because of low availability and high cost.

Measurement of the *QKD interval* is another option for evaluating regional arterial stiffness.[30] A special apparatus for ambulatory BP measurement (Diasys Integra, Novacor) employs the auscultatory method for measuring BP and determines the time between the QRS wave on the electrocardiogram and detection of the last Korotkoff sound (QKD interval) during deflation of a cuff placed on the arm. This time is equal to the sum of the pre-ejection time and the time of pulse transmission from the aorta to the brachial artery, so it evaluates the stiffness of the ascending aorta, aortic arch, left subclavian, and brachial artery.[30]

22.2.2.3 Measurement of Systemic Arterial Stiffness

Several methods attempt to estimate the overall "elastic behaviour" of the arterial network. These techniques do not measure stiffness in a direct manner but require the adoption of models for the circulation.

The *ratio of stroke volume to pulse pressure* has been used as a gross index of systemic compliance,[31] but its use is limited owing to the lack of accurate noninvasive methods for measurement of stroke volume.

Total systemic arterial compliance can be calculated using the "*area method*."[32] According to this technique, total peripheral resistance R is calculated by measuring total blood flow in the aorta using a Doppler probe placed at the suprasternal notch. The carotid pressure waveform is recorded with a sensitive tonometer (Millar instrument). The central carotid end-systolic and end-diastolic pressures (P_{es}, DP) are obtained after calibration of the carotid waveform with oscillometric diastolic and mean pressures measured in the brachial artery, as mean and DP barely change along the arterial tree. After calculating the area under the diastolic part of the carotid waveform A_d, systemic compliance is calculated using the formula $A_d/[R\ (P_{es} - DP)]$ (Fig. 22.3).[32]

Systemic compliance can also be measured using the HDI/PulseWave CR-2000 System (Hypertension Diagnostics Inc.). According to this technique, the arterial pressure waveform is recorded in the radial artery and analysis of the diastolic decaying sinusoidal wave follows.[33] The method is based on a modified Windkessel model, which assumes that two basic compliance components are present in parallel: the systemic *large artery (capacitive) compliance* C_1, which is BP-dependent, and the *small artery (oscillatory) compliance* C_2, which is mainly dependent on the function of small resistance arteries.[33]

Analysis of the pulse waveform of central arteries (aorta, CCA) may provide information on systemic stiffness.[2-5,8] The central pulse waveform is influenced not only by local (central) stiffness, but also by the elastic properties of the whole arterial network. The aortic waveform can be estimated noninvasively with the SphygmoCor system (AtCor Medical), which employs the principle of applanation tonometry of the radial artery with sensitive tonometers (Millar Instruments). The radial waveform is calibrated with the brachial systolic and diastolic BP, as there is negligible pressure amplification between the brachial and radial arteries, and the mean BP is computed automatically by numerically averaging the area under the radial pressure waveform. Then, the aortic waveform is reconstructed automatically with the use of a generalized transfer function.[2-5,8,34] The transfer function may not apply to all clinical circumstances, and there is concern as to whether it accurately estimates the central waveform.[35] However, although the transfer function is an approximation, it may be useful clinically as transfer function-derived parameters have been shown to predict outcomes in a recent large epidemiological study and a randomized trial.[8,36] In contrast, the carotid waveform can be recorded with a tonometer placed directly over the CCA, and no transfer function is needed.[19,20,34] The central pressure waveforms are calibrated with the mean and diastolic BP from the peripheral arteries, as these pressures remain almost constant throughout the whole arterial tree, and the *central* (*aortic or carotid*) *pressures* are calculated. The point where the incident wave merges with the reflected wave (*reflection point*) is recognized in the central waveform, and *augmentation pressure* (*AP*), which represents the pressure added to the incident wave by the returning reflected one is computed (Fig. 22.3). *Augmentation index* (*AIx*) is then calculated as AP divided by central pulse pressure. Although AIx is not an index of stiffness in a strict sense, it comprises a composite measure of the magnitude of wave reflections (that mainly depends on the tone of the resistance arteries) and PWV (arterial stiffness), which affects the timing of wave reflections. AIx also depends on other variables such as heart rate, left ventricular contractility, and somatometric characteristics. For similar heart rate and effective length of the arterial system, larger values of AIx indicate increased wave reflections from the periphery and/or earlier return of the reflected wave as a result of increased PWV (owing to increased arterial stiffness), and vice versa.[2-5,8,34] The *arrival time* (Δt) of reflected waves at aorta is the time from the foot of the pressure wave to the reflection point and represents the time needed for pressure waves to travel from the aorta to peripheral arterial sites and return back to the aorta owing to wave reflections (Fig. 22.3). A lower Δt indicates a shorter travel time of the pressure waves and a higher arterial stiffness.[3,4]

Ambulatory arterial stiffness index is a new measure of arterial elasticity that is derived from ambulatory BP monitoring.[37] The diastolic BP is plotted against systolic BP, and the regression slope is then calculated. Ambulatory arterial stiffness index is defined as one minus this regression slope. The stiffer the arterial tree, the closer the regression slope to zero and the ambulatory arterial stiffness index is to one.

22.2.3 Determinants and Predictive Value of Arterial Elasticity

Large epidemiological studies including the Anglo-Cardiff Collaborative Trial and the Framingham Heart Study have shown that age is the single most important determinant of arterial elasticity.[38,39] However, other factors such as gender and race also play a role.

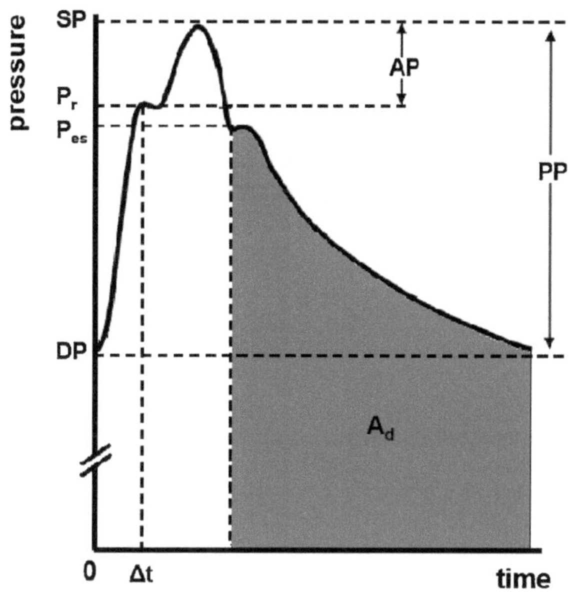

Fig. 22.3 The central (aortic or carotid) pressure waveform is recorded either indirectly by applanation tonometry of the radial artery and reconstruction of the aortic waveform with the use of transfer functions, or directly by tonometry of the common carotid artery (CCA). After calibrating the central waveforms with the mean and diastolic pressure from the peripheral arteries, central systolic pressure (Ps), diastolic pressure (DP), and pulse pressure (PP) is calculated. Augmentation pressure (AP) is calculated as the pressure difference from the reflection point (P_r) to the late systolic peak (SP). Augmentation index (AIx) is the ratio of AP to central PP. The arrival time (Δt) is the time from the foot of the waveform to the reflection point and represents an index of stiffness. Systemic compliance is calculated from the area under the diastolic curve A_d, end-systolic pressure (P_{es}), and end-diastolic pressure (DP) of the carotid waveform

Data from the large bi-racial atherosclerosis risk in communities (ARIC) Study and Bogalusa Heart Study showed that large artery stiffening may be more accelerated in African Americans than their white counterparts.[40,41] Arterial elastic properties are also influenced by hemodynamic parameters (i.e., distending arterial pressure), anthropomorphics, and presence of traditional CV risk factors.[1-3] Table 22.2 lists the main demographic, lifestyle, and clinical characteristics that influence arterial stiffness.

Several longitudinal studies have shown that arterial elastic properties may independently predict CV morbidity and mortality. The Framingham Study has shown that even crude indices of arterial stiffness such as the brachial artery pulse pressure is a predictor of the risk of coronary artery disease in the general population.[42] Recent studies have shown that more sophisticated indices of arterial stiffness may independently predict outcomes. For example, in the Strong Heart Study of American Indians, central (aortic) pulse pressure, an indirect index of central artery stiffness, predicted left ventricular hypertrophy, prevalent carotid atherosclerosis, and future CV events, independently of brachial pulse pressure.[36] PWV has been

Table 22.2 Determinants and clinical correlations of arterial elastic properties

1. Age
2. Gender
3. Cardiovascular (CV) risk factors
hypertension
dyslipidemia (hypercholesterolemia, hypertriglyceridemia)
diabetes mellitus/impaired glucose tolerance
smoking
obesity
4. Cardiovascular diseases
coronary artery disease
peripheral artery disease
heart failure
cardiac syndrome X
5. Endocrinology and metabolic diseases
metabolic syndrome
hyperhomocysteinemia
hypothyroidism
6. Nutrition and lifestyle
coffee, caffeine
chronic alcohol consumption
sedentary lifestyle
resistance training
7. Genetic factors
family history of atherosclerotic disease
polymorphisms of the renin–angiotensin system genes
polymorphisms of the extracellular matrix proteins genes
8. Gynecological disorders
menopause
pre-eclampsia
polycystic ovaries syndrome
9. Other disorders
inflammation
acute inflammation (i.e., acute infections)
chronic inflammatory diseases (i.e., rheumatoid arthritis)
subclinical, low-grade inflammation
end-stage renal disease
sleep apnea

also related to outcomes in several populations. This was initially shown in high-risk patient groups (end-stage renal disease),[43-47] but later studies demonstrated that stiffness may predict outcomes in subjects with risk factors (hypertensives, diabetics)[30,48-52] and even in the general population.[53-59] Indeed, the Rotterdam Study of elderly subjects[57] and another large study in the general population[58] showed that aortic PWV is a predictor of future development of CV disease. Table 22.3 summarizes the reports relating arterial stiffness indices (except pulse pressure) to outcomes according to population, variable measured, and site of measurement.

Table 22.3 Longitudinal studies indicating the value of arterial elastic properties as predictors of cardiovascular outcomes

First author, year	Study population	Index of stiffness	Follow-up	Clinical end point
Blacher, 1998[43]	79 ESRD pts	Common carotid incremental elastic modulus	25 ± 7 months	All cause and CV mortality
Blacher, 1999[44]	241 ESRD pts	Aortic PWV	72 ± 41 months	All cause and CV mortality
De Simone, 1999[31]	294 hypertensive pts	SV/PP	10 years	CV mortality
Stefanadis, 2000[97]	54 CAD pts	Aortic stiffness constant[a] Aortic root distensibility	3 years	Acute coronary syndrome
Fagard, 2001[48]	192 hypertensive pts	SV/PP[a]	16.5 years	CV events and mortality
Laurent, 2001[49]	1,980 hypertensive pts	Aortic PWV	112 ± 53 months	All cause and CV mortality
London, 2001[45]	180 ESRD pts	Carotid AIx	52 ± 36 months	All cause and CV mortality
Meaume, 2001[98]	141 elderly subjects	Aortic PWV	30 months	CV mortality
Boutouyrie, 2002[50]	1,045 hypertensive pts	Aortic PWV	5.7 years	Primary coronary events
Cruickshank, 2002[52]	397 diabetic pts	Aortic PWV	10.7 years	All cause and CV mortality
Guerin, 2002[46]	150 ESRD pts	Aortic PWV	72 ± 41 months	↑ all cause and CV mortality in pts with absence of PWV decrease with antihypertensive treatment
Blacher, 2003[47]	242 ESRD pts	Aortic PWV index (measured –theoretical PWV)	78 ± 46 months	All cause and CV mortality
Grey, 2003[99]	419 subjects, mixed populations	Small artery compliance (C2)	1–7 years	CV events
Laurent, 2003[51]	1 715 hypertensive pts	Aortic PWV	7.9 ± 5.7 years	Fatal stroke
Chirinos, 2005[100]	297 men undergoing diagnostic catheterization	Aortic AP[a]	1,186 ± 424 days	CV events and mortality
Gosse, 2005[30]	412 hypertensive pts	QKD interval	65 ± 36 months	Combined CV outcomes
Matsuoka, 2005[53]	298 elderly subjects	Brachial-ankle PWV	1,227 days	CV mortality
Shokawa, 2005[101]	492 members of ethnic minority	Aortic PWV	10 years	CV mortality
Sutton-Tyrrell, 2005[54]	2,488 elderly subjects	Aortic PWV	4.6 years	All cause and CV mortality
Tomiyama, 2005[102]	215 pts with acute coronary syndrome	Brachial-ankle PWV	26 ± 10 months	Major CV event

(continued)

Table 22.3 (continued)

First author, year	Study population	Index of stiffness	Follow-up	Clinical end point
Weber, 2005[103]	262 pts undergoing PCI	Aortic AIx	2 years	CV events and mortality
Dolan, 2006[55]	11,291 subjects, general population	AASI	5.3 years	CV mortality
Hansen, 2006[56]	1,829 subjects, general population	AASI	9.4 years	Incidence of fatal and nonfatal stroke
Mattace-Raso, 2006[57]	2,835 elderly subjects	Aortic PWV	4.1 ± 0.8 years	Incidence of CAD and stroke
Willum-Hansen, 2006[58]	1,678 subjects, general population	Aortic PWV	9.4 years	Combined CV outcomes
Kikuya, 2007[59]	1,542 subjects, general population	AASI	13.3 years	CV and stroke mortality

[a]Studies employing invasive (intra-arterial) measurements

AASI ambulatory arterial stiffness index; AIx augmentation index; *AP* augmentation pressure; *CAD* coronary artery disease; *CV* cardiovascular; *ESRD* end-stage renal disease; *PP* pulse pressure; *PWV* pulse wave velocity; *QKD* time between the QRS and detection of the last Korotkoff sound; *SV* stroke volume

From a therapeutic standpoint, recent studies with antihypertensive medications support the role of arterial elastic properties as a worthwhile treatment target. In patients with end-stage renal disease, the administration of antihypertensive treatment was clinically beneficial only in subjects who experienced a decrease of aortic PWV, and this effect was independent of the pressure changes.[46] Furthermore, in the recent Conduit Artery Function Evaluation (CAFÉ) Study, a substudy of the Anglo-Scandinavian Cardiac Outcomes Trial (ASCOT), a higher treatment-related decrease of central pulse pressure, that was not evident in brachial pressure measurements, was independently associated with clinical benefit and reduced CV events in hypertensive patients.[8] There is no doubt that in hypertensives, the peripheral (brachial) BP and its changes comprise a major independent determinant of CV prognosis and antihypertensive treatment-related clinical benefit.[60] However, as we discussed previously, the BP readings we obtain in the brachial artery do not reliably reflect the central BP, because of the pressure amplification. Therefore, the above-mentioned studies[8,46] suggest that the evaluation of central hemodynamics is clinically useful and may add to decision-making.

22.3 Carotid Intima-Media Thickness (IMT)

Although atherosclerosis is a diffuse process, certain arteries have a predilection for development of atherosclerotic lesions. The carotid arteries and especially the carotid bifurcations (BIF) are affected early by the atherosclerotic process.[61] An increase in the thickness of the carotid wall intima-media complex is considered a marker of generalized subclinical atherosclerosis and reflects the overall atherosclerotic burden. Carotid wall thickening tends to progress proximally along the CCA and distally to the proximal portion of the internal carotid artery (ICA).[62]

22.3.1 Methods for Evaluation of Carotid IMT

Carotid IMT is measured noninvasively using two-dimensional ultrasonic B-mode imaging that visualizes a double-line pattern on both walls of the carotid artery on the longitudinal image. This double-line pattern is formed by two parallel lines that represent the leading edges of the lumen–intima and media–adventitia interfaces.[63] Atherosclerotic plaque is defined as a focal structure encroaching into the arterial lumen that measures at least 0.5 mm in thickness, or 50% of the surrounding IMT value, or demonstrates a thickness of 1.5 mm in total when measured from the media–adventitia interface to the lumen–intima interface (Fig. 22.4).[64]

Clinical trials utilizing carotid IMT measurements have used a variety of imaging protocols. These protocols differ in (1) the number and location of carotid artery segments imaged; (2) whether far wall or near wall or both are imaged; (3) the number of imaging angles; (4) capture of video images (loops or electrocardiogram-gated still images);

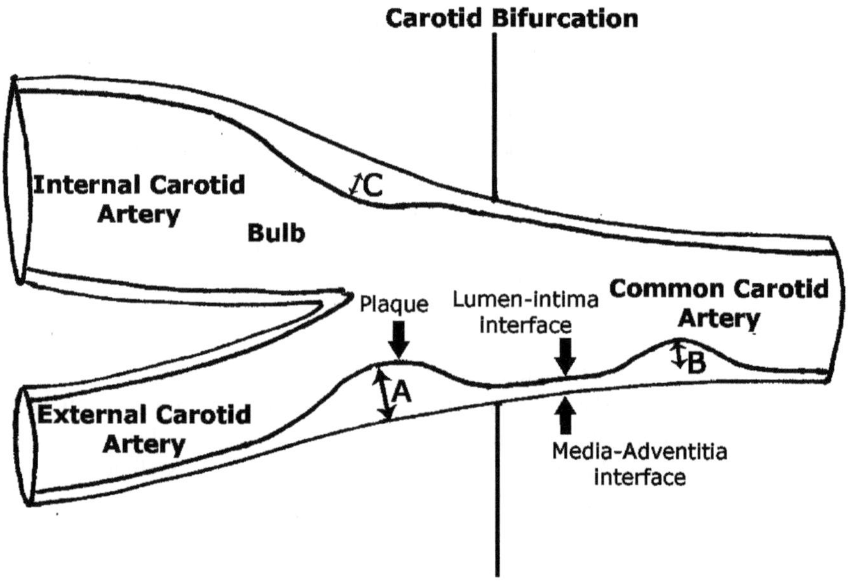

Fig. 22.4 Schematic diagram representing the carotid arterial tree to define carotid plaque. Plaque is defined as a focal structure by either of three criteria: (**a**): Thickness > 1.5 mm; (**b**): Thickness >50% of surrounding IMT value, and (**c**): Encroachment into the lumen by >0.5 mm

(5) manual or automated measurement; and (6) which measurements are finally reported (maximum or mean).[65] Implementation of standardized protocols and strict quality control can make ultrasonic measurements of carotid IMT highly reliable in large multinational clinical trials.[66] Herein follows a description of the widely used protocols for measuring carotid IMT.

22.3.1.1 Measurement in Multiple Carotid Sites

This approach consists of measuring IMT in the near and far walls of the three main segments of the carotid artery – the CCA, the BIF, and the ICA.[67] In each segment, the IMT is measured in multiple directions for both the near and far walls, and the maximal IMT recorded. The final IMT is then calculated from the average of the maximal IMTs obtained in the near and far walls of 12 sites (near and far walls of CCA, BIF, and ICA bilaterally), 6 sites (far wall of the CCA, BIF, and ICA), or even 4 sites (near and far walls of CCA or far wall of CCA and BIF bilaterally).[68] The IMT is measured manually or semi-automatically by placing the cursor on the lumen–intima and media–adventitia interfaces of the digitized ultrasonic image.[67] The multiple carotid sites approach tends to provide a comprehensive assessment of the carotid vasculature and aims to incorporate plaque thickness into the IMT value, which is a better marker for early carotid atherosclerosis.[69]

22.3.1.2 Measurement in the Far Wall of the Common Carotid Artery

This approach consists of restricting the measurement of carotid IMT to just the far wall of the CCA, which is a straight segment of the vessel that is easy to visualize. The ultrasonic video image is frozen in end-diastole by electrocardiogram triggering to avoid the confounding effect of pulsatile deformation of the arterial wall and hence alteration in arterial wall thickness. Computerized edge-detection software is used for accurate delineation of the lumen–intima and media–adventitia interfaces. IMT is automatically measured at nearly 100 points over 1 cm of a longitudinal length of the vessel, giving an averaged final IMT value with reduced variability, higher precision, and reproducibility.[70] Because IMT measurement is limited to the CCA, which often develops plaque at a later stage when compared with the BIF, the IMT value obtained by this method serves as a marker for early arterial wall change rather than early carotid atherosclerosis.[69]

22.3.2 Determinants and Predictive Value of Carotid IMT

Age is an important determinant of carotid IMT.[71-73] IMT increases by almost 0.03 mm/year.[74] IMT is also influenced by gender, CV risk factors (smoking, hypertension, diabetes), and the genetic background.[71-76] Body mass index, abdominal adiposity, and physical inactivity are also associated with increased carotid IMT.[71,72,77]

Recent data support that IMT is also related to ethnicity and lifestyle characteristics. Data from large studies indicate that IMT in blacks exceeds IMT in whites in both genders.[78,79] Furthermore, it has been shown that Japanese Americans have a higher IMT than native Japanese of similar characteristics, presumably due to adoption of western lifestyle and a higher prevalence of metabolic disorders.[80] Interestingly, the ARIC study showed that even the socioeconomic status may affect the rate of IMT progression.[78]

Low-grade inflammatory activation also plays a role to the progression of subclinical atherosclerosis. Levels of inflammatory markers such as high-sensitivity C-reactive protein and ICAM-1 have been found to predict IMT in several studies,[81-83] although this is not a consistent finding in all studies.[84] Other factors such as oxidant stress[85,86] and depletion of endothelial progenitor cells[87] may also play a role.

Increased carotid IMT can also be used as a surrogate marker for the presence of atherosclerosis in other arterial beds. For example, IMT has been associated with the presence and extent of coronary artery disease as assessed by angiography.[88] In patients presenting with cardiomyopathy of unknown origin, those with increased IMT are more likely to suffer from severe coronary artery disease, whereas a normal IMT is almost diagnostic of a nonischemic etiology.[89] The MultiEthnic Study of Atherosclerosis showed that even in populations without overt CV disease, an increased IMT is related to impaired systolic and diastolic myocardial function.[90]

Finally, a number of longitudinal studies have examined the relationship between carotid IMT and future clinical CV events, including coronary and cerebrovascular

Table 22.4 Longitudinal studies indicating the value of carotid intima-media thickness (IMT) and plaques as predictors of CV outcomes

Study, Year	Study population	Site of measurement	Follow-up	IMT: relation to clinical end point	Plaque: definition and relation to clinical end point
KIHD 1991[73]	1,275 men, general population	CCA (FW)	3 years	MI: HR of 2.17 for IMT ≥ 1 mm	Echogenic or focal protrusion into lumen; HR 4.15
ARIC 1997[6]	5,552 men and 7,289 women, age 45–64, no clinical CV disease	Mean max of 6 sites (FW of CCA, BIF, and ICA)	4–7 years	MI: HR of 1.85 in males and 5.07 in females with IMT ≥ 1 mm.	No data
Rotterdam 1997[92]	7,983 subjects, age > 55, general population	CCA (FW right side only)	3 years	HR of 1.43 for MI and 1.41 for stroke per 0.16 mm (1 SD) i ncrease in IMT	No data
CHS 1999[7]	4,476 subjects, no clinical CV disease	Mean max of 8 sites (NW+FW of CCA and ICA)	6 years	HR for MI and stroke combined: CCA 0.87–0.96 mm: 1.61 0.97–1.05 mm: 1.44 1.06–1.17 mm: 2.04 ≥1.18 mm: 2.85 ICA 0.91–1.10 mm: 1.52 1.11–1.39 mm: 2.15 1.40–1.80 mm: 2.13 ≥1.81 mm: 3.2	No data
ARIC 2000[91]	6,349 men and 7,865 women, age 45–64, no clinical CV disease	Mean max of 6 sites (FW of CCA, BIF, and ICA)	6–9 years	Stroke: HR of 3.6 in males and 8.5 in females with IMT ≥1 mm.	No data

CAFES-CAVE 2001[95]	13,221 low risk asymptomatic subjects	BIF (FW)	10 years	Combined rate for CV events and death: 8.6% for IMT >1 mm and/or increased echogenicity in deep layers; 0.13% when IMT is normal	1. Nonobstructing plaque (IMT >1 mm, increased echogenicity in all layers, no hemodynamic disturbance): 39.3% 2. Obstructing plaque (stenosis >50%): 81.1%
MDCS 2005[104]	5,163 middle-aged subjects, general population	CCA (FW right side only)	7 years	HR of 1.31 for stroke per 0.15 mm (1 SD) increase in IMT	Focal IMT > 1.2 mm; HR of 3.44 for stroke.
CAPS 2006[93]	5,056 subjects, general population	Mean max of 6 sites (FW of CCA, wBIF, and ICA)	4.2 years	HR for MI: 2.18 for CCA-IMT ≥0.79 mm and 2.15 for BIF-IMT: ≥0.98 mm	No data

CCA common carotid artery; *BIF* carotid bifurcation; *FW* far wall; *HR* hazard ratio; *ICA* internal carotid artery; *MI* myocardial infarction; *NW* near wall events, and found that IMT predicts future CV risk (Table 22.4). In the ARIC Study that followed a large cohort of subjects without CV disease, the mean IMT was an independent predictor of the future incidence of myocardial infarction and stroke.[6,91] IMT was also predictive of myocardial infarction and stroke in the elderly population of the Rotterdam Study[92] and in other large longitudinal studies.[73,75,93] A recent meta-analysis showed that IMT has a slightly better predictive value for stroke than for myocardial infarction.[94] Finally, the Carotid And FEmoral ultrasound morphology Screening and CArdioVascular Events (CAFES-CAVE) Study showed that IMT also predicts CV mortality in low-risk asymptomatic subjects.[95] Although it is often assumed that the relationship between carotid IMT and CV risk is linear, a recent meta-analysis demonstrated that it is in fact nonlinear in most populations, especially for myocardial infarction.[94] The risk associated with a specific increase in IMT is age-dependent, the risk being higher in younger individuals (<50 years) with increased IMT.[93]

IMT has been also used as an end-point in intervention studies. Although treatments that reduce IMT usually benefit prognosis, this clinical benefit most possibly is not accounted for by the decrease in IMT per se. Rather, it is derived from favorable effects on other factors, such as blood pressure, lipids, and glucose metabolism.

22.4 Epilogue

Arterial elastic properties and carotid wall thickness are important determinants of global CV performance. There is mounting evidence that these arterial characteristics are also important and independent predictors of CV risk in several populations, and may even be used to monitor the clinical benefit of treatment strategies. Recent guidelines underscore the value of both arterial stiffness and wall thickness as tools for stratification of risk in hypertensive patients.[9] Recently, attempts to define the normal values for arterial characteristics[39,96] and to standardize the available techniques[4,64] have been forthcoming. These efforts will facilitate the implementation of the concept of "vascular age" in everyday clinical practice. However, there continues to be a need to further refine available methods for measuring arterial elastic properties and wall thickness and standardization of the techniques before assessment of these characteristics is introduced into everyday clinical practice.

References

1. Cohn JN, Quyyumi AA, Hollenberg NK, Jamerson KA. Surrogate markers for cardiovascular disease: functional markers. *Circulation*. 2004;109(25 suppl 1):IV31-IV46.
2. Vlachopoulos C, Aznaouridis K, Stefanadis C. Clinical appraisal of arterial stiffness: the Argonauts in front of the golden fleece. *Heart*. 2006;92:1544-1550.
3. Nichols WW, O'Rourke MF. *McDonald's Blood Flow in Arteries*. London: Arnold; 2005.

4. Laurent S, Cockcroft J, Van Bortel L, et al. European network for non-invasive investigation of large arteries. Expert consensus document on arterial stiffness: methodological issues and clinical applications. *Eur Heart J*. 2006;27:2588-2605.
5. Agabiti-Rosei E, Mancia G, O'Rourke MF, et al. Central blood pressure measurements and antihypertensive therapy: a consensus document. *Hypertension*. 2007;50:154-160.
6. Chambless LE, Heiss G, Folsom AR, et al. Association of coronary heart disease incidence with carotid arterial wall thickness and major risk factors: the Atherosclerosis Risk in Communities (ARIC) Study, 1987–1993. *Am J Epidemiol*. 1997;146:483-494.
7. O'Leary DH, Polak JF, Kronmal RA, et al. Carotid-artery intima and media thickness as a risk factor for myocardial infarction and stroke in older adults. Cardiovascular health study collaborative research group. *N Engl J Med*. 1999;340:14-22.
8. Williams B, Lacy PS, Thom SM, et al. CAFE investigators, for the anglo-scandinavian cardiac outcomes trial investigators. Differential impact of blood pressure-lowering drugs on central aortic pressure and clinical outcomes: principal results of the Conduit Artery Function Evaluation (CAFE) study. *Circulation*. 2006;113:1213-1225.
9. Mancia G, De Backer G, Dominiczak A, et al. Management of arterial hypertension of the European society of hypertension; European society of cardiology. 2007 guidelines for the management of arterial hypertension: The task force for the management of arterial hypertension of the European society of hypertension (ESH) and of the European society of cardiology (ESC). *J Hypertens*. 2007;25:1105-1187.
10. Zieman SJ, Melenovsky V, Kass DA. Mechanisms, pathophysiology, and therapy of arterial stiffness. *Arterioscler Thromb Vasc Biol*. 2005;25:932-943.
11. O'Rourke MF, Staessen JA, Vlachopoulos C, et al. Clinical applications of arterial stiffness; definitions and reference values. *Am J Hypertens*. 2002;15:426-444.
12. Wilkinson IB, MacCallum H, Cockcroft JR, Webb DJ. Inhibition of basal nitric oxide synthesis increases aortic augmentation index and pulse wave velocity in vivo. *Br J Clin Pharmacol*. 2002;53:189-192.
13. Laurent S, Boutouyrie P, Lacolley P. Structural and genetic bases of arterial stiffness. *Hypertension*. 2005;45:1050-1055.
14. Stefanadis C, Stratos C, Vlachopoulos C, et al. Pressure-diameter relation of the human aorta. A new method of determination by the application of a special ultrasonic dimension catheter. *Circulation*. 1995;92:2210-2219.
15. Bank AJ, Wilson RF, Kubo SH, et al. Direct effects of smooth muscle relaxation and contraction on in vivo human brachial artery elastic properties. *Circ Res*. 1995;77:1008-1016.
16. Van der Heijden-Spek JJ, Staessen JA, Fagard RH, et al. Effect of age on brachial artery wall properties differs from the aorta and is gender dependent: a population study. *Hypertension*. 2000;35:637-642.
17. Stefanadis C, Stratos C, Boudoulas H, et al. Distensibility of the ascending aorta: comparison of invasive and non-invasive techniques in healthy men and in men with coronary artery disease. *Eur Heart J*. 1990;11:990-996.
18. Ikonomidis I, Lekakis J, Stamatelopoulos K, et al. Aortic elastic properties and left ventricular diastolic function in patients with Adamantiades Behcet's disease. *J Am Coll Cardiol*. 2004;43:1075-1081.
19. Roman MJ, Ganau A, Saba PS, et al. Impact of arterial stiffening on left ventricular structure. *Hypertension*. 2000;36:489-494.
20. Van Bortel LM, Balkestein EJ, van der Heijden-Spek JJ, et al. Non-invasive assessment of local arterial pulse pressure: comparison of applanation tonometry and echo-tracking. *J Hypertens*. 2001;19:1037-1044.
21. Van der Meer RW, Diamant M, Westenberg JJ, et al. Magnetic resonance assessment of aortic pulse wave velocity, aortic distensibility, and cardiac function in uncomplicated type 2 diabetes mellitus. *J Cardiovasc Magn Reson*. 2007;9:645-651.
22. Wiesmann F, Petersen SE, Leeson PM, et al. Global impairment of brachial, carotid, and aortic vascular function in young smokers: direct quantification by high-resolution magnetic resonance imaging. *J Am Coll Cardiol*. 2004;44:2056-2064.

23. Schmidt-Trucksäss A, Grathwohl D, Schmid A, et al. Assessment of carotid wall motion and stiffness with tissue Doppler imaging. *Ultrasound Med Biol.* 1998;24:639-646.
24. Herrington DM, Brown WV, Mosca L, et al. Relationship between arterial stiffness and subclinical aortic atherosclerosis. *Circulation.* 2004;110:432-437.
25. Zambanini A, Cunningham SL, Parker KH, et al. Wave-energy patterns in carotid, brachial, and radial arteries: a noninvasive approach using wave-intensity analysis. *Am J Physiol Heart Circ Physiol.* 2005;289:H270-H276.
26. Lehmann ED, Hopkins KD, Rawesh A, et al. Relation between number of cardiovascular risk factors/events and noninvasive Doppler ultrasound assessments of aortic compliance. *Hypertension.* 1998;32:565-569.
27. Jiang B, Liu B, McNeill KL, Chowienczyk PJ. Measurement of pulse wave velocity using pulse wave Doppler ultrasound: comparison with arterial tonometry. *Ultrasound Med Biol.* 2008;34:509-512.
28. Sandor GG, Hishitani T, Petty RE, et al. A novel Doppler echocardiographic method of measuring the biophysical properties of the aorta in pediatric patients. *J Am Soc Echocardiogr.* 2003;16:745-750.
29. Patrianakos AP, Karakitsos DN, de Groot E, et al. Alteration of proximal aorta biophysical properties in patients with end stage renal disease. *Heart.* 2006;92:228-232.
30. Gosse P, Lasserre R, Minifié C, et al. Arterial stiffness evaluated by measurement of the QKD interval is an independent predictor of cardiovascular events. *Am J Hypertens.* 2005;18(4 Pt 1):470-476.
31. De Simone G, Roman MJ, Koren MJ, et al. Stroke volume/pulse pressure ratio and cardiovascular risk in arterial hypertension. *Hypertension.* 1999;33:800-805.
32. McGrath BP, Liang YL, Teede H, et al. Age-related deterioration in arterial structure and function in postmenopausal women: impact of hormone replacement therapy. *Arterioscler Thromb Vasc Biol.* 1998;18:1149-1156.
33. Cohn JN, Finkelstein S, McVeigh G, et al. Noninvasive pulse wave analysis for the early detection of vascular disease. *Hypertension.* 1995;26:503-508.
34. Segers P, Rietzschel E, Heireman S, et al. Carotid tonometry versus synthesized aorta pressure waves for the estimation of central systolic blood pressure and augmentation index. *Am J Hypertens.* 2005;18(9 Pt 1):1168-1173.
35. Hope SA, Meredith IT, Cameron JD. Arterial transfer functions and the reconstruction of central aortic waveforms: myths, controversies and misconceptions. *J Hypertens.* 2008;26:4-7.
36. Roman MJ, Devereux RB, Kizer JR, et al. Central pressure more strongly relates to vascular disease and outcome than does brachial pressure: the strong heart study. *Hypertension.* 2007;50:197-203.
37. Li Y, Wang JG, Dolan E, et al. Ambulatory arterial stiffness index derived from 24-hour ambulatory blood pressure monitoring. *Hypertension.* 2006;47:359-364.
38. Mitchell GF, Guo CY, Benjamin EJ, et al. Cross-sectional correlates of increased aortic stiffness in the community: the Framingham heart study. *Circulation.* 2007;115:2628-2636.
39. McEniery CM, Yasmin, Hall IR, et al, on behalf of the ACCT Investigators. Normal vascular aging: differential effects on wave reflection and aortic pulse wave velocity. The Anglo-Cardiff Collaborative Trial (ACCT). *J Am Coll Cardiol.* 2005;46:1753-1760.
40. Din-Dzietham R, Couper D, Evans G, et al. Arterial stiffness is greater in African Americans than in whites. Evidence from the Forsyth County, North Carolina, ARIC cohort. *Am J Hypertens.* 2004;17:304-313.
41. Chen W, Srinivasan SR, Bond MG, et al. Nitric oxide synthase gene polymorphism (G894T) influences arterial stiffness in adults. The Bogalusa Heart Study. *Am J Hypertens.* 2004;17:553-559.
42. Franklin SS, Khan SA, Wong ND, et al. Is pulse pressure useful in predicting risk for coronary heart disease: the Framingham Heart Study. *Circulation.* 1999;100:354-360.
43. Blacher J, Pannier B, Guerin AP, et al. Carotid arterial stiffness as a predictor of cardiovascular and all-cause mortality in end-stage renal disease. *Hypertension.* 1998;32:570-574.
44. Blacher J, Guerin AP, Pannier B, et al. Impact of aortic stiffness on survival in end-stage renal disease. *Circulation.* 1999;99:2434-2439.

45. London GM, Blacher J, Pannier B, et al. Arterial wave reflections and survival in end-stage renal failure. *Hypertension.* 2001;38:434-438.
46. Guerin AP, Blacher J, Pannier B, et al. Impact of aortic stiffness attenuation on survival of patients in end-stage renal failure. *Circulation.* 2001;103:987-992.
47. Blacher J, Safar ME, Guerin AP, et al. Aortic pulse wave velocity index and mortality in end-stage renal disease. *Kidney Int.* 2003;63:1852-1860.
48. Fagard RH, Pardaens K, Staessen JA, Thijs L. The pulse pressure-to-stroke index ratio predicts cardiovascular events and death in uncomplicated hypertension. *J Am Coll Cardiol.* 2001;38:227-231.
49. Laurent S, Boutouyrie P, Asmar R, et al. Aortic stiffness is an independent predictor of all-cause and cardiovascular mortality in hypertensive patients. *Hypertension.* 2001;37:1236-1241.
50. Boutouyrie P, Tropeano AI, Asmar R, et al. Aortic stiffness is an independent predictor of primary coronary events in hypertensive patients: a longitudinal study. *Hypertension.* 2002;39:10-15.
51. Laurent S, Katsahian S, Fassot C, et al. Aortic stiffness is an independent predictor of fatal stroke in essential hypertension. *Stroke.* 2003;34:1203-1206.
52. Cruickshank K, Riste L, Anderson SG, et al. Aortic pulse-wave velocity and its relationship to mortality in diabetes and glucose intolerance: an integrated index of vascular function? *Circulation.* 2002;106:2085-2090.
53. Matsuoka O, Otsuka K, Murakami S, et al. Arterial stiffness independently predicts cardiovascular events in an elderly community – Longitudinal Investigation for the Longevity and Aging in Hokkaido County (LILAC) study. *Biomed Pharmacother.* 2005;59(Suppl 1):S40-S44.
54. Sutton-Tyrrell K, Najjar SS, Boudreau RM, et al, for the Health ABC Study. Elevated aortic pulse wave velocity, a marker of arterial stiffness, predicts cardiovascular events in well-functioning older adults. *Circulation.* 2005;111:3384-3390.
55. Dolan E, Thijs L, Li Y, et al. Ambulatory arterial stiffness index as a predictor of cardiovascular mortality in the Dublin Outcome Study. *Hypertension.* 2006;47:365-370.
56. Hansen TW, Staessen JA, Torp-Pedersen C, et al. Ambulatory arterial stiffness index predicts stroke in a general population. *J Hypertens.* 2006;24:2247-2253.
57. Mattace-Raso FU, van der Cammen TJ, Hofman A, et al. Arterial stiffness and risk of coronary heart disease and stroke: the Rotterdam Study. *Circulation.* 2006;113:657-663.
58. Willum-Hansen T, Staessen JA, Torp-Pedersen C, et al. Prognostic value of aortic pulse wave velocity as index of arterial stiffness in the general population. *Circulation.* 2006;113:664-670.
59. Kikuya M, Staessen JA, Ohkubo T, et al. Ambulatory arterial stiffness index and 24-hour ambulatory pulse pressure as predictors of mortality in Ohasama, Japan. *Stroke.* 2007;38:1161-1166.
60. Staessen JA, Wang JG, Thijs L. Cardiovascular protection and blood pressure reduction: a meta-analysis. *Lancet.* 2001;358:1305-1315.
61. Javid H. Development of carotid plaque. *Am J Surg.* 1979;138:224-227.
62. Solberg LA, Eggen DA. Localization and sequence of development of atherosclerotic lesions in the carotid and vertebral arteries. *Circulation.* 1971;43:711-724.
63. Pignoli P, Tremoli E, Poli A, et al. Intimal plus medial thickness of the arterial wall: a direct measurement with ultrasound imaging. *Circulation.* 1986;74:1399-1406.
64. Touboul PJ, Hennerici MG, Meairs S, et al. Mannheim carotid intima-media thickness consensus (2004-2006). An update on behalf of the Advisory Board of the 3rd and 4th Watching the Risk Symposium, 13th and 15th European Stroke Conferences, Mannheim, Germany, 2004, and Brussels, Belgium, 2006. *Cerebrovasc Dis.* 2007;23:75-80.
65. Mitchell CKC, Aeschlimann SE, Korcarz CE. Carotid intima-media thickness testing: technical considerations. *J Am Soc Echocardiogr.* 2004;17:690-692.
66. Tang R, Hennig M, Thomasson B, et al. Baseline reproducibility of B-mode ultrasonic measurement of carotid artery intima-media thickness: the European Lacidipine Study on Atherosclerosis (ELSA). *J Hypertens.* 2000;18:197-201.
67. Bond MG, Barnes RW, Riley WA, et al. High-resolution B-mode ultrasound scanning methods in the Atherosclerosis Risk in Communities Study (ARIC). The ARIC Study Group. *J Neuroimaging.* 1991;1:68-73.

68. Howard G, Sharrett AR, Heiss G, et al. Carotid artery intimal-medial thickness distribution in general populations as evaluated by B-mode ultrasound. *ARIC Investigators Stroke.* 1993;24:1297-1304.
69. Simon A, Gariepy J, Chironi G, et al. Intima-media thickness: a new tool for diagnosis and treatment of cardiovascular risk. *J Hypertens.* 2002;20:159-169.
70. Selzer RH, Hodis HN, Kwong-Fu H, et al. Evaluation of computerized edge tracking for quantifying intima-media thickness of the common carotid artery from B-mode ultrasound images. *Atherosclerosis.* 1994;111:1-11.
71. Gnasso A, Irace C, Mattioli PL, Pujia A. Carotid intima-media thickness and coronary heart disease risk factors. *Atherosclerosis.* 1996;119:7-15.
72. Gariepy J, Salomon J, Denarie N, et al. Sex and topographic differences in associations between large-artery wall thickness and coronary risk profile in a French working cohort: the AXA Study. *Arterioscler Thromb Vasc Biol.* 1998;18:584-590.
73. Salonen JT, Salonen R. Ultrasonographically assessed carotid morphology and the risk of coronary heart disease. *Arterioscler Thromb.* 1991;11:1245-1249.
74. Kullo IJ, Malik AR. Arterial ultrasonography and tonometry as adjuncts to cardiovascular risk stratification. *J Am Coll Cardiol.* 2007;49:1413-1426.
75. Crouse JR, Goldbourt U, Evans G, et al. Risk factors and segment-specific carotid arterial enlargement in the Atherosclerosis Risk in Communities (ARIC) cohort. *Stroke.* 1996;27:69-75.
76. Salonen R, Salonen JT. Progression of carotid atherosclerosis and its determinants: a population-based ultrasonography study. *Atherosclerosis.* 1990;81:33-40.
77. Folsom AR, Eckfeldt JH, Weitzman S, et al. Relation of carotid artery wall thickness to diabetes mellitus, fasting glucose and insulin, body size, and physical activity. Atherosclerosis Risk in Communities (ARIC) Study Investigators. *Stroke.* 1994;25:66-73.
78. Ranjit N, Diez-Roux AV, Chambless L, et al. Socioeconomic differences in progression of carotid intima-media thickness in the atherosclerosis risk in communities study. *Arterioscler Thromb Vasc Biol.* 2006;26:411-416.
79. D' Agostino RB, Burke G, O'Leary D, et al. Ethnic differences in carotid wall thickness. The Insulin Resistance Atherosclerosis Study. *Stroke.* 1996;27:1744-1749.
80. Watanabe H, Yamane K, Equsa G, et al. Influence of westernization of lifestyle on the progression of IMT in Japanese. *J Atheroscler Thromb.* 2004;11:330-334.
81. Wang TJ, Nam B-H, Wilson PWF, et al. Association of C-reactive protein with carotid atherosclerosis in men and women: the Framingham Heart Study. *Arterioscler Thromb Vasc Biol.* 2002;22:1662-1667.
82. Hashimoto H, Kitagawa K, Hougaku H, et al. Relationship between C-reactive protein and progression of early carotid atherosclerosis in hypertensive subjects. *Stroke.* 2004;35:1625-1630.
83. van der Meer IM, de Maat MPM, Bots ML, et al. Inflammatory mediators and cell adhesion molecules as indicators of severity of atherosclerosis: the Rotterdam Study. *Arterioscler Thromb Vasc Biol.* 2002;22:838-842.
84. Lorenz MW, Karbstein P, Markus HS, Sitzer M. High-sensitivity C-reactive protein is not associated with carotid intima-media progression: the carotid atherosclerosis progression study. *Stroke.* 2007;38:1774-1779.
85. Ashfaq S, Abramson JL, Jones DP, et al. The relationship between plasma levels of oxidized and reduced thiols and early atherosclerosis in healthy adults. *J Am Coll Cardiol.* 2006;47:1005-1011.
86. Hulthe J, Fagerberg B. Circulating oxidized LDL is associated with subclinical atherosclerosis development and inflammatory cytokines (AIR Study). *Arterioscler Thromb Vasc Biol.* 2002;22:1162-1167.
87. Fadini GP, Coracina A, Baesso I, et al. Peripheral blood CD34+KDR+ endothelial progenitor cells are determinants of subclinical atherosclerosis in a middle-aged general population. *Stroke.* 2006;37:2277-2282.
88. Graner M, Varpula M, Kahri J, et al. Association of carotid intima-media thickness with angiographic severity and extent of coronary artery disease. *Am J Cardiol.* 2006;97:624-629.

89. Androulakis AE, Andrikopoulos GK, Richter DJ, et al. The role of carotid atherosclerosis in the distinction between ischaemic and non-ischaemic cardiomyopathy. Eur Heart J. 2000;21:919-926.
90. Fernandes VR, Polak JF, Edvardsen T, et al. Subclinical atherosclerosis and incipient regional myocardial dysfunction in asymptomatic individuals: the Multi-Ethnic Study of Atherosclerosis (MESA). J Am Coll Cardiol. 2006;47:2420-2428.
91. Chambless LE, Folsom AR, Clegg LX, et al. Carotid wall thickness is predictive of incident clinical stroke: the Atherosclerosis Risk in Communities (ARIC) study. Am J Epidemiol. 2000;151:478-487.
92. Bots ML, Hoes AW, Koudstaal PJ, et al. Common carotid intima-media thickness and risk of stroke and myocardial infarction: the Rotterdam Study. Circulation. 1997;96:1432-1437.
93. Lorenz MW, von Kegler S, Steinmetz H, et al. Carotid intima-media thickening indicates a higher vascular risk across a wide age range: prospective data from the Carotid Atherosclerosis Progression Study (CAPS). Stroke. 2006;37:87-92.
94. Lorenz MW, Markus HS, Bots ML, et al. Prediction of clinical cardiovascular events with carotid intima-media thickness: a systematic review and meta-analysis. Circulation. 2007;115:459-467.
95. Belcaro G, Nicolaides AN, Ramaswami G, et al. Carotid and femoral ultrasound morphology screening and cardiovascular events in low risk subjects: a 10-year follow-up study (the CAFES-CAVE study(1)). Atherosclerosis. 2001;156:379-387.
96. Wojciechowska W, Staessen JA, Nawrot T, et al. European Project on Genes in Hypertension (EPOGH) Investigators. Reference values in white Europeans for the arterial pulse wave recorded by means of the SphygmoCor device. Hypertens Res. 2006;29:475-483.
97. Stefanadis C, Dernellis J, Tsiamis E, et al. Aortic stiffness as a risk factor for recurrent acute coronary events in patients with ischaemic heart disease. *Eur Heart J.* 2000;21:390-396.
98. Meaume S, Benetos A, Henry OF, et al. Aortic pulse wave velocity predicts cardiovascular mortality in subjects >70 years of age. *Arterioscler Thromb Vasc Biol.* 2001;21:2046-2050.
99. Grey E, Bratteli C, Glasser SP, et al. Reduced small artery but not large artery elasticity is an independent risk marker for cardiovascular events. *Am J Hypertens.* 2003;16:265-269.
100. Chirinos JA, Zambrano JP, Chakko S, et al. Aortic pressure augmentation predicts adverse cardiovascular events in patients with established coronary artery disease. *Hypertension.* 2005;45:980-985.
101. Shokawa T, Imazu M, Yamamoto H, et al. Pulse wave velocity predicts cardiovascular mortality: findings from the Hawaii-Los Angeles-Hiroshima study. *Circ J.* 2005;69:259-264.
102. Tomiyama H, Koji Y, Yambe M, et al. Brachial-ankle pulse wave velocity is a simple and independent predictor of prognosis in patients with acute coronary syndrome. *Circ J.* 2005;69:815-822.
103. Weber T, Auer J, O'Rourke MF, et al. Increased arterial wave reflections predict severe cardiovascular events in patients undergoing percutaneous coronary interventions. *Eur Heart J.* 2005;26:2657-2663.
104. Rosvall M, Janzon L, Berglund G, et al. Incidence of stroke is related to carotid IMT even in the absence of plaque. *Atherosclerosis.* 2005;179:325-331.

Chapter 23
The Metabolic Syndrome

Benjamin D. Mackie, David G. Harrison, and A. Maziar Zafari

23.1 Introduction

Metabolic syndrome (MetS) is a clustering of risk factors, mostly of metabolic origin that taken together enhance the predictive ability of diseases, specifically cardiovascular disease (CVD) and type 2 diabetes mellitus (T2DM). The risk factors comprising MetS are abdominal obesity, dyslipidemia, hypertension, and hyperglycemia or insulin resistance. Of these, abdominal obesity and insulin resistance are central to the pathogenesis and development of MetS, while genetic and environmental factors also play a role. Additionally, MetS is closely associated with both a proinflammatory and prothrombotic state, which at least partially mediates insulin resistance and the associated increased risk of CVD. There remains some debate regarding whether MetS is an entity in its own right and if it has greater predictive ability of CVD risk and T2DM than its individual components. New data suggest that it does improve predictive ability; this combined with the mounting body of evidence to support a distinct pathophysiological process is beginning to earn MetS a place of its own. MetS has been, and continues to be, an area of great interest and robust research and is generally accepted as a paradigm identifying patients at risk for CVD and T2DM, allowing for aggressive risk factor modification at an early stage of disease.

23.2 Definition

In 1988, Gerald Reaven proposed insulin resistance as the root cause of MetS.[1] Since that time, the definition of MetS has evolved, currently including anthropometric, pathophysiological, and clinical criteria. All definitions consist of similar compo-

B.D. Mackie, D.G. Harison, and A.M. Zafari (✉)
Division of Cardiology, Emory University School of Medicine,
1639 Pierce Drive, 319 WMB, Atlanta, GA, 30322, USA
e-mail: azafari@emory.edu

D. Abraham et al. (eds.), *Advances in Vascular Medicine*,
DOI 10.1007/978-1-84882-637-3_23, © Springer-Verlag London Limited 2010

nents; however, they vary in their detail. The three most widely accepted and utilized definitions are those put forth by the World Health Organization (WHO),[2] the 2001 National Cholesterol Education Program (Adult Treatment Panel III) (NCEP:ATP3), updated in 2005,[3] and the International Diabetes Federation (IDF).[4] Other definitions include those proposed by the American Association of Clinical Endocrinologists (AACE),[5] and The Group for the Study of Insulin Resistance (EGIR).[6] Variations between these definitions are listed in Table 23.1. The WHO definition requires formal testing to establish glucose intolerance or insulin resistance. It is widely used in Europe, and is felt to be somewhat better suited for research purposes than the NCEP:ATP3 definition. The NCEP:ATP3 definition is the simplest definition, and the most widely used in the United States, and more clinically applicable. The IDF definition is a compromise between the WHO and NCEP:ATP3 definitions including ethnicity-specific values for central obesity. These ethnicity-specific values are included to address the observed increased risk of T2DM associated with smaller degrees of visceral obesity in certain patient populations, particularly Asians.

When these definitions are applied to large study populations such as the Diabetes Epidemiology: Collaborative Analysis of Diagnostic Criteria in Europe (DECODE)[7] and The National Health and Nutrition Examination Survey III (NHANES IIII)[8] in the United States, one identifies a similar percentage of individuals with MetS; however, the definitions do not provide for concordant selection of the same

Table 23.1 Comparison of metabolic syndrome (MetS) definitions [102]

Guide-line	Glucose/insulin abnormality	Obesity/central adiposity	Dyslipidemia	Hypertension (with or without medication)	Other	Minimum criteria for diagnosis
WHO (1999)[3]	Type 2 diabetes, impaired fasting glucose (FBG ≥6.1 mmol/l), impaired glucose tolerance (2 h PPG ≥7.8 mmol/l), or lowest 25% for hyperinsulinemic euglycemic clamp-glucose uptake	Waist-to-hip ratio >0.9 (M) or >0.85 (F) and/or BMI >30 kg/m^2	Triglycerides ≥1.7 mmol/l and/or HDL-C <0.9 mmol/l (M) or <1.0 mmol/l (F)	BP ≥140/90 mmHg (and/or medication)	Microalbuminuria (≥20 μg/min albumin excretion rate or albumin: creatinine ratio ≥30 mg/g)	Glucose intolerance/ insulin resistance, plus two other features
EGIR[a] (1999)[6]	Insulin resistance: hyperinsulinemia (nondiabetic fasting insulin in top 25%) and impaired fasting glucose (FBG ≥6.1 mmol/l)	Waist circumference ≥94 cm (M) or ≥80 cm (F)	Triglycerides >2 mmol/l and/or HDL-C <1.0 mmol/l	BP ≥140/90 mmHg (and/or medication)	–	Insulin resistance, plus two other features
NCEP ATP III (2001)[4]	Impaired fasting glucose (FBG ≥6.1 mmol/l)	Waist circumference >102 cm (M) or >88 cm (F)	Triglycerides ≥1.69 mmol/l, HDL-C <1.04 mmol/l (M) or <1.29 mmol/l (F)	BP ≥130/≥85 mmHg (and/or medication)	–	Any three features
AACE (2003)[7]	Glucose intolerance (FBG ≥6.1 mmol/l or 2 h PPG >7.8 mmol/l)	BMI ≥25 kg/m^2	Triglycerides ≥1.69 mmol/l, HDL-C <1.04 mmol/l (M) or <1.29 mmol/l (F)	BP ≥130/≥85 mmHg[b]	Family history of or high-risk ethnic group for type 2 diabetes, hypertension or CVD; polycystic ovarian syndrome; sedentary lifestyle; advancing age	Clinical judgment based on all features
IDF (2005)[5]	Glucose intolerance (FBG ≥5.6 mmol/l) or pre-existing diabetes	Waist circumference: European ≥94 cm (M) or ≥80 cm (F); South Asian and Chinese ≥90 cm (M) or ≥80 cm (F); Japanese ≥85 cm (M) or ≥90 cm (F)	Triglycerides ≥1.7 mmol/l, HDL-C <1.0 mmol/l (M) or <1.3 mmol/l (F)	BP ≥130/≥85 mmHg (and/or medication)	–	Central adiposity plus two other features

[a]Individuals without diabetes only. [b]Without medication only. Abbreviations: AACE, the American College of Endocrinology; BP, blood pressure; CVD, cardiovascular disease; EGIR, European Group for the Study of Insulin Resistance; F, female; FBG, fasting blood glucose; HDL-C, HDL cholesterol; IDF, International Diabetes Federation; M, male; NCEP ATP III, National Cholesterol Education Program Adult Treatment Panel III; PPG, postprandial glucose.

individuals.[9,10] Even though different individuals are identified with different definitions, all appear to confer similar predictive values for the development of CVD and T2DM.[11,12] The IDF definition will most likely become the consensus definition given that it incorporates, what most consider, the most useful aspects of the WHO and NCEP:ATP3 definitions. This will serve to have globally applicable criteria enabling uniform inclusion criteria for clinical and epidemiological research.

23.3 Epidemiology

The prevalence of MetS is increasing worldwide paralleling the observed obesity epidemic. MetS is no longer solely an adult disease state as a large percentage of obese adolescents currently meet the criteria for MetS.[13] The increased prevalence of MetS is attributed to excessive macronutrient intake in the setting of physical inactivity and if not intervened on will have profound negative effects worldwide. When applied to the NHANES III data, the WHO and NCEP:ATPIII definitions, respectively, identify 25.1 and 23.9% of individuals as meeting MetS criteria.[14] A similar percentage of individuals (24%) are identified in the Framingham Offspring Study.[9] Perhaps, most striking is the positive association between age and prevalence with only 6.7% of affected individuals being 20–29 years of age. This figure increases to 44% of individuals aged 60–69 years.[14] Additional variability is seen between males and females and different ethnicities. As stated above, the overall prevalence for MetS in the US population is 24%; this figure is the same for nonHispanic whites (24%); however, the prevalence is lower in African Americans (22%) and is higher in Mexican Americans (32%), which is in accordance with the higher prevalence of obesity in the Mexican American population.[15] The lower prevalence of MetS in African Americans is interesting given the well-accepted increased risk of hypertension and T2DM in this population.[16] Globally, South Asian populations exhibit an increased prevalence of MetS, however not approaching the exceedingly high prevalence observed in the United States. Another commonly observed phenomenon is the increased prevalence of MetS among those residing in the urban setting when compared with more rural populations. While the reasons for the above variations in prevalence are not well understood, they can likely be discovered through a more detailed understanding of the pathophysiology involved in MetS, which is further discussed.

23.4 Pathophysiology

23.4.1 Inflammation and Obesity

MetS is most commonly associated with visceral obesity, a surrogate marker for insulin resistance. The association of visceral obesity with impaired glucose homeostasis is well accepted, with current evidence linking insulin resistance to the chronic inflammatory state observed in obese subjects. In this paradigm, visceral adipose tissue functions as an endocrine organ mediating both systemic and local effects through bioactive proteins such as pro-inflammatory cytokines, chemokines, and adipocyte-derived hormones collectively known as adipokines.[17] The most prominent of these being adiponectin, tumor necrosis factor-alpha (TNF-α), and interleukin-6 (IL-6), all of which are involved in the development of insulin resistance and thus are paramount to the pathophysiology of hypertension, dyslipidemia, and the hypercoagulability associated with MetS. Other important bioactive proteins increased in the obese state include angiotensinogen, monocyte chemoattractant protein-1 (MCP-1), and plasminogen activator factor-1 (PAI-1). It is unclear if the majority of these bioactive proteins originate from macrophages that have infiltrated visceral adipose tissue or from adipocytes themselves. The infiltration of visceral adipose tissue with macrophages has been shown experimentally; however, the importance of macrophages in the inflammatory cascade of MetS is just beginning to be understood.[18] Some evidence suggests an important role for preadipocytes since they exhibit macrophage-like properties in the proliferative phase of development.[19] Macrophages also increase the expression of adhesion molecules in adipose tissue and endothelial cells by activated T cells through antigen presentation.[20] The exact importance of these mechanisms in regard to the pathophysiology of MetS has not been determined, but promises to be an active area of investigation in the future to further discern the origins of obesity-induced inflammation.

Current literature suggests that the initiation of obesity-induced inflammation begins with excessive intake of macronutrients leading to increased adipocyte size and volume causing increased activation and recruitment of macrophages ultimately leading to increased systemic inflammation. Clearly, macronutrient intake is a regulator of oxidative stress and comprehensive systemic inflammation. This has been shown experimentally with increased levels of inflammatory transcription factors and reactive oxygen species (ROS) found in the setting of macronutrient challenge.[21] Furthermore, the levels of inflammatory mediators and ROS are decreased with a reduction of macronutrient intake in obese individuals.[22] However, increased macronutrient intake alone cannot explain the observed chronic inflammatory state seen with visceral obesity. As previously mentioned, activated macrophages appear central to this process, and multiple theories have been proposed to explain the initiating events. One of these is that free fatty acids (FFA) induce endoplasmic reticular stress directly stimulating nuclear factor kappa B (NFκB), inhibitor of nuclear factor kappa B kinase beta subunit (IKKB), and C-Jun nuclear kinase (JNK) leading to increased levels of inflammatory cytokines. It is also plausible that FFAs directly activate toll-like receptors triggering the observed inflammatory cascade.[23] These processes appear to have a positive relationship with the degree of visceral obesity and hence increased adipocyte size. Activated macrophages increase levels of TNF-α and IL-6 leading to initiation of pro-inflammatory downstream signaling involving NFκB and JNK.[18] NFκB, JNK, along with TNF-α and IL-6 impair insulin signaling causing insulin resistance. They also participate in positive feedforward signaling loops causing further increases in inflammatory biomarkers. This vicious cycle of systemic inflammation forms the pathophysiological foundation of MetS (Fig. 23.1).

In addition to a role for macrophages, recent evidence points to T cells as contributing to the inflammation associated with obesity. When T cells are activated by antigenic challenge, they express homing markers that target them to tissues, including the chemokine receptor 5 (CCR5), the hyaluronan receptor (CD44), and receptors for selectins such as CD62L. These activated T cells then enter tissues that harbor docking molecules for such homing signals. The ligand for CCR5, known as regulated on activation and normally T cell expressed and secreted (RANTES), is one such marker that is highly expressed by visceral fat cells and its expression correlates with the degree of obesity.[24] As body mass index (BMI) increases, the accumulation of activated T cells (expressing CCR5) in visceral adipose tissues increases. It has also been recently shown that newly activated effector T cells enter visceral adipose tissue in high numbers.[25] On resolution of the antigenic challenge, these "rested" T cells disappear from peripheral stores and can only be found in lymphoid tissues. On repeat antigenic challenge, newly activated effector T cells emerge from lymphoid tissues and reappear in peripheral tissues such as visceral fat. Recently, it has been recognized that the visceral adipose tissue surrounding vessels, so-called

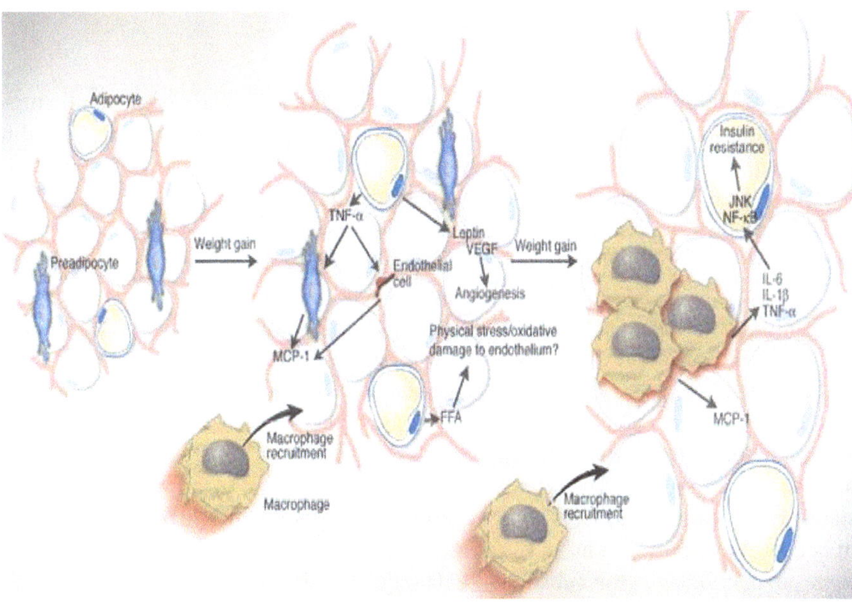

Fig. 23.1 Inflammation and obesity. Above is a schematic representation of the interplay between preadipocytes, weight gain, FFA, macrophages, and cytokines in the pathogenesis of obesity and inflammation.[18] Adipocytes and inflammation. *FFA* free fatty acid; *IL* interleukin; *JNK* jun N-terminal kinase; *MCP* monocyte chemotactic protein; *NF* nuclear factor; *TNF* tumor necrosis factor; *VEGF* vascular endothelial growth factor

perivascular fat, is particularly prone to express such homing molecules and to serve as a site of inflammation.[26,27] In the perivascular fat, T cells release cytokines, such as the T_{H1} cytokine TNF-α, which have myriad roles including activation of the NADPH oxidase in the adjacent vessel, causing endothelial dysfunction and promoting lesion development and hypertension.[26] Of note, T_{H1} cytokines have been implicated in decreasing skeletal muscle utilization of glucose by decreasing glucose receptors, altering hepatocyte lipid synthesis, and damaging pancreatic beta cells. In this fashion, the entry of T cells into visceral adipose tissue and in particular perivascular fat might play an important role in the genesis of the MetS.

These considerations regarding T cell activation might also explain the common co-existence of chronic low-grade infections or inflammation with components of the MetS that have been poorly understood. For example, there is a correlation between the severity of chronic peridontitis and hypertension.[28] Up to 70% of patients with rheumatoid arthritis (a T cell mediated disease) have hypertension[29] and there is a relationship between the severity of psoriasis (another T cell mediated disease) and blood pressure elevation.[30] It is conceivable that these chronic inflammatory diseases lead to a condition where recently activated effector T cells are constantly entering perivascular fat and promoting the injurious responses mentioned above.

23.4.2 *Insulin Resistance*

Insulin is a pleiotropic hormone that has multiple effects on various metabolic processes and numerous organs throughout the body. The principal actions of insulin are to suppress hepatic gluconeogenesis and production of very low-density lipoprotein (VLDL), facilitate glucose uptake into skeletal muscle and adipose tissues, and to suppress the release of FFAs from adipose tissue. The pathophysiology of insulin resistance is complex, and is at least partially mediated through the pro-inflammatory state associated with visceral obesity.[31]

Insulin resistance is generally defined as a state in which normal concentrations of insulin do not achieve the expected biological and metabolic effects in target tissues. This results in a hyperinsulinemic state that downregulates insulin receptors in adipose, muscle, and hepatic tissues causing further increases in insulin secretion and perpetuation of insulin resistance.[32] The initiating events leading to insulin resistance are thought to be related to elevated levels of FFAs, a pro-inflammatory state, and in some individuals a strong genetic predisposition. Elevated FFAs result in activation of protein kinase C causing phosphorylation of serine and threonine sites on insulin receptor substrate-1 (IRS-1), thus impairing insulin signaling.[33] Increased levels of FFAs have also been correlated with elevated levels of ceramides, which induce many inflammatory pathways involving NFκB and JNK, also leading to abnormal phosphorylation of IRS-1.[34] Recent research has shown that in addition to FFAs, many pro-inflammatory cytokines are directly involved with the development of insulin resistance.

Hotamisligil and colleagues were the first to describe a molecular pathway linking inflammation to insulin resistance, specifically focusing on the pro-inflammatory cytokine TNF-α.[35] Since that time, many other bioactive substances have been found to play a role in the inflammation associated with insulin resistance, including IL-6, interleukin-1 beta (IL-1β), MCP-1, resistin, leptin, and adiponectin among many others. Of these, adiponectin is the only substance whose levels are decreased with higher levels of obesity, which is most likely mediated through increased levels of TNF-α and IL-6.[36] Adiponectin is the predominant adipokine, with many of its beneficial effects being lost in the setting of MetS since levels are inversely proportional to BMI. In healthy nonobese individuals, adiponectin dominates over the pro-inflammatory cytokines and functions to maintain metabolic harmony and contributes to healthy vascular endothelial function. This is accomplished through stimulation of nitric oxide production in vascular endothelial cells, induction of peroxisome proliferator-activated receptor-alpha (PPAR-α) leading to modulation of FFA oxidation in muscle tissue, activation of adenosine monophosphate (AMP) kinase, and thus decreased hepatic gluconeogenesis and promotion of glucose transporter-4 (GLUT4) -mediated glucose uptake in muscle tissue.[37,38] To the contrary, resistin and leptin, two other adipokines, are increased in subjects with MetS and are pro-inflammatory. Leptin increases macrophage activation and release of TNF-α and IL-6, while resistin conveys resistance to insulin and also induces inflammation via activation of NFκB.[39,40]

At increased levels, pro-inflammatory cytokines and chemokines increase local and systemic oxidative stress, increase levels of FFAs, induce apoptosis of pancreatic β-cells, and mediate insulin resistance.[41] Experimentally, IL-6 and TNF-α induce insulin resistance through interference with tyrosine phosphorylation of IRS-1, thus preventing the translocation of GLUT4 to the plasma membrane.[21] Macrophages in adipose tissue play a key role in the production of the pro-inflammatory mediators found in MetS through their increased binding of NFκB, a key pro-inflammatory transcription factor, which is inhibited by adiponectin.[42] The combination of increased macrophage activation and recruitment, elevated levels of TNF-α and IL-6, increased activity of pro-inflammatory transcription factors, and elevation in serum FFA concentrations accompanied by a hyperinsulinemic state all serve to drive the insulin resistance, forming what many feel is the foundation of MetS.

23.4.3 Dyslipidemia

The dyslipidemia in MetS has been termed "atherogenic" dyslipidemia because of decreased levels of high-density lipoprotein cholesterol (HDL-C), increased levels of small dense low-density lipoprotein cholesterol (LDL-C), and an increase in serum triglycerides, the combination of which is associated with increased atherogenesis.[43] The mechanism of dyslipidemia in MetS is largely the result of impaired insulin signaling having many deleterious downstream effects. These include (1) decreased activity of insulin-dependent lipase in adipocytes leading to increased lipolysis and release of FFAs, (2) increased activity of hepatic lipase leading to abnormal metabolism of HDL-C and LDL-C, and (3) impaired degradation of apoliporotein B (apo B) leading to increased levels of VLDL and triglycerides[44] (Fig. 23.2). In MetS, FFAs are elevated and synthesized to triglycerides before being packaged into apo B proteins and ultimately released into the circulation as VLDL particles. Normally, apo B is degraded in the liver via an insulin-dependent pathway involving phosphoinositide-3 kinase (PI3K). This pathway is inhibited in the insulin-resistant state, further increasing serum levels of VLDL.[45] Additionally, the peripheral clearance of triglyceride-rich VLDL is impaired given the decreased levels of lipoprotein lipase in the insulin-resistant state, which also contributes to the elevation of serum triglycerides. Once circulating, VLDL particles transfer triglycerides to HDL-C particles via the action of cholestryl ester transfer protein (CETP) resulting in triglyceride-rich HDL-C particles. Triglyceride-rich HDL-C particles are the preferred substrate of hepatic lipase, the activity of which is increased in insulin resistance. Increased hepatic lipase activity causes increased hydrolyzation of triglycerides from HDL-C particles resulting in decreased HDL-C diameter. This is problematic because renal clearance of HDL-C is inversely proportional to HDL-C size, thus more HDL-C particles are removed from the circulation and low serum levels ensue. Additionally, the morphology of LDL-C particles is altered resulting in small dense particles that are more atherogenic than their larger more buoyant counterparts.[46] Another cardioprotective lipid, ApoA-I, is also decreased in the insulin-resistant and pro-inflammatory state. Its gene expression appears to be directly downregulated by TNF-α and other inflammatory cytokines. It is possible that pro-inflammatory cytokines also have direct effects on HDL-C and LDL-C; however, these effects have not been well described.

23.4.4 Endothelial Dysfunction and Hypercoagulability

The vascular endothelium is a dynamic structure, the dysfunction of which is mediated through pro-inflammatory and prothrombotic proteins. Many of these proteins are increased in MetS and the insulin-resistant state potentially increasing the risk of atherothrombosis. Elevated levels of TNF-α, IL-6, PAI-1, C-reactive protein (CRP), and tissue factor (TF) are integral in this process.[47] TNF-α, through its upregulation of NFκB, leads to increased levels of vascular adhesion molecules and decreases the activity of endothelial nitric oxide synthase (NOS3), thus leading to decreased levels of nitric oxide (NO) and impaired endothelial relaxation. Decreased NO levels also promote platelet adhesion and aggregation in vessel walls favoring a state of thrombosis.[48,49] Insulin itself has many beneficial properties beyond its integral role in glucose metabolism. Insulin decreases levels of PAI-1, TF, matrix metalloproteinases, suppresses activity of NFκB, directly increases NOS3 expression and NO release from vascular endothelial cells, and suppresses production of ROS through inhibition of NADPH oxidases, all of which improve

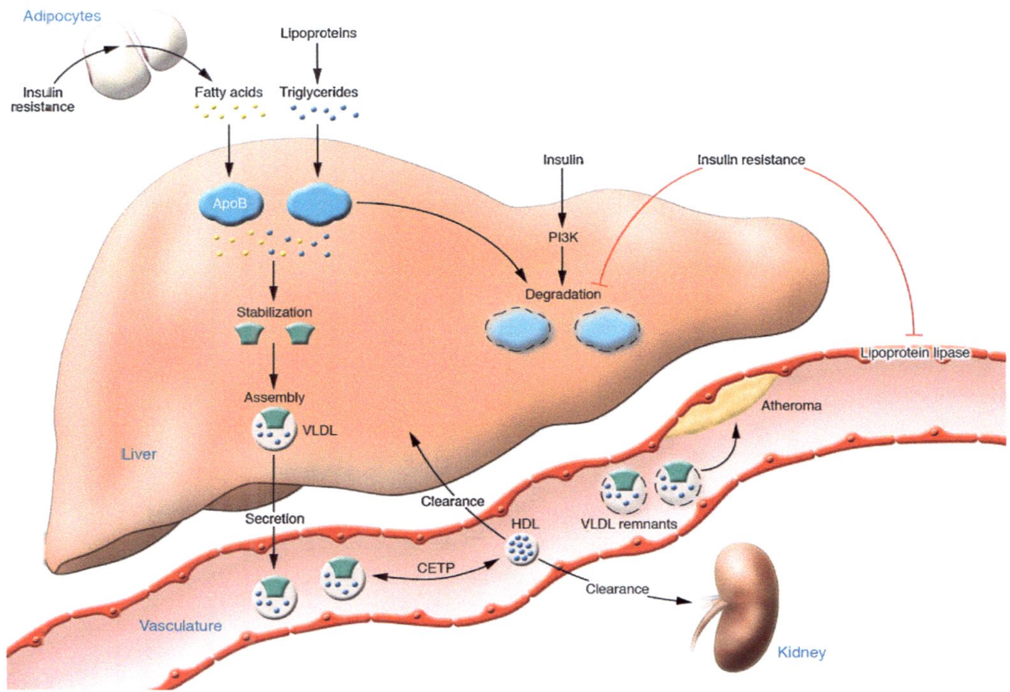

Fig. 23.2 Dyslipidemia in MetS. Above is an illustrative representation of the pathogenesis of dyslipidemia as found in MetS.[44] *ApoB* apolipoprotein B; *VLDL* very low-density lipoprotein cholesterol; *PI3K* phosphoinositide-3 kinase; *CETP* cholesteryl ester transfer protein; *HDL* high-density lipoprotein; *LDL* low-density lipoprotein

endothelial function or mitigate thrombosis [21,50]. In the insulin-resistant state, all of these beneficial functions are lost leading to endothelial dysfunction and a pro-thrombotic environment. Insulin also causes the release of endothelin-1, a potent vasoconstrictor, which in the insulin-resistant state is favored over the production of NO leading to increased vasoconstriction.[51] The combination of increased ROS, a hypercoagulable state, decreased NO availability, and increased vascoconstriction serves to alter normal hemostasis and endothelial function, most likely contributing to the increased risk of CVD and stroke in subjects with MetS.

23.4.5 Hypertension

Insulin resistance, obesity, a Western lifestyle and diet, and a pro-inflammatory state, all contribute to the pathophysiology of hypertension in MetS. Insulin resistance and abdominal obesity are associated with increased adrenergic tone, activation of the renin–angiotensin system (RAS), and microvascular dysfunction.[51,52] Activation of the RAS is directly mediated through adipocytes and increased systemic sympathetic activity. Increased expression of genes encoding for vasoactive proteins involved in the RAS is observed in adipocytes of abdominally obese subjects resulting in secretion of angiotensinogen.[52,53] Elevated levels of endothelin-1, the major effector peptide of the endothelin family, are found in insulin-resistant subjects causing increased vasoconstriction in resistance vessels, thus contributing to the development of hypertension through microvascular dysfunction.[51,54] In addition to vasoactive substances, many pro-inflammatory cytokines, particularly TNF-α, have also been associated with insulin resistance and hypertension. TNF-α impairs the vasodilatory effects of insulin through activation of JNK, in addition to directly decreasing skin capillary recruitment. Additionally, FFAs, also elevated in MetS, can directly lead to vasoconstriction partially explaining why hypertension has long been associated with the typical Western diet.[55,56] Hypertension in the obese insulin-resistant individual is complex and multifactorial with vasoactive substances, pro-inflammatory cytokines, and FFAs all playing a role. Given the heritability of essential hypertension, one might expect to find a strong genetic component to hypertension in subjects with MetS; however, hypertension is not one of the MetS traits that has strong evidence to support its inheritance as is further discussed below.

23.5 Genetics

The increased prevalence in obesity-related MetS is undoubtly a result of environmental factors such as physical inactivity, overeating, and consumption of the traditional Western diet; however, there is evidence for a genetic predisposition to MetS. In 1962, James Neel proposed the *thrifty gene hypothesis* suggesting that in primitive humans natural selection favored energy-conserving genotypes given the frequency of malnutrition and the scarcity of food supply. In times of nutritional abundance, these genes become unfavorable and quite likely predispose to conditions such as MetS and T2DM.[57] Neel's hypothesis helped to fuel the movement to identify a genetic basis for the development of T2DM that has now carried over to MetS. The first step to establish a genetic basis for these disease states is through a heritability assessment utilizing concordance statistics comparing monozygotic to dizygotic twins or through family studies. No such studies have been performed for MetS specifically; however, all the individual traits comprising the MetS do appear to have some genetic basis with glucose intolerance, obesity, and low HDL-C having the strongest evidence for genetic influence.[58,59] This is well described in the Pima Indian population who on migrating to the United States and adopting a Western lifestyle manifest a high prevalence of obesity (85%) and T2DM (50%) by the age of 35 years. In contrast, Pima Indians of the same lineage living in Mexico with a traditional lifestyle have a much lower prevalence of these conditions.[60] With the completion of the human genome project, the ability to search for candidate genes predisposing to MetS and T2DM has now become possible. Some candidate genes have been identified and are discussed below.

Firm evidence exists for linkages between chromosomes 2, 3, and 16, specifically bands 3q27, 16p13-pter, and 2 at 240-cm, with the development of MetS. Additionally, suggestive linkages have been observed with chromosomes 7, 12, 14, and 15.[59] Identification of a proline to alanine (Pro12Ala) mutation in the PPAR-γ, a commonly targeted receptor in the treatment of T2DM, at 3p25 has been associated with many traits of MetS. Two other mutations involving the PPAR-γ gene, Val290Met and Pro467leu, have been identified in patients who developed insulin resistance, T2DM, and hypertension at an early age.

Another candidate gene is the beta-3 adrenergic receptor that is expressed in visceral fat. Reportedly, a missense mutation at Trp64Arg confers increased abdominal obesity, insulin resistance, and high blood pressure in individuals who are carriers for this genetic variant when compared with individuals who are homozygous for the wild-type Trp64Trp.[61] However, this observation has not been reproduced by other investigators.[62] In T2DM, there is increased frequency of genetic polymorphisms in the IRS-1 when compared with controls. This finding increases the likelihood that mutations leading to changes in the insulin receptor substrates may confer a genetic predisposition to MetS. Thus, there appears to be some evidence to support a genetic predisposition to MetS; however, this data is in its infancy and more research is needed to further clarify a genetic cause-and-effect relationship.

23.6 Cardiovascular Risk

It is well accepted that each component of MetS confers increased CVD risk on its own. Thus, it makes sense that MetS is positively associated with CVD risk, CVD-related death, and overall mortality. The more appropriate question is whether or not MetS confers greater CVD risk than would be expected by the sum of its components alone. Some recent evidence suggests that it does[63] however, other studies have shown that it does not,[64,65] and this continues to be a topic of great debate.[66] Multiple studies have shown an increased risk of CVD in subjects with MetS after adjusting for age, smoking status, LDL-C, and race. Additionally, MetS confers an increased risk of CVD mortality and death in those with known coronary atherosclerotic disease and prior myocardial infarction.[67] Interestingly, the risk of CVD conferred by MetS appears to be slightly higher for women than men.[68,69] The reason for this is not well understood, but thought to be related to deficiencies in sex hormones in the postmenopausal state contributing to endothelial dysfunction and atherogenesis.

In European cohorts, the relative risk (RR) of CVD mortality in men with MetS are 2.09, 1.72, and 1.51 based on WHO, NCEP:ATP3, and IDF definitions, respectively.[70] These results are relatively consistent with other studies assessing CVD risk and MetS.[12] When compared with the Framingham Risk Score, the MetS category is inferior for the prediction of CVD risk at 10 and 20 years; however, the predictive power of MetS is increased at 20 years compared with 10 years. This is quite likely because MetS does not take into account factors such as age, gender, total cholesterol, and smoking history.[71] MetS is also an independent predictor for the development of heart failure after adjusting for the presence of established etiologies.[72] This is thought to reflect the chronic inflammatory state of MetS and possibly the presence of sleep-disordered breathing, which is more commonly found in obese individuals. The risk of stroke is also increased in MetS; however, as with CVD, the Framingham Risk Score is a superior predictor. Using the diagnosis of MetS as a clinical tool to identify patients at high risk for CVD may be easier than calculating the Framingham Risk Score and potentially more clinically applicable. However, MetS is not a robust predictor of short-term CVD risk, thus recognition of MetS should prompt the physician to aggressively risk stratify individuals with proven CVD risk prediction models and then treat them appropriately.

23.7 Risk of Type 2 Diabetes

It is not surprising that MetS is a strong predictor for future development of T2DM given the inclusion of insulin resistance and abdominal obesity in its definition, both of which have been independently associated with an increased risk of T2DM.[73,74] Applying the NCEP:ATP3 definition to the Framingham Offspring Study showed a 6.9-fold increased risk for the development of T2DM over an 8-year period. This increased RR was similar among men and women.[12] When compared with the Framingham risk score, MetS was superior in predicting future development of T2DM; this is not surprising given that the Framingham risk score does not incorporate either abdominal obesity or insulin resistance in its calculation. When taken together, the risk of developing either CVD or T2DM rises as the number of MetS components affecting an individual increases. In using MetS as a clinical tool, it has the added benefit of providing a quantifiable risk without requiring more labor-intensive laboratory testing to identify individuals with glucose intolerance and insulin resistance. The incorporation of abdominal obesity into the definition of MetS is a surrogate for insulin resistance, thus making MetS a very powerful tool for identifying individuals who are at high risk of developing T2DM.

23.8 Treatment

23.8.1 Lifestyle Modification

The clinical care of individuals with MetS should focus on decreasing risk of CVD and the risk of developing T2DM. Lifestyle interventions, i.e. smoking cessation, weight loss, increased physical activity, and diet modifications are the cornerstones of therapy for individuals with MetS and have been shown to lower blood pressure, increase insulin sensitivity, improve lipid profiles, and decrease systemic inflammation.[75] Additionally, weight loss and physical activity are independently associated with decreased CVD risk and risk of developing T2DM.[76,77] Weight loss can be accomplished through increased physical activity, diet modifications, surgical therapy, drug therapy, or a combination of these interventions. Physical activity is the preferred method of weight loss given its additional beneficial effects and inverse association with cardiovascular risk and progression from insulin resistance to T2DM. Currently, the American Heart Association recommends 30 min or more of moderate intensity physical activity at least five times per week.[78] The Diabetes Prevention Program "lifestyle exercise" has shown promise with multiple short bouts of exercise showing similar benefits to a single day exercise session.[79] For some, this should increase compliance. Adopting a Mediterranean diet high in fruits, vegetables, fiber, and monounsaturated fats reduces the prevalence of MetS by up to 25%, improves endothelial function, and reduces serum levels of inflammatory markers.[80,81] Taken together, physical activity, weight loss, and diet modification can drastically reduce the prevalence of MetS in affected individuals.

23.8.2 Pharmacologic and Surgical Interventions

In addition to lifestyle modification, other weight loss therapies, such as pharmacologic and surgical interventions, show promise in the treatment of MetS. Clinical trials assessing the efficacy of gastric bypass surgery show drastic reductions in MetS up to 2 years out from surgery.[82] In these trials, the average amount of weight lost was 55–75% of their presurgical weight. This weight loss correlated with 85–95% of subjects showing resolution of MetS at 2-year follow-up. This data is promising; however, longer follow-up periods are needed before firm conclusions can be made. The same benefit is not reproduced with subcutaneous surgical fat removal correlating with the hypothesis that the accumulation of visceral adipose tissue is responsible for the development of MetS.

Additional research has shown promise with the selective cannabinoid-1 receptor blocker, rimonabant, showing modest sustained weight and waist circumference reductions with favorable effects on cardiometabolic risk factors and statistically significant reductions in the prevalence of MetS among treated individuals.[83] Rimonabant appears to have an advantage over other pharmacologic weight loss options, orlistat: a lipase inhibitor, and sibutramine: a norepinephrine, serotonin, and dopamine re-uptake inhibitor that acts as an appetite suppressant, given its improved tolerability and increased quantity of weight loss.[84] Orlistat and sibutramine are currently approved for the treatment of morbid obesity in the United States; however, rimonabant is currently only approved for use in Europe.

23.8.3 Insulin Resistance

The treatment of Insulin resistance, also known as prediabetes or glucose intolerance in MetS, is of utmost importance given its key role in the pathogenesis of this syndrome. The Diabetes Prevention Program compared physical activity and weight loss to oral metformin therapy and placebo for the prevention of progression to T2DM in patients with glucose intolerance. Over a 2.8-year follow-up, lifestyle modifications yielding a 7% weight loss was the most efficacious therapy with the development of 4.8 cases of T2DM per 100 person-years, the metformin group had 7.8 cases per 100 person-years, while the placebo group had 11 cases per 100 person-years.[85] A study in Finland showed similar results.[86] The same interventions were shown to decrease the incidence of MetS defined by NCEP:ATP3 criteria by 41 and 17% in the lifestyle modification and metformin groups, respectively.[87] Additionally, metformin is the only oral hypoglycemic drug that has been shown to decrease the incidence of macrovascular complications associated withT2DM, which is a beneficial effect of this agent.

Therapy with thiazolidinediones (TZD) has also been assessed in prediabetic populations. Recently, TZDs have shown a decrease in the progression of prediabetes to T2DM in glucose-intolerant individuals, with 25% of the placebo group and only 10.6% of the study group receiving rosiglitazone 8 mg daily, progressing to T2DM.[88] TZD's also decrease blood pressure and have more recently been proposed to be a first-line therapy for individuals with MetS because of the previously described beneficial effects.[86] Caution must be taken in utilizing this class of drugs because of their association with an increased incidence of congestive heart failure.[88] There are no official recommendations for pharmacological treatment of insulin resistance; however, this may change in the near future.

23.8.4 Hypertension

Current recommendations for the treatment of elevated blood pressure for individuals with MetS remain in concert with those proposed by the seventh Report of the Joint National Commission.[89] These along with other treatment recommendations are shown in Table 23.2. Dietary modifications, specifically the *D*ietary *A*pproaches to *S*top *H*ypertension (DASH) trial diet reduce blood pressure and improve the metabolic risk profile of individuals with MetS.[90] Additionally, aggressive weight reduction is associated with a decrease of 5–20 mmHg in systolic blood pressure for every 10 kg of weight loss.[89] Regular aerobic exercise also reduces blood pressure yielding a 4–6 mmHg decrease in systolic blood pressure.[91] In addition to dietary and lifestyle

Table 23.2 Therapy of metabolic risk factors for prevention of CVD or treatment of type 2 diabetes

Therapeutic target and goals of therapy	Therapeutic recommendations
Lifestyle risk factors	Long-term prevention of CVD and prevention (or treatment) of type 2 diabetes mellitus (T2DM)
Abdominal obesity Reduce body weight by 7–10% during year 1 of therapy. Continue weight loss thereafter to extent possible with goal to ultimately achieve desirable weight (BMI <25 kg/m^2)	Consistently encourage weight maintenance/reduction through appropriate balance of physical activity, caloric intake, and formal behavior-modification programs when indicated to maintain/achieve waist circumference of <40 inches in men and <35 inches in women. Aim initially at slow reduction of &7–10% from baseline weight. Even small amounts of weight loss are associated with significant health benefits.
Physical inactivity Regular moderate-intensity physical activity; at least 30 min of continuous or intermittent (and preferably >60 min) 5 d/wk, but preferably daily	In patients with established CVD, assess risk with detailed physical activity history and/or an exercise test, to guide prescription. Encourage 30–60 min of moderate-intensity aerobic activity: brisk walking, preferably daily, supplemented by increase in daily lifestyle activities (e.g., pedometer step tracking, walking breaks at work, gardening, housework). Longer exercise times can be achieved by accumulating exercise throughout day. Encourage resistance training 2 d/wk. Advise medically supervised programs for high-risk patients (e.g., recent acute coronary syndrome or revascularization, CHF).
Atherogenic diet Reduced intake of saturated fat, *trans*fat, cholesterol	Recommendations: saturated fat <7% of total calories; reduce *trans* fat; dietary cholesterol <200 mg/dL; total fat 25–35% of total calories. Most dietary fat should be unsaturated; simple sugars should be limited.
Metabolic risk factors	Shorter-term prevention of CVD or treatment of T2DM
Atherogenic dyslipidemia Primary target: elevated LDL-C (per NCEP:ATP3 Guidelines) Secondary target: elevated non-HDL-C High-risk patients[a]: <130 mg/dL (3.4 mmol/L) (optional: <100 mg/dL [2.6 mmol/L] for very high-risk patients[b]) Moderately high-risk patients[c]: <160 mg/dL (4.1 mmol/L) Therapeutic option: <130 mg/dL (3.4 mmol/L) Moderate-risk patients[d]: <160 mg/dL (4.1 mmol/L) Lower-risk patients[e]: <190 mg/dL (4.9 mmol/L)	Follow NCEP:ATP3 guidelines for LDL lowering therapy. First option to achieve non-HDL-C goal: Intensify LDL-lowering therapy Second option to achieve non-HDL-C goal: Add fibrate (preferably fenofibrate) or nicotinic acid if non-HDL-C remains relatively high after LDL-lowering drug therapy Give preference to adding fibrate or nicotinic acid in high-risk patients Give preference to avoiding addition of fibrate or nicotinic acid in moderately high-risk or moderate-risk patients All patients: If TG is ≥500 mg/dL, initiate fibrate or nicotinic acid (before LDL-lowering therapy; treat non-HDL-C to goal after TG-lowering therapy) Maximize lifestyle therapies: weight reduction and increased physical activity Consider adding fibrate or nicotinic acid after LDL-C-lowering drug therapy as outlined for elevated non-HDL-C

(continued)

Table 23.2 (continued)

Therapeutic target and goals of therapy	Therapeutic recommendations
Elevated BP Reduce BP to at least achieve BP of <140/90 mmHg (or <130/80 mmHg if diabetes present). Reduce BP further to extent possible through lifestyle changes.	For BP≥120/80 mmHg: Initiate or maintain lifestyle modification in all patients with MetS: weight control, increased physical activity, alcohol moderation, sodium reduction, and emphasis on increased consumption of fresh fruits, vegetables, and low-fat dairy products For BP≥140/90 mmHg (or ≥130/80 mmHg for individuals with chronic kidney disease or diabetes): As tolerated, add BP medication as needed to achieve goal BP
Elevated glucose For IFG, delay progression to T2DM. For diabetes, hemoglobin A_{1C} <7.0%	For IFG, encourage weight reduction and increased physical activity. For T2DM, lifestyle therapy, and pharmacotherapy, if necessary, should be used to achieve near-normal HbA1C (<7%). Modify other risk factors and behaviors (e.g., abdominal obesity, physical inactivity, elevated BP, lipid abnormalities).
Prothrombotic state Reduce thrombotic and fibrinolytic risk factors	High-risk patients: Initiate and continue low-dose aspirin therapy; in patients with ASCVD, consider clopidogrel if aspirin is contraindicated. Moderately high-risk patients: Consider low-dose aspirin prophylaxis
Proinflammatory state	Recommendations: no specific therapies beyond lifestyle therapies

TG indicates triglycerides; *BP* blood pressure; *CVD* cardiovascular disease; *CHF* congestive heart failure; *BMI* body mass index; *IFG* impaired fasting glucose; and *ASCVD* atherosclerotic cardiovascular disease

[a] High-risk patients are those with established ASCVD, diabetes, or 10-year risk for coronary heart disease >20%. For cerebrovascular disease, high-risk condition includes TIA or stroke of carotid origin or >50% carotid stenosis

[b] Very high-risk patients are those who are likely to have major CVD events in next few years, and diagnosis depends on clinical assessment. Factors that may confer very high risk include recent acute coronary syndromes, and established coronary heart disease + any of following: multiple major risk factors (especially diabetes), severe and poorly controlled risk factors (especially continued cigarette smoking), and MetS

[c] Moderately high-risk patients are those with 10-year risk for coronary heart disease 10–20%. Factors that favor therapeutic option of non-HDL-C <100 mg/dL are those that can raise individuals to upper range of moderately high risk: multiple major risk factors, severe and poorly controlled risk factors (especially continued cigarette smoking), MetS, and documented advanced subclinical atherosclerotic disease (e.g., coronary calcium or carotid intimal-medial thickness >75th percentile for age and sex)

[d] Moderate-risk patients are those with 2+ major risk factors and 10-year risk <10%

[e] Lower-risk patients are those with 0 or 1 major risk factor and 10-year risk <10%

Modified from ref.[78]

modifications, multiple pharmacologic options exist for the treatment of hypertension. Angiotensin-converting enzyme inhibitors (ACE-I) and angiotensin receptor blockers have favorable effects on glucose profiles and thus should be considered early in the treatment of hypertensive individuals with MetS.[92]

23.8.5 *Dyslipidemia*

Atherogenic dyslipidemia is intimately associated with insulin resistance. As previously discussed, increased triglycerides, increased number of small dense LDL-C particles, and decreased levels of HDL-C are the manifestations of the atherogenic lipid profile in MetS. Importantly, the quantitative increase in smaller LDL-C particles seen in insulin-resistant and diabetic individuals may not be accurately depicted by conventional lipid profile assays. Recent clinical trials comparing lipid measurements taken by nuclear magnetic resonance spectroscopy and conventional lipid profile show an increase in small LDL-C particle number that is not well represented in the conventional lipid profile.[93] The implications of this are not yet well established. Currently, the goals of treatment of patients with MetS are the same as for any other patient population and are based on the NCEP:ATP3 guidelines[78] (see Table 23.2). Many pharmacological options are currently available including 3-hydroxy-3-methylglutaryl (HMG) coenzyme A reductase inhibitors (statins), fibric acid derivatives, and niacin.

Statins have proven efficacy in both the primary and secondary prevention of CVD in patients both with and without the MetS.[94,95] Some data suggests a potentially greater benefit for secondary prevention of coronary heart disease (CHD) in those with the atherogenic lipid profile associated with MetS.[94] Another study looking at secondary prevention of CVD-related events in patients with MetS showed improved risk reduction with high-dose vs. low-dose statin therapy; this data was strongest for individuals with three or more components of MetS.[96] This data supports the current recommendation of an optional, but reasonable LDL-C target of <70 mg/dL set forth by the American Heart Association and the American College of Cardiology.[97] In addition to reducing LDL-C levels, statins also have beneficial effects on HDL-C and triglycerides, which are of particular importance in individuals with MetS.

Fibric acid derivatives have long been an option for the treatment of dyslipidemia and have recently been evaluated in those with MetS showing beneficial results. Fibrates have beneficial effects on serum LDL-C, HDL-C, and triglycerides. European studies show significant reductions in CHD events and sudden death.[98] This data is best shown through subgroup analyzes when individuals with MetS are looked at specifically. It appears that of all patient populations, those with MetS benefit the most from fibric acid therapy. These benefits are most likely the result of increased HDL-C levels, a shift from smaller dense to larger and softer LDL-C particles, and decreased triglycerides all translating to decreased atherogenesis. Trials are currently underway to assess the benefits of combination therapy with fibrates, statins, and niacin in individuals with MetS.

Niacin, also known as nicotinic acid or vitamin B3, has been shown in multiple clinical trials to increase serum HDL-C levels, decrease LDL-C levels, increase LDL-C particle size, and decrease triglycerides. Immediate release niacin appears to be slightly more effective in modifying the lipid profile than its intermediate-release counterpart; however, only the latter is realistically tolerated in the clinical setting.[99] Recent data suggests that niacin in combination with statins actually causes regression of carotid intima media thickness in contrast to statin therapy alone.[100] Importantly, niacin does cause a slight increase in fasting blood glucose levels; however, this can be adequately controlled by hypoglycemic therapy. Uric acid levels can also be increased with niacin therapy; therefore, caution should be taken in individuals with gout. Unfortunately, niacin is poorly tolerated given its cutaneous side effects, especially a flushing sensation. To improve tolerability, a 325 mg aspirin can be taken 30 min prior to niacin dosing to reduce frequency and intensity of these reactions.[101] Additionally, it is reasonable to implement daily anti-platelet therapy in patients with MetS given the increased risk of CVD. Given the beneficial effects of niacin on the atherogenic lipid profile, decrease in CVD events and regression of atherosclerosis when combined with statin therapy, niacin is a valuable pharmacological agent for the treatment of dyslipidemia in MetS.

23.9 Summary

MetS has emerged as a unique entity conferring risk for development of CVD and T2DM. Prevalence of MetS is reaching epidemic proportions and will undoubtedly have a profound impact on the development of CVD and T2DM for generations to come. The origins of MetS appear to be multifactorial with visceral obesity and insulin resistance central to its pathogenesis. Environmental and lifestyle factors are also paramount in this disease process with genetic predisposition most likely contributing to a lesser degree. The increased risk of CVD and T2DM is well documented in those with MetS and is the reason for the early identification and aggressive treatment of these individuals. Standard therapeutic lifestyle modifications are the first-line treatment for MetS. Currently, pharmacological therapy is directed by current guidelines for the treatment of each individual component of MetS. It remains to be seen if any of these recommendations are specifically curtailed toward the treatment of individuals with MetS in the future.

References

1. Reaven GM (1988) Banting lecture 1988. Role of insulin resistance in human disease. Diabetes 37:1595–1607
2. Alberti KG, Zimmet PZ (1998) Definition, diagnosis, and classification of diabtes mellitus and its complications. Part 1: diagnosis and classification of diabetes mellitus, provisional report of WHO consultation. Diabet Med 15:539–553

3. Grundy SM, Cleeman JI, Merz CN et al (2004) Implications of recent clinical trials for the National cholesterol education program adult treatment panel III guidelines. Circulation 110:227–239
4. Alberti KG, Zimmet PZ, Shaw J (2005) The metabolic syndrome-a new worldwide definition. Lancet 355:1059–1062
5. Einhorn D, Reaven GM, Cobin RH et al (2003) American college of endocrinology position statement on the insulin resistance syndrome. Endocr Pract 9:237–252
6. Balkau B, Charles MA (1999) Comment on the provisional report from the WHO consultation. European group for the study of insulin resistance (EGIR). Diabet Med 16:442–443
7. Balkau B (2000) The DECODE study. Diabetes epidemiology: collaborative analysis of diagnostic criteria in Europe. Diabetes Metab 26:282–286
8. Ford ES, Giles WH, Dietz WH (2002) Prevalence of the metabolic syndrome among US adults: findings from the third national health and nutrition examination survey. JAMA 287:356–359
9. Meigs JB, Wilson PW, Nathan DM et al (2003) Prevalence and characteristics of the metabolic syndrome in the San Antonio heart and Framingham offspring studies. Diabetes 52:2160–2167
10. Pyorala K, Qiao Q, Gao WG et al, for the DECODE Study Group. Prevalence of the metabolic syndrome in non-diabetic Europeans: relation to cardiovascular mortality. *Arch Intern Med.* 2004;164:1066-1076.
11. Hanley AJ, Karter AJ, Williams K et al (2005) Prediciton of type 2 diabetes Mellitus with alternative definitions of the metabolic syndrome: the insulin resistance atherosclerosis study. Circulation 112:3713–3721
12. Wilson PW, D'Agostino RB, Parise H et al (2005) Metabolic syndrome as a precursor of cardiovascular disease and type 2 diabetes mellitus. Circulation 112:3066–3072
13. Zafari AM, Mackie BD. Physical activity and the metabolic syndrome. *Hosp Physician.* 2006;42:26-38 <http://www.turnerwhite.com/memberfile.php?PubCode=hp_sep06_metabolic.pdf> Accessed January 2008.
14. Ford ES, Giles WH (2003) A comparison of the prevalence of the metabolic syndrome using two proposed definitions. Diabetes Care 2693:575–581
15. Kolovou GD, Anagnostopoulou KK, Salpea KD et al (2007) The prevalence of metabolic syndrome in various populations. Am J Med Sci 333:362–371
16. Mainous AG III, King DE, Garr DR et al (2004) Race, rural residence, and control of diabetes and hypertension. Ann Fam Med 2:563–568
17. Kershaw EE, Flier JS (2004) Adipose tissue as an endocrine organ. J Clin Endocrinol Metab 89:2548–2556
18. Wellen KE, Hotamisligil GS (2003) Obesity-induced inflammatory changes in adipose tissue. J Clin Invest 112:1785–1788
19. Gustafson B, Hammarstedt A, Andersson CX et al (2007) Inflamed adipose tissue: a culprit underlying the metabolic syndrome and atherosclerosis. Arterioscler Thromb Vasc Biol 27:2276–2283
20. Shoelson SE, Lee J, Goldfine AB (2006) Inflammation and insulin resistance. J Clin Invest 116:1793–1801
21. DandonaP AA, Chaudhuri A et al (2005) Metabolic syndrome: a comprehensive perspective based on interactionsbetween obesity, diabetes, and inflammation. Circulation 111:1448–1454
22. Dandona P, Mohanty P, Ghanim H et al (2001) The suppressive effect of dietary restriction and weight loss in the obese on a generation of reactive oxygen species by leukocytes, lipid peroxidation, and protein carbonylation. J Clin Endocrinol Metab 86:355–362
23. Lee JY, Sohn KH, Rhee SH et al (2001) Saturated fatty acids, but not unsaturated fatty acids, induce the expression of cyclooxygenase-2 mediated through Toll-like receptor 4. J Biol Chem 276:16683–16689
24. Wu H, Ghosh S, Dai Perrard X et al (2007) T-Cell accumulation and regulated on activation, normal T cell expressed and secreted upregulation in adipose tissue in obesity. Circulation 115:1029–1038
25. Agrewala JN, Brown DM, Lepak NM et al (2007) Unique ability of activated CD4+ T cells but not rested effectors to migrate to non-lymphoid sites in the absence of inflammation. J Biol Chem 282:6106–6115

26. Guzik TJ, Hoch NE, Brown KA et al (2007) Role of the T cell in the genesis of angiotensin II induced hypertension and vascular dysfunction. J Exp Med 204:2449–2460
27. Guzik TJ, Marvar PJ, Czesnikiewicz-Guzik M et al (2007) Perivascular adipose tissue as a messenger of the brain-vessel axis: role in vascular inflammation and dysfunction. J Physiol Pharmacol 58:591–610
28. Angeli F, Verdecchia P, Pellegrino C et al (2003) Association between periodontal disease and left ventricle mass in essential hypertension. Hypertension 41:488–492
29. Panoulas VF, Douglas KM, Milionis HJ et al (2007) Prevalence and associations of hypertension and its control in patients with rheumatoid arthritis. Rheumatology (Oxford, England) 46:1477–1482
30. Neimann AL, Shin DB, Wang X et al (2006) Prevalence of cardiovascular risk factors in patients with psoriasis. J Am Acad Dermatol 55:829–835
31. Eckel RH, Grundy SM, Zimmet PZ (2005) The metabolic syndrome. Lancet 365:1415–1428
32. Lann D, LeRoith D (2007) Insulin resistance as the underlying cause for the metabolic syndrome. Med Clin N Am 91:1063–1077
33. Shulman GI (2000) Cellular mechanisms of insulin resistance. J Clin Invest 106:171–176
34. Straczkowski M, Kowalska I, Nikolajuk A et al (2004) Relationship between insulin sensitivity and sphingomyelin signaling pathway in human skeletal muscle. Diabetes 53:1215–1221
35. Hotamisligil GS, Shargill NS, Spiegelman BM (1993) Adipose expression of tumor necrosis factor-direct role in obesity-linked insulin resistance. Science 259:87–91
36. Mehta S, Farmer JA (2007) Obesity and inflammation: a new look at an old problem. Curr Atheroscler Rep 9:134–138
37. Ahima RS (2006) Adipose tissue as an endocrine organ. Obesity 14:242S–249S
38. Pittas AG, Joseph NA, Greenberg AS (2004) Adipocytokines and insulin resistance. J Clin Endocrinol Metab 89:447–452
39. Guzik TJ, Managalat D, Korbut R (2006) Adipocytokines – novel link between inflammation and vascular function? J Physiol Pharmacol 57:505–528
40. Steppan CM, Bailey ST, Bhat S et al (2001) The hormone resistin links obesity to diabetes. Nature 409:307–312
41. Zafari AM, Khan AQ, Sola S, Khan BV. The metabolic syndrome: inflammation and endothelial dysfunction. *Hosp Physician*. 2006;42:26-37. <http://www.turnerwhite.com/memberfile.php?PubCode=hp_aug06_matabolic.pdf> Accessed January 2008.
42. Ghanim H, Aljada A, Hofmeyer D et al (2004) The circulating mononuclear cells in the obese are in the pro-inflammatory state. Circulation 110:1564–1571
43. Szapary PO, Rader DJ (2004) The triglyceride-high density lipoprotein axis: an important target of therapy? Am Heart J 148:211–221
44. Semenkovich CF (2006) Insulin resistance and atherosclerosis. J Clin Invest 116:1813–1822
45. Smith DA (2007) Treatment of the dyslipidemia of insulin resistance. Med Clin North Am 91:1185–1210
46. Berneis KK, Kraus RM (2002) Metabolic origins and clinical significance of LDL heterogeneity. J Lipid Res 43:1363–1379
47. Koh KK, Han SH, Quon MJ (2005) Inflammatory markers and the metabolic syndrome: insights from therapeutic interventions. J Am Coll Cardiol 46:1978–1985
48. Ross R (1999) Atherosclerosis – an inflammatory disease. N Engl J Med 340:115–126
49. Vincent MA, Montagnani M, Quon MJ (2003) Molecular and physiologic actions of insulin related to production of nitric oxide in vascular endothelium. Curr Diab Rep 3:279–288
50. Kim JA, Montagnani M, Koh KK et al (2006) Reciprocal relationships between insulin resistance and endothelial dysfunction: molecular and pathophysiological mechanisms. Circulation 113:1888–1904
51. Serne EH, Renate TD, Etto C et al (2007) Microvascular dysfunction: a potential pathophysiological role in the metabolic syndrome. Hypertension 50:204–211
52. Barton M, Carmona R, Ortmann J et al (2003) Obesity-associated activation of angiotensin and endothelin in the cardiovascular system. Int J Biochem Cell 35:826–837
53. Alihoud G (2006) Adipose tissue as a secretory organ: from adipogenesis to the metabolic syndrome. C R Biologies 329:570–577

54. Cardillo C, Campia U, Iantorno M et al (2004) Enhanced vascular activity of endogenous endothelin-1 in obese hypertensive patients. Hypertension 43:36–40
55. Svetkey LP, Simons-Morton D, Vollmer WM et al (1999) Effects of dietary patterns on blood pressure: subgroup analysis of the dietary approaches to stop hypertension (DASH) randomized clinical trial. Arch Intern Med 159:285–293
56. Tripathy D, Mohanty P, Dhindsa S et al (2003) Elevation of free fatty acids induces inflammation and impairs vascular reactivity in healthy subjects. Diabetes 52:2882–2887
57. Neel JV (1962) Diabetes mellitus: a "thrifty" genotype rendered detrimental by "progress"? Am J Hum Genet 14:353–362
58. Groop L, Orho-Melander M (2001) The dysmetabolic syndrome. J Intern Med 250:105–120
59. Tang W, Miller MB, Rich SS et al (2003) Linkage analysis for the multiple metabolic syndrome: The national heart, lung, and blood institute family study. Diabetes 52:2840–2847
60. Ravissun E, Valencia ME, Esparza J et al (1994) Effects of a traditional lifestyle on obesity in Pima Indians. Diabetes Care 17:1067–1074
61. Widen E, Lehto M, Kanninen T et al (1995) Association of a polymorphism in the beta 3-adrenergic-receptor gene with features of the insulin resistance syndrome in Finns. N Eng J Med 333:348–351
62. Shihara N, Yasuda K, Mortani T et al (1999) The association between Trp64Arg polymorphism of the beta3-adrenergic receptor and autonominc nervous system activity. J Clin Endocrinol Metab 84:1623–1627
63. Gami AS, Witt BJ, Howard DE et al (2007) Metabolic syndrome and risk of incident cardiovascular events and death: a systematic review and meta-analysis of longitudinal studies. J Am Coll Cardiol 49:403–414
64. Lawlor DA, Smith GD, Ebrahim S (2006) Does the new International Diabetes Federation definition of the metabolic syndrome predict CHD any more strongly than older definitions? Findings from the British women's heart and health study. Diabetologia 49:41–48
65. McNeil AM, Rosamond WD, Girman CJ et al (2005) The metabolic syndrome and 11-year risk of incident cardiovascular disease in the atherosclerosis risk in communities study. Diabetes Care 28:385–390
66. Kahn R, Buse J, Ferrannini E et al (2005) The metabolic syndrome: time for a critical appraisal: joint statement from the American diabetes association and the European association for the study of diabetes. Diabetes Care 28:2289–2304
67. Levantesi G, Macchia A, Marfisi R, Investigators GISSI-Prevenzione et al (2005) Metabolic syndrome and risk of cardiovascular events after myocardial infarction. J Am Coll Cardiol 46:277–283
68. Hunt KJ, Resendez RG, Williams K et al (2004) National cholesterol education program versus world health organization metabolic syndrome in relation to all-cause and cardiovascular mortality in the San Antonio Heart Study. Circulation 110:1251–1257
69. Kip KE, Marroquin OC, Kelley DE et al (2004) Clinical importance of obesity versus the metabolic syndrome in cardiovascular risk in women: a report from the Women's Ischemia Syndrome Evaluation (WISE) Study. Circulation 109:706–713
70. Graham I, Atar D, Borch-Johnsen K (2007) European guidelines on cardiovascular disease prevention in clinical practice: full text. Fourth joint task force of the European society of cardiology and other societies on cardiovascular disease prevention in clinical practice (constituted by representatives of nine societies and by invited experts). Eur J Cardiovasc Prev Rehabil 14(suppl 2):S1–S113
71. Wannamethee SGP, Sharper AGF, Lennon LM et al (2005) Metabolic syndrome versus Framingham risk score for prediction of coronary heart disease, stroke, and type 2 diabetes mellitus. Arch Intern Med 165:2644–2650
72. Ingelsson E, Arnlov J, Lind L et al (2006) Metabolic syndrome and risk for heart failure in middle aged men. Heart 92:1409–1413
73. Edelstein SL, Knowler WC, Bain RP et al (1997) Predictors of progression from impaired glucose tolerance to NIDDM: an analysis of six prospective studies. Diabetes 46:701–710
74. Ohlson LO, Larsson B, Svardsudd K et al (1985) The influence of body fat distribution on the incidence of diabetes mellitus. 13.5 years of follow-up of the participants in the study of men born in 1913. Diabetes 34:1055–1058

75. Ford ES (2002) Does exercise reduce inflammation? Physical activity and C-reactive protein among U.S. adults. Epidemiology 13:561–568
76. Manson JE, Greenland P, LaCroix AZ et al (2002) Walking compared to vigorous exercise for the prevention of cardiovascular disease in women. N Engl J Med 347:716–725
77. Tupper T, Gopalakrishnan G (2007) Prevention of diabetes development in those with the metabolic syndrome. Med Clin North Am 91:1091–1105
78. Grundy SM, Cleeman JI, Daniels SR et al (2005) Diagnosis and management of the metabolic syndrome: an American heart association/national heart, lung, and blood institute scientific statement: executive summary. Circulation 112:e285–e290
79. Klein S, Sheard NF, Pi-Sunyer X et al (2004) Weight management through lifestyle modification for the prevention and management of type 2 diabetes: rationale and strategies: a statement of the American diabetes association, the North American association for the study of obesity, and the American society for clinical nutrition. Diabetes Care 27:2067–2073
80. Esposito K, Marfella R, Citola M et al (2004) Effect of a Mediterranean-style diet on endothelial dysfunction and markers of vascular inflammation in the metabolic syndrome: a randomized trial. JAMA 292:1440–1446
81. Esposito K, Citola M, Giugliano D et al (2007) Mediterranean diet and the metabolic syndrome. Mol Nutr Food Res 51:1268–1274
82. Lee WJ, Huang MT, Wang W et al (2004) Effects of obesity surgery on the metabolic syndrome. Arch Surg 139:1088–1092
83. Pi-Sunyer FX, Aronne LJ, Heshmati HM et al (2006) Effect of rimonabant, a cannabinoid-1 receptor blocker, on weight and cardiometabolic risk factors in overweight or obese patients: RIO-North America: a randomized controlled trial. JAMA 295:761–775
84. Bray GA, Ryan DH (2007) Drug treatment of the overweight patient. Gastroenterology 132:2239–2252
85. Knowler WC, Barrett-Connor E, Fowler SE et al (2002) Reduction in the incidence of type 2 diabetes with lifestyle intervention or metformin. N Engl J Med 346:393–403
86. Brietzke SA (2007) Controversy in diagnosis and management of the metabolic syndrome. Med Clin N Am 91:1041–1061
87. Orchard TJ, Temprosa M, Goldberg R et al, Diabetes Prevention Program Research Group (2005) The effect of metformin and intensive lifestyle intervention on the metabolic syndrome: the diabetes prevention program randomized trial. Ann Intern Med 142:611–619
88. DREAM Trial Investigators (2006) Effects of rosiglitazone on the frequency of diabetes in patients with impaired glucose tolerance or impaired fasting glucose. A randomized contolled trial. Lancet 368:1096–1105
89. Chobanian AV, Bakris GL, Black HR et al (2003) Seventh report of the Joint National Committee on prevention, detection, evaluation, and treatment of high blood pressure. Hypertension 42:1206–1252
90. Azadbakht L, Mirmiran P, Esmaillzadeh A et al (2005) Beneficial effects of dietary approaches to stop hypertension eating plan on features of the metabolic syndrome. Diabetes Care 28:2823–2831
91. Whelton SP, Chin A, Xin X et al (2002) Effect of aerobic exercise on blood pressure: a meta-analysis of randomized, controlled trials. Ann Intern Med 136:493–503
92. Abuissa H, Jones PG, Marso SP et al (2005) Angiotensin-converting enzyme inhibitors or angiotensin receptor blockers for prevention of type 2 diabetes: a meta-analysis of randomized clinical trials. J Am Coll Cardiol 46:821–826
93. Garvey WT, Kwon S, Zheng D et al (2003) Effects of insulin resistance and type 2 diabetes on lipoprotein subclass particle size and concentration determined by nuclear resonance. Diabetes 52:453–462
94. Ballantyne CM, Olsson AG, Cook TJ et al (2001) Influence of low high-density cholesterol and elevated triglyceride on coronary heart disease events and response to simvastatin therapy in 4 S. Circulation 104:3046–3051
95. Shepherd J, Cobbe SM, Ford I et al (1995) Prevention of coronary heart disease with pravastatin in men with hypercholesterolemia. West Scotland coronary prevention study group. N Engl J Med 333:1301–1307

96. Deedwania P, Barter P, Carmen R et al (2006) Reduction of low-density lipoprotein cholesterol in patients with coronary heart disease and the metabolic syndrome: analysis of the treating to new targets study. Lancet 368:919–928
97. Smith SC Jr, Allen J, Blair SN et al (2006) AHA/ACC guidelines for secondary prevention for patients with coronary and other atherosclerotic vascular disease: 2006 update: endorsed by the national heart, lung, and blood institute. Circulation 113:2363–2372
98. Frick MH, Elo O, Haapa K et al (1987) Helsinki heart study: primary prevention trial with gemfibrozil in middle-aged men with dyslipidemia. Safety of treatment, changes in risk factors, and incidence of coronary heart disease. N Engl J Med 317:1237–1245
99. Knopp RH (2000) Evaluating niacin in its various forms. Am J Cardiol 86:51L–56L
100. Taylor AJ, Lee HJ, Sullenberger LE (2006) The effect of 24 months of combination statin and extended-release niacin on carotid intima-media thickness: ARBITER 3. Curr Med Res Opin 22:2243–2250
101. Whelan AM, Price SO, Fowler SF et al (1992) The effect of aspirin on niacin-induced cutaneous reactions. J Fam Pract 34:165–168
102. Pollex RL, Hegele RA (2006) Genetic determinants of the metabolic syndrome. Nat Clin Pract Cardiovasc Med 3:482–489

Index

F

G

H

I

K

L

M

MIX
Papier aus verantwortungsvollen Quellen
Paper from responsible sources
FSC® C105338

If you have any concerns about our products,
you can contact us on
ProductSafety@springernature.com

In case Publisher is established outside the EU,
the EU authorized representative is:
Springer Nature Customer Service Center GmbH
Europaplatz 3, 69115 Heidelberg, Germany

Printed by Libri Plureos GmbH
in Hamburg, Germany